Study Guide for Atkins and Jones's

CHEMICAL PRINCIPLES

The Quest for Insight

John Krenos and Joseph Potenza

Rutgers, the State University of New Jersey

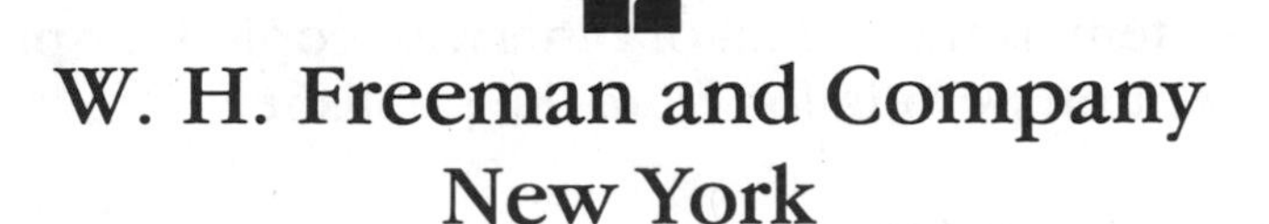

W. H. Freeman and Company
New York

ISBN: 0-7167-3357-9

Printed in the United States of America

First Printing 1999

CONTENTS

PREFACE

This S*tudy Guide* accompanies the textbook *Chemical Principles: The Quest for Insight* by Peter Atkins and Loretta Jones – an authoritative and thorough introduction to chemistry for students anticipating careers in science or engineering disciplines. We have followed the order of topics in the textbook chapter for chapter. The parallel between the symbols, concepts, and style of the *Study Guide* and the textbook enable the reader to easily move back and forth between the two. Although we have not provided a chapter dealing explicitly with the Fundamentals section of the text (pages F1 to F115), we refer to this basic material in many instances as the need arises in subsequent chapters. For example, the several definitions of an acid and a base, introduced in the Fundamentals section of the text, are incorporated in Chapter 10 of the *Study Guide*, Acids and Bases.

Following this preface, a **Brief Review of Symbols, Units, and Number Conventions** is provided. This section is written in summary/tabular form and is intended to be a resource to the student when reading the text or solving problems. We suggest that students examine this material before using the *Study Guide* or the text.

Much of the *Study Guide* is presented in outline style with material highlighted by bullets, arrows, and tables. In general, bullets offset major items of importance for each section and arrows provide explanatory material, descriptive material, or examples to reinforce a concept. This telegraphic style should help the student obtain a broad perspective of large blocks of material in a relatively short period of time and may prove particularly useful in preparing for examinations. We believe that the *Study Guide* will be most useful following a careful reading of the text.

Most chapters begin with an **Overview** section, which provides a summary of the main ideas and concepts to be introduced and, in some instances, the **Goals** of the chapter. Each major topic in a chapter begins with a list of **Key Concepts**. After reading the text and the *Study Guide*, students should understand and be able to define these concepts, most of which are defined in the guide or in the glossary at the back of the text. Chapter **Sections** follow those in the text and contain descriptive material (bullets and arrows) as well as numerous examples (problems). Most examples are worked out in detail. In some instances, to reinforce a given concept, a given worked-out example is followed by a similar one for which only the answer is shown. **Examples** are numbered by section rather than in numerical order so that if students need to brush up on a particular topic, it should be relatively easy to do so. While we have covered most of the material in the text, we have not attempted to be encyclopedic. To help obtain a broad overview of the material, important equations are highlighted in boxes. Students may find this aspect of the guide particularly useful before exams. Where appropriate, we have introduced supplementary tables either to amplify material in the text, to clarify it further, or to summarize a body of material. Sprinkled throughout the guide are **Notes**; in the main, these are designed to point out common pitfalls.

The usefulness of any study guide depends on its accuracy and clarity. We welcome feedback regarding what you like about the guide or how you feel it might be improved. We would particularly appreciate learning of any errors in the text. Our mailing address is: Department of Chemistry, Rutgers, the State University of New Jersey, 610 Taylor Road, Piscataway, NJ 08854-8087. Our e-mail addresses are krenos@rutchem.rutgers.edu and potenza@rutchem.rutgers.edu. Our fax number is 732-445-5312.

ACKNOWLEDGMENTS

We are indebted to many individuals who have made significant contributions to this *Study Guide*.

First and foremost, we wish to thank Beth Van Assen for a thorough and incisive reading of the drafts. Her critical suggestions led to improved readability and scientific accuracy of the *Study Guide*. Many examples and important sections of descriptive text were clarified and expanded with her help. Her encouragement, patience, and persistence were essential to the completion of this project.

We also thank two of our colleagues at Rutgers for critically reading three chapters. Harvey Schugar carefully reviewed Chapter16. Our knowledge and appreciation of inorganic chemistry were broadened greatly by his comments and analysis. Spencer Knapp critiqued Chapters 18 and 19 and gave us a quick tutorial on modern organic chemistry. Many of the chemical structures were prepared with his help.

David Becker, author of the study guide accompanying the textbook *Chemistry: Molecules, Matter, and Change* by Peter Atkins and Loretta Jones, provided inspiration for the layout and format of our worked-out examples. We were also guided by his **Pitfalls** sections that alert students to commonly encountered misperceptions and traps. We attempted to incorporate some of his insights into our comments (**Notes**) for students.

The copy editors, Jodi Simpson and Alice Allen, did an excellent job in making the material clear, complete, and in harmony with the textbook. In particular, we thank Jodi Simpson for help in developing a workable, consistent format for the telegraphic style.

We also wish to acknowledge the help and encouragement of the staff at W. H. Freeman and Company. We especially thank Michelle Russel Julet (chemistry editor) for asking us to author this guide, Jodi Isman (project editor) for guiding us through the laborious process of developing a manuscript from scratch and also for helping with the editing, and Matthew Fitzpatrick (supplements editor) for spearheading the project and trying to keep us (sometimes successfully) to a schedule.

Finally, we extol the efforts of the authors of the textbook, Peter Atkins and Loretta Jones, in creating a new approach to teaching general chemistry. Beginning an introductory chemistry textbook with quantum theory is a logical, but challenging approach. The authors succeed by starting the book with a section on the fundamentals of chemistry (an "as needed" review) and by not covering a great deal of material (mostly historical) presented in other books. In our opinion, this approach is justified because it leads to the development of all the major topics in chemistry enlightened and enlivened in the first instance by the molecular viewpoint.

Brief Review of Symbols, Units, and Number Conventions

Mathematical Symbols

$=$ (equal to)	$\equiv$ (identically equal to)
$\neq$ (not equal to)	$\propto$ (proportional to)
$\approx$ (approximately equal to)	$\cong$ (approximately identical to)
$>$ (greater than)	$<$ (less than)
$\geq$ (greater than or equal to)	$\leq$ (less than or equal to)
$+$ (add, plus)	$-$ (subtract, minus)
$\times$ (multiply)	/ or $\div$ (divide)
Δ (finite difference)	d or ∂ (infinitesimal difference)
$\sum$ (series summation)	$\int$ (integration symbol)
$\prod$ (series multiplication)	powers (numerical superscripts)

Note: A *bar* above a variable or a *pair* of surrounding angle brackets is used to denote the average (mean) value of a variable.

$$\bar{x} = \langle x \rangle = \sum_{i=1}^{n} \left(\frac{x_i}{n} \right), \quad \text{where } i = 1,\ 2,\ 3,\ \ldots,\ n$$

Mathematical Functions

Special functions in normal type: log, ln, exp, sin, cos, tan, d, δ,…

General functions in italic type: $f(x)$, $F(x,y)$, …

Note: The exponential function can be written in alternative forms.

$$y = \exp(x) = \mathrm{e}^x \qquad \ln y = x \qquad \ln \equiv \log_\mathrm{e}$$

Property Symbols

Normally in italic type: v (particle speed), m (particle mass), T (temperature), E (energy)

Greek letters: ν (frequency: nu), λ (wavelength: lambda), θ and ϕ (spherical polar coordinate angles: theta and phi)

Commonly used alternative symbols for frequency are f (normal) and *f* (italic).

Some authors use Greek letters in italic type: compare θ and ϕ above.

Note: Greek letter nu (ν) and italic letter v (v) are similar in appearance.

Note: Some property symbols appear in several variations.

One example is the set of symbols used to denote <u>directions</u> in space.

x, y, z (positions in Cartesian coordinate space)
u, v, w (corresponding speeds in the same coordinate system)

Symbols for Numbers and Constants

Mathematical infinity symbol ∞
Symbol (Greek letter pi) $\pi = 3.14159265359\ldots$
(ratio of circumference to diameter of a circle)
Symbol for base of natural logarithms $\mathrm{e} = 2.71828182846\ldots$

Note: Do not confuse e with the fundamental elementary charge e.

Physical constants follow the convention for property symbols:

c (speed of light in vacuum)
h (Planck constant)
ε_0 (vacuum permittivity) [Note the use of a number subscript.]

An important constant in chemistry is denoted by a special symbol.

N_A or L (Avogadro constant) = $6.022\,136\,7 \times 10^{23}$ "objects" mol^{-1}

Note: Some constants are given special symbols that are not universal.

$\mathfrak{R}$ (Rydberg constant) [See Symbol font used on personal computers.]

Conventions for Units : International System of Units (SI)

The Seven SI Base Units

Physical Quantity	Symbol(s)	Name of SI Unit	Symbol for SI Unit
length	l,x,y,z	meter	m
mass	m	kilogram	kg
time	t	second	s
electric current	I	ampere	A
thermodynamic temperature	T	kelvin	K
amount of substance	n	mole	mol
luminous intensity	I_v	candela	cd

Notes: A symbol for a quantity is *italicized*, whereas a symbol for a unit is <u>not</u>. The symbol for the quantity of mass is m; whereas the SI unit of length is the meter m. The symbol for various constants in chemistry is K.

Common SI Derived Units

Physical Quantity (Common Symbol)	Name of SI Unit	Symbol for SI Unit	Expression in Terms of SI Base Units	
frequency (ν, f)	hertz	Hz	s^{-1}	
force (F)	newton	N	$kg{\cdot}m{\cdot}s^{-2}$	
pressure (p, P)	pascal	Pa	$N{\cdot}m^{-2}$	$= kg{\cdot}m^{-1}{\cdot}s^{-2}$
energy (E)	joule	J	$N{\cdot}m$	$= kg{\cdot}m^2{\cdot}s^{-2}$
power (P)	watt	W	$J{\cdot}s^{-1}$	$= kg{\cdot}m^2{\cdot}s^{-3}$
electric charge (Q)	coulomb	C	$A{\cdot}s$	
electric potential (V, ϕ)	volt	V	$J{\cdot}C^{-1}$	$= kg{\cdot}m^2{\cdot}s^{-3}{\cdot}A^{-1}$
magnetic flux density (B)	tesla	T	$V{\cdot}s{\cdot}m^{-2}$	$= kg{\cdot}s^{-2}{\cdot}A^{-1}$
magnetic flux (Φ)	weber	Wb	$V{\cdot}s$	$= kg{\cdot}m^2{\cdot}s^{-2}{\cdot}A^{-1}$

Common SI Derived Units for Other Quantities

Physical Quantity (Common Symbol)	Expression in Terms of SI Base Units	
area (A)	m^2	
volume (V)	m^3	
speed (v, u, w, c)	$m{\cdot}s^{-1}$	
acceleration (a)	$m{\cdot}s^{-2}$	
density (d, ρ)	$kg{\cdot}m^{-3}$	
molar energy ($\overline{E}, E_m$)	$J{\cdot}mol^{-1}$	$= kg{\cdot}m^2{\cdot}s^{-2}{\cdot}mol^{-1}$
heat capacity (C), entropy (S)	$J{\cdot}K^{-1}$	$= kg{\cdot}m^2{\cdot}s^{-2}{\cdot}K^{-1}$
energy density (ρ, w)	$J{\cdot}m^{-3}$	$= kg{\cdot}m^{-1}{\cdot}s^{-2}$
surface tension (γ, σ)	$N{\cdot}m^{-1} = J{\cdot}m^{-2}$	$= kg{\cdot}s^{-2}$
amount concentration (c)	$mol{\cdot}m^{-3}$	

Common Units in Use with the SI

Physical Quantity (Symbol)	Name of Unit	Symbol for Unit	Value in SI Units
time (t)	minute	min	60 s
time (t)	hour	h	3600 s
time (t)	day	d	86,400 s
plane angle (θ, ϕ)	degree	°	(π/180) rad
length (l)	ångström	Å	10^{-10} m
volume (V)	liter	L	10^{-3} m^3
pressure (p, P)	bar	bar	10^5 Pa (10^5 $N{\cdot}m^{-2}$)
energy (E)	electronvolt	eV (= $e \times$ V)	$\approx 1.602\,18 \times 10^{-19}$ J
mass (m)	unified atomic mass unit	u (= $m_a(^{12}C)/12$)	$\approx 1.660\,54 \times 10^{-27}$ kg

Note: The last two units are not SI exact, because they depend on the values of physical constants. The unified mass unit is also called the dalton (symbol Da).

Common SI prefixes

$f \equiv 10^{-15}$ (femto)	$p \equiv 10^{-12}$ (pico)	$n \equiv 10^{-9}$ (nano)	$\mu \equiv 10^{-6}$ (micro)	$m \equiv 10^{-3}$ (milli)	$c \equiv 10^{-2}$ (centi)	$k \equiv 10^{3}$ (kilo)	$M \equiv 10^{6}$ (mega)	$G \equiv 10^{9}$ (giga)

Note: Do not confuse prefixes with unit or constant symbols.
mm (millimeter) cm (centimeter) c (speed of light) k (Boltzmann constant)
Physicians/pharmacists use mc in place of the Greek prefix μ (mu),
for example, 1 μg = 1 mcg.

Conventions for Numbers

Arabic integers: 1 (one), 2 (two), 3 (three), …
Roman numerals: I, II, III, IV, V, VI, VII, VIII, IX, X, XI, XII, …

A number is usually in normal type; however, one that is used to label a symbol in the form of a superscript and/or a subscript is sometimes *italicized* to avoid confusion.

n (principal quantum number): n_1 or n_{1} (labeled quantum number for energy state 1)
n_1^2 or n_{1}^2 (*square* of labeled quantum number): $n_1^2 = n_1 \times n_1$ (the power 2 is *not* a label)

Note: photoelectric effect E_0 (threshold energy) particle in a box E_1 (zero-point energy) hydrogen atom E_1 (ground-state energy)

Chapter 1 Atoms: The Quantum World

Observing Atoms (Sections 1.1-1.6)

Key Concepts

electronic structure, classical mechanics, quantum mechanics, spectroscopy, electromagnetic radiation, oscillation, cycle, frequency, amplitude, intensity, wavelength, speed of light, visible light, ultraviolet radiation, infrared radiation, black body, black-body radiation, Wien's law, Stefan-Boltzmann law, ultraviolet catastrophe, quanta, Planck constant, photoelectric effect, photons, diffraction, constructive interference, destructive interference, duality of electromagnetic radiation, duality of matter, linear momentum, de Broglie relation, Heisenberg uncertainty principle (complementarity of location and momentum), trajectories, wavefunction, Schrödinger equation, Born interpretation, node of the wavefunction, probability of finding a particle in some region of space, differential equation, particle in a box, standing wave, quantum number ***n***, quantized (discrete) energy values, zero-point energy, spectral lines, energy levels, Bohr frequency condition, Rydberg constant

1.1 The Characteristics of Electromagnetic Radiation

- **Oscillating amplitude of electric and magnetic field →** Wave characterized by *wavelength* and *frequency*

Distance behavior (fixed time *t*) **Time behavior (fixed position *x*)**

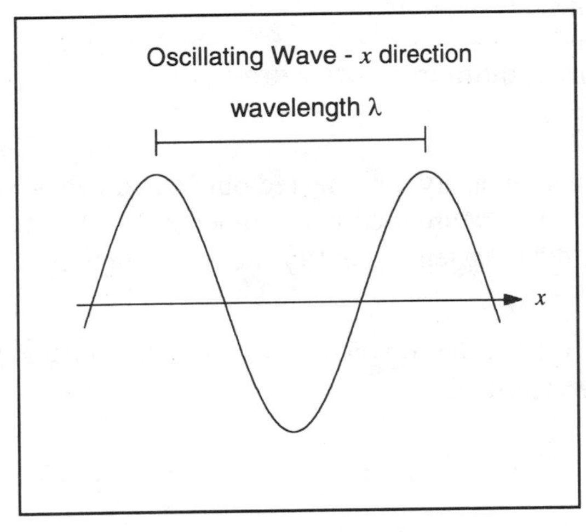

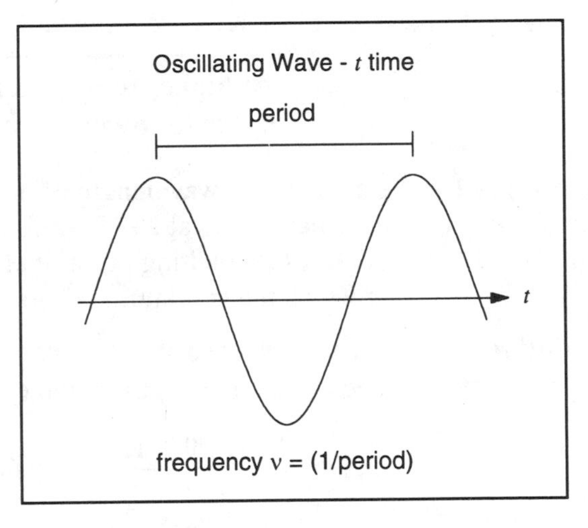

Speed of light (distance/time) = wavelength / period = wavelength × frequency

$c = \lambda\nu$	speed of light	= wavelength	× frequency	
SI units:	$(\text{m}\cdot\text{s}^{-1})$	(m)	(s^{-1})	[1 Hz (hertz) = 1 s^{-1}]

Note: The speed of light depends on the medium in which the radiation is traveling. In vacuum, c is given an optional zero subscript c_0. The effect of the medium (for example, air compared with a vacuum) on wavelengths in the visible region is small (beyond three significant figures). $c_0 = 2.997\,924\,58 \times 10^8\ \text{m}\cdot\text{s}^{-1} \approx 3.00 \times 10^8\ \text{m}\cdot\text{s}^{-1}$

Example 1.1a Calculate the range in frequency and period that corresponds to the wavelength limits of vision, 400 to 700 nm. (See *Brief Review* Section for SI prefix definitions)

Solution First calculate each frequency by solving for ν in the equation $\lambda\nu = c$. Then calculate the period, which is 1 over the frequency.

$$\nu = \frac{c}{\lambda} = \frac{3.00\times10^{8}\ \text{m}\cdot\text{s}^{-1}}{400\times10^{-9}\ \text{m}} = 7.50\times10^{14}\ \text{s}^{-1} \quad \& \quad \text{period} = \frac{1}{\nu} = 1.33\times10^{-15}\ \text{s} = 1.33\ \text{fs}$$

$$\nu = \frac{c}{\lambda} = \frac{3.00\times10^{8}\ \text{m}\cdot\text{s}^{-1}}{700\times10^{-9}\ \text{m}} = 4.29\times10^{14}\ \text{s}^{-1} \quad \& \quad \text{period} = \frac{1}{\nu} = 2.33\times10^{-15}\ \text{s} = 2.33\ \text{fs}$$

Frequency range = $7.50 \times 10^{14}\ \text{s}^{-1}$ to $4.29 \times 10^{14}\ \text{s}^{-1}$ (decreases as λ increases)
Period range = 1.33 fs to 2.33 fs

Example 1.1b Calculate the range in frequency and period that corresponds to the wavelength limits of the infrared region, 1000 nm to 3.00 mm.

Answers Frequency range = $3.00 \times 10^{14}\ \text{s}^{-1}$ to $1.00 \times 10^{11}\ \text{s}^{-1}$
Period range = 3.33 fs to 10.0 ps

1.2 Quanta and Photons

- **Black body**
  - → Solid that is a perfect absorber and emitter of radiation
  - → Intensity of radiation for a series of temperatures
  - → Wavelength corresponding to maximum intensity λ_{max}
  - → Wien's law: $T\lambda_{max} = \text{constant} = 2.88\times10^{-3}\ \text{K}\cdot\text{m}$

    At higher temperature, maximum intensity of radiation shifts to lower wavelength.

Example 1.2a Calculate the wavelength of maximum intensity for a heated object that behaves as a black body at 298 K (reference temperature used in thermochemistry) and 3680 K (the melting point of elemental tungsten). Identify the wavelength region in which the maximum occurs.

Solution Wien's law is easily rearranged to evaluate the wavelength of maximum intensity. Then, the wavelength region can be identified.

$$\lambda_{max} = \frac{2.88\times10^{-3}\ \text{K}\cdot\text{m}}{298\ \text{K}} = 9.66\times10^{-6}\ \text{m} = 9.66\ \mu\text{m} \quad \text{Infrared: 1000 nm to 3 mm}$$

$$\lambda_{max} = \frac{2.88\times10^{-3}\ \text{K}\cdot\text{m}}{3680\ \text{K}} = 7.82\times10^{-7}\ \text{m} = 782\ \text{nm} \quad \text{Visible: 400 nm to 800 nm}$$

Example 1.2b Calculate the wavelength of maximum intensity for a heated object that behaves as a black body at 194.5 K (dry ice temperature) and 6000 K (Sun's "color" temperature). Identify the wavelength region in which the maximum occurs.

Answers $\lambda_{max} = 1.48\times10^{-5}\ \text{m} = 14.8\ \mu\text{m}$ Infrared: 1000 nm to 3 mm
$\lambda_{max} = 4.80\times10^{-7}\ \text{m} = 480\ \text{nm}$ Visible: 400 nm to 800 nm (appears yellow)

- **Energy of a quantum (packet) of light (later called a photon)**
  → Postulated by Max Planck to explain black body radiation
  → Quantization of electromagnetic radiation

$E = h\nu$	photon energy = Planck constant × photon frequency		
SI units:	(J)	($h = 6.6261 \times 10^{-34}$ J·s)	(s^{-1})

Example 1.2c Calculate the photon energy range in Example 1.1a for one photon and one mole of photons.

Solution Use the frequencies calculated in Example 1.1a in the Planck equation for single photon energies. Then multiply by the Avogadro constant to calculate energies for one mole of photons.

Note: Energy changes in chemical reactions are usually given in $\text{kJ}\cdot\text{mol}^{-1}$.

$E = h\nu = (6.63 \times 10^{-34}\ \text{J}\cdot\text{s})(7.50 \times 10^{14}\ \text{s}^{-1}) = 4.97 \times 10^{-19}\ \text{J}$ for single photon (400 nm)

$E \times N_A = (4.97 \times 10^{-19}\ \text{J})(6.02 \times 10^{23}\ \text{mol}^{-1}) = 2.99 \times 10^{5}\ \text{J}\cdot\text{mol}^{-1} = 299\ \text{kJ}\cdot\text{mol}^{-1}$

$E = h\nu = (6.63 \times 10^{-34}\ \text{J}\cdot\text{s})(4.29 \times 10^{14}\ \text{s}^{-1}) = 2.84 \times 10^{-19}\ \text{J}$ for single photon (700 nm)

$E \times N_A = (2.84 \times 10^{-19}\ \text{J})(6.02 \times 10^{23}\ \text{mol}^{-1}) = 1.71 \times 10^{5}\ \text{J}\cdot\text{mol}^{-1} = 171\ \text{kJ}\cdot\text{mol}^{-1}$

Thus, the photon energy range of vision (400 to 700 nm) is 299 $\text{kJ}\cdot\text{mol}^{-1}$ to 171 $\text{kJ}\cdot\text{mol}^{-1}$.

Example 1.2d Calculate the photon energy range in Example 1.1b for one photon and one mole of photons.

Answers $E = 1.99 \times 10^{-19}$ J for single photon (1000 nm) & 120 $\text{kJ}\cdot\text{mol}^{-1}$
$E = 6.63 \times 10^{-23}$ J for single photon (3.00 mm) & 39.9 $\text{J}\cdot\text{mol}^{-1}$

Thus, the photon energy range of the infrared (1000 nm to 3.00 mm) is about 120 $\text{kJ}\cdot\text{mol}^{-1}$ to 40 $\text{J}\cdot\text{mol}^{-1}$.

- **Photoelectric effect**
  → Ejection of electrons from the surface of a metal exposed to electromagnetic radiation (photons) of sufficient energy
  → Light behaves as a particle

$$E_K = h\nu - E_0$$

E_K is the kinetic energy of the ejected electron, E_0 is the threshold energy required to eject the electron from the surface of the metal, and $h\nu$ is the photon energy.

Note: $h\nu \geq E_0$ required for electron ejection

Note: Multiple photon absorption with high-powered lasers can eject electrons even if a single photon energy value is below threshold.

- **Wave behavior of light** → Diffraction and interference effects of superimposed waves (*constructive* and *destructive*)

1.3 The Wave-Particle Duality of Matter

- **Wave properties**
  - → Matter with mass m and velocity v
  - → Matter behaving as a wave has a characteristic wavelength
  - → Proposed by Louis de Broglie

$$\boxed{\lambda = \frac{h}{mv}}$$ de Broglie wavelength for a particle with linear momentum $p = mv$

Example 1.3a Calculate the de Broglie wavelength for an electron formed by photoelectric emission from a metal surface ($E_0 = 7.20 \times 10^{-19}$ J) exposed to light of wavelength 260 nm.

Solution First find the energy of the photon. Then find the kinetic energy of the electron from the photoelectric effect. Use the definition of kinetic energy to find the electron speed. Calculate the de Broglie wavelength.
Note: $1\text{ J} = 1\text{ kg·m}^2\text{·s}^{-2}$ and $E_K = (1/2)mv^2$, where $m = m_e$ = mass of electron

$$E = h\nu = \frac{hc}{\lambda} = \frac{(6.63 \times 10^{-34}\text{ J·s})(3.00 \times 10^{8}\text{ m·s}^{-1})}{(260 \times 10^{-9}\text{ m})} = 7.65 \times 10^{-19}\text{ J (photon energy)}$$

$$E_K = h\nu - E_0 = (7.65 \times 10^{-19}\text{ J}) - (7.20 \times 10^{-19}\text{ J}) = 4.5 \times 10^{-20}\text{ J (electron kinetic energy)}$$

$$v = \sqrt{\frac{2E_K}{m_e}} = \sqrt{\frac{2(4.5 \times 10^{-20}\text{ kg·m}^2\text{·s}^{-2})}{(9.1094 \times 10^{-31}\text{ kg})}} = \sqrt{9.88 \times 10^{10}\text{ m}^2\text{·s}^{-2}} = 3.1 \times 10^{5}\text{ m·s}^{-1}\text{(speed)}$$

$$\lambda = \frac{h}{m_e v} = \frac{(6.63 \times 10^{-34}\text{ J·s})}{(9.1094 \times 10^{-31}\text{ kg})(3.1 \times 10^{5}\text{ m·s}^{-1})} = 2.3 \times 10^{-9}\text{ m} = 2.3\text{ nm}$$

Example 1.3b Repeat Example 1.3a for a different light source, but in reverse. Suppose that the wavelength of the ejected electron is determined to be 232 pm. What is the wavelength of a photon that ejects the electron from the same surface?

Answer 53.6 nm

1.4 The Uncertainty Principle

- **Complementarity of location (x) and momentum (p)**
  - → Uncertainty in x is Δx
  - → Uncertainty in p is Δp
  - → Heisenberg uncertainty principle
  - → Limitation of knowledge

$$\boxed{\Delta p \Delta x \geq \hbar/2}$$ Heisenberg uncertainty principle, where $\hbar = h/2\pi$

→ $\hbar$ is called "h bar" $\hbar = 1.0546 \times 10^{-34}$ J·s

→ Refutes classical physics on the atomic scale

1.5 Wavefunctions and Energy Levels

- **Classical trajectories** → Precisely defined paths
- **Wavefunction** ψ → Probable position of particle with mass m
- **Born interpretation** → Probability of finding particle in a region proportional to ψ^2
- **Schrödinger equation** → Allows calculation of ψ by solving a differential equation
- **Particle in a box**
  → Mass m confined between two rigid walls a distance L apart
  → $\psi = 0$ outside the box and at the walls (boundary condition)

$$\boxed{\psi_n(x) = \left(\frac{2}{L}\right)^{1/2} \sin\left(\frac{n\pi x}{L}\right) \qquad n = 1,\ 2,\ \ldots}$$

Wavefunction that satisfies the Schrödinger equation between the box limits. n is a *quantum number*.

Note: A node is a point in the box where $\psi = 0$ and ψ *changes* sign. $\boxed{\psi^2 \geq 0, \text{ always}}$

$$\boxed{E_n = \frac{n^2h^2}{8mL^2}}$$

Allowed energy values of a particle in a box
Note: $n = 1$ gives the zero-point energy E_1
$E_1 \neq 0$ implies residual motion

$$\boxed{\Delta E = E_{n+1} - E_n = \frac{(2n+1)h^2}{8mL^2}}$$

Energy difference between two neighboring levels

Example 1.5a Calculate the probability of finding a particle in the box at $x = L/2$ for $n = 2$. How many nodes occur in the wavefunction?

Solution Find ψ^2 for $n = 2$. Find values of x where $\psi = 0$ and ψ changes sign within the box.

$$\psi_2^2\left(x = \frac{L}{2}\right) = \left(\frac{2}{L}\right)\sin^2\left(\frac{2\pi x}{L}\right) = \left(\frac{2}{L}\right)\sin^2\left(\frac{2\pi(L/2)}{L}\right) = \left(\frac{2}{L}\right)\sin^2(\pi) = \left(\frac{2}{L}\right)(0) = 0 \text{ (Node)}$$

$\psi = 0$, and ψ changes sign in the center of the box. There is only one node ($x = L/2$).

Example 1.5b Repeat Example 1.5a for $n = 3$.

Answer
$$\psi_3^2\left(x = \frac{L}{2}\right) = \left(\frac{2}{L}\right)(-1)^2 = \frac{2}{L}$$
(Not a node, because $\psi^2 \neq 0$)

$\psi = 0$, and ψ changes sign at $x = \dfrac{L}{3}$ and $\dfrac{2L}{3}$ (2 nodes)

Note: A pattern is suggested, in which $(n - 1)$ nodes exist for ψ_n. The existence of locations of zero particle probability is a *purely* quantum effect.

1.6 Atomic Spectra and Energy Levels

- **Spectral lines** → Discharge lamp of hydrogen

  H_2 + electrical energy → $H + H^*$ [* ≡ asterisk denotes excitation]

  $H^* \rightarrow H^{(*)} + h\nu$ [$^{(*)}$ ≡ denotes a less excited atom]

- **Lines form a discrete pattern** → Discrete energy levels
- **Bohr frequency condition**

$$h\nu = E_{\text{upper}} - E_{\text{lower}}$$

Relates the photon energy to the difference in energy between two energy levels in an atom.

- **Hydrogen atom spectral lines**

  → Johann Rydberg's general equation

  → $n_2 = n_{\text{upper}}$ and $n_1 = n_{\text{lower}}$

$$\nu = \mathfrak{R}\left(\frac{1}{n_1^2} - \frac{1}{n_2^2}\right) \quad n_1 = 1, 2, \ldots \quad n_2 = n_1 + 1, n_1 + 2, \ldots$$

$\mathfrak{R} = 3.29 \times 10^{15}$ Hz (Rydberg constant)

Rydberg expression reproducing the pattern of lines in the hydrogen atom emission spectrum. The value of $\mathfrak{R}$ is obtained from an empirical fitting procedure.

Note: Lines with a common n_1 can be grouped into a series and some have special names:

n_1 = 1 (Lyman), 2 (Balmer), 3 (Paschen), 4 (Brackett), 5 (Pfund)

Example 1.6a Calculate the wavelength of light emitted when a hydrogen atom in the third excited state reverts to the ground state (member of Lyman series). (See *Study Guide* Section 1.10 for a presentation of ground and excited states.)

Solution The third excited state corresponds to $n_2 = 4$, the ground state to $n_1 = 1$.

Thus, combining $\nu = \mathfrak{R}\left[\frac{1}{n_1^2} - \frac{1}{n_2^2}\right]$ and $\lambda\nu = c$, we obtain

$$\lambda = \frac{c}{\nu} = \frac{c}{\mathfrak{R}\left[\frac{1}{n_1^2} - \frac{1}{n_2^2}\right]} = \frac{3.00 \times 10^8\,\text{m}\cdot\text{s}^{-1}}{\left(3.29 \times 10^{15}\,\text{s}^{-1}\right)\left[\frac{1}{1^2} - \frac{1}{4^2}\right]} = 9.73 \times 10^{-8}\,\text{m} = 97.3\text{ nm}$$

Example 1.6b Calculate the wavelength of light emitted when a hydrogen atom in the third excited state undergoes a transition to the second excited state (member of the Paschen series).

Answers $n_2 = 4$, $n_1 = 3$, and $\lambda = 1.88\ \mu\text{m}$

Models of Atoms (Sections 1.7-1.10)

Key Concepts

principal quantum number, Coulomb potential energy, energy level/state, ionization, ground state, excited state, atomic orbital, radial wavefunction, angular wavefunction, Bohr radius, shell, subshell, orbital angular momentum, quantum number, magnetic quantum number, boundary surface, node, electron spin, spin magnetic quantum number

1.7 The Principal Quantum Number

See *Brief Review* section of the *Study Guide* for presentation of SI unit conventions.

- **Charge on an electron** → $e = -1.60217733 \times 10^{-19}$ C
- **Vacuum permittivity** → $\varepsilon_0 = 8.854187817 \times 10^{-12}\ \mathrm{C^2{\cdot}N^{-1}{\cdot}m^{-2}}$
- **Coulomb potential energy** $\boxed{V(r) = \dfrac{e_1 e_2}{4\pi\varepsilon_0 r}}$

$$V(r) = \frac{\text{product of charges on particles}}{4\pi \times \text{permittivity of free space} \times \text{distance between charges}} \qquad \text{(SI system)}$$

Units of $V(r)$: $\mathrm{J} = \dfrac{\mathrm{C^2}}{(\mathrm{C^2{\cdot}N^{-1}{\cdot}m^{-2}})\ \mathrm{m}} = \mathrm{N{\cdot}m} = (\mathrm{kg{\cdot}m{\cdot}s^{-2}})\ \mathrm{m} = \mathrm{kg{\cdot}m^2{\cdot}s^{-2}}$

Example 1.7a Calculate the coulomb potential energy for two electrons separated by a distance of 10^{-6} m.

Solution Insert values into the equation for $V(r)$ with $e_1 = e_2 =$ the charge of the electron and $r = 10^{-6}$ m. Thus,

$$V(r) = \frac{e_1 e_2}{4\pi\varepsilon_0 r} = \frac{(-1.602 \times 10^{-19}\ \mathrm{C})^2}{4\pi(8.854 \times 10^{-12}\ \mathrm{C^2{\cdot}N^{-1}{\cdot}m^{-2}})(10^{-6}\ \mathrm{m})} = 2.307 \times 10^{-22}\ \mathrm{J}.$$

The energy is positive indicating a repulsive interaction, as expected.

Hint: When evaluating expressions with many terms, particularly ones containing exponents, it is often useful to estimate the result first. This estimation can be accomplished by factoring the exponents and estimating the pre-exponent and exponent terms separately. With practice, the estimation can be done quickly and in one's head. Thus, in the present case, the expression above factors to

$\left(\dfrac{(-1.602)^2}{4 \cdot \pi \cdot 8.854}\right)\left(\dfrac{10^{-38}}{10^{-18}}\right) \approx 0.03 \times 10^{-20} = 3 \times 10^{-22}$, which is a good estimate of the correct answer.

Example 1.7b Calculate the coulomb potential energy for four electrons at the corner of a square of edge 10^{-6} m.

Solution

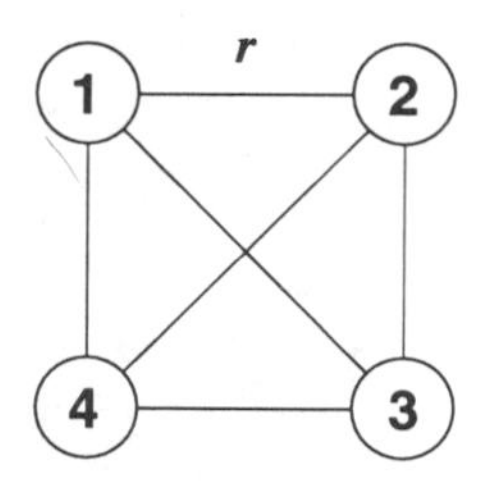

Label the electrons 1 through 4 as shown in the diagram above. All pairwise interactions must be considered. There are 4 interactions (r_1, r_2, r_3, r_4) with $r = 10^{-6}$ m and two diagonal interactions (r_5, r_6) with $r = 1.414 \times 10^{-6}$ m. Now,

$$V(r) = \frac{e_1 e_2}{4\pi\varepsilon_0}\sum_{i=1}^{6}\frac{1}{r_i} = \left(\frac{(-1.602\times10^{-19})^2}{(4\pi)(8.854\times10^{-12})}\right)\left(\frac{4}{10^{-6}}+\frac{2}{1.414\times10^{-6}}\right) = 1.249\times10^{-21}\text{ J.}$$

Notice that this repulsive energy is more than four times that in Example 1.7a because of the cross terms. Electron-electron repulsions are important in understanding the energy levels of atoms and molecules.

- **H atom energy levels**

→ Solutions to the Schrödinger equation

$$E_n = \frac{-h\Re}{n^2} \qquad n = 1, 2, 3, \ldots \qquad (n \text{ is dimensionless})$$

$$\Re = \frac{m_e e^4}{8h^3\varepsilon_0^2} = 3.29\times10^{15}\text{ Hz} \qquad \text{Units: } \frac{\text{kg}\cdot\text{C}^4}{(\text{J}\cdot\text{s})^3(\text{C}^2\cdot\text{N}^{-1}\cdot\text{m}^{-2})} = \text{s}^{-1} \equiv \text{Hz}$$

→ All quantum numbers are dimensionless. With $\Re$ in units of frequency, the H atom energy-level equation has the same form as the Planck equation, $E = h\nu$.

→ E_n is the equation for the energy levels (states) of the H atom. Note the *negative* sign.

→ $\Re$ is the Rydberg constant, calculated exactly, using Bohr theory.

→ n is the principal quantum number.

→ As n increases, energy increases, and the atom becomes less stable and energy states become more closely spaced (more dense).

→ Integer n varies from 1 (ground state) to higher integers (excited states) to ∞ (ionization).

→ Energies of H atom states vary from $-h\Re$ ($n = 1$) to 0 ($n = \infty$). States with $E > 0$ are possible. In these states, the electron is free (ionized atom) and the energy > 0 corresponds to the kinetic energy of the electron.

1.8 Atomic Orbitals (AOs)

- **Definition of AO**
  - → Wavefunction (ψ, psi) describes an electron in an atom
  - → Orbital (ψ^2) holds 0, 1, or 2 electrons
  - → Orbital can be viewed as a *cloud* within which the point density represents the *probability* of finding the electron at that point
  - → Orbital specified by *three* quantum numbers (n, ℓ, m_ℓ; see below)

- **Wavefunction**
  - → Fills all space
  - → Depends on the *three* spherical coordinates: r, θ, ϕ
  - → Written as a product of a radial [$R(r)$] and an angular [$Y(\theta,\phi)$] wavefunction
    Mathematically, $\psi(r,\theta,\phi) = R(r)Y(\theta,\phi)$

- **Wavefunction for the H atom 2*s* orbital** → $n = 2$, $\ell = 0$, and $m_\ell = 0$

$$R(r) = \frac{\left(2 - \frac{r}{a_0}\right)e^{-r/2a_0}}{(2a_0)^{3/2}} \qquad Y(\theta,\phi) = (4\pi)^{-1/2} \qquad a_0 = \frac{4\pi\varepsilon_0\hbar^2}{m_e e^2} = \left\{\begin{matrix} 5.29177 \times 10^{-11}\ \text{m} \\ \text{(Bohr radius)} \end{matrix}\right\}$$

$$\text{Units:}\ R(r) = \text{m}^{-3/2} \qquad Y(\theta,\phi) = \text{none} \qquad a_0 = \frac{(\text{C}^2\cdot\text{N}^{-1}\cdot\text{m}^{-2})(\text{J}\cdot\text{s})}{\text{kg}\cdot\text{C}^2} = \text{m}$$

Example 1.8a The H atom 2*s*-orbital has one *nodal surface* [surface for which $\psi(r,\theta,\phi) = 0$]. Using the wavefunction given above, determine the location and geometrical form of this surface.

Solution $\psi_{2s}(r,\theta,\phi) = 0$, if $R_{2s}(r) = 0$, which occurs when $r = 2a_0 = 2(5.29 \times 10^{-11}\ \text{m}) = 1.06 \times 10^{-10}$ m. The locus of points for which r is constant defines the surface of a sphere. The nodal surface is located 1.06×10^{-10} m (106 pm) from the proton.

Three Quantum Numbers [n, ℓ, m_ℓ] Specify an Atomic Orbital

Symbol	Name	Allowed Values	Constraints
n	Principal quantum number	$= 1, 2, 3, \ldots$	Positive integer
ℓ	Orbital angular momentum quantum number	$= 0, 1, 2, \ldots, n-1$	Each value of n corresponds to n allowed values of ℓ.
m_ℓ	Magnetic quantum number	$= \ell, \ell-1, \ell-2, \ldots, -\ell$ $= 0, \pm1, \pm2, \ldots, \pm\ell$	Each value of ℓ corresponds to $(2\ell+1)$ allowed values of m_ℓ.

- **Terminology (nomenclature)**

shell: AOs with the same n value

subshell: AOs with the same n and ℓ values;

$\ell = 0, 1, 2, 3$ equivalent to s-, p-, d-, f-subshell, respectively, or

s-orbital $\Rightarrow$ $\ell = 0$ $m_\ell = 0$

p-orbital $\Rightarrow$ $\ell = 1$ $m_\ell = -1, 0,$ or $+1$

d-orbital $\Rightarrow$ $\ell = 2$ $m_\ell = -2, -1, 0, +1,$ or $+2$

f-orbital $\Rightarrow$ $\ell = 3$ $m_\ell = -3, -2, -1, 0, +1, +2,$ or $+3$

- **Physical significance of the wavefunction $\psi(r,\theta,\phi)$**

$\psi^2(r,\theta,\phi)$ is proportional to the probability of finding the electron at a point with coordinates r,θ,ϕ. We can also regard $\psi^2(r,\theta,\phi)$ as the electron density at point r,θ,ϕ. These statements are true for all atoms and molecules.

Example 1.8b For the H atom $2s$ orbital, calculate the electron density at $r = a_0$, $\theta = \phi = \pi$ (radians).

Solution The electron density (probability of finding the electron) at a point is given by the square of the wavefunction.

$$\psi_{2s}^2 = R_{2s}^2 Y_{2s}^2 = \frac{\left(2 - \frac{r}{a_0}\right)^2 e^{-r/a_0}}{4\pi(2a_0)^3}$$

With $r = a_0$, this becomes $\psi_{2s}^2 = \dfrac{e^{-1}}{4\pi(2a_0)^3} = 2.47 \times 10^{28}$ electron·m^{-3}.

Example 1.8c For the H atom $2s$-orbital, calculate the electron density at $r = 3a_0$, $\theta = \phi = 0$.

Solution For $r = 3a_0$, the value is $\psi_{2s}^2 = \dfrac{e^{-3}}{4\pi(2a_0)^3} = 4.95 \times 10^{27}$ electron·m^{-3}.

Note: These values of the point electron density are huge. Expressing them on a scale more nearly comparable to the size of the atom (a_0) gives the values

Example 1.8b $\Rightarrow 3.66 \times 10^{-3} \dfrac{\text{electron}}{a_0^3}$ and Example 1.8c $\Rightarrow 4.95 \times 10^{-4} \dfrac{\text{electron}}{a_0^3}$.

Note: Integrating over all space: $\int_0^{2\pi}\int_0^{\pi}\int_0^{\infty} \psi_{2s}^2 r^2 \sin\theta \, dr \, d\theta \, d\phi = 1$ electron,

where the volume element $dx\,dy\,dz = r^2 \sin\theta \, dr \, d\theta \, d\phi$.

- **Plot of electron density for the $2s$-orbital of hydrogen**

→ A computer program using a random number generator produces an array of points in two dimensions, which are then selected or rejected in comparison with the square of the wavefunction (probability) for an electron in the $2s$-orbital. The higher the probability, the greater the number of points selected in a given region.

→ The resulting points can be plotted as a dot diagram representing electron density. The following diagram shows the points for the $2s$-orbital of the H atom.

Electron density dot diagram for the 2*s*-orbital of the hydrogen atom. The nucleus is at the center of the square and the density of dots is proportional to the probability of finding the electron. Notice the location of the spherical node.

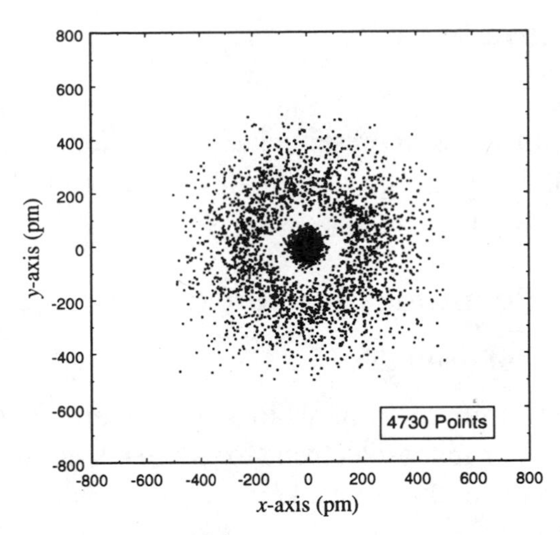

- **Concept of AOs** → Two interpretations are useful:

  1 Visualize AO as a *cloud of points*, with the density of points in a given volume of space proportional to probability of finding the electron in that volume.

  2 Visualize AO as a *surface* (boundary surface) within which there is a given probability of finding the electron. For example, the 95% boundary surface, within which the probability of finding an electron is 95%, defines the surface of a sphere for an *s*-orbital.

- **Boundary surfaces** → *Shapes* of atomic orbitals

s-orbital	*spherical*
p-orbital	*dumbbell* or *peanut*
d-orbital	*four-leaf clover* or *dumbbell* with equatorial *torus* (doughnut)

- **Number and type of orbitals** → Follows from allowed values of quantum numbers

s-orbitals *s* means that $l = 0$; if $l = 0$, then $m_l = 0$. So, for each value of n, there is *one s*-orbital.

one for each n : 1*s*, 2*s*, 3*s*, 4*s*, ...

p-orbitals *p* means that $l = 1$; if $l = 1$, then $m_l = -1, 0, +1$, and $n > 1$. So, for each value of $n > 1$, there are *three p*-orbitals, p_x, p_y, p_z.

three for each $n > 1$: $2p_x, 2p_y, 2p_z$; $3p_x, 3p_y, 3p_z$; ...

The *np*-orbitals are referred to collectively as the 2*p*-orbitals, 3*p*-orbitals, and so on.

d-orbitals *d* means that $l = 2$; if $l = 2$, then $m_l = -2, -1, 0, +1, +2$, and $n > 2$. So, for each value of $n > 2$, there are *five d*-orbitals, d_{xy}, d_{xz}, d_{yz}, $d_{x^2-y^2}$ and d_{z^2}.

five for each $n > 2$: $3d_{xy}$, $3d_{xz}$, $3d_{yz}$, $3d_{x^2-y^2}$ and $3d_{z^2}$;
$4d_{xy}$, $4d_{xz}$, $4d_{yz}$, $4d_{x^2-y^2}$ and $4d_{z^2}$; ...

The *nd*-orbitals are referred to collectively as the 3*d*-orbitals, 4*d*-orbitals, and so on.

1.9 Electron Spin

- **Spin states**
  → *Two* spin states, represented as ↑ and ↓ or α and β
  → Described by the spin magnetic quantum number, m_s
  → Only two allowed values of m_s: $+\frac{1}{2}, -\frac{1}{2}$

1.10 The Electronic Structure of Hydrogen

- **Degeneracy of orbitals**

  In the H atom, the orbitals in a given shell are *degenerate* [have the same energy].
  Note: For many-electron atoms, this is not true and the energy of a given orbital depends on *n* and *l*.

- **Ground and excited states**

  In the ground state, $n = 1$, and the electron is in the 1*s*-orbital. The first excited state corresponds to $n = 2$, and the electron occupies one of the *five* possible orbitals with $n = 2$ (1*s*, 2*s*, $2p_x$, $2p_y$, or $2p_z$). Similar considerations hold for higher excited states.

- **Ionization**

  Absorption of a photon with energy $\geq h\Re$ ionizes the atom, creating a free electron and a free proton. Energy in excess of $h\Re$ is utilized as kinetic energy of the system.

- **Summary**

  The state of an electron in a hydrogen atom is defined by four quantum numbers $\{n, \ell, m_\ell, m_s\}$. As the value of *n* increases, the size of the hydrogen atom increases.

The Structures of Many-Electron Atoms (Sections 1.11-1.14)

Key Concepts

effective nuclear charge, penetration, shielding, building-up principle, Pauli exclusion principle, spin pairing, electron configuration, closed shell, valence electron(s), valence shell, Hund's rule, excited state, periodic table, period, long period, group, block

1.11 Orbital Energies

- **Many-electron atoms**
  → Atoms with more than one electron
  → Coulomb potential energy equals the sum of *nucleus-electron* **attractions** and *electron-electron* **repulsions**

→ Schrödinger equation cannot be solved exactly
→ Accurate wavefunctions obtained numerically by using computers

Example 1.11a Write an expression for the coulomb potential energy of the Li atom (3 protons in the nucleus plus 3 electrons). The atomic number Z is equal to the number of protons in the nucleus. Which terms are attractive? Which are repulsive?

Solution This is a 4-body problem similar to the one solved in Example 1.7b. Label the *nuclear charge Ze* ($Z = 3$) and the *electrons* 1, 2, and 3 (charge $= -e$). The potential energy is given as:

$$V(r) = \sum \frac{(Ze)e}{4\pi\varepsilon_0 r_i} + \sum \frac{e^2}{4\pi\varepsilon_0 r_{ij}}$$

$$= -\frac{(Ze)e}{4\pi\varepsilon_0 r_1} - \frac{(Ze)e}{4\pi\varepsilon_0 r_2} - \frac{(Ze)e}{4\pi\varepsilon_0 r_3} + \frac{e^2}{4\pi\varepsilon_0 r_{12}} + \frac{e^2}{4\pi\varepsilon_0 r_{13}} + \frac{e^2}{4\pi\varepsilon_0 r_{23}}$$

Here, r_i is the distance of electron i from the nucleus and r_{ij} is the distance between electrons i and j. The *electron-nucleus* terms (the first three terms in the equation above) are attractive whereas the *electron-electron* terms (the last three terms) are repulsive.

- **Variation of energy of orbitals** → For orbitals in the same shell but in different subshells, a combination of *nucleus-electron* attraction and *electron-electron* repulsion influences the orbital energies.
- **Shielding and penetration** → Qualitative understanding of orbital energies in atoms
- **Shielding** → Each electron in an atom is *attracted* by the nucleus and *repelled* by all the other electrons. In effect, each electron feels a *reduced* nuclear charge (Z_{eff} = effective nuclear charge). The electron (orbital) is *shielded* to some extent from the nuclear charge and its energy is raised accordingly.
- **Penetration** → The *s*-, *p*-, *d*-, ... orbitals have different shapes and different electron density distributions. For a given *shell* (same value of the principal quantum number *n*), s-electrons tend to be closer to the nucleus than p-electrons, which are closer than d-electrons. We say that s-electrons are more *penetrating* than p-electrons, and p-electrons are more *penetrating* than d-electrons.

Note: Shielding and penetration can be understood qualitatively on the basis of the nucleus-electron potential energy term: $-(Ze)e/4\pi\varepsilon_0 r$, where Ze is the nuclear charge and r the nucleus-electron distance. *Shielded* electrons have the equivalent of a reduced Z value ($Z_{eff} < Z$) and therefore *higher* energy; *penetrating* electrons have the equivalent of a reduced value of r and therefore *lower* energy.

- **Review** → Z has three equivalent meanings:
  Nuclear charge (actually, Ze)
  Atomic number
  Number of protons in the nucleus

- **Consequences of *shielding* and *penetration* in many-electron atoms**
  → Orbitals with the same n and different l values have *different* energies.
  → For a given *shell* (n), *subshell* energies *increase* in the order: $ns < np < nd < nf$.
  Example: A 3*s*-electron is lower in energy than a 3*p*-electron, which is lower in energy than a 3*d*-electron.
  → Orbitals within a given *subshell* have the *same* energy.
  Example: The five 3*d*-orbitals are degenerate for a given atom.
  → *Penetrating* orbitals of higher shells may be lower in energy than less *penetrating* orbitals of lower shells.
  Example: A penetrating 4*s*-electron may be lower in energy than a less penetrating 3*d*-electron.

1.12 The Building-Up Principle

- **Pauli exclusion principle**
  → No more than *two* electrons per orbital
  → *Two* electrons occupying a single orbital must have paired spins: ↑↓
  → *Two* electrons in an atom may *not* have the same *four* quantum numbers.

- **Terminology**

closed shell: Shell with maximum number of electrons allowed by the exclusion principle

valence electrons: Electrons in the outermost occupied shell of an atom; they occupy the shell with the largest value of n; they are used to form chemical bonds.

electron configuration: List of all occupied *subshells* or *orbitals*, with the number of electrons in each indicated as a numerical superscript.

Note: There are several different ways of writing electron configurations. Examples 1.12a-d show four of them.

Example 1.12a Determine the electron configuration of carbon (atomic number $Z = 6$).

Solution Carbon has six electrons in the neutral atom. The order of filling subshells is 1*s*, 2*s*, 2*p*. The *s*-subshells contain a maximum of two electrons and the *p*-subshells each contain a maximum of six (two electrons in each of three orbitals). The electron configuration is then $1s^2 2s^2 2p^2$. In the ground state, carbon has two electrons in the 1*s* orbital, two in the 2*s* orbital and two in the 2*p* subshell. This notation does *not* show the individual orbitals in subshells with $\ell \neq 0$.

- **Building-Up (*Aufbau*) Principle** → Order in which electrons are added to *subshells* and *orbitals* to yield the *electron configuration* of atoms

Note: As the elements are built up in order of increasing atomic number, (or when atoms are ionized), subtle differences in shielding and penetration can change the order of the energy of subshells. Energy-level order is a function of both the nuclear charge and the number of electrons in an atom. *Thus, energy ordering of subshells is not fixed absolutely, but may vary from atom to atom, or from an atom to its ion.*

$(n + \ell)$ rule → Order of filling subshells in *neutral atoms* is determined by filling those with the *lowest* values of $(n + \ell)$ first. Subshells in a group with the same value of $(n + \ell)$ are filled in the order of increasing n.

The rule is summarized by the following diagram

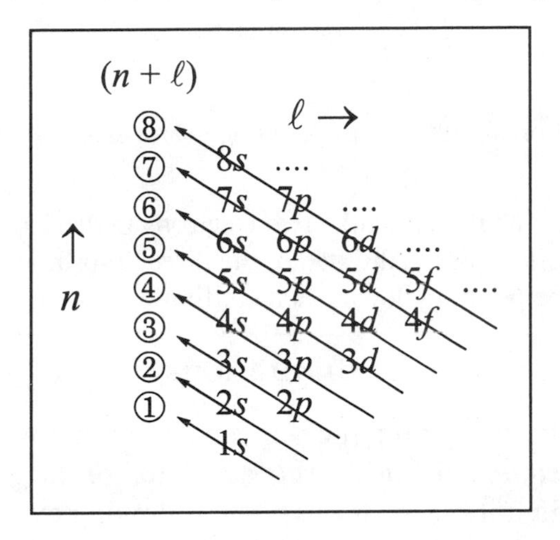

Start with 1s and read diagonally from the bottom to the top of each diagonal arrow with increasing $(n + \ell)$

- **Filling order of subshells** → $1s < 2s < 2p < 3s < 3p < 4s < 3d < 4p < 5s < 4d < \ldots$

- **Orbitals within a subshell** → Hund's rule:

  Electrons add to *different* orbitals of a subshell with spins *parallel* until the subshell is half full.

Example 1.12b Write the electron configuration of C ($Z = 6$) using *noble-gas abbreviations* for the core electrons.

Solution From Example 1.12a, the configuration is $1s^2 2s^2 2p^2$. The configuration of the nearest noble gas atom of lower Z is [He] = $1s^2$. Therefore, the abbreviated configuration is $[\text{He}]2s^2 2p^2$. This shorthand notation separates the *core* electrons ([He] = $1s^2$) from the *valence-shell* electrons. Core electrons are not involved in bonding and are relatively uninteresting.

Example 1.12c Write the electron configuration of C ($Z = 6$) showing orbital occupancy.

Solution The $2p$-subshell has two electrons. According to Hund's Rule, they should be placed in different orbitals. Therefore, the electron configuration showing orbital occupancy is: $1s^2 2s^2 2p_x^1 2p_y^1$ *or* $[\mathrm{He}]2s^2 2p_x^1 2p_y^1$.

This notation shows *orbital occupancy* (n, ℓ, m_ℓ quantum numbers). You may choose any two p-orbitals (x and y, x and z, or y and z).

Example 1.12d Write the electron configuration of C ($Z = 6$) showing orbital occupancy and the spin states of each electron. Use ↑ for $m_s = +\frac{1}{2}$ and ↓ for $-\frac{1}{2}$.

Solution The $2p$-subshell has two electrons. According to Hund's rule, they should be placed in different orbitals with parallel spins. Therefore, the electron

configuration is

↑↓	↑↓	↑	↑	__	or
$1s$	$2s$	$2p_x$	$2p_y$	$2p_z$	

↑↓	↑↓	↓	↓	__	or four other possibilities.
$1s$	$2s$	$2p_x$	$2p_y$	$2p_z$	

This notation shows *orbital occupancy* and *electron spin* (n, ℓ, m_ℓ, m_s quantum numbers).

Note: In accordance with Hund's rule, the two electrons in the $2p$-subshell occupy *separate* orbitals with *parallel* spins. Any two of the three $2p$-orbitals may be occupied and both spins could also be pointing down; such configurations are all *degenerate*.

1.13 Ground-State Electron Configurations

- **Applicability** → The building-up principle in combination with the *Pauli exclusion principle* and *Hund's rule* accounts for the *ground-state* electron configurations of atoms. The following examples show that the principle is generally valid, but there are exceptions.

Example 1.13a Write the ground-state electron configuration (*subshells* only) of germanium, Ge. How many *unpaired* electrons does Ge have?

Solution Ge has atomic number 32 and therefore has 32 protons and 32 electrons. The electrons will fill the subshells in the order given above: $1s$, $2s$, $2p$, $3s$, $3p$, $4s$, $3d$, $4p$, $5s$, $4d$, ... Recognizing that s-, p-, and d-subshells can hold 2, 6, and 10 electrons, respectively, we fill the subshells from lowest to highest energy with 32 electrons.

Ge: $1s^2 2s^2 2p^6 3s^2 3p^6 4s^2 3d^{10} 4p^2$ filling order
or $1s^2 2s^2 2p^6 3s^2 3p^6 3d^{10} 4s^2 4p^2$ final energy order
or $[\mathrm{Ar}]\ 3d^{10} 4s^2 4p^2$ noble-gas abbreviations

Note: The valence electrons are in the shell with $n = 4$. From the electron configuration, only the $4p$-subshell is partially filled and therefore only it can contain unpaired electrons. Using Hund's rule, the two electrons in the $4p$-subshell should occupy separate orbitals with spins parallel. So, Ge has two unpaired electrons:

↑	↑	__ .
$4p_x$	$4p_y$	$4p_z$

Example 1.13b In the ground state, how many *unpaired* electrons does selenium, Se, have?

Solution Se, atomic number 34, has two more electrons than Ge. The additional two electrons will enter the 4*p*-subshell. The first will enter the vacant $4p_z$-orbital and the second must pair with one of the three. So the 4*p*-subshell will appear as

$\underline{\uparrow\downarrow}$ $\underline{\uparrow}$ $\underline{\uparrow}$, so the Se atom also has two unpaired electrons in the ground state.
$4p_x$ $4p_y$ $4p_z$

Example 1.13c An *exception* to the rule. Determine the ground-state electron configuration of chromium, Cr.

Solution Cr has 24 electrons, and we expect the electron configuration to be

Cr: $1s^2 2s^2 2p^6 3s^2 3p^6 4s^2 3d^4$ filling order

However, the correct configuration is

Cr: $1s^2 2s^2 2p^6 3s^2 3p^6 4s^1 3d^5$ *or* $[\mathrm{Ar}]3d^5 4s^1$ final energy order

Note: To understand Example 1.13c, look at the orbitals of the outermost two subshells, 4*s* and 3*d*. We expect configuration **(A)** with four unpaired electrons in the 3*d* subshell; we find configuration **(B)** with six unpaired electrons, one in the 4*s*-orbital and 5 in the 3*d*-subshell.

(A) $\underline{\uparrow\downarrow}$ $\underline{\uparrow}$ $\underline{\uparrow}$ $\underline{\uparrow}$ $\underline{\uparrow}$ $\underline{\ \ }$
$4s$ $3d_{xy}$ $3d_{xz}$ $3d_{yz}$ $3d_{x^2-y^2}$ $3d_{z^2}$

(B) $\underline{\uparrow}$ $\underline{\uparrow}$ $\underline{\uparrow}$ $\underline{\uparrow}$ $\underline{\uparrow}$ $\underline{\uparrow}$
$4s$ $3d_{xy}$ $3d_{xz}$ $3d_{yz}$ $3d_{x^2-y^2}$ $3d_{z^2}$

We can rationalize configuration **(B)** by assuming that *half-filled* subshells have additional stability and lower energy. Configuration **(B)** has two *half-fill*ed subshells (4*s*, 3*d*) while configuration **(A)** has none. There are a number of such anomalies in the periodic table. (See Section 1.14.)

1.14 Electronic Structure and the Periodic Table

- **Order of filling subshells**
  - → Understanding the organization of the periodic table
  - → Straightforward determination of (most) electron configurations
- **Terminology** → Given in the following two tables

groups:	Columns in the periodic table, labeled 1–18 horizontally
***s*-block elements:**	Groups 1, 2; *s*-subshell fills
***p*-block elements:**	Groups 13–18; *p*-subshell fills
***d*-block elements:**	Groups 3–12; *d*-subshell fills
transition elements:	Groups 3–11; *d*-subshell fills
main-group elements:	Groups 1, 2 and 13–18
lanthanides:	4*f*-subshell fills
actinides:	5*f*-subshell fills

Note: A transition element has a partially filled *d*-subshell either as the element or in any commonly occurring oxidation state. By this definition, Group 12 metals (Zn, Cd, Hg) are not transition elements.

valence shell:	Outermost occupied shell (highest *n*)
period:	Row of the periodic table
period number:	Principal quantum number of valence shell
Period 1:	H, He; 1*s*-subshell fills
Period 2:	Li through Ne; 2*s*-, 2*p*-subshells fill
Period 3:	Na through Ar; 3*s*-, 3*p*-subshells fill
Period 4: (first long period)	K through Kr; 4*s*-, 3*d*-, 4*p*-subshells fill

- **Periodic table** → Two forms displayed below:
  One shows the elements, the other the *final* subshell filled.
  Look at the tables to understand the terminology.

Periodicity of Elements in the Periodic Table

Period ↓ **Group (1–18)** →

Period	1	2	3	4	5	6	7	8	9	10	11	12	13	14	15	16	17	18
1	H																	He
2	Li	Be											B	C	N	O	F	Ne
3	Na	Mg											Al	Si	P	S	Cl	Ar
4	K	Ca	Sc	Ti	V	*Cr*	Mn	Fe	Co	Ni	*Cu*	Zn	Ga	Ge	As	Se	Br	Kr
5	Rb	Sr	Y	Zr	*Nb*	*Mo*	Tc	*Ru*	*Rh*	***Pd***	*Ag*	Cd	In	Sn	Sb	Te	I	Xe
6	Cs	Ba	Lu	Hf	Ta	W	Re	Os	Ir	*Pt*	*Au*	Hg	Tl	Pb	Bi	Po	At	Rn
7	Fr	Ra	Lr	Rf	Db	Sg	Bh	Hs	Mt	Uun	Uuu	Uub	Uut					

↑

Lanthanides	*La*	*Ce*	Pr	Nd	Pm	Sm	Eu	*Gd*	Tb	Dy	Ho	Er	Tm	Yb
Actinides	*Ac*	***Th***	*Pa*	*U*	*Np*	Pu	Am	*Cm*	Bk	Cf	Es	Fm	Md	No

- **17 italicized elements**
  → Exceptions to the ($n + \ell$) rule
  Differ by the placement of *one* electron
  → See Examples 1.13c and 1.14c
- **2 bold, italicized elements** → Differ by the placement of *two* electrons (Pd, Th)

Pd We expect $[Kr]4d^8 5s^2$, but *actually* find $[Kr]4d^{10}$
No electrons in the 5*s*-subshell

Th We expect $[Rn]5f^2 7s^2$, but *actually* find $[Rn]6d^2 7s^2$
No electrons in the 5*f*-subshell

Order of Filling Subshells in the Periodic Table

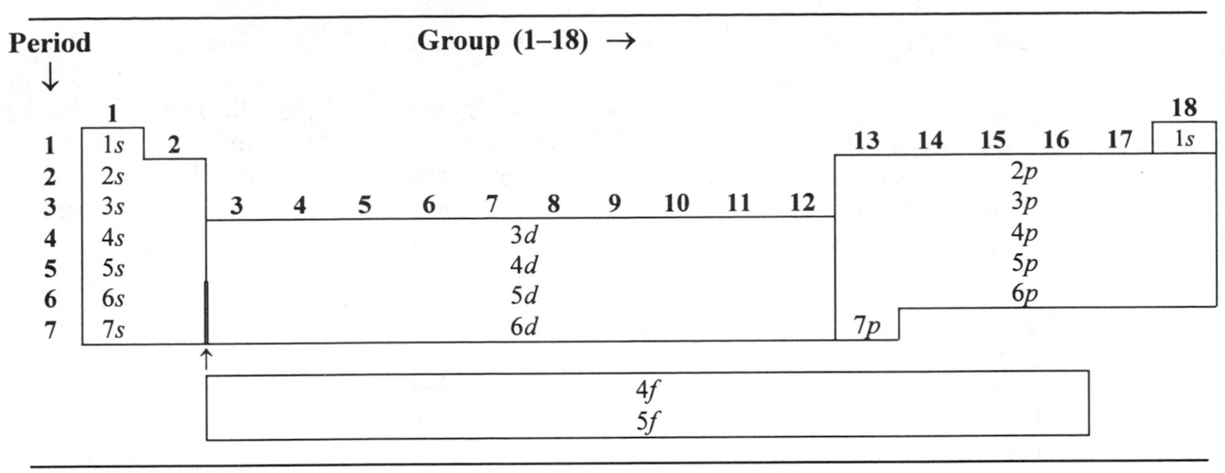

Example 1.14a Write the ground-state electron configuration (*core* plus *valence* electrons) of arsenic, As ($Z = 33$).

Solution The element As in Period 4 has the Ar ($Z = 18$) core and additional subshells $4s$, $3d$, and $4p$. According to its position in Group 15, the $4s$- and $3d$-subshells are filled and the $4p$-subshell has 3 electrons. It is not an exception. Thus, the configuration is

As: [Ar] $4s^2\,3d^{10}\,4p^3$ (filling order) *or* [Ar] $3d^{10}\,4s^2\,4p^3$ (final energy order)

The $4s$- and $3d$-subshells are filled whereas the $4p$-subshell is half filled, with one electron in each of the $4p_x$-, $4p_y$- and $4p_z$-orbitals. The filled $3d$-subshell becomes part of the core, as shown in the second configuration. There are *five* valence electrons.

Example 1.14b Write the ground-state electron configuration (*core* plus *valence* electrons) of zirconium, Zr ($Z = 40$).

Solution The element Zr is in Period 5 and Group 4. It has the Kr ($Z = 36$) core and is in the period with additional subshells $5s$, $4d$, and $5p$. It is *not* an exception. Thus, we have

Zr: [Kr] $5s^2\,4d^2$ (filling order) *or* [Kr] $4d^2\,5s^2$ (final energy order)

The $5s$-subshell is filled; the two $4d$-electrons occupy separate orbitals, with spins parallel according to Hund's rule. The second configuration shows the correct energy pattern *after* the subshells are filled.

Example 1.14c Write the ground-state electron configuration (*core* plus *valence* electrons) of gold, Au ($Z = 79$).

Solution The element Au is in Period 6 and Group 11. It has the Xe ($Z = 54$) core and is in the period with additional subshells $6s$, $5d$ and $6p$. We also observe that the $4f$ subshell (corresponding to the lanthanides) fills between Ba and Hf. Finally, we note that Au is a one-electron exception to the subshell-filling rule. Using the rule first, we find

Au: [Xe] $6s^2\,4f^{14}\,5d^9$, which is incorrect.

The correct configuration is

Au: [Xe] $6s^1\,4f^{14}\,5d^{10}$ (filling order) *or* [Xe] $4f^{14}\,5d^{10}\,6s^1$ (final energy order)

is obtained by moving one electron from the 6*s*- to the 5*d*-subshell. Thus, gold has a complete 5*d*-subshell and a half-filled 6*s*-subshell. The 4*f*-electrons are lower in energy (penetration) and are not involved in the chemistry of gold. The other two coinage metals, silver (Ag) and copper (Cu), show similar one-electron anomalies in their electron configurations.

The Periodicity of Atomic Properties (Sections 1.15-1.20)

Key Concepts

atomic radius, covalent radius, ionic radius, isoelectronic atoms and ions, ionization energy, second ionization energy, metallic character, inert-pair effect, diagonal relationship, electron affinity

1.15 Atomic Radius (*r*)

- **Definition of atomic radius**
  - → Half the distance between the centers of neighboring atoms (nuclei)
  - → For metallic elements, *r* is determined for the solid.

    Example: For solid Zn, 274 pm between nuclei, so *r* (atomic radius) = 137 pm
  - → For nometallic elements, *r* is determined for diatomic molecules (covalent bond).

    Example: For I_2 molecules, 266 pm between nuclei, so *r* (covalent radius) = 133 pm

- **General trends in radius with atomic number**
  - → *r* decreases from *left* to *right* across a period (*effective* nuclear charge increases)
  - → *r* increases from *top* to *bottom* down a group (change in valence electron principal quantum number and size of valence shell)

1.16 Ionic Radius

- **An ion's share of the distance between neighbors in an ionic solid (cation to anion)**
  - → Distance between the nuclei of a neighboring cation and anion is the sum of two ionic radii
  - → Radius of the oxide anion (O^{2-}) is 140 pm

    Example: Distance between Zn and O nuclei in zinc oxide is 223 pm; therefore Zn^{2+} is 83 pm. [$r\,(Zn^{2+}) = 223\text{ pm} - r\,(O^{2-})$]
  - → Cations are smaller than parent atoms; for example, Zn (133 pm) and Zn^{2+}(83 pm)
  - → Anions are larger than parent atoms; for example, O (66 pm) and O^{2-}(140 pm)
- **Isoelectronic atoms and ions** → Atoms and ions with the same number of electrons

Example 1.16 Consider the isoelectronic species of the elements sulfur, chlorine, argon, potassium and calcium with 18 electrons. Determine the charge on each species and arrange them in order of decreasing size.

Solution The atomic numbers of S, Cl, Ar, K, and Ca are 16, 17, 18, 19, and 20. In this problem, all species have 18 electrons. Sulfur with 16 protons has a charge of –2 (sulfide anion). Chlorine with 17 protons has a charge of –1 (chloride anion). Argon with 18 protons is neutral. Potassium with 19 protons has a charge of +1. Calcium with 20 protons has a charge of +2. Ionic size decreases as nuclear charge increases, therefore the expected order with decreasing size is

$$r\,(S^{2-}) > r\,(Cl^-) > r\,(Ar) > r\,(K^+) > r\,(Ca^{2+})$$

Note: The atomic radius of noble-gas atoms can be obtained for low-temperature solids (high pressure for helium). These atoms are held together by weak forces, so their size defined in this way is actually *larger* than the covalent radius of the neighboring halogen atom. There are other definitions of size. If *all* atomic sizes are defined from *atomic wavefunctions*, then each noble-gas atom is *smaller* than its neighboring halogen atom.

(See the Periodic Table web site: http://www.shef.ac.uk/~chem/web-elements/)

1.17 Ionization Energy (I)

- **Symbol** → Number subscripts denote removal of successive electrons.
- ***I*** → Energy needed to remove an electron from a gas-phase atom in its lowest energy state.

Example 1.17 Consider the first three ionization energies (I_1, I_2, and I_3) of the argon atom (Group 18 noble gases). Arrange the ionization energies in order of increasing magnitude. Look up the values on the Periodic Table web site: http://www.shef.ac.uk/chemistry/web-elements/Ar/key.html

Solution All ionization energies are positive numbers, and each successive value is larger than the preceding one. Therefore, the order is $I_3 > I_2 > I_1$. The reactions and values taken from the web site are

first ionization energy: $Ar(g) \rightarrow Ar^+(g) + e^-(g)$ $I_1 = 1520.6\text{ kJ·mol}^{-1}$
second ionization energy: $Ar^+(g) \rightarrow Ar^{2+}(g) + e^-(g)$ $I_2 = 2665.8\text{ kJ·mol}^{-1}$
third ionization energy: $Ar^{2+}(g) \rightarrow Ar^{3+}(g) + e^-(g)$ $I_3 = 3931\text{ kJ·mol}^{-1}$

- **General periodic table trends in I_1 for the main-group elements**
  → Increases from *left* to *right* across a period (Z$_{eff}$ increases)
  → Decreases from *top* to *bottom* down a group (change in principal quantum number *n* of valence electron)

  Exceptions: For Groups 2 and 15 in Periods 2–4, I_1 is *larger* than the neighboring main-group element in Groups 13 and 16, respectively. Repulsions between electrons occupying the same orbital and/or extra stability of completed and half-completed subshells are responsible for these exceptions to the general trends.

- **For metals found toward the lower left of the periodic table**
  → Low ionization energy
  → Atoms readily lose electrons

- **For nonmetals found toward the upper right of the periodic table**
  → High ionization energy
  → Atoms do not readily lose electrons

1.18 The Inert-Pair Effect

- **Tendency to form ions two units lower in charge than expected from the group number relation**
  → Due in part to the different energies of the valence *p*- and *s*-electrons
  → Important for the *lower* two members of Groups 13, 14, and 15
  → Valence *s*-electrons are called a "lazy pair"

Example 1.18 List the elements in the main-group that exhibit the inert-pair effect.

Answer			
	Group 13	In^{+} and In^{3+}	Tl^{+} and Tl^{3+}
	Group 14	Sn^{2+} and Sn^{4+}	Pb^{2+} and Pb^{4+}
	Group 15	Sb^{3+} and Sb^{5+}	Bi^{3+} and Bi^{5+}

1.19 Diagonal Relationships

- **Diagonally related pairs of elements often show similar chemical properties**
  → Diagonal band of metalloids dividing metals from nonmetals
  → Similarity of Li and Mg (react directly with N_2 to form nitrides)
  → Similarity of Be and Al (both react with acids and bases)

1.20 Electron Affinity (E_{ea})

- **Energy *released* when an electron is added to a gas-phase atom**

Example 1.20a Write equations showing the chemical reactions occurring when two electrons are added *sequentially* to an atom of sulfur, S. Show the energy change, ΔE, for each reaction and the value of E_{ea}.

Solution Unlike ionization energy values, electron affinity values may be positive or negative numbers. Energy *release* requires a change in sign: $E_{ea} = -\Delta E$. If the electron-attachment reaction gives off energy, then $E_{ea} > 0$ (anion stable). If the electron-attachment reaction consumes energy, then $E_{ea} < 0$ (anion unstable). The values taken from Fig. 1.49 for sulfur are

first electron affinity: $S(g) + e^-(g) \rightarrow S^-(g)$ $\quad E_{ea} = -\Delta E = +200\ kJ{\cdot}mol^{-1}$

second electron affinity: $S^-(g) + e^-(g) \rightarrow S^{2-}(g)$ $\quad E_{ea} = -\Delta E = -532\ kJ{\cdot}mol^{-1}$

The anion, S^-, is stable in the gas phase, whereas the anion, S^{2-}, is unstable.

Example 1.20b Write an expression that combines the two reactions in Example 1.20a. Calculate the value of E_{ea} for the attachment of *two* electrons to S in one step.

Solution Add the two reactions together to obtain the overall reaction for the addition of two electrons to S. Combine the energy changes for the two reactions as well.

combined reaction: $S(g) + 2e^-(g) \rightarrow S^{2-}(g)$ $\quad E_{ea} = 200 + (-532) = -332\ kJ{\cdot}mol^{-1}$

Notes: Only mononegative anions are stable in the gas phase.

Some mononegative anions are always unstable: nitrogen, the Group 2 elements, and the Group 18 elements are the major examples.

In ionic solids, many *polynegative* anions are stable: O^{2-}(oxide), S^{2-}(sulfide), N^{3-}(nitride), C^{4-}(carbide: methanide). There are other carbides: C_2^{2-} (carbide: acetylide).

- **General periodic table trends in E_{ea} for the main-group elements**
  - → Increases from *left* to *right* across a period (Z_{eff} increases)
  - → Decreases from *top* to *bottom* down a group (change in principal quantum number n of valence electron)
  - → Generally, the same as for I_1 with the major exceptions given in Example 1.20b

- **Summary of trends in r, I_1, and E_{ea}**
  - → All depend on Z_{eff} and n of outer subshell electrons.
  - → Recall that there are *many* exceptions to the general trends, as noted earlier.

Main-Group Elements in the Periodic Table

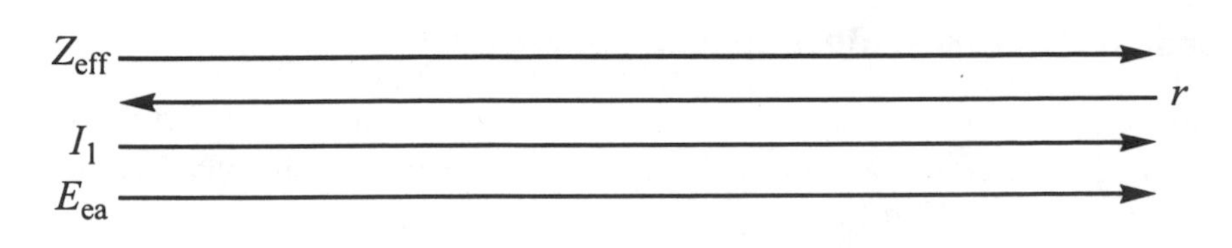

H							He
Li	Be	B	C	N	O	F	Ne
Na	Mg	Al	Si	P	S	Cl	Ar
K	Ca	Ga	Ge	As	Se	Br	Kr
Rb	Sr	In	Sn	Sb	Te	I	Xe
Cs	Ba	Tl	Pb	Bi	Po	At	Rn

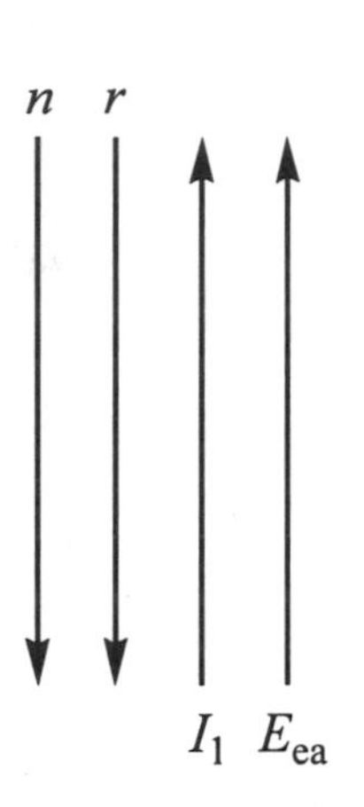

The Impact on Materials (Sections 1.21-1.22)

Key Concepts

metals, metalloids, nonmetals, *s*-block elements, *p*-block elements, *d*-block elements, transition metals

1.21 The Main-Group Elements (Groups 1–2, 13–18)

- ***s*-block elements (Groups 1 and 2)** → All are reactive metals (except H); they form *basic* oxides (O^{2-}), peroxides (O_2^{2-}), or superoxides (O_2^{-}).

Example 1.21 List the chemical reactions of sodium metal with water and oxygen.

Solution Sodium metal reacts violently with water to form a basic solution, a metal cation, and hydrogen gas. Sodium metal reacts with gaseous oxygen to form the peroxide salt. Write the overall reactions and balance by inspection.

Answer

$2\,Na(s) + 2\,H_2O(l) \rightarrow 2\,Na^+(aq) + 2\,OH^-(aq) + H_2(g)$ (See Section H.1)

$2\,Na(s) + O_2(g) \rightarrow Na_2O_2(s)$ [O_2^{2-} (basic peroxide ion)]

- **Compounds of *s*-block elements are ionic, except for beryllium.**

Notes: *Hydrogen* (a nonmetal) is placed by itself or *more usually* in Group 1, because its electronic configuration, $1s^1$, is similar to those of the alkali metals: [noble gas] ns^1.

Helium, with electronic configuration $1s^2$, is placed in Group 18, because its properties are similar to those of neon, argon, krypton, and xenon.

- ***p*-block elements (Groups 13–18)** → Members are metals, metalloids, and nonmetals.

Metals: Group 13 (Al, Ga, In, Tl)
Group 14 (Sn, Pb)
Group 15 (Bi)

Note: These metals have relatively low I_1, but *larger* than those of the *s* block and *d* block.

Metalloids: Group 14 (Si, Ge)
Group 15 (As, Sb)
Group 16 (Te, Po)

Note: Metalloids have the *physical* appearance and properties of *metals*, but behave *chemically* as *nonmetals*.

Nonmetals: Group 13 (B)
Group 14 (C)
Group 15 (N, P)
Group 16 (O, S, Se)
Group 17 (All)
Group 18 (All)

Notes: High E_{ea} in Groups 13–17; these atoms tend to *gain* electrons to complete their subshells.

Group 18 noble-gas atoms are generally *nonreactive*, except for Kr and Xe, which form a few compounds.

1.22 The Transition Metals

- ***d*-block elements (Groups 3–12)** → All are metals (*most* of them are transition metals), with properties intermediate to those of *s*-block and *p*-block metals.

Note: *All* the cations formed by Group 12 members retain the filled *d*-subshell. For this reason, these elements are *not* considered to be transition metals.

- **Characteristically, many transition metals form compounds with a variety of oxidation states (oxidation numbers).**

Example 1.22 Chromium forms oxides with the oxidation states of +2, +3, and +6. Write the formula of each oxide compound. Show the electron configuration of each cation.

Solution Use the oxidation number rules and procedures in Section K.2. The electron configuration of Cr is $[Ar]3d^5 4s^1$. When forming cations, remove the outer 4*s* electron first.

Answer

Oxidation state	Oxide compound	Cation configuration
II	CrO	Cr^{2+} $[Ar]3d^4$
III	Cr_2O_3	Cr^{3+} $[Ar]3d^3$
VI	CrO_3	Cr^{6+} $[Ar]$

Note: Common anion salts incorporating chromium (VI) are CrO_4^{2-} (chromate) and $Cr_2O_7^{2-}$ (dichromate).

- ***f*-block elements (lanthanides and actinides)**
  → Rare on Earth
  → Lanthanides are incorporated in *superconducting* materials
  → Actinides are all *radioactive* elements

Chapter 2 Chemical Bonds

Ionic Bonds (Sections 2.1-2.4)

Key Concepts

chemical bonds, valence electrons, ionic bonds, ionic model, ionic solids, crystalline solids, lattice energy, Madelung constant, refractory material, octet of electrons, duplet of electrons, Lewis symbols, variable valence

Overview

- **Chemical bond** → *Link* between atoms
- **Valence electrons** → Electrons in the *outermost shells*

2.1 The Formation of Ionic Bonds

- **Ionic bond** → Electrostatic attraction (coulombic) of *oppositely* charged ions

Example 2.1a Calculate the energy change for the reaction $K(g) + Cl(g) \rightarrow K^+(g) + Cl^-(g)$.

Solution Break the reaction into two parts: ionization and electron gain. Use the ionization energy and electron affinity data from Chapter 1 in the text.

Ionization:	$K(g) \rightarrow K^+(g) + e^-(g)$	$I_1 =$	$418\ kJ \cdot mol^{-1}$
Electron gain:	$Cl(g) + e^-(g) \rightarrow Cl^-(g)$	$-E_{ea} =$	$-349\ kJ \cdot mol^{-1}$
Overall:	$K(g) + Cl(g) \rightarrow K^+(g) + Cl^-(g)$		$\Delta E = 1\ mol \times (I_1 - E_{ea}) = 69\ kJ$

Example 2.1b Given that I_2 for Ca(g) is $1150\ kJ \cdot mol^{-1}$, calculate the energy change for the reaction:

$$Ca(g) + 2\,Br(g) \rightarrow Ca^{2+}(g) + 2\,Br^-(g)$$

Answer $\Delta E = 1\ mol \times (I_1 + I_2) - (2\ mol \times E_{ea}) = [590 + 1150 - (2 \times 325)]\ kJ = 1090\ kJ$

- **Ionic model** → Energy required for the *formation* of ionic bonds is supplied *mainly* by the coulombic attraction between *oppositely* charged ions. The ionic model is a good description of bonding between metals (particularly *s*-block metals) and nonmetals.

- **Ionic solids**
  - → Assembly of cations and anions arranged in a regular array
  - → The ionic bond is *nondirectional*; each ion is bound to *all* its neighbors.
  - → Ionic solids have *high* melting and boiling points.
  - → Form electrolyte solutions if they dissolve in water

- **Crystalline solids** → Atoms, molecules, or ions arranged in a regular pattern

2.2 Interactions Between Ions

- **Coulomb potential energy**

$$V_{12} = \frac{(z_1 e) \times (z_2 e)}{4\pi\varepsilon_0 r_{12}} = \frac{2.307\, z_1 z_2}{r_{12}} \times 10^{-19}\,\text{J·nm}$$

e is the elementary charge; z_1 and z_2 are the charges on the two ions; r_{12} is the distance in nanometers (nm) between the ions; ε_0 is the vacuum permittivity.

Example 2.2a Calculate the energy change for the reactions: (A) $K^+(g) + Cl^-(g) \rightarrow KCl(g)$ and (B) $KCl(g) \rightarrow K(g) + Cl(g)$, where $r_{12} = 0.267$ nm for KCl(g).

Note: The bonding is considered to be totally ionic and only the coulomb energy between ions is considered. The repulsive energy and other sources of attractive energy are ignored. In (B), use the results of (A) and Example 2.1a.

Solution (A) The energy between two ions at infinite separation is 0. The energy change ΔE for one molecule is then V_{12}(product) – V_{12}(reactants).

$$\Delta E = V_{12}(r_{12} = 0.267\text{ nm}) - V_{12}(r_{12} = \infty) = \frac{2.307(+1)(-1)}{(0.267\text{ nm})} \times 10^{-19}\,\text{J·nm} - 0 = -8.64 \times 10^{-19}\,\text{J}$$

or $N_A \times \Delta E = -520$ kJ for 1 mol of KCl molecules

Note: Compare $K^+(g) + Cl^-(g) \rightarrow KCl(s)$, where $N_A \times \Delta E = -717$ kJ

(B) Break the reaction into two parts:

$KCl(g) \rightarrow K^+(g) + Cl^-(g)$	–(–520 kJ)	(A)
$K^+(g) + Cl^-(g) \rightarrow K(g) + Cl(g)$	–(69 kJ)	Example 2.1a
$KCl(g) \rightarrow K(g) + Cl(g)$	451 kJ	

The experimentally derived value for this reaction is 429 kJ.

Example 2.2b Calculate the energy change for the reactions (A) $Ca^{2+}(g) + 2\,Br^-(g) \rightarrow CaBr_2(g)$ and (B) $CaBr_2(g) \rightarrow Ca(g) + 2\,Br(g)$, where $r_{12} = 0.296$ nm (sum of the ionic radii of the cation and anion) for $CaBr_2(g)$, in which the bonding is considered to be totally ionic and the molecule is assumed to be linear.

Consider the coulomb interaction for each cation-anion pair and single anion-anion pair separately. As before, the energy at infinite separation is 0. *Beware* of the cation charge and the anion-anion distance. In (B), use the results of (A) and Example 2.1b.

Answer (A) $N_A \times \Delta E = -1642$ kJ for one mol of $CaBr_2$ molecules

(B) –(–1642 kJ) –1090 kJ = 552 kJ

Note: The ionic radii are obtained from ionic solids. In Example 2.2a, the actual bond distance for a gas-phase KCl molecule is used.

- **Lattice energy** → Potential energy difference:
  ions in solid *minus* ions infinitely far apart

- **Madelung constant (A)** $\boxed{V = -A \times \dfrac{|z_1 z_2| N_A e^2}{4\pi\varepsilon_0 d}}$ A is a positive numerical integer; d is the distance between the centers of nearest neighbors in the solid crystal.

→ Depends on the *arrangement* of ions in the solid

Values of A → 1.747 56 (NaCl structure) [$d = 0.2798$ nm for NaCl (actual value at 0 K)]
1.762 67 (CsCl structure) [$d = 0.351$ nm for CsCl (sum of ionic radii)]

Example 2.2c Calculate the ratio of the lattice energy of KCl to that of RbCl. Use the ionic radii of the cations and anions in Chapter 1 of the text to estimate *d*. Notice that both KCl and RbCl have the NaCl crystal structure.

Solution The Madelung constants and ion charges are the same for KCl and RbCl. Therefore, the ratio of lattice energies is simply

$$\frac{V_{KCl}}{V_{RbCl}} = \frac{d_{RbCl}}{d_{KCl}} = \frac{149 \text{ pm} + 181 \text{ pm}}{138 \text{ pm} + 181 \text{ pm}} = 1.03$$

Example 2.2d Calculate the ratio of the lattice energy of CsCl to that of CsI. Both solids have the CsCl crystal structure.

Answer $(170 + 220)/(170 + 181) = (390/351) = 1.11$

- **Refractory material** → Substance that can withstand high temperatures; MgO, for example.

- **Coulomb interaction between ions in a solid** → *Large* when ions are highly charged and small

2.3 The Electron Configurations of Ions

- **Cations** → Remove outermost electrons in the order *np*, *ns*, $(n-1)d$

  Example: Iron (II), Fe^{2+}, $[Ar]3d^6$ and iron (III), Fe^{3+}, $[Ar]3d^5$

- **Anions** → Add electrons until the next noble-gas configuration is reached.

  Example: Carbide (methanide), C^{4-}, $[He]2s^2 2p^6$ or [Ne] {octet}
  Hydride, H^-, $1s^2$ or [He] {duplet}

- **Formulas of compounds composed of monatomic ions**
  → Formulas are predicted by assuming that atoms forming cations lose all valence electrons and those forming anions gain electrons in the valence subshell(s) until each ion has an octet of electrons or a duplet in the case of H, He, and Be.
  → For cations with variable valence, the Stock number, or oxidation number, is used (see Fundamentals section of text).
  → Relative numbers of cations and anions are chosen to achieve electrical neutrality, using the smallest possible integers as subscripts.

Example 2.3 Determine the formula of calcium bromide.

Answer Ca, $[Ar]4s^2$, loses two 4s-electrons to become Ca^{2+}, [Ar].
Br, $[Ar]3d^{10}4s^24p^5$, gains an electron to become Br^-, $[Ar]3d^{10}4s^24p^6$ or [Kr].
Electrical neutrality: combining two Br^- ions for each Ca^{2+} gives $CaBr_2$, an ionic solid.

2.4 Lewis Symbols (for Atoms and Ions)

- **Valence electrons** → Depicted as dots
  → Pair of dots represents two paired electrons

Example 2.4a Write the Lewis symbols for the neutral atoms in the first two periods of the Periodic Table.

Answer H• He: Li• Be: •B• •C• :N• :O• :F• :Ne:

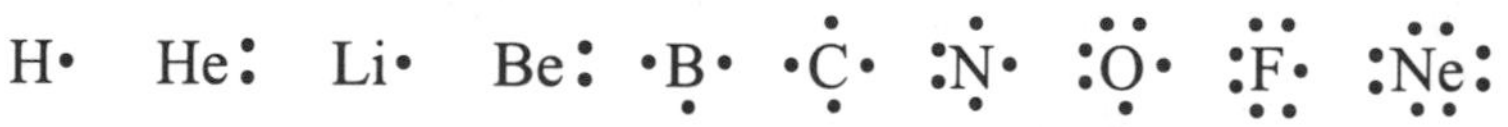

Note: The ground-state structures for B and C are :B• and :C• .

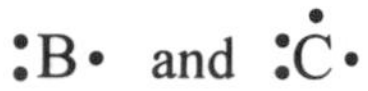

Example 2.4b Write the Lewis symbols for the lithium cation, beryllium cation, nitride anion, oxide anion, and the fluoride anion.

Answer Li^+ Be^{2+} $[:N:]^{3-}$ $[:O:]^{2-}$ $[:F:]^-$

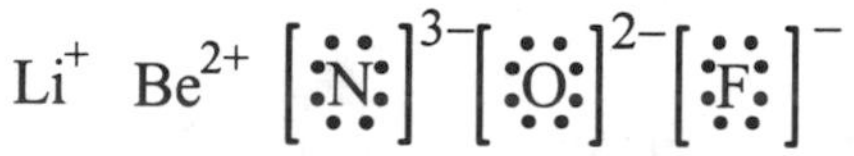

Example 2.4c Write the Lewis symbols for the reaction between calcium atoms and oxygen atoms to form calcium oxide.

Answer Ca: + :O• ⟶ $Ca^{2+}[:O:]^{2-}$

Covalent Bonds (Sections 2.5-2.9)

Key Concepts

covalent bond, octet rule, valence, lone pairs, Lewis structure, single bond, double bond, triple bond, multiple bonds, bond order, resonance, resonance hybrid, delocalization, Kekulé structure, formal charge, structure plausibility

2.5 The Nature of the Covalent Bond

- **Covalent bond**
  → Pair of electrons *shared* between two atoms
  → Lies between two neighboring atoms and *binds* them together

  Example: Nonmetallic elements H_2, N_2, O_2, F_2, Cl_2, Br_2, I_2, P_4, S_8

2.6 Lewis Structures

- **Rules**
  → Atoms attempt to complete their duplets or octets by sharing electron pairs.
  → Valence of an atom is the number of bonds it can form.
  → A line (–) represents a shared pair of electrons.
  Example: H–H, (single bond) duplet on each atom (valence = 1) and **:N≡N:,** (triple bond and lone pair) octet on each atom (valence = 3)

2.7 Lewis Structures for Polyatomic Species

- **Rules**
  → Count the total number of valence electrons in the species.
  → Arrange the atoms next to bonded neighbors.
  → Use the minimum number of electrons to make all single bonds.
  → Count the number of nonbonding electrons required to satisfy octets.
  → Compare to the actual number of electrons left.
  → If lacking a sufficient number of electrons to satisfy octets, make *one extra bond* for each deficit pair of electrons.
  → Octet or duplet is completed by *sharing* pairs of electrons with a neighbor.
  → Each shared pair counts as one *covalent bond* (line).

One shared pair	**single bond** (–)	
Two shared pairs	**double bond** (=)	(multiple bond)
Three shared pairs	**triple bond** (≡)	(multiple bond)

- **Bond order** → Number of bonds that link a specific pair of atoms
- **Terminal atom** → Bonded to only one other atom
- **Central atom** → Bonded to at least two other atoms
- **Molecular ions** → Consist of *covalently* bonded atoms: NH_4^+, Hg_2^{2+}, SO_4^{2-}
- **Rules of thumb**
  → Usually the element with the lowest I_1 is a central atom, but electronegativity is a better indicator (see Section 2.14).
  For example, in HCN, carbon has the lowest I_1 and is the central atom. It is also less electronegative than nitrogen.
  → Usually, there is a symmetrical arrangement about the central atom.
  For example, in SO_2, OSO is symmetrical, with S as the central atom and two terminal O atoms.
  → Oxoacids have H atoms bonded to O atoms; H_2SO_4 is actually $(HO)_2SO_2$.

Example 2.7a Determine the Lewis structure of ethyne (acetylene), C_2H_2.

Solution Count the number of valence electrons [2(4) + 2(1) = 10 valence electrons or 5 electron pairs]. Draw a symmetrical arrangement of atoms with C atoms as central atoms, since H atoms are terminal. There are a minimum of three single bonds [H–C–C–H] accounting for three pairs of electrons. Four electron pairs are required to complete the octets on carbon atoms, but there are only two pairs left. Make two additional bonds. Since H atoms make only one bond, the two additional bonds are located between the carbon atoms.

Answer The final Lewis structure is H–C≡C–H with one triple bond between the carbon atoms and no lone pair (nonbonding) electrons.

Example 2.7b Determine the Lewis structure of hydrogen cyanide, HCN.

Solution Count the number of valence electrons [1 + 4 + 5 = 10 valence electrons or 5 electron pairs]. Draw an arrangement of atoms with the C atom as the central atom, since C has the lowest I_1 and the H atom is terminal. There are a minimum of two single bonds [H–C–N] accounting for two pairs of electrons. Five electron pairs are required to complete the octets on the carbon and nitrogen atoms, but there are only three pairs left. One of the three pairs remaining is assigned as a lone pair and the other two pairs make additional bonds. Since H atoms make only one bond, the two additional bonds are located between the carbon and nitrogen atoms. The lone pair is assigned to the nitrogen atom.

Answer The final Lewis structure is H–C≡N: with one triple bond between the carbon and nitrogen atoms and a lone pair (nonbonding) of electrons on the nitrogen atom.

Example 2.7c Determine the Lewis structure of the ammonium ion, NH_4^+.

Answer Valence electrons = 5 + 4(1) – 1 (electron removed) = 8 or four pairs.

```
[    H    ]+
[    |    ]
[ H—N—H ]
[    |    ]
[    H    ]
```

NH_4^+ has the same Lewis structure as methane, CH_4.

2.8 Resonance

- **Electron delocalization** → Many molecules can be represented by different Lewis structures in which the *location* of the electrons, but *not* the nuclei, vary.
- **Blending of structures** → Depicted by double-headed arrows (⟷)
- **Blended structure** → **Resonance hybrid** of the contributing Lewis structures
- **Delocalized electrons** → Distributed over several pairs of atoms

Example: N_2O, nitrous oxide, has 2(5) + 6 = 16 valence electrons or eight pairs.

:N̈=N=Ö: ⟷ :N≡N–Ö: (with three lone pairs on O) Blending of two structures

Note: A triple bond to an O atom is only found in BO^-, CO, NO^+, and O_2^{2+}.

Note: Central N atom is the least electronegative atom (Section 2.14).

Example: C_6H_6, benzene, has 6(4) + 6(1) = 30 valence electrons or 15 pairs.

Kekulé structures Resonance hybrid

Note: Carbon atoms are at the corners of the hexagon (the six C–H bonds radiating from each corner are not shown). The last structure depicts six valence electrons *delocalized* around the ring.

2.9 Formal Charge

- **Formal charge** → An atom's number of valence electrons (*V*) minus the number of electrons assigned to it in a Lewis structure
- **Electron assignment**
  → Atom possesses all of its lone pair electrons (*L*) and half of its bonding electrons (*S*) [*S* means shared electrons].
  → Formal charge = $V - (L + \frac{1}{2}S)$
- **Contribution of individual Lewis structures to the resonance hybrid** → The structures with individual formal charges as close to 0 as possible usually have the lowest energy and are major contributors.

Example 2.9a Calculate the formal charges on each atom in the first Lewis structure of N_2O given above, namely :N̈=N=Ö: . Also, calculate the formal charges in the second structure with a triple bond between the N atoms and in one with a triple bond between N and O.

Solution Consider the first structure.

Left N atom: formal charge = 5 – [4 + (0.5)4] = 5 – 6 = –1
Central N atom: formal charge = 5 – [0 + (0.5)8] = 5 – 4 = +1
Right N atom: formal charge = 6 – [4 + (0.5)4] = 6 – 6 = 0
The sum of the formal charges equals the charge on the species.

Answer Finally, :N̈=N=Ö: (formal charges –1, +1, 0). For the second structure, :N≡N–Ö: with three lone pairs on O (formal charges 0, +1, –1).

For the structure with a triple bond between N and O, :N̈–N≡O: with three lone pairs on the left N (formal charges –2, +1, +1). In this case, the formal charges are not as close to 0 as in the previous two structures. The last structure does not contribute appreciably to the resonance hybrid and is normally not included.

- **Plausibility of isomers** → Isomer with lowest formal charges is usually preferred.

Example 2.9b Write the Lewis structures for hydrogen cyanide, HCN, and hydrogen isocyanide, HNC. Calculate the formal charges and predict the isomer that is favored.

Answer H–C≡N: (formal charges 0, 0, 0) and H–N≡C: (formal charges 0, +1, –1) Lower formal charges on HCN, which is favored.

Example 2.9c Write the Lewis structure with two double bonds for nitrous oxide, NNO, and its isomer, NON. Calculate the formal charges and predict the isomer that is favored.

Answer

$$\underset{-1}{:\ddot{N}}=\underset{+1}{N}=\underset{0}{\ddot{O}:} \quad \text{and} \quad \underset{-1}{:\ddot{N}}=\underset{+2}{O}=\underset{-1}{\ddot{N}:}$$

Lower formal charges on NNO, which is energetically more favorable. Note that there are usually only one or two bonds to an oxygen atom except for BO^-, CO, NO^+, O_2^{2+}, and O_3 (central O atom).

- **Summary**
  - → Formal charge gives an indication of the extent to which atoms have gained or lost electrons in a Lewis structure (covalent bonding).
  - → Structures with the lowest formal charges usually have the lowest energy (major contributors to the resonance hybrid).
  - → Isomers with lower formal charges are usually favored.

Exceptions to the Octet Rule (Sections 2.10-2.11)

Key Concepts

radicals, antioxidant, biradicals, expanded valence shell, variable covalence

2.10 Radicals and Biradicals

- **Radicals**
  - → Species having electrons with unpaired spins
  - → All species with an odd number of electrons are radicals.
  - → Highly reactive; cause rancidity in foods and degradation of plastics in sunlight, and perhaps human aging
  - → CH_3 (methyl radical), OH (hydroxyl), OOH (hydrogenperoxyl) NO (nitric oxide), NO_2 (nitrogen dioxide), O_3^- (ozonide ion)

- **Biradicals**
  - → Two unpaired electrons
  - → For example, O (atom) and CH_2 (methylene) [on single atom]
    O_2 (oxygen) and larger organic molecules [on different atoms]

- **Antioxidant** → Additive that reacts rapidly with radicals before they have a chance to do damage

Example 2.10 Write the Lewis structure(s) for the ozonide ion, O_3^-.

Solution First determine the number of valence electrons. Locate them in the molecular ion to satisfy the octet rule, if possible. Because this is a radical, one atom will be unsatisfied. If more than one structure is possible, show the blending of Lewis structures (resonance hybrid).

Answer Valence electrons = 6 + 6 + 6 + 1 (charge on anion) = 19. The all-single-bond structure is O–O–O with 15 electrons unaccounted for. A total of 16 electrons are needed to satisfy the octet rule, but only 15 remain. Three choices for locating the odd electron are $[:\dot{\ddot{O}}-\ddot{O}-\ddot{O}:]^- \longleftrightarrow [:\ddot{O}-\dot{\ddot{O}}-\ddot{O}:]^- \longleftrightarrow [:\ddot{O}-\ddot{O}-\dot{\ddot{O}}:]^-$.

2.11 Expanded Valence Shells

- **Expanded valence shells**
  - → More than 8 electrons on an atom in a Lewis structure (requires empty *d*-orbitals)
  - → Electrons may be present as lone pairs or as bonding pairs
  - → Characteristic of nonmetal atoms in Period 3 or higher
- **Variable covalence**
  - → Ability to form different numbers of covalent bonds
  - → Some elements showing variable covalence:

Valence Shell Occupancy

Elements	8 electrons	10 electrons	12 electrons
P	PCl_3 / PCl_4^+	PCl_5	PCl_6^-
S	SF_2	SF_4	SF_6
I	IF	IF_3	IF_5

Example 2.11 Write the major Lewis structures for sulfuric acid, H_2SO_4.

Solution Count the number of valence electrons and first produce a structure satisfying the octet rule. Then consider octet expansion, which is allowed for sulfur atoms.

Answer The number of valence electrons = 2(1) + 6 + 4(6) = 32. A minimum structure requires 6 single bonds or 12 electrons. The remaining 20 electrons may occupy lone pair sites, thus satisfying the octet/duplet rule. Recall that oxoacids have H atoms bonded to O atoms, and S is a central atom. When the structure is drawn with only single bonds, the formal charge on S is +2. This charge can be reduced to +1 by forming an *additional* bond from a lone pair on one terminal O atom. It can be further reduced to 0 by forming an *additional* bond from a lone pair on the other terminal O atom. Three representative structures are

```
   :Ö:                    :Ö                    :Ö
    |                      ||                    ||
H−Ö−S−Ö−H   ←→   H−Ö−S−Ö−H   ←→   H−Ö−S−Ö−H
    |                      |                     ||
   :Ö:                    :Ö:                   :Ö
```

Note: In the structure on the right, all atoms have a formal charge of 0. There is one additional Lewis structure with a single double bond not shown. Other examples similar to this one include SO_4^{2-}, SO_2, and S_2O (S central atom).

Lewis Acids and Bases (Sections 2.12-2.13)

Key Concepts:

coordinate covalent bond, Lewis acid, Lewis base, Lewis acid-base complex

2.12 The Unusual Structures of Group 13 Halides

- **Incomplete octet**
  → Fewer than eight valence electrons on an atom in a Lewis structure; for example, BF_3 :F–B–F: (with :F: below B)
  → Other compounds with incomplete octets are BCl_3 and $AlCl_3$, which is a vapor at very high temperature.
- **Coordinate covalent bond** → Donation of a lone pair by one atom to create a bond that completes the octet of another atom with an incomplete octet

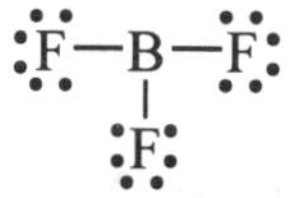

Example 2.12a Write Lewis structures for the reaction of BF_3 and NH_3.

Answer $BF_3 + :NH_3 \longrightarrow F_3B\text{—}NH_3$

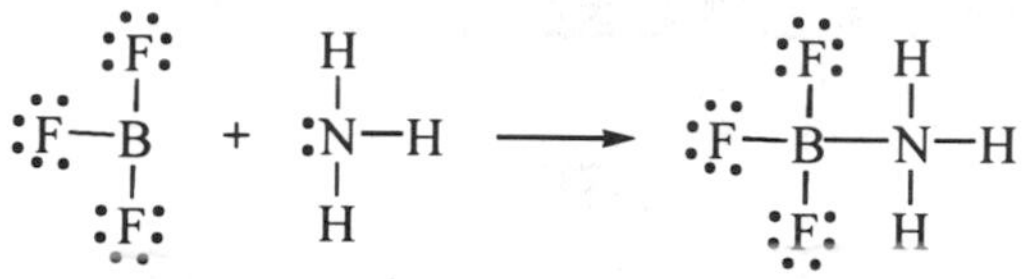

Example 2.12b Write the Lewis structure for the dimer Al_2Cl_6 molecule (vapor at high temperatures) that is formed from the reaction of $AlCl_3$ and $AlCl_3$.

Solution Each coordinate covalent bond is formed by the donation of a lone pair of electrons on Cl to complete the octet on Al. The resulting Al–Cl–Al bridging bonds (**bold**) are equivalent.

Answer $Cl_2Al(\mu\text{-}Cl)_2AlCl_2$ (two Al atoms bridged by two Cl atoms, each Al bearing two terminal Cl)

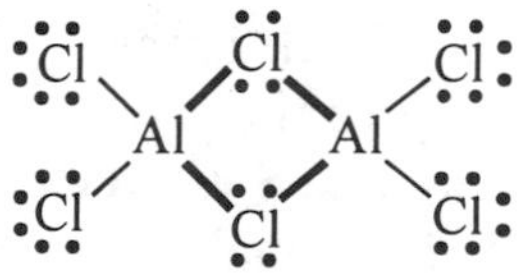

2.13 Lewis Acid-Base Complexes

- **Lewis base**
  → Electron pair donor
  → NH_3, OH^-, F^-, O^{2-}

- **Lewis acid**
  → Electron pair acceptor
  → BF_3, H^+ (also metal cations: Mg^{2+}, Al^{3+})

- **Lewis acid-base complex**
  → Product of reaction between a Lewis acid and a Lewis base
  → Acid + :base → complex

Example 2.13 Write Lewis structures for the reaction of BF_3 and F^-.

Answer

$$BF_3 + [:\ddot{F}:]^- \longrightarrow [BF_4]^-$$

Note: BF_4^- (tetrafluoroborate ion) is a Lewis acid-base complex. The bond formed is coordinate covalent, but, as in the Al_2Cl_6 bridging bonds (Example 2.12b), the four B–F bonds are equivalent.

- **Reactions of H^+, OH^-, H_2O, and H_3O^+**
  → Consider the reaction between a Lewis acid-base complex (H_3O^+) and the Lewis base (OH^-): $H_3O^+ + OH^- \rightarrow 2\,H_2O$

  The Lewis acid H^+ in the H_3O^+ complex reacts with the Lewis base OH^- to form two molecules of a Lewis acid-base complex H_2O.
  → Now consider the reverse reaction: $H_2O + H_2O \rightarrow H_3O^+ + OH^-$
  The Lewis acid H^+ in one H_2O complex reacts with the Lewis base H_2O.

Ionic versus Covalent Bonds (Sections 2.14-2.15)

Key Concepts

partial charges, polar covalent bond, electric dipole, electric dipole moment, debye, electronegativity, ionic character, polarizable, polarizing power

2.14 Correcting the Covalent Model: Electronegativity

- **Bonds** → All molecules are resonance hybrids of pure covalent and ionic structures.
  Consider H_2, the structures are $H{-}H \longleftrightarrow H^+H^- \longleftrightarrow H^-H^+$.

- **Partial charges**
  → Ionic structures are not energetically equivalent when the bonded atoms differ.
  → Unequal sharing of electrons results in a **polar covalent bond** between atoms bearing partial charges.

Consider HF, the structures are H—F $\longleftrightarrow$ H^+F^- $\longleftrightarrow$ H^-F^+.
(lone pairs not shown)

Note: $E_{ea}(F) > E_{ea}(H) \Rightarrow {}^{\delta+}H\text{–}F^{\delta-}$ (partial charges, $|\delta+| = |\delta-|$)
$E(H^+F^-) \ll E(H^-F^+) \Rightarrow H^-F^+$ is a *minor* contributor

- **Electric dipole** → A partial *positive* charge separated from an equal but *negative* partial charge
- **Electric dipole moment (μ)**
  → Size of an electric dipole: partial charge times distance between charges
  → Units: debye (D)
  → 4.80 D ≡ an electron (–) separated by 100 pm from a proton (+)
  → $\boxed{\mu = (4.80\ \text{D}) \times \delta \times (\text{distance in pm} / 100\ \text{pm})}$

Example 2.14a In the HF molecule, the measured dipole moment is 1.82 D and the distance between partial charges (bond distance) is 91.7 pm. Estimate the partial charges on the HF molecule by rearranging the above relationship to solve for δ.

Answer

$$\delta = \frac{\mu}{(4.80\ \text{D}) \times (\text{distance in pm}/100\ \text{pm})} = \frac{1.82\ \text{D}}{(4.80\ \text{D}) \times (91.7\ \text{pm}/100\ \text{pm})}$$

$$\delta = 0.413$$

- **Electronegativity (χ)**
  → Electron-attracting power of an atom when it is part of a bond
  → Mulliken scale: $\chi = \frac{1}{2}(I_1 + E_{ea})$
  Follows same periodic table trends as I_1 and E_{ea}
  → Pauling numerical scale based on bond energies [χ(F) = 4.0] used in text (qualitatively similar to Mulliken's scale).
  See Section 2.14 in the text for the tablulation of values.
  → $\boxed{\mu = (\chi_A - \chi_B)\ \text{D}}$

Example 2.14b Estimate the dipole moment of the HF molecule by using the electronegativity values of H and F. Compare with the measured value of 1.82 D.

Answer $\mu = (\chi_F - \chi_H)\ \text{D} = (4.0 - 2.2)\ \text{D} = 1.8\ \text{D}$, very close to the measured value. The HF bond has considerable *ionic character*. It is a *polar* covalent bond.

- **Rough rules of thumb**

$(\chi_A - \chi_B) \geq 2$	bond is *essentially* ionic
$0.5 \leq (\chi_A - \chi_B) \leq 1.5$	bond is *polar* covalent
$(\chi_A - \chi_B) \leq 0.5$	bond is *essentially* covalent

2.15 Correcting the Ionic Model: Polarizability

- **Ionic bonds** → All have *some* covalent character.
  A cation's positive charge attracts the anion's electrons in the direction of the cation (*distortion* of spherical electron cloud).
- **Highly *polarizable* atoms and ions** → Readily undergo a *large* distortion.
  For example, large anions and atoms I^-, Br^-, Cl^-, I, Br, and Cl.
- **Polarizing power** → Property of ions (and atoms) that cause large distortions.
  For example, the small and highly charged cations Li^+, Be^{2+}, Mg^{2+}, and Al^{3+} have significant polarizing power.
- **Significant covalent bonding character**
  → The bonds in compounds composed of highly polarizing cations and highly polarizable anions have significant covalent character.
  → The Be^{2+} cation is highly polarizing and the Be–Cl bond is significantly covalent (χ difference of 1.6).
  → In the series AgCl to AgI, the bonds become more covalent as the polarizability of the anion increases ($Cl^- < Br^- < I^-$).

The Strengths and Lengths of Covalent Bonds (Sections 2.16-2.18)

Key Concepts

dissociation energy, bond strength, bond length, covalent radius

2.16 Bond Strengths

- **Dissociation energy (*D*)**
  → *Strength* of a chemical bond
  The *greater* the dissociation energy, the *stronger* the bond.
  → *D* is defined exactly for diatomic (two-atom) molecules.
  For polyatomic (greater than two-atom) molecules, *D* also depends on the other bonds in the molecule.
- ***D* for several diatomic molecules**

H–H(g)	→ H(g) + H(g)	$D = 432\ kJ{\cdot}mol^{-1}$	(strong)
H–F(g)	→ H(g) + F(g)	$D = 562\ kJ{\cdot}mol^{-1}$	(strong)
F–F(g)	→ F(g) + F(g)	$D = 155\ kJ{\cdot}mol^{-1}$	(weak)
I–I(g)	→ I(g) + I(g)	$D = 149\ kJ{\cdot}mol^{-1}$	(weak)
O=O(g)	→ O(g) + O(g)	$D = 494\ kJ{\cdot}mol^{-1}$	(strong)
N≡N(g)	→ N(g) + N(g)	$D = 942\ kJ{\cdot}mol^{-1}$	(very strong)

2.17 The Variation of Bond Strength

- **Bond strength** → For polyatomic molecules, bond strength is defined as the *average* dissociation energy for one type of bond found in different molecules.

  For example, the tabulated C–H single bond value is the *average* strength of such bonds in a selection of organic molecules, such as methane (CH_4), ethane (C_2H_6), and ethene (C_2H_4).

  Note: The values of average dissociation energy given in Table 2.2 in the text are actually ones for the thermodynamic property of *bond enthalpy* introduced in Section 6.21. The thermodynamic values of *bond enthalpy* are measured at a temperature of 298.15 K. The bond *dissociation energies* of diatomic molecules given in Section 2.16 apply at a temperature of 0 K (absolute zero).

- **Factors influencing bond strength**
  - → Bond multiplicity (C≡C > C=C > C–C)
  - → Resonance (C=C > C⋯C (benzene) > C–C)
  - → Lone pairs on neighboring atoms (F–F < H–H)
  - → Atomic radii (HF > HCl > HBr > HI)
    (the smaller the radius, the stronger the bond)

- **Assessing energy changes (ΔE)** → In gas-phase chemical reactions, energy change (ΔE) is equal to the sum of the dissociation energies (D) of bonds broken minus those of bonds formed.

$$\Delta E = \sum nD \text{ (bonds broken)} - \sum nD \text{ (bonds formed)}$$

Notes: Keep track of the number of moles n of each bond broken and each one formed.
Dissociation energy has the units of kJ per mol of bonds.

Example 2.17a Use Tables 2.2 and 6.6 in the text to calculate the energy change for the reaction (oxidation of ethanol to acetic acid and water):

$$CH_3CH_2OH + O_2 \rightarrow CH_3COOH + H_2O$$

Solution First draw Lewis structures for each reactant and product. List the moles of bonds of a given type that are broken and formed. If there is no net change in bonds of a given type, these can be ignored. Finally, calculate the energy change by the method discussed above.

Answer The reaction written in terms of Lewis structures is

```
  H  H                               H :Ö
  |  |   ..                          |  ‖    ..              ..
H-C--C--O-H   +   :Ö=Ö:   ──►     H-C--C--O-H    +    H-O-H
  |  |   ..                          |       ..              ..
  H  H                               H
```

Bonds broken	Bonds formed	Net change
5 C–H	3 C–H	2 C–H broken
1 C–C	1 C–C	no change in C–C
1 C–O	1 C–O	no change in C–O
1 O–H	3 O–H	2 O–H formed
1 O=O	no O=O	1 O=O broken
no C=O	1 C=O	1 C=O formed

Net E of bonds broken: 2 (C–H) + 1 (O=O) = 2 mol (412 kJ·mol^{-1}) + 496 kJ = 1320 kJ
Net E of bonds formed: 2 (O–H) + 1 (C=O) = 2 mol (463 kJ·mol^{-1}) + 743 kJ = 1669 kJ
ΔE = (bonds broken – bonds formed) = 1320 kJ – 1669 kJ = –349 kJ (–439 kJ actual)

Note: Table 6.6 is used for consistency, since the dissociation energies given in Table 2.2 and Table 6.7 are bond enthalpy values (see Section 6.21).

Example 2.17b Use Table 2.2 in the text to calculate the energy change for the reaction (decarboxylation of acetic acid):

$$CH_3COOH \rightarrow CH_4 + CO_2$$

Answer ΔE = (bonds broken – bonds formed) = 16 kJ (–36 kJ actual)

Note: The value of D for C=O has a substantial range depending on which molecule the bond is a constituent of. A more appropriate value in CO_2 is 804 kJ·mol^{-1}. Then, ΔE = –106 kJ.

Note: The values of ΔE are not quantitative but are a good approximation to the correct values.

2.18 Bond Lengths

- **Bond length**
  - → Distance between the centers of two atoms joined by a covalent bond
  - → Helps determine the overall size and shape of a molecule
  - → Evaluated by spectroscopy or x-ray diffraction of solids
  - → Length is *inversely* proportional to strength
- **Factors influencing bond length**
  - → Bond multiplicity (C≡C < C=C < C–C)
  - → Resonance (C=C < C┅C (benzene) < C–C)
  - → Lone pairs on neighboring atoms (F–F > H–H)
  - → Atomic radii (HF < HCl < HBr < HI)
    (the smaller the radius, the shorter the bond)

  Note: These trends are the opposite of the ones for bond strength.

- **Covalent radius**
  → *Half* the distance between the centers of neighboring atoms joined by a covalent bond (*for like atoms*)
  See Section 1.15 (nonmetallic elements).
  → Atom's contribution to the length of a covalent bond (*for unlike atoms*)
  → Covalent radii are added together to estimate the lengths of bonds in molecules.
  → Tabulated values are *averages* of radii in polyatomic molecules.

Example 2.18a Use the covalent radii displayed in Fig. 2.26 in the text to estimate the various bond lengths in the acetic acid molecule, CH_3COOH.

Solution First write a Lewis structure for acetic acid. Add together the covalent radii of the atoms in each bond to estimate its length.

Answer Lewis structure:

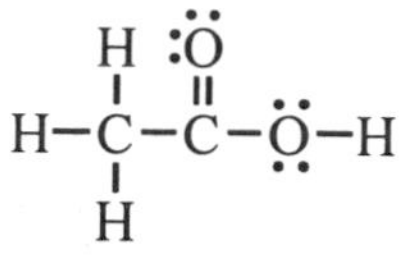

```
  H :Ö
  |  ||  ..
H−C−C−O−H
  |      ..
  H
```

Bond type	Bond length estimate	Actual
C–H	77 pm + 37 pm = 114 pm	≈109 pm
C–C	77 pm + 77 pm = 154 pm	≈150 pm
C–O	77 pm + 74 pm = 151 pm	134 pm
C=O	67 pm + 60 pm = 127 pm	120 pm
O–H	74 pm + 37 pm = 111 pm	97 pm

Example 2.18b Use the covalent radii displayed in Fig. 2.26 in the text to estimate the various bond lengths in the hydroxyl amine molecule, NH_2OH.

Answer N–H, 112 pm N–O, 149 pm O–H, 111 pm

Note: The H atom covalent radius of 37 pm obtained from H_2 overestimates bond lengths in other molecules. A better value is 30 pm for –H bonds except H–H.

Chapter 3 Molecular Shape and Structure

The Shapes of Molecules and Ions (Sections 3.1-3.3)

Key Concepts

molecular formula, structural formula, space-filling model, ball-and-stick model, bond angle, valence-shell electron-pair repulsion (VSEPR) model, bonding pairs, lone pairs, central atom, electron arrangement about central atom, multiple *versus* single bonds, polar bond, nonpolar bond; molecular shapes: linear, angular (bent), trigonal planar, trigonal pyramidal, tetrahedral, square pyramidal, trigonal bipyramidal, octahedral, T-shaped, seesaw, square pyramidal, square planar

Representation of Molecules

- **Molecular formula** → Number and types of atoms present in the molecule or ion
- **Structural formula** → Atoms represented by chemical symbols, linked by lines
- **Lewis structure** → Atoms represented by chemical symbols, valence electrons shown as lines (*bonds*) or dots (*lone pairs*)
- **Space-filling model** → Shaded or colored shapes (atoms) that fit into one another
- **Ball-and-stick model** → Shaded or colored balls (atoms) linked by sticks (bonds)

Examples for Ammonia

Type	Example	Advantage(s)	Disadvantage(s)
Molecular formula	NH_3	Shows chemical composition	No structural information
Structural formula	H—N—H H	Shows chemical connectivity (which atoms are bonded together)	Little three-dimensional structural information
Lewis formula	H—$\ddot{N}$—H H	Shows chemical connectivity and all valence electrons involved in bonding, reactivity	Little three-dimensional structural information
Space-filling model		Three-dimensional contour of molecule clearly visible	Bonds difficult to see, particularly for interior atoms
Ball-and-stick model		Atoms, connectivity, and general shape visible	Poor representation of actual bonds and atoms

3.1 The Valence-Shell Electron-Pair Repulsion (VSEPR) Model

- **Valence electrons about central atom(s)** → Controls shape of a molecule
- **Lewis structure** → Shows distribution of *valence* electrons in bonding pairs (bonds) and as lone pairs
- **Bonds, lone pairs** → Regions of high electron density that repel each other (Coulomb's law) by rotating about the central atom, thereby maximizing their separation distance
- **Bond angle(s)** → Angle(s) between bonds joining atom centers
- **Multiple bonds** → Treated as a *single* region of high electron concentration
- **Electron arrangement** → Ideal locations of bond pairs and lone pairs about the central atom (angles between electron pairs define the geometry)

Typical Electron Arrangements of Molecules with *One* Central Atom*

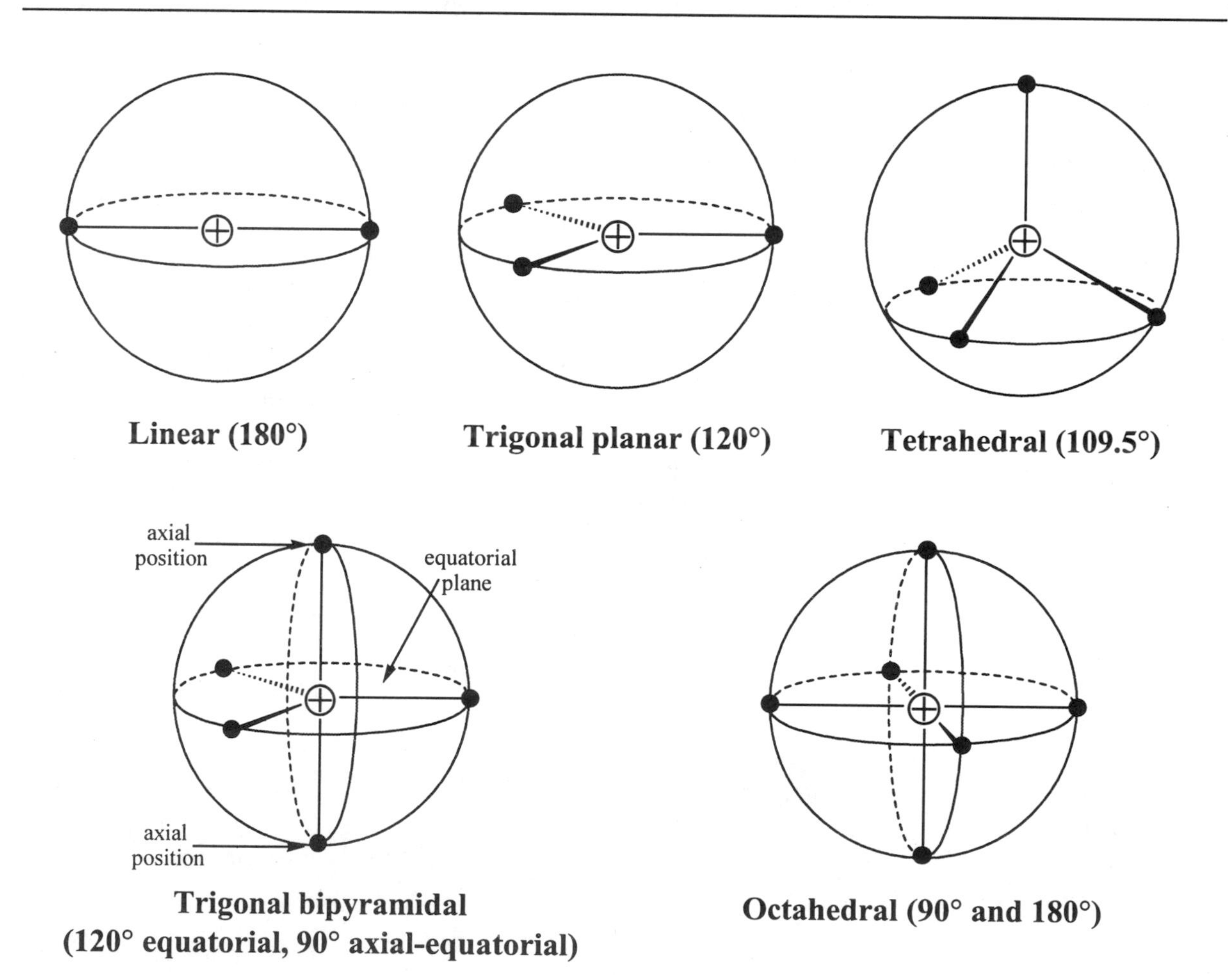

Linear (180°) **Trigonal planar (120°)** **Tetrahedral (109.5°)**

Trigonal bipyramidal (120° equatorial, 90° axial-equatorial) **Octahedral (90° and 180°)**

* Pentagonal bipyramidal is not shown.

Typical Shapes of Molecules with *One* Central Atom*

(Ball-and-Stick Models)

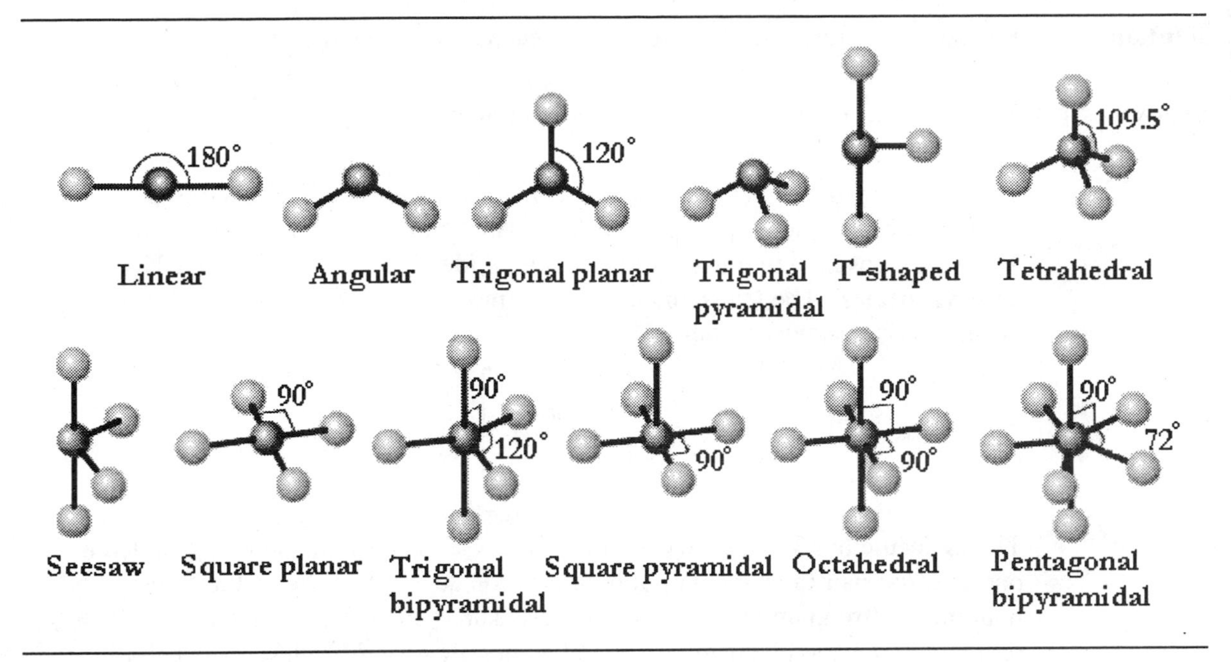

* These figures depict the locations of the atoms only, *not* lone pairs.

- **VSEPR formula, AX_nE_m** (See Section 3.2 in the text)
  - → A = central atom
  - → X_n = *n* atoms (*same* or *different*) bonded to central atom
  - → E_m = *m* lone pairs on central atom
  - → Useful for generalizing types of structures

Note: The shape of the molecule is defined by the location of the *atoms alone*.
The bond angles are $\angle$ X–A–X.

- **VSEPR method** (See Toolbox 3.1 in the text)
  - → Write the Lewis structure(s).
    If there are resonance structures, pick *any* one.
  - → Count the number of electron pairs (bonding and nonbonding) around the central atom(s).
    Treat a multiple bond as a *single* unit of high electron density.
  - → Identify the *electron arrangement.*
    Place electron pairs as far apart as possible.
  - → Locate the *atoms* and classify the *shape* of the molecule.
  - → Optimize *bond angles* for molecules with *lone pairs* on the central atom(s) with the concept in mind that the repulsions are in the order
    lone pair-lone pair > lone pair-bonding pair > bonding pair-bonding pair.

Example 3.1a Use the VSEPR model to predict the *shape* of (1) fluoromethane, CH_3F; (2) the phosphate anion, PO_3^{3-}; (3) sulfur hexafluoride, SF_6; and (4) tetrafluoroethene (tetrafluorethylene), C_2F_4. All are molecules with no lone pairs ($m = 0$).

Solution For each molecule, follow the general procedure outlined above.

Answer (1) The Lewis structure of CH_3F (C is the central atom) is

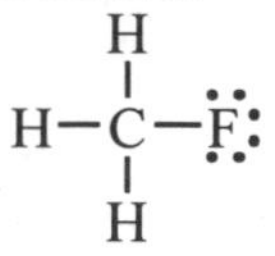

```
   H
   |  ..
H—C—F:
   |  ''
   H
```

The molecule has 14 valence electrons. There are four bonding electron pairs around carbon. The shape is *tetrahedral*. The bond angles will *not* be exactly 109.5°, because the terminal atoms are *not* all identical. The molecule CH_3F is an AX_4 species. The lone pairs on the fluorine atom are not considered because it is a *terminal* atom.

(2) One Lewis structure of PO_4^{3-} is

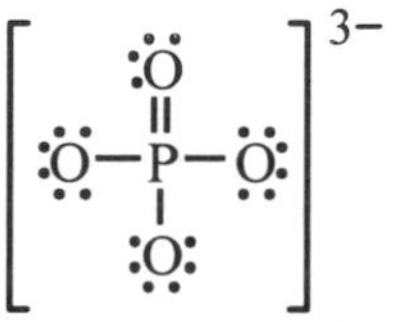

```
[      ..       ]3−
[     :O        ]
[  ..  ||   ..  ]
[ :O—P—O:       ]
[  ''  |    ''  ]
[     :O:       ]
[      ''       ]
```

The molecule has 32 valence electrons. Because the multiple bond is counted as one electron pair (a *single* unit of high electron density), there are then four bonding electron pairs around phosphorus. The shape is again *tetrahedral*. The terminal atoms are identical. The bond angles are *exactly* 109.5°. The anion PO_4^{3-} is an AX_4 species. The Lewis structure is an example of octet expansion on the central P atom. An equally acceptable Lewis structure with four single bonds between P and O gives the same answer.

(3) The Lewis structure of SF_6 is

```
 ..  ..  ..
:F  :F:  F:
 ''\ | /''
     S
 ../ | \..
:F  :F:  F:
 ''  ''  ''
```

The molecule has 48 valence electrons. There are six bonding electron pairs around sulfur. The shape is *octahedral*. The terminal atoms are identical. The bond angles are exactly 90° and 180°. The molecule SF_6 is an AX_6 species. The Lewis structure is an example of octet expansion on the central S atom.

(4) The Lewis structure of C_2F_4 (each C is a central atom) is

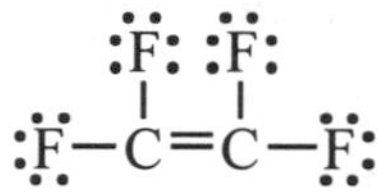

```
      ..  ..
     :F: :F:
 ..   |   |   ..
:F—C=C—F:
 ''           ''
```

The molecule has 36 valence electrons. Because the multiple bond is counted as one electron pair, there are three bonding electron pairs around each carbon atom. The shape is *trigonal planar* on each end of the molecule. The molecule C_2F_4 is an AX_3AX_3 species. As before, the lone pairs on the fluorine atoms are not considered because they are on *terminal* atoms.

Note: According to the VSEPR method, one might expect a *twisted* structure that minimizes electron repulsion for molecule (4). Experiments show, however, that this and similar molecules containing double bonds are planar overall. Its inability to predict planarity in this instance is a failing of the VSEPR method.

Example 3.1b Name the type of species according to the VSEPR formula (AX_nE_m) and predict the shapes of the following molecules (1) carbon disulfide, CS_2, and (2) ethane, C_2H_6. Both are molecules with no lone pairs ($m = 0$).

Answer (1) AX_2, linear; (2) AX_4AX_4, tetrahedral at each C atom

3.2 Molecules with Lone Pairs on the Central Atom

- **VSEPR method for molecules with lone pairs ($m \neq 0$)**
  - → Lone pair electrons have preferred positions in certain electron arrangements.
  - → In the *trigonal bipyramidal* arrangement, lone pairs prefer the *equatorial* positions, in which electron repulsions are minimized.
  - → In the *octahedral* arrangement, all positions are *identical*. The preferred electron arrangement for *two lone pairs* is the occupation of *opposite* corners of the octahedron.

Example 3.2a Use the VSEPR model to predict the shape of (1) trichloroamine, NCl_3; (2) xenon tetrafluoride, XeF_4; (3) xenon difluoride, XeF_2; and (4) sulfur tetrafluoride, SF_4. All are molecules with lone pairs ($m \neq 0$).

Solution Follow the general procedure used in Section 3.2 for each molecule, but consider the modifications indicated above.

Answer (1) The Lewis structure of NCl_3 (N is the central atom) is

:C̤̈l—N̈—C̤̈l:
 |
:C̤̈l:

The molecule has 26 valence electrons. There are four electron pairs around nitrogen (three bonding pairs and one lone pair). The electron arrangement is *tetrahedral*. The shape is *trigonal pyramidal*. The lone pair-bonding pair repulsion is greater than the bonding pair-bonding pair value. Therefore, the bond angles [∠ Cl–N–Cl] will be smaller than 109.5°. The molecule NCl_3 is an AX_3E species. The lone pairs on the chlorine atoms are not considered because they are on *terminal* atoms.

(2) The Lewis structure of XeF_4 is

:F̤̈ ⟍ ·· ⟋ F̤̈:
Xe
:F̤̈ ⟋ ·· ⟍ F̤̈:

The molecule has 36 valence electrons. There are six electron pairs around xenon (four bonding pairs and two lone pairs). The electron arrangement is *octahedral*. The shape is *square planar* because the lone pairs prefer opposite corners of the octahedron. The bond angles will be 90° and 180° in the plane of the molecule. The molecule XeF_4 is an AX_4E_2 species. The lone pairs on the fluorine atoms are not considered because they are on *terminal* atoms.

(3) The Lewis structure of XeF_2 is :F̤̈—Ẍ̤e:—F̤̈:

The molecule has 22 valence electrons. There are five electron pairs around xenon (two bonding pairs and three lone pairs). The electron arrangement is *trigonal bipyramidal*. The shape is *linear* because the lone pairs prefer the *equatorial* positions. The bond angle will be 180°. The molecule XeF_2 is an

AX_2E_3 species. The lone pairs on the fluorine atoms are not considered because they are on *terminal* atoms.

(4) The Lewis structure of SF_4 is 

The molecule has 34 valence electrons. There are five electron pairs around sulfur (four bonding pairs and one lone pair). The electron arrangement is *trigonal bipyramidal*. The shape is *seesaw* because the lone pair prefers one of the *equatorial* positions. The bond angles will be less than 120° in the equatorial plane and less than 90° between the axial and equatorial positions (greater repulsions between bonding and lone pair electrons than between bonding electron pairs). The molecule SF_4 is an AX_4E species. The lone pairs on the fluorine atoms are not considered because they are on *terminal* atoms.

Example 3.2b Name the type of species according to the VSEPR formula (AX_nE_m) and predict the shapes of the following molecules: (1) sulfur dioxide, SO_2; (2) the triiodide anion, I_3^-; (3) iodine pentafluoride, IF_5; (4) methyl alcohol, CH_3OH. All are molecules with lone pairs ($m \neq 0$).

Answer (1) AX_2E, angular; (2) AX_2E_3, linear; (3) AX_5E, square pyramidal; (4) AX_4, tetrahedral at C, AX_2E_2, angular at the O atom.

3.3 Polar Molecules (Review Section 2.14)

- **Polar molecule** → Molecule with a *nonzero* dipole moment
- **Polar bond** → Bond with a *nonzero* dipole moment
- **Molecular dipole moment** → Vector sum of bond dipole moments
- **Representation** → Use arrow with head pointed toward the *positive* end of dipole.
  Example: Lewis structure H—C̤̈l: and polar (dipole) representation H⟵Cl
- **Predicting molecular polarity**
  → Determine the *shape* of the molecule using VSEPR theory
  → Estimate electric dipole (bond) moments (Section 2.14 in the text) or use Fig. 3.10 in the text to decide whether the *symmetry* of the molecule leads to a *cancellation* of bond moments (*nonpolar molecule*, no dipole moment) or not (*polar molecule*, nonzero dipole moment).

Example 3.3a Identify which, if any, of the following molecules are polar: (1) nitrogen triiodide, NI_3; (2) xenon difluoride, XeF_2; (3) carbon tetrachloride, CCl_4. All three molecules have polar bonds.

Solution Determine the shape of the molecule by using VSEPR theory. Decide whether the symmetry of the molecule leads to a nonzero dipole moment.

Answer (1) The Lewis structure of NI_3 (N is the central atom) is 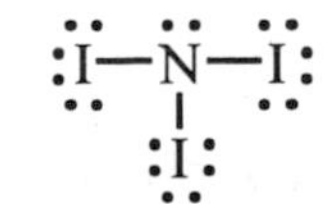

The molecule has 26 valence electrons. The electron arrangement is *tetrahedral*. The shape is *trigonal pyramidal*. The molecule NI_3 is an AX_3E species. The bond dipoles do *not* cancel and the molecule is *polar*.

(2) The Lewis structure of XeF_2 is :F–Xe–F:

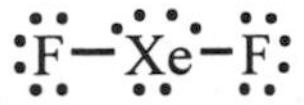

The molecule has 22 valence electrons. The electron arrangement is *trigonal bipyramidal*. The shape is *linear* because the lone pairs prefer the equatorial positions. The molecule XeF_2 is an AX_2E_3 species. The bond angle will be 180°, and the bond dipoles cancel. The molecule is *nonpolar*. F→Xe←F

(3) The Lewis structure of CCl_4 is

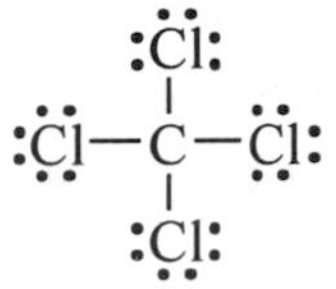

The molecule has 32 valence electrons. The electron arrangement and shape are both *tetrahedral*. The molecule CCl_4 is an AX_4 species. The bond angles of 109.5° point to four of the eight corners of a cube in such a way that the bond dipoles cancel. The molecule is *nonpolar*.

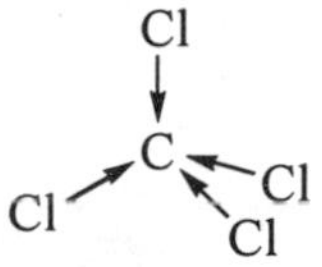

Note: For *identical* terminal atoms, all VSEPR formulas with $m = 0$ yield *nonpolar* molecules. Assuming that the *bonds* are all *polar*, the formulas with $m \neq 0$ all yield *polar* molecules, except for AX_2E_3 and AX_4E_2, which are *nonpolar*.

Example 3.3b Identify which, if any, of the following molecules are polar: (1) water, H_2O; (2) methane, CH_4; (3) iodine trichloride, ICl_3.

Answer (1) AX_2E_2, angular, *polar;* (2) AX_4, tetrahedral, *nonpolar;* (3) AX_3E_2, T-shaped, *polar.*

Note: The C–H *bond* is essentially *nonpolar*. Molecules containing only C and H that have nonsymmetrical arrangements of atoms are essentially *nonpolar*.

Valence-Bond (VB) Theory (Sections 3.4-3.8)

Key Concepts

valence-bond theory, σ-bond, π-bond, nodal surface, overlap of atomic orbitals, single bond, double bond, triple bond, promotion, hybridization of atomic orbitals, hybrid atomic orbitals, hybridization schemes and relation to geometry, hydrocarbons, alkanes, alkenes, alkynes, properties of double bonds

Overview

- **Two-center covalent bond** → Overlap of atomic orbitals (AO) or hybridized atomic orbitals (HO), each containing an electron of *opposite spin* on an adjacent atom
- **Bond strength** → As the degree of overlap of adjacent orbitals (AO or HO) increases, bond strength increases.

3.4 Sigma (σ) and Pi (π) Bonds

- **Two major types of *bonding* orbitals** → σ and π
- **σ-orbital**
  → Has no *nodal surface* containing the interatomic (bond) axis
  → Cylindrical or "sausage" shape
  → Formed from the overlap of two *s*-orbitals, an *s*-orbital and a *p*-orbital end-to-end, two *p*-orbitals end-to-end, a certain hybrid orbital and an *s*-orbital, or a *p*-orbital end-to-end with a certain hybrid orbital
- **Example of σ-orbital formation** → Two 1*s*-orbitals of H atoms combine to form a σ-orbital of H_2.

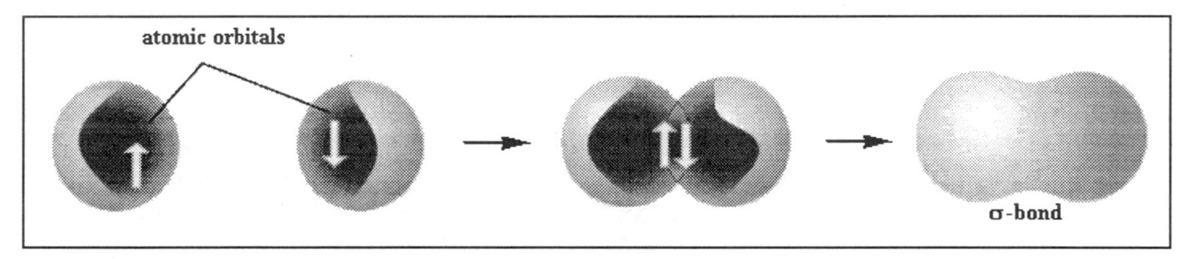

- **π-orbital**
  → *Nodal plane* containing the interatomic (bond) axis
  → Two cylindrical shapes, one above and the other below the nodal plane
  → Formed from side-by-side overlap of two *p*-orbitals
- **Example of π-orbital formation** → Two $2p_x$-orbitals overlap side-by-side.

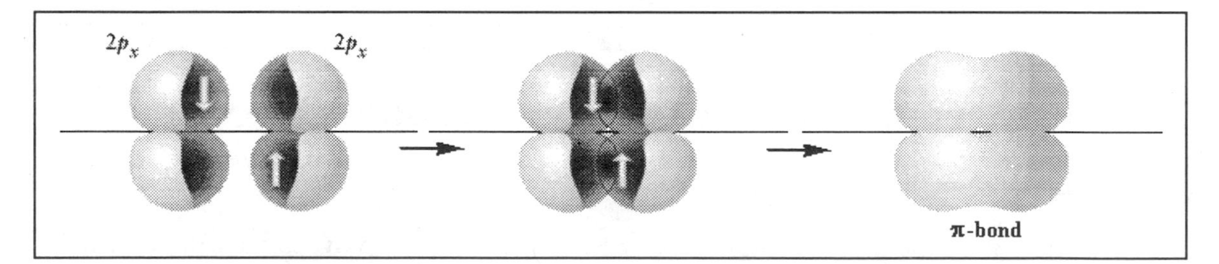

- **Electron occupancy**
  → σ- and π-orbitals can hold 0, 1, or 2 electrons, corresponding to no bond, a half-strength bond, or a full covalent bond, respectively.
  → Spins of two electrons must be *paired* (opposite direction of arrows).

- **Types of bonds according to valence-bond theory**
  → **Single bond** is a σ-bond [H_2, for example]
  → **Double bond** is a σ-bond plus one π-bond [O_2, for example]
  → **Triple bond** is a σ-bond plus two π-bonds [N_2, for example]

Example 3.4a Use the valence-bond theory to describe the bonding in (1) H_2; (2) HCl; (3) O_2.

Answer (1) In H_2, each H atom has one unpaired electron in a 1*s*-orbital. Two H atoms with *opposite* spins are brought together to form H_2. The two 1*s*-orbitals overlap to form *one σ-bond* (paired electrons).

(2) In HCl, the H atom has one unpaired electron in a 1*s*-orbital; the Cl atom has one unpaired electron in a 3*p*-orbital. The two orbitals overlap end-to-end to form *one σ-bond* (paired electrons).

(3) In O_2, the O atom ($1s^2 2s^2 2p_x^2 2p_y^1 2p_z^1$) has two unpaired electrons. The half-occupied $2p_z$-orbitals overlap end-to-end with each other to form *one σ-bond.* The half-occupied $2p_y$-orbitals overlap side-by-side to form *one π-bond.* Thus, O_2 has a *double bond* consisting of *one σ-bond* and *one π-bond* (an electron pair in each bond). The two electrons in a bonding pair have opposite spins.

Note: The *z*-axis is customarily chosen to be the bond axis.

Note: The *molecular orbital theory* has an improved description of the bonding in the O_2 molecule, which is a *biradical* (see Sections 2.10 and 3.10).

Example 3.4b Use the valence-bond theory to describe the bonding in (1) CO; (2) NO^-; (3) Li_2 (gas-phase, high temperature molecule).

Answer
(1) triple bond: *one σ-bond* and *two π-bonds* [isoelectronic with N_2]
(2) double bond: *one σ-bond* and *one π-bond* [isoelectronic with O_2]
(3) single bond: *one σ-bond* [two 2*s*-orbitals overlapping]

3.5 Hybridization of Orbitals

- **Hybrid orbitals**
  → Produced by mixing (hybridizing) orbitals of a central atom
  → A construct consistent with observed shapes and bonding in molecules

Hybridization Schemes

Electron Arrangement Around the Central Atom	Hybrid Orbitals (Number)	Angle(s) Between σ-Bonds	Example(s) [Underlined Atom]
linear	*sp* (two)	180°	<u>Be</u>Cl_2, <u>C</u>O_2
trigonal planar	sp^2 (three)	120°	<u>B</u>F_3, <u>C</u>H_3^+
tetrahedral	sp^3 (four)	109.5°	<u>C</u>Cl_4, <u>N</u>H_4^+
trigonal bipyramidal	sp^3d (five)	120°, 90°, and180°	<u>P</u>Cl_5
octahedral	sp^3d^2 (six)	90° and 180°	<u>S</u>F_6

- **Determination of hybridization schemes**
  → Write the Lewis structure(s). Pick *any* one, if there is resonance.
  → Determine the number of lone pairs and σ-bonds on the central atom.
  → Determine the number of orbitals required for hybridization on the central atom to accommodate the lone pairs and σ-bonds.
  → Determine the hybridization scheme from the above table.
  → Use the same rules to determine the *shape* of the molecule and the *bond angle(s)* as used in the VSEPR model. (π-bonds are formed from the overlap of atomic orbitals [unhybridized] in a side-by-side arrangement).

Example 3.5a Use the valence-bond theory and hybridization to describe the bonding in BF_3.

Solution The ground-state electron configuration of B is $[He]2s^2 2p_x^1$. The molecule BF_3 has a *trigonal-planar* shape with three equal B–F single bonds (using the VSPER technique). To account for this with the VB theory, *promote* one 2*s*-electron on B to a vacant 2*p*-orbital $\{2s^2 2p_x^1 \rightarrow 2s^1 2p_x^1 2p_y^1\}$. *Mix* or *hybridize* the resulting half-filled valence orbitals (one 2*s*- and two 2*p*-orbitals) to get *three* sp^2 hybrid orbitals, each with one unpaired electron. Now overlap *one* hybrid orbital with one 2*p*-orbital of F (end-to-end). Overlap a *second* hybrid orbital of B with a 2*p*-orbital of another F. Repeat with the *third* hybrid orbital and the last F atom. One obtains three σ-bonds [B(sp^2)–F(2*p*)].

Example 3.5b Use the valence-bond theory and hybridization to describe the bonding in CH_4.

Answer There are four σ-bonds [C(sp^3)–H(1*s*)]. See Section 3.5 in the text.

Note: Valence atomic orbitals that are empty or that are needed to make π-*bonds* are *not* hybridized, unless they are needed to produce coordinate covalent σ-*bonds* (as in BF_4^-).

Example 3.5c Use the VB theory to predict the hybridization, shape, and bond angle(s) in (1) trichloroamine, NCl_3; (2) xenon tetrafluoride, XeF_4; (3) xenon difluoride, XeF_2; (4) sulfur tetrafluoride, SF_4. See the Lewis structures in Example 3.2a.

Solution Follow the general hybridization scheme for each molecule.

Answer (1) According to its Lewis structure, NCl_3 has one lone pair and *three* σ-bonds to the central N atom. Four valence atomic orbitals on N must be hybridized to form four sp^3 hybrid orbitals. Three of the hybrid orbitals overlap end-to-end with three 3*p*-orbitals on Cl to form three σ-bonds. The remaining hybrid orbital on N contains the lone-pair electrons. The electron arrangement is *tetrahedral*. The shape is *trigonal pyramidal*. The lone pair-bonding pair repulsion is greater than the bonding pair-bonding pair value. Therefore, the bond angles will be smaller than 109.5°. The same adjustments to the bond angle(s) are used as in the VSEPR model.

(2) According to its Lewis structure, XeF_4 has two lone pairs and four σ-bonds to the central Xe atom. Six valence atomic orbitals on Xe must be hybridized to form six sp^3d^2 hybrid orbitals. Four of the hybrid orbitals overlap end-to-end with four 2*p*-orbitals on F to form four σ-bonds. The remaining two hybrid

orbitals on Xe contain the two lone pairs. The electron arrangement is *octahedral*. The shape is *square planar* because the lone pairs prefer opposite corners of the octahedron. The bond angles will be 90° and 180° in the plane of the molecule.

(3) According to its Lewis structure, XeF_2 has three lone pairs and two σ-bonds to the central Xe atom. Five valence atomic orbitals on Xe must be hybridized to form five sp^3d hybrid orbitals. Two of the hybrid orbitals overlap end-to-end with two 2*p*-orbitals on F to form two σ-bonds. The remaining three hybrid orbitals on Xe contain the three lone pairs. The electron arrangement is *trigonal bipyramidal*. The shape is *linear* because the lone pairs prefer the equatorial positions. The bond angle will be 180°.

(4) According to its Lewis structure, SF_4 has one lone pair and four σ-bonds to the central S atom. Five valence atomic orbitals on S must be hybridized to form five sp^3d hybrid orbitals. Four of the hybrid orbitals overlap end-to-end with four 2*p*-orbitals on F to form four σ-bonds. The last hybrid orbital contains the lone pair. The electron arrangement is *trigonal bipyramidal*. The shape is *seesaw*, because the lone pair prefers one of the equatorial positions. The bond angles will be less than 120° in the equatorial plane and less than 90° between the axial and equatorial positions (greater repulsions between bonding and lone pair electrons than between bonding electron pairs).

Example 3.5d Use the VB theory to predict the hybridization, shape, and bond angle(s) in (1) sulfur dioxide, SO_2; (2) the triiodide anion, I_3^-; (3) iodine pentafluoride, IF_5; (4) methyl alcohol, CH_3OH. See Example 3.2b.

Answer (1) **S**: sp^2 (three), angular, $<120°$; (2) **I**: sp^3d (five), linear, 180°; (3) I: sp^3d^2 (six), square pyramidal, $<90°$ and $<180°$; (4) **C**: sp^3 (four), tetrahedral, ~109.5° and **O**: sp^3 (four), angular, $<109.5°$.

3.6 Hybridization in More Complex Molecules

- **Central atoms →** Each central atom can be represented in the hybridization scheme.
  See Example 3.5d and Example 3.6a,b.

Example 3.6a Use the VB theory to describe the geometry and bonding in acetaldehyde, CH_3CHO, a molecule with two central atoms.

Solution The Lewis structure shows two central C atoms: one with a *tetrahedral* electron arrangement (sp^3 hybridization, methyl carbon atom) and one with a *trigonal-planar* electron arrangement (sp^2 hybridization, aldehyde carbon atom). The Lewis structure and three-dimensional representation are

```
    H :Ö:              H      ·Ö·
    |  ||               \     //
H—C—C—H                C—C
    |                  ◁◀     \
    H                  HH      H
```

The C–C σ-bond forms from the overlap of an sp^3 hybrid orbital on the methyl C atom with an sp^2 hybrid orbital on the aldehyde C atom: σ-bond $[C(sp^3)–C(sp^2)]$.

The C=O σ-bond forms from the overlap of an sp^2 hybrid on the aldehyde C atom with one 2*p*-orbital on the oxygen atom: σ-bond [C(sp^2)–O(2*p*)].

Each of three C–H single bonds forms from the overlap of an sp^3 hybrid on the methyl C atom with an H 1*s*-orbital: three σ-bonds [C(sp^3)–H(1*s*)].

One C–H single bond forms from the overlap of an sp^2 hybrid orbital on the aldehyde C atom with an H 1*s*-orbital: one σ-bond [C(sp^2)–H(1*s*)].

The C=O π-bond is formed by the side-by-side overlap of a 2*p*-orbital on C with a 2*p*-orbital on O (unhybridized orbitals): one π-bond [C(2*p*)–C(2*p*)].

The shape of acetaldehyde is *tetrahedral* at the methyl C atom and *trigonal planar* at the aldehyde C-atom. Bonding consists of a C–C single (σ) bond, four C–H single (σ) bonds and a C=O double bond (σ plus π) formed from the atomic and hybrid orbitals given above.

The bond angles are three ∠ H–C–H ~109.5°, three ∠ H–C–C ~109.5°, one ∠ C–C=O ~120°, one ∠ O=C–H ~120°, and one ∠ C–C–H ~120°.

Note: In acetaldehyde, the methyl group is free to rotate with respect to the aldehyde group about the C–C σ-bond.

Example 3.6b Describe the geometry, hybridization, and bonding in methylacetylene, $HCCCH_3$, a molecule with *three* central atoms.

Answer The *shape* at C atoms: *linear, linear, tetrahedral. Hybridization* at C atoms: *sp, sp,* sp^3. *Bonding:* four C–H σ-bonds, one C≡C triple bond (one σ plus two π), one C–C σ-bond. *Bond angles:* one ∠ H–C≡C =180°, one ∠ C≡C–C =180°, three ∠ C–C–H ~109.5°, three ∠ H–C–H ~109.5°.

3.7 Bonding in Hydrocarbons

- **Alkanes (C_nH_{2n+2}, n = 1, 2, 3, ...)**
  - → Examples include methane, CH_4; and propane, $CH_3CH_2CH_3$
  - → Characteristics: tetrahedral geometry at C atoms; sp^3 hybridization at all C atoms; all C–C and C–H single (σ) bonds; allowed rotation about C–C single bonds
- **Alkenes (C_nH_{2n}, n = 2, 3, 4, ...)**
  - → Examples include ethene (ethylene), $H_2C{=}CH_2$; and propylene, $CH_3CH{=}CH_2$
  - → Characteristics: one C=C double bond (σ plus π); other C–C bonds and all C–H bonds single (σ); trigonal-planar geometry at double-bonded C atoms; sp^2 hybridization at double-bonded C atoms; tetrahedral geometry at other C (sp^3) atoms; disallowed rotation about double C=C bond; allowed rotation about C–C single bonds
- **Alkynes (C_nH_{2n-2}, n= 2, 3, 4, ...)**
  - → Examples include ethyne (acetylene), HC≡CH; 2-butyne (dimethylacetylene), $H_3CC{\equiv}CCH_3$
  - → Characteristics: one C≡C triple bond (σ plus two π); other C–C bonds and all C–H bonds (σ); linear geometry at triple-bonded C atoms; *sp* hybridization at triple-bonded C atoms; tetrahedral geometry at other C (sp^3) atoms

- **Benzene (C_6H_6)**
  - → Characteristics: *Planar* molecule with hexagonal C framework; trigonal-planar geometry at C atoms; sp^2 hybridization at C atoms; formation of σ framework yields C–C and C–H single bonds; 2*p*-orbitals (one from each C atom) overlap in a side-by-side fashion to form *three* π-bonds (*six* π-electrons) *delocalized over the entire ring* of six C atoms
  - → See Lewis resonance structures in Section 2.8.

3.8 Characteristics of Double Bonds

- **Double bonds**
  - → Consist of *one* σ- and *one* π-bond
  - → Always *stronger* than a single σ-bond:
    C=C is *weaker* than two single C–C σ-bonds
    N=N is *stronger* than two single N–N σ-bonds
    O=O is *stronger* than two single O–O σ-bonds
  - → Formed readily by Period 2 elements (C, N, O)
    Rarely found in Period 3 and higher period elements
  - → Impart rigidity to molecules and influence molecular shape

Example 3.8a Describe the structure of the carbonate anion, CO_3^{2-}, in terms of hybrid orbitals, bond angles, and σ- and π-bonds.

Solution The Lewis structure given below implies a trigonal-planar geometry with sp^2 hybridization at C and 120° ∠ O–C–O bond angles. The nonhybridized C 2*p*-orbital can overlap with O 2*p*-orbitals to form a π-bond in each of three ways corresponding to the resonance structures below. Therefore, each carbon-oxygen bond can be viewed as one σ-bond and one-third of a π-bond.

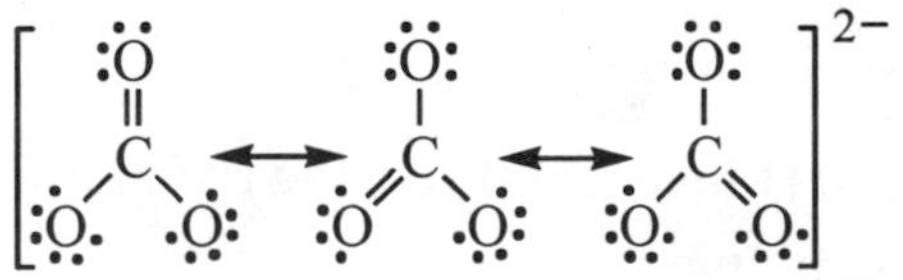

[:Ö=C(–:Ö:)(–:Ö:) ⟷ :Ö–C(=Ö)(–:Ö:) ⟷ :Ö–C(–:Ö:)(=Ö)]$^{2-}$

Example 3.8b Describe the structure of the nitrite anion, NO_2^-, in terms of hybrid orbitals, bond angles, and σ- and π-bonds.

Answer There is sp^2 hybridization at N and somewhat less than a 120° ∠ O–N–O bond angle. The nonhybridized N 2*p*-orbital can overlap with O 2*p*-orbitals to form a π-bond in each of two ways. Each nitrogen-oxygen bond can be viewed as one σ-bond and one-half of a π-bond.

Molecular Orbital (MO) Theory (Sections 3.9-3.13)

Key Concepts

electron-deficient compound, paramagnetism, diamagnetism, molecular orbital (MO), linear combination of atomic orbitals (LCAO), bonding orbital, antibonding orbital, nonbonding orbital, molecular orbital energy-level diagram (correlation diagram), σ-orbital, π-orbital, homonuclear diatomic molecule, bond order, heteronuclear diatomic molecule, delocalized electrons, highest occupied molecular orbital (HOMO), lowest unoccupied molecular orbital (LUMO), electronic conductor, metallic conductor, semiconductor, superconductor, insulator, band theory, conduction band, valence band, band gap, doping, n-type semiconductor, p-type semiconductor, p-n junctions

Overview

- **Valence Bond (VB) deficiencies**
  - → Cannot explain *paramagnetism* of O_2
  - → Difficulty treating electron-deficient compounds such as diborane, B_2H_6
  - → No simple explanation for spectroscopic properties of compounds such as color
- **Molecular Orbital (MO) advantages**
  - → Addresses all of the above shortcomings of VB theory
  - → Provides a deeper understanding of the electron-pair bond
  - → Accounts for the structure and properties of metals and semiconductors
  - → More facile for computer calculations than VB theory

3.9 Molecular Orbitals (MOs)

- **MO theory** → Electrons occupy *molecular orbitals* that are *delocalized* over the *entire* molecule, whereas in Valence Bond theory, the bonding electrons are *localized* between the two bonded atoms.
- **Molecular orbitals (MO)**
  - → Formed by superposition (**linear combination**) of atomic orbitals **(LCAO-MO)**
  - → **Bonding orbital** (constructive interference):
    Increased amplitude or electron density between atoms
  - → **Antibonding orbital** (destructive interference):
    Decreased amplitude or electron density between atoms
- **Types of Molecular Orbitals**
  - → **Bonding MO:** energy *lower* than that of the constituent AOs
  - → **Nonbonding MO:** energy *equal* to that of the constituent AOs
  - → **Antibonding MO:** energy *greater* than that of the constituent AOs
- **Rules for combining AOs to obtain MOs** → *N* atomic orbitals (AOs) yield *N* molecular orbitals (MOs)

Note: There is little overlapping of inner shell AOs, and these MOs are usually **nonbonding.**

- **MO energy-level diagrams** → The *relative* energies of the original AOs and the resulting MOs are shown schematically in energy-level diagrams. Often the location of the electrons in the separated atoms and the molecule are indicated by arrows to show electron spin. The following simple examples illustrate the appearance of **LCAO-MOs** and the **MO energy-level diagrams.**

Example 3.9a Show schematically the overlap of two 2*s* AOs on adjacent Li atoms to form a 2*s*σ-bonding orbital (constructive interference) and a 2*s*σ*-antibonding orbital (destructive interference). Draw an energy-level diagram to show that the bonding MO is lower in energy than the atomic 2*s*-orbitals and the antibonding MO is higher in energy by about the same amount. Consider the Li_2 molecule. Show the locations of the two valence electrons in both the *separated* atomic orbitals and the molecular orbitals in the energy-level diagram.

Solution

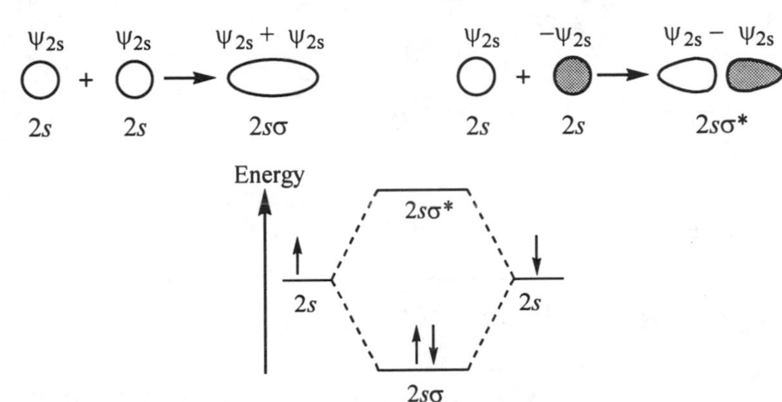

Example 3.9b Show schematically the overlap of two 1*s* AOs on adjacent H atoms to form a 1*s*σ-bonding orbital (constructive interference) and a 1*s*σ*-antibonding orbital (destructive interference). Draw an energy-level diagram to show that the bonding MO is lower in energy than the atomic 1*s*-orbitals and the antibonding MO is higher in energy by about the same amount. Consider the H_2^- molecular anion. Show the locations of the three valence electrons in both the *separated* atomic orbitals and the molecular orbitals in the energy-level diagram.

Solution

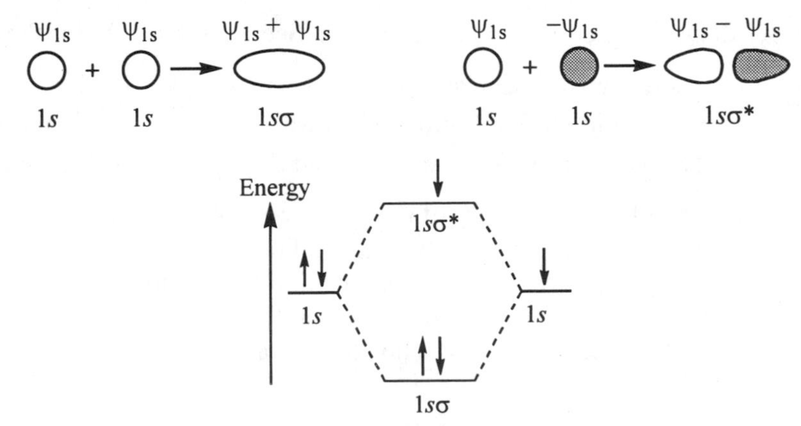

3.10 The Electron Configurations of Diatomic Molecules

- **Procedure for determining the electronic configuration of diatomic molecules**
  → Construct all possible MOs from *valence-shell* AOs
  → Place *valence electrons* in the lowest energy, unoccupied MOs. Follow the **Pauli exclusion principle** and **Hund's rule** as for AOs: electrons have *spins paired* in a fully occupied molecular orbital and enter different *degenerate* orbitals with *parallel spins*, if the orbitals are initially unoccupied.

- **Valence-shell MOs for Period 1 and Period 2 *homonuclear* diatomic molecules**

Correlation Diagrams

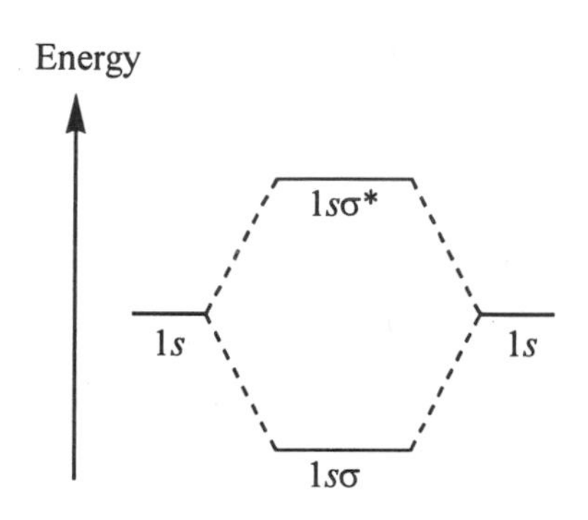

Schematic order of MO energy levels for H_2, He_2

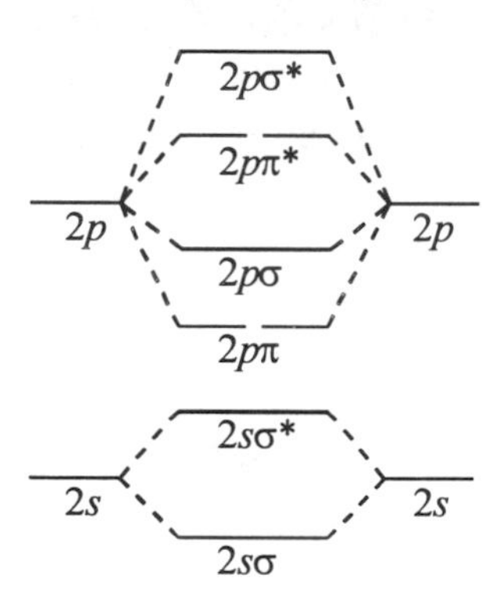

Schematic order of MO energy levels for Li_2, Be_2, B_2, C_2, N_2

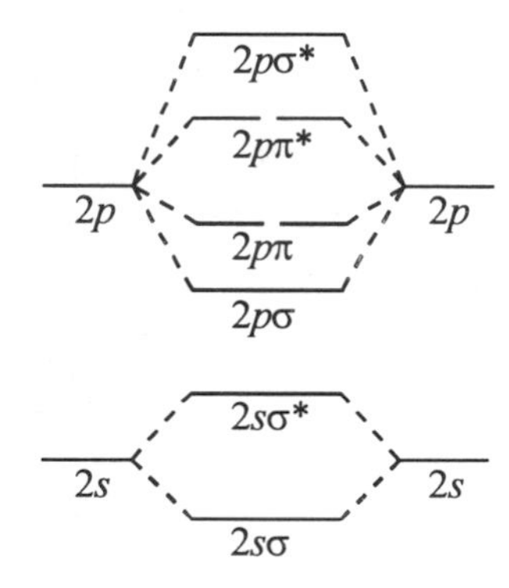

Schematic order of MO energy levels for O_2, F_2, Ne_2

- **Bond order (BO)** → $\text{BO} = \frac{1}{2}(B - A)$

where B = total number of electrons in *bonding* MOs and A = total number of electrons in *antibonding* MOs.

Example 3.10a Determine the bond order in Li_2. See Example 3.9a for the correlation diagram.

Solution There are two electrons in the $2s\sigma$-bonding MO, so $B = 2$. There are no electrons in the $2s\sigma^*$-antibonding MO, so $A = 0$. The bond order is $\frac{1}{2}(2 - 0) = 1$. This picture is similar to the VB theory with a single σ-bond between the Li atoms.

Example 3.10b Determine the bond order in H_2^-. See Example 3.9b for the correlation diagram.

Solution There are two electrons in the $1s\sigma$-bonding MO, so $B = 2$. There is one unpaired electron in the $1s\sigma^*$-antibonding MO, so $A = 1$. The bond order is $\frac{1}{2}(2 - 1) = 1/2$. The MO theory predicts a net bond that is *half* a single bond value.

Note: The ultimate stability of molecular anions requires a more detailed treatment. The H_2^- anion is actually unstable in the gas phase, as is Li_2^-. The ion H_2^- *is* bound with respect to its dissociation to a neutral hydrogen atom and an atomic hydrogen anion. The ion is not stable, however, with respect to the neutral hydrogen molecule, which is formed after the extra electron detaches [$H_2^- \rightarrow H_2 + e^-$]. The consequence of this behavior is that the electron affinity

E_{ea} of H_2 is less than zero. The homonuclear diatomic molecules in Period 2 with *positive* values of E_{ea} (stable molecular anions) are C_2, O_2, and F_2.

Example 3.10c Determine the ground-state, valence-shell electron MO configuration, number of unpaired electrons and bond order for (1) Li_2^+ and (2) C_2.

Solution (1) The molecular ion Li_2^+ has *one* valence electron, which should be placed in the $2s\sigma$, which is a bonding MO. The valence electron configuration is $(2s\sigma)^1$, which represents one unpaired electron. $B = 1$ and $A = 0$. Bond order $= \frac{1}{2}(1 - 0)$ $= 1/2$. The molecular ion is stable in the gas phase.

(2) The molecule C_2 has *eight* valence electrons, which should be placed in the $2s\sigma$, $2s\sigma^*$, and two *degenerate* $2p\pi$ MOs. The valence electron configuration is $(2s\sigma)^2\,(2s\sigma^*)^2\,(2p\pi)^4$, which represents all paired electrons. $B = 6$ and $A = 2$. Bond order $= \frac{1}{2}(6 - 2) = 2$. The MO theory predicts an <u>unusual</u> double bond: *two* π-bonds and *no* σ-bond. The molecule is very stable and is found in flames and interstellar space.

Example 3.10d Determine the ground-state, valence-shell electron MO configuration, number of unpaired electrons and bond order for (1) O_2^+ and (2) Be_2.

Solution (1) The molecular ion O_2^+ has *eleven* valence electrons, which should be placed in the $2s\sigma$, $2s\sigma^*$, $2p\sigma$, two *degenerate* $2p\pi$, and one of the *degenerate* $2p\pi^*$ MOs. The valence electron configuration is $(2s\sigma)^2\,(2s\sigma^*)^2\,(2p\sigma)^2\,(2p\pi)^4\,(2p\pi^*)^1$, which represents one unpaired electron. $B = 8$ and $A = 3$. Bond order $= \frac{1}{2}(8 - 3)$ $= 5/2$. The MO theory predicts unusual bonding: *one* σ-bond and *one* and *one-half* π-bonds. The bond order is greater than in O_2 [BO = 2], which has *two* unpaired electrons and is *paramagnetic*. The molecular ion is stable in the gas phase.

(2) The molecule Be_2 has *four* valence electrons, which are placed in the $2s\sigma$ and $2s\sigma^*$ MOs. The valence electron configuration is $(2s\sigma)^2\,(2s\sigma^*)^2$, which represents full pairing of electrons. $B = 2$ and $A = 2$. Bond order $= \frac{1}{2}(2 - 2) = 0$. The MO theory predicts *no* chemical bond.

3.11 Bonding in Heteronuclear Diatomic Molecules

- **Bond properties**
  - → Polar in nature
  - → The atomic orbital from the *more* electronegative atom is *lower* in energy.
  - → The *bonding* orbital has a *greater* contribution from the *more* electronegative atom.
  - → The *antibonding* orbital has a *greater* contribution from the *less* electronegative atom.

Example 3.11 Using a MO energy-level diagram, describe the bonding in HCl and give the electron configuration of its *valence* electrons.

Solution The valence orbitals are H [1*s*-orbital], Cl [3*s*- and three 3*p*-orbitals]. The more electronegative atom is Cl, so we expect the 3*s*- and 3*p*-orbitals of Cl to be lower in energy than the 1*s*-orbital of H. The polar covalent bond of H–Cl is formed from the overlap of the H 1*s*-orbital and the Cl $3p_z$-orbital. (The *z*-axis is customarily chosen to be the bond axis.) The correlation diagram shows the resulting σ-bonding orbital and σ*-antibonding orbital. The electronic configuration is $(\text{Cl } 3s)^2\, \sigma^2\, (\text{Cl } 3p_x)^2\, (\text{Cl } 3p_y)^2$.

Note: The Cl 3*s*-orbital is a *nonbonding* σ MO, and the Cl $3p_x$- and Cl $3p_y$-orbitals are *nonbonding* π MOs.

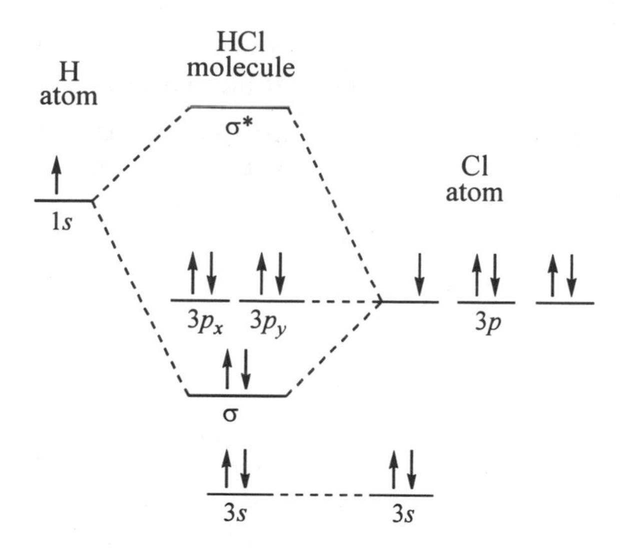

3.12 Orbitals in Polyatomic Molecules

- **Bond properties**

→ All atoms in the molecule are encompassed by the MOs.

→ Electrons in MOs bond all the atoms in a molecule.

→ Electrons are *delocalized* over all the atoms.

→ Energies of the MOs are obtained by spectroscopy.

Note: Although MO theory requires MOs to be *delocalized* over all atoms, sometimes electrons are *effectively localized* on an atom or group of atoms.

See Example 3.11 and text example: The O $2p_x$ *nonbonding* orbital in H_2O is localized on O.

Note: In describing the bonding in complex molecules, VB and MO theory are sometimes jointly invoked to explain the bonding.

See text example: In the bonding in benzene, we use *localized* sp^2 hybridization (hybrids of C 2*s*-, C $2p_x$-, and C $2p_y$-orbitals on each C) to describe the σ-bonding framework of the ring, and *delocalized* MOs (LCAOs of six C $2p_z$-orbitals) to describe the π-bonding.

- **Triumphs of MO theory**

→ Accounts for paramagnetism of O_2

→ Accounts for bonding in electron-deficient molecules. In diborane (B_2H_6), *six* electron pairs bond *eight* atoms using *six* MOs. VB theory requires *eight* electron pairs and fails to describe the bonding.

→ Provides a framework for interpreting visible and ultraviolet spectroscopy

Example 3.12a Consider the *delocalized* π-electron system in benzene. The ultraviolet absorption spectrum of benzene is interpreted as the excitation of an electron from the *highest occupied molecular orbital* **(HOMO)** to the *lowest unoccupied molecular orbital* **(LUMO)**. In benzene, the HOMO is a *bonding* π MO and the LUMO is an *antibonding* π* MO. The bond order of the π-system changes from 3 to 2. The schematic energies and electron occupancies of the π and π* MOs are

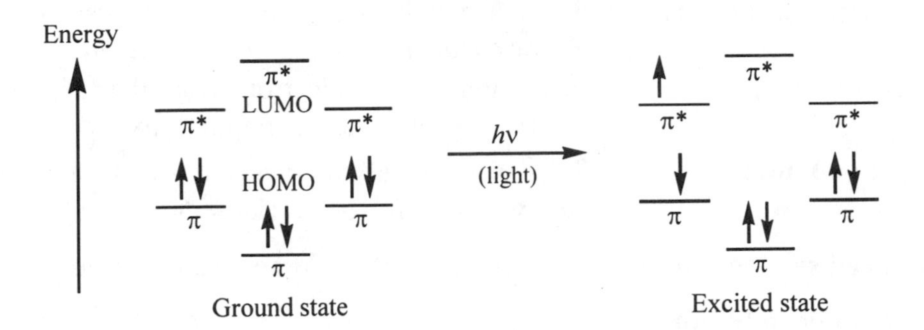

Example 3.12b Consider the HCl molecule described in Example 3.11. As in the case of benzene, the ultraviolet absorption spectrum is interpreted as the excitation of an electron from a HOMO to a LUMO. In the case of HCl, the HOMO is a *nonbonding* π MO (Cl $3p_x$ or Cl $3p_y$) and the LUMO is an *antibonding* σ* MO. The bond order of the molecule changes from 1 to 0.5.

3.13 The Impact on Materials: The Band Theory of Solids

- **MO theory** → Accounts for electrical properties of metals and semiconductors
- **Electronic conductors** → Metals, semiconductors, and superconductors

  Insulator: Does not conduct electricity

  Metallic conductor: Current carried by *delocalized* electrons in *bands*
  Conductivity *decreases* with *increasing* temperature

  Semiconductor: Current carried by *delocalized* electrons in *bands*
  Conductivity *increases* with *increasing* temperature

  Superconductor: *Zero* resistance (infinite conductivity) to an electric current below a definite transition temperature
  An example is the ionic solid $K_3(C_{60})$ below 19.3 K. The ions in the solid are K^+ and $(C_{60})^{3-}$. The molecule C_{60} is buckminsterfullerene (buckyball).

- **Bands**
  → Groups of MOs having closely spaced energy levels (see Fig. 3.46 in the text)
  Formed by the spatial overlap of a very large number of AOs
  → Separate into two groups (one mostly *bonding*, the other mostly *antibonding*)

- **Conduction band** → Empty or partially filled band of MOs
  An example is the formation of a conduction band in Na by the overlap of 3*s*-orbitals as discussed in the text.

- **Valence band** → Completely filled band of MOs
  Examples include insulators (molecular solids).
- **Band gap** → An energy range between bands in which there are *no* orbitals

  Metallic conductor: Substance with a partially filled *conduction band* (gap *not* relevant)

  Semiconductor: Substance with a full *valence band* close in energy to an empty *conduction band* (small *band gap*)
  Excitation of some electrons from the *valence* to the *conduction band* yields conductivity.

  Insulator: Substance with a full *valence band* far in energy from an empty *conduction band* (large *band gap*)
- **Enhanced semiconductors** → *Doping*, that is, adding a small amount of impurity
- **n-type semiconductor** → Adding some *valence electrons* to the *conduction band* (n = electrons added)
  An example is a small amount of As (Group 15) added to Si (Group 14). Extra electrons are transferred from As into the previously empty *conduction band* of Si]
- **p-type semiconductor** → Removing some *valence electrons* from the *valence band* (p = positive "holes" in the *valence band*)
  An example is a small amount of In (Group 13) added to Si (Group 14). A valence electron deficit is created in the previously full *valence band* of Si, which thereby becomes a *conduction band.*
- **p-n junctions**
  → p-type semiconductor in contact with an n-type semiconductor
  → Solid-state electronic devices
  One example is a type of transistor that is a p-n sandwich with two separate wires (*source* and *emitter*) connected to the n-side. It acts as a *switch* in the following way. When a positive charge is applied to a polysilicon contact (*base* wire) positioned between the *source* and *emitter* wires, a current flows from the *source* to the *emitter.*
  Other examples include diodes and integrated circuits.

Chapter 4 The Properties of Gases

Significant Figure Convention

At least one *extra* significant figure is displayed in all *intermediate* calculations. The *final* answer is expressed with the correct number of significant figures.

The Nature of Gases (Sections 4.1-4.3)

Key Concepts

compressibility, pressure, barometer, pascal (Pa), manometer, gauge pressure, mmHg (Torr), atmosphere (atm), bar

4.1 Observing Gases

- **Atmosphere** → Not uniform in composition, temperature (T), or density (d)
- **Physical properties** → Most gases behave similarly at low pressure
- **Compressibility**
  → Ease with which a gas undergoes a volume (V) decrease
  → Gas molecules are widely spaced
- **Expansivity**
  → The ability of a gas to rapidly fill the space available to it
  → Atomic and molecular gases move rapidly and respond quickly to changes in the available volume

4.2 Pressure

- **Pressure** $\equiv \dfrac{\text{force}}{\text{area}}$ *or* $P = \dfrac{F}{A}$ Opposing force from a confined gas
- **Barometer**
  → A glass tube, sealed at one end, filled with liquid mercury, and inverted into a beaker also containing liquid mercury (Torricelli)
  → The height (h) of the liquid column is a measure of the atmospheric pressure (P) expressed as the length of the column for a particular liquid.
  Common pressure units: mmHg, cmHg, cmH_2O

$$P = \frac{F}{A} = \frac{mg}{A} = \frac{(dV)g}{A} = \frac{(dhA)g}{A} = dhg$$

where m = mass of liquid, g = acceleration of gravity ($9.806\,65\,\text{m·s}^{-2}$ Earth's surface), and d = density of liquid.

Example 4.2a Suppose the height of the column of mercury in a barometer is exactly 760 mm. Given that the density of mercury at 0°C is 13 595.1 kg·m^{-3}, calculate the atmospheric pressure.

Solution $P = dhg = (13\,595.1\ \text{kg·m}^{-3})(0.760\ \text{m})(9.806\,65\ \text{m·s}^{-2}) = 101\,325\ \text{kg·m}^{-1}\text{·s}^{-2}$

$= 101\,325\ \text{Pa} = 101.325\ \text{kPa}$

Note: 1 Pa (pascal) = 1 kg·m^{-1}·s^{-2} SI unit of pressure

Example 4.2b At the same atmospheric pressure obtained in Example 4.2a, calculate the height of liquid in a barometer that employs water at 20°C. [$d\,(H_2O) = 0.998$ g·cm^{-3}]

Answer $h = 10.4$ m

- **Manometer (see Fig. 4.5 in the text)**
  - → U-shaped tube filled with liquid connected to an experimental system
  - → "open-tube" pressure (system) = atmosphere when levels are equal
  - → "closed-tube" pressure (system) = difference in level heights

Example 4.2c The height of the system-side column in an *open-tube* mercury manometer was 15 mm above that of the open side when the atmospheric pressure was 762 mmHg at 20°C. Calculate the pressure in the system in units of mmHg and kPa.

Solution In this case, the pressure in the system is equal to atmospheric pressure minus the difference in height of the two columns. The system is at lower pressure than the atmosphere (open end). The conversion to kPa requires the density of Hg at 20°C, which is 13 546 kg·m^{-1}·s^{-2}.

$P = 762 - 15 = 747\ \text{mmHg} = 0.747\ \text{mHg}$
$P = dhg = (13\,546\ \text{kg·m}^{-1}\text{·s}^{-2}) \times (0.747\ \text{m}) \times (9.806\,65\ \text{m·s}^{-1}) = 9.92 \times 10^4\ \text{Pa}$
$= 99.2\ \text{kPa}$

Example 4.2d Calculate the pressure inside a system when a *closed-tube* mercury manometer shows a height difference of 12 cm (higher on the closed side) at 20°C.

Answer $P = 120$ mmHg and 15.9 kPa

- **Gauge pressure** → Difference between the pressure inside a vessel containing a compressed gas (a tire, for example) and the atmospheric pressure

4.3 Alternative Units of Pressure

- **Atmosphere** → 1 atm = 760 Torr (exactly)
  1 atm = 101 325 Pa (exactly)
  1 bar = 10^5 Pa = 10^2 kPa
  1 Torr = 1 mmHg (exactly)

Example 4.3 Convert an atmospheric pressure of 740 mmHg into each of the pressure units given above.

Solution $P = 740\ \text{Torr} = 0.974\ \text{atm} = 9.87 \times 10^4\ \text{Pa} = 98.7\ \text{kPa} = 0.987\ \text{bar}$

The Gas Laws (Sections 4.4-4.10)

Key Concepts

Boyle's law, isotherm, hyberbola, Charles's law, extrapolation, Kelvin scale, molar volume, Avogadro's principle, combined gas law, ideal gas law, gas constant, equation of state, ideal gas, limiting law, standard temperature and pressure (STP), molar concentration, density, amount-to-volume conversions in reacting gases, gas mixtures, partial pressure, Dalton's law of partial pressures, vapor pressure

4.4 Boyle's Law (fixed amount of gas *n* at constant temperature *T*)

- **Gas behavior** → Boyle's law describes an isothermal system: $\Delta T = 0$ (constant T).

$$\text{Volume} \propto \frac{1}{\text{pressure}} \quad \text{or} \quad V \propto \frac{1}{P} \quad \text{or} \quad PV = \text{constant}$$

The functional form is a *hyperbola*.

- **Changes between two conditions (1 and 2)**

$$P_1V_1 = \text{constant} = P_2V_2 \quad \text{or} \quad \frac{P_2}{P_1} = \frac{1/V_2}{1/V_1} = \frac{V_1}{V_2}$$

- **Deviations** → At low temperature

4.5 Charles's Law (fixed amount of gas *n* at constant pressure *P*)

- **Gas behavior** → Charles's law describes an isobaric system: $\Delta P = 0$ (constant P).

$$\text{Volume} \propto \text{temperature} \quad \text{or} \quad V \propto T \quad \text{or} \quad V/T = \text{constant}$$

The functional form is *linear*.

- **Kelvin temperature scale** → Two *fixed* points define the absolute temperature scale:

  $T \equiv 273.16$ K The triple point of water, where ice, liquid, and vapor are in equilibrium (see Chapter 8).

  $T \equiv 0$ K Defined from the limiting behavior of Charles's law. V approaches 0 at fixed *low* pressure (extrapolation).

- **Celsius temperature scale** → $t(°\text{C}) \equiv T(\text{K}) - 273.15$ (exactly)

Note: The freezing point of water at 1 atm external pressure is 0°C or 273.15 K. (The symbol t is also used for time.)

- **Changes between two conditions (1 and 2)**

$$V_1/T_1 = \text{constant} = V_2/T_2 \quad \text{or} \quad \frac{V_2}{V_1} = \frac{T_2}{T_1}$$

- **Deviations** → At high pressure

- **Another aspect of gas behavior (fixed amount of gas n at constant volume V)**

  → Pressure ∝ temperature or $P \propto T$ or P/T = constant

  → This relation describes an isochoric system: $\Delta V = 0$ (constant V). The functional form is *linear*.

- **Changes between two conditions (1 and 2)**

$$P_1/T_1 = \text{constant} = P_2/T_2 \quad \text{or} \quad \frac{P_2}{P_1} = \frac{T_2}{T_1}$$

- **Deviations** → For small volume

4.6 Avogadro's Principle (gas at constant pressure P and temperature T)

- **Gas behavior** Volume ∝ amount or $V \propto n$ or V/n = constant

- **Avogadro's principle** → A given amount of gas molecules at a particular P and T occupies the same volume regardless of chemical identity

- **Molar volume**

$$V_m = \frac{V}{n} = \text{constant} \qquad \text{for } \Delta P = 0 \text{ and } \Delta T = 0$$

- **Changes between two conditions (1 and 2)**

$$V_1/n_1 = \text{constant} = V_2/n_2 \quad \text{or} \quad \frac{V_2}{V_1} = \frac{n_2}{n_1}$$

- **Deviations** → At high pressure and/or low temperature

- **Another aspect of gas behavior (gas at constant volume V and temperature T)**

  Pressure ∝ amount or $P \propto n$ or P/n = constant

- **Changes between two conditions (1 and 2)**

$$P_1/n_1 = \text{constant} = P_2/n_2 \quad \text{or} \quad \frac{P_2}{P_1} = \frac{n_2}{n_1}$$

- **Deviations** → For small volume and/or low temperature

4.7 The Ideal Gas Law

- **Derivation**

$PV = \text{constant}_1$ $\qquad V = \text{constant}_2 \times T$ $\qquad P = \text{constant}_3 \times n$

$PV = (\text{constant}_3 \times n)(\text{constant}_2 \times T) = nRT$ ***equation of state***

R = gas constant = $8.314\,51 \text{ L·kPa·K}^{-1}\text{·mol}^{-1} = 8.205\,78 \times 10^{-2} \text{ L·atm·K}^{-1}\text{·mol}^{-1}$

- **Gas behavior** → As P approaches 0, all gases behave *ideally* (*limiting law*).

Example 4.7a A 2.6-μL ampule of xenon has a pressure of 2.00 Torr at 15°C. How many xenon atoms are present?

Solution Convert the volume units into liters, the pressure units into atmospheres, and the temperature units into kelvins. Solve for the number of moles of xenon, using the ideal gas law, and then use Avogadro's number to determine the number of atoms.

$V = 2.6 \times 10^{-6}$ L $\qquad P = (2.00/760)$ atm $= 2.632 \times 10^{-3}$ atm
$T = 15 + 273.15 = 288.2$ K

$$n = \frac{PV}{RT} = \frac{(2.632 \times 10^{-3}\text{ atm})(2.6 \times 10^{-6}\text{ L})}{(8.205\,78 \times 10^{-2}\text{ L·atm·K}^{-1}\text{·mol}^{-1})(288.2\text{ K})} = 2.9 \times 10^{-10}\text{ mol}$$

$$n \times N_A = 1.7 \times 10^{14}\text{ atoms}$$

Example 4.7b A 20-L flask at 200 K and 20 Torr contains nitrogen. What mass of nitrogen is present (in grams)?

Answer 0.89 g

4.8 Applications of the Ideal Gas Law

- **Molar volume** $\boxed{V_m = \dfrac{RT}{P}}$

- **Standard temperature and pressure (STP)** → 0°C and 1 atm [or 273.150 K and 1.013 25 bar]

- **Molar volume at STP**

$$\rightarrow V_m = \frac{RT}{P} = \frac{(8.205\,78 \times 10^{-2}\text{ L·atm·K}^{-1}\text{·mol}^{-1})(273.150\text{ K})}{(1\text{ atm})} = 22.4141\text{ L·mol}^{-1}$$

- **Molar concentration** $\boxed{\dfrac{n}{V} = \dfrac{1}{V_m} = \dfrac{P}{RT}}$

- **Molar concentration at STP**

$$\rightarrow \frac{n}{V} = \frac{1}{V_m} = \frac{1}{22.4141\text{ L·mol}^{-1}} = 4.461\,48 \times 10^{-2}\text{ mol·L}^{-1}$$

- **Density of a gas** $\boxed{d = \dfrac{m}{V} = \dfrac{nM}{V} = \dfrac{MP}{RT}}$

- **Density of a gas at STP** → $d = \dfrac{m}{V} = \dfrac{nM}{V} = (4.461\,48 \times 10^{-2}\text{ mol·L}^{-1}) \times M$

Example 4.8a The density of an unknown hydrocarbon gas (molecules containing only carbon and hydrogen atoms) is found to be 0.875 g·L^{-1} at 570 Torr and 20°C. Calculate the molar mass of the gas and identify it.

Solution Convert the pressure units into atmospheres and the temperature units into kelvin. Solve for the unknown molar mass. Compare with molar masses of simple hydrocarbons to identify the gas.

$P = (570/760)$ atm $= 0.7500$ atm and $T = 20 + 273.15 = 293.2$ K

$$M = \frac{dRT}{P} = \frac{(0.875\ \text{g}\cdot\text{L}^{-1})(8.20578\times10^{-2}\ \text{L}\cdot\text{atm}\cdot\text{K}^{-1}\cdot\text{mol}^{-1})(293.2\ \text{K})}{(0.7500\ \text{atm})} = 28.1\ \text{g}\cdot\text{mol}^{-1}$$

If the gas contains 2 moles of carbon atoms (24.02 g·mol^{-1}) and 4 moles of hydrogen atoms (4.03 g·mol^{-1}), then $M = 28.1$ g·mol^{-1}. Gas is ethene C_2H_4.

Example 4.8b Calculate the density of methane gas at STP.

Answer $d = 0.715$ g·L^{-1}

- **Changes between two conditions (1 and 2)**

$$\frac{P_1V_1}{n_1T_1} = \frac{P_2V_2}{n_2T_2}$$

Combined gas law

Note: The more general form of the combined gas law includes the case where the amount of gas may change. The form of the combined gas law given in the text assumes that $n_1 = n_2$.

Example 4.8c A gas sample is compressed to half of its original volume and the absolute temperature is increased by 15%. What is the pressure change?

Solution Let indices 1 and 2 represent the initial and final states, respectively. Note the changes, if any, in amount, volume, and temperature of the gas. Use the combined form of the ideal gas law given above to solve for the pressure change.

Change in the amount of gas: $n_2 = n_1$
50% decrease in the volume: $V_2 = V_1 - 0.50V_1 = 0.50V_1$
15% increase in the temperature: $T_2 = T_1 + 0.15T_1 = 1.15T_1$

$$P_2 = P_1\left(\frac{V_1}{V_2}\right)\left(\frac{n_2}{n_1}\right)\left(\frac{T_2}{T_1}\right) = P_1\left(\frac{V_1}{0.50V_1}\right)\left(\frac{n_1}{n_1}\right)\left(\frac{1.15T_1}{T_1}\right) = P_1(2)(1)(1.15) = 2.30P_1$$

Because $P_2 = P_1 + 1.30P_1$, the pressure has increased by 130%.

Example 4.8d A gas sample has its pressure quadrupled, absolute temperature doubled, and amount doubled. Calculate the volume change.

Answer $V_2 = V_1$ (no change in volume)

Note: The actual units used in manipulating ratios of quantities are *not* important. Consistency in units is essential. *Absolute* temperature, however, *must* be used in these equations.

4.9 The Stoichiometry of Reacting Gases

- **Accounting for reaction volumes** → Use mole-to-mole calculations and convert to gas volume by using the ideal gas law.

Example 4.9a Calculate the volume of each gas produced when 225 g of NH_4NO_3 (ammonium nitrate) decomposes completely to N_2O and H_2O gases at atmospheric pressure (1.00 atm) and 300°C (573 K). [Note the convention for significant figures described at the beginning of this chapter.]

Solution Balance the chemical equation for the decomposition of ammonium nitrate solid. Calculate the number of moles of solid and use mole-to-mole calculations to determine the number of moles of each gas formed. Use the ideal gas law to calculate the volume of each gas formed at 1.00 atm and 573 K.

$$NH_4NO_3(s) \rightarrow N_2O(g) + 2H_2O(g)$$

$(225 \text{ g } NH_4NO_3)/(80.05 \text{ g·mol}^{-1}) = 2.811 \text{ mol}$

$1 \text{ mol } NH_4NO_3 \rightarrow 1 \text{ mol } N_2O$ and $2.811 \text{ mol } NH_4NO_3 \rightarrow 2.811 \text{ mol } N_2O$

$1 \text{ mol } NH_4NO_3 \rightarrow 2 \text{ mol } H_2O$ and $2.811 \text{ mol } NH_4NO_3 \rightarrow 5.622 \text{ mol } H_2O$

$V_{N_2O} = nRT/P = (2.811 \text{ mol})(0.082\,057\,8 \text{ L·atm·K}^{-1}\text{·mol}^{-1})(573 \text{ K})/(1.00 \text{ atm}) = 132 \text{ L}$

$V_{H_2O} = nRT/P = (5.622 \text{ mol})(0.082\,057\,8 \text{ L·atm·K}^{-1}\text{·mol}^{-1})(573 \text{ K})/(1.00 \text{ atm}) = 264 \text{ L}$

Example 4.9b An initial volume of 352 L of butane at 250°C and 3.42 atm is burned with excess oxygen: $2C_4H_{10}(g) + 13O_2(g) \rightarrow 8CO_2(g) + 10H_2O(g)$. Calculate the mass of carbon dioxide released to the atmosphere.

Answer $m = 4.94 \times 10^3$ g

4.10 Mixtures of Gases

- **Behavior** → Mixtures of nonreacting gases behave as if only a single component were present.
- **Partial pressure** → Pressure a gas would exert if it occupied the container alone
- **Law of partial pressures** → Total pressure of a gas mixture is the sum of the partial pressures of its components [John Dalton]
  $P = P_A + P_B + \ldots$ for mixture containing A, B, ...
- **Components at the same *T* and *V*** → $n = n_A + n_B + \ldots$ = total amount of mixture

$$P = \frac{nRT}{V} \quad \text{and} \quad P_A = \frac{n_A RT}{V}$$

$$x_A = \frac{n_A}{n} = \frac{P_A}{P} = \text{mole fraction of A in the mixture}$$

Thus, $P_A = x_A P$

Example 4.10a A 1.00-L sample of 100% humid air from the atmosphere is taken at 15°C. An analysis reveals the following composition by mass: 934 mg of nitrogen (N_2), 285 mg of oxygen (O_2), 12.8 mg of water vapor (H_2O), 15.9 mg of argon (Ar), and 0.658 mg of carbon dioxide (CO_2). Calculate the mole fraction and partial pressure of each component of air in the sample.

Solution Convert the mass of each component into amount (n_A, n_B, ...) by using the appropriate molar mass. Sum the amounts of each component to obtain the total amount n. Take the ratio of the amount of each component to the total amount to obtain the mole fraction. The total pressure P (atm) is obtained from the total number of moles, the volume of the sample, and the absolute temperature by the ideal gas law. The partial pressures are then obtained from Dalton's law ($P_A = x_A P$, ...).

$$n_{N_2} = 934\text{ mg}\left(\frac{1\text{ g}}{1000\text{ mg}}\right)\left(\frac{1\text{ mol}}{28.02\text{ g}}\right) = 3.333\times10^{-2}\text{ mol}$$

$$n_{O_2} = 285\text{ mg}\left(\frac{1\text{ g}}{1000\text{ mg}}\right)\left(\frac{1\text{ mol}}{32.00\text{ g}}\right) = 8.906\times10^{-3}\text{ mol}$$

$$n_{H_2O} = 12.8\text{ mg}\left(\frac{1\text{ g}}{1000\text{ mg}}\right)\left(\frac{1\text{ mol}}{18.016\text{ g}}\right) = 7.105\times10^{-4}\text{ mol}$$

$$n_{Ar} = 15.9\text{ mg}\left(\frac{1\text{ g}}{1000\text{ mg}}\right)\left(\frac{1\text{ mol}}{39.95\text{ g}}\right) = 3.980\times10^{-4}\text{ mol}$$

$$n_{CO_2} = 0.658\text{ mg}\left(\frac{1\text{ g}}{1000\text{ mg}}\right)\left(\frac{1\text{ mol}}{44.01\text{ g}}\right) = 1.495\times10^{-5}\text{ mol}$$

$$n = n_{N_2} + n_{O_2} + n_{H_2O} + n_{Ar} + n_{CO_2} = 4.336\times10^{-2}\text{ mol}$$

$$x_{N_2} = n_{N_2}/n = 0.7687; \quad x_{O_2} = n_{O_2}/n = 0.2054; \quad x_{H_2O} = n_{H_2O}/n = 0.01639;$$

$$x_{Ar} = n_{Ar}/n = 0.009\,179; \quad x_{CO_2} = n_{CO_2}/n = 0.000\,344\,8$$

$$P = nRT/V = (0.043\,33\text{ mol})(0.082\,057\,8\text{ L·atm·K}^{-1}\text{·mol}^{-1})(288.2\text{ K})/(1.00\text{ L}) = 1.025\text{ atm}$$

$$P_{N_2} = x_{N_2}P = 0.788\text{ atm}; \quad P_{O_2} = x_{O_2}P = 0.211\text{ atm}; \quad P_{H_2O} = x_{H_2O}P = 0.0168\text{ atm};$$

$$P_{Ar} = x_{Ar}P = 0.00941\text{ atm}; \quad P_{CO_2} = x_{CO_2}P = 0.000\,353\text{ atm}$$

Example 4.10b A sample of damp air in a 1.00-L container exerts a pressure of 762.0 Torr at 20°C; but when it is cooled to –10.0°C, the pressure falls to 607.1 Torr as the water condenses. What mass of water was present?
Assume the vapor pressure of ice at –10°C is negligible.

Answer $m = 8.44 \times 10^{-2}$ g

Molecular Motion (Section 4.11)

Key Concepts

diffusion, effusion, Graham's law

4.11 Diffusion and Effusion

- **Diffusion** → Gradual dispersal of one substance through another substance
- **Effusion**
  → Escape of a gas through a small hole (or assembly of microscopic holes) into a vacuum
  → Depends on molar mass and temperature of gas
- **Graham's law** → Rate of effusion of a gas *at constant temperature* is *inversely* proportional to the square root of its molar mass M.

For two gases A and B

$$\frac{\text{Rate of effusion of A}}{\text{Rate of effusion of B}} = \sqrt{\frac{M_B}{M_A}}$$

→ Rate of effusion is proportional to the average speed of the molecules in a gas *at constant temperature.*

$$\frac{\text{Average speed of A molecules}}{\text{Average speed of B molecules}} = \sqrt{\frac{M_B}{M_A}}$$

→ Rate of effusion is *inversely* proportional to the time taken for given volumes or amounts of gas to effuse *at constant temperature.*

$$\frac{\text{Rate of effusion of A}}{\text{Rate of effusion of B}} = \frac{\text{time for B to effuse}}{\text{time for A to effuse}} = \sqrt{\frac{M_B}{M_A}}$$

Example 4.11a It takes a certain amount of hydrogen gas 45 s to effuse through a porous barrier. How long, in minutes, does it take the same amount of uranium hexafluoride (UF_6) to effuse under the same conditions?

Solution First determine the molar mass of H_2 and UF_6. The time for effusion is proportional to the square root of the molar mass. Because UF_6 has a higher molar mass, it takes longer to effuse.

$M_{H_2} = 2(1.0079) = 2.0158\ \text{g·mol}^{-1}$; $M_{UF_6} = 238.03 + 6(19.00) = 352.03\ \text{g·mol}^{-1}$

$$(\text{time for } UF_6 \text{ to effuse}) = (\text{time for } H_2 \text{ to effuse})\sqrt{\frac{M_{UF_6}}{M_{H_2}}} = (45\text{ s})\sqrt{\frac{352.03}{2.0158}}$$

$$= (45\text{ s})\sqrt{174.635} = (45\text{s})\,(13.215) = 5.9 \times 10^2\text{ s} = 9.9\text{ min}$$

Example 4.11b It takes 50.0 mL of nitrogen 55.0 s to effuse through a porous barrier. An unknown compound takes 164 s to effuse through the same barrier under the same conditions. What is the molar mass of this compound?

Answer M of unknown compound = 248 g·mol^{-1}

- **Effusion at different temperatures** → Rate of effusion and average speed increase as the square root of the temperature.

$$\frac{\text{Rate of effusion at } T_2}{\text{Rate of effusion at } T_1} = \frac{\text{average speed of molecules at } T_2}{\text{average speed of molecules at } T_1} = \sqrt{\frac{T_2}{T_1}}$$

Example 4.11c How much does the rate of effusion (and average speed) in H_2 change when the temperature is decreased from 25°C to −120°C?

Solution Convert temperature to the absolute scale before using the above relationship.

$T_1 = 25 + 273.15 = 298.2$ K and $T_2 = -120 + 273.15 = 153.2$ K

$$\frac{\text{Rate of effusion (average speed) at 153 K}}{\text{Rate of effusion (average speed) at 298 K}} = \sqrt{\frac{153.2\text{ K}}{298.2\text{ K}}} = \sqrt{0.5137} = 0.717$$

Example 4.11d The average speed of molecules in a gas increases by 50% when the temperature is increased from an initial value of 320 K. What is the final temperature of the gas?

Answer $T_2 = 720$ K

- **Combined relationship** → Average speed of molecules is *directly* proportional to the square root of the temperature and *inversely* proportional to the square root of the molar mass.

$$\text{Average speed of molecules in a gas} \propto \sqrt{\frac{T}{M}}$$

Example 4.11e Which has the higher average speed, N_2O molecules at 50°C or HCN molecules at 25°C?

Solution Calculate the molar mass of N_2O and HCN. Convert Celsius temperature to the absolute scale. Determine which gas has the larger value of the square root of the temperature over the molar mass.

$M_{N_2O} = 2(14.01) + 16.00 = 44.02 \text{ g·mol}^{-1}$

$M_{HCN} = 1.0079 + 12.01 + 14.01 = 27.0279 \text{ g·mol}^{-1}$

$T_{N_2O} = 50 + 273.2 = 323.2$ K and $T_{HCN} = 25 + 273.2 = 298.2$ K

Average speed of $N_2O \propto (323.2/44.02)^{1/2} = 2.71$

Average speed of HCN $\propto (298.2/27.0279)^{1/2} = 3.32$

Therefore, HCN has the greater average speed under these conditions.

Example 4.11f What ratio of temperatures is required for methane (CH_4) and helium (He) to have the same average speed?

Answer

$$\frac{T_{CH_4}}{T_{He}} = \frac{M_{CH_4}}{M_{He}} = \frac{16.0416}{4.00} = 4.01$$

The Kinetic Model of Gases (Sections 4.12-4.13)

Key Concepts

kinetic model, pressure-volume product proportional to square of speed, root mean square speed, Maxwell distribution of speeds

Kinetic model assumptions

1. Gas molecules are in continuous random motion.
2. Gas molecules are infinitesimally small particles.
3. Particles move in straight lines until they collide.
4. Molecules do not influence one another except during collisions.

4.12 The Pressure of a Gas (see text for detailed derivation)

- **Wall collisions** → Consider molecules traveling only in one dimension x with an average square-velocity of $\langle v_x^2 \rangle$

- **Pressure on wall** → $P = \dfrac{Nm\langle v_x^2 \rangle}{V}$, where N is the number of molecules, m is the mass of each molecule, and V is the volume of the container

- **Mean square speed** → $c^2 = \langle v^2 \rangle = \langle v_x^2 + v_y^2 + v_z^2 \rangle = \langle v_x^2 \rangle + \langle v_y^2 \rangle + \langle v_z^2 \rangle = 3\langle v_x^2 \rangle$

  Note: The symbol c is also used for the *speed of light* in Chapter 1.

- **Pressure on wall** → $P = \dfrac{Nmc^2}{3V}$ and $N = nN_A$, where N_A is the Avogadro constant

  $P = \dfrac{nN_A mc^2}{3V} = \dfrac{nMc^2}{3V}$, where $M = nN_A$ is the molar mass

  $$\boxed{PV = \frac{nMc^2}{3} = nRT}$$ The term nRT is from the ideal gas law.

- **Solve for c** $$\boxed{c = \sqrt{\frac{3RT}{M}}}$$ c is the **root mean square** (*rms*) speed of the particle.

Note: *Beware of units.* Use R in units of $J{\cdot}K^{-1}{\cdot}mol^{-1}$ and convert M to $kg{\cdot}mol^{-1}$. The resulting units of speed will then be $m{\cdot}s^{-1}$.

Example 4.12a Calculate the root mean square speed of helium and radon gases at 300 K and 1200 K.

Solution Look up the molar masses of helium and radon and convert them to kg·mol^{-1}. Use the above expression for root mean square speed to calculate c at the two temperatures. Be careful to use the gas constant in the proper SI units.

$M_{He} = (4.00 \text{ g·mol}^{-1}) \times (10^{-3} \text{ kg·g}^{-1}) = 4.00 \times 10^{-3} \text{ kg·mol}^{-1}$
$M_{Rn} = (222 \text{ g·mol}^{-1}) \times (10^{-3} \text{ kg·g}^{-1}) = 2.22 \times 10^{-1} \text{ kg·mol}^{-1}$

$$c = \sqrt{\frac{3RT}{M}} = \sqrt{\frac{3(8.314\,51 \text{ J·K}^{-1}\text{·mol}^{-1})(300 \text{ K})}{4.00\times10^{-3} \text{ kg·mol}^{-1}}} = \sqrt{1.871\times10^{6} \text{ m}^2\text{·s}^{-2}} = 1.37\times10^{3} \text{ m·s}^{-1}$$

The ratio of temperatures is (1200/300) = 4. Because $c \propto \sqrt{T}$, the speed at 1200 K will be $\sqrt{4} = 2$ times that at 300 K The final speed for helium is therefore $c = 2.74 \times 10^3$ m·s^{-1} at 1200 K.

For radon, the same procedure yields
$c = 184$ m·s^{-1} at $T = 300$ K and $c = 368$ m·s^{-1} at $T = 1200$ K

Example 4.12b Calculate the root mean square speed for ethene (ethylene, C_2H_4) gas at 298 K. At what temperature will ethyne (acetylene, C_2H_2) gas have the same root mean square speed?

Answer $c = 515$ m·s^{-1} and $T = 277$ K for ethyne gas

Molar kinetic energy $= \left(\frac{1}{2}mc^2\right)N_A = \left(\frac{1}{2}m\right)\left(\frac{3RT}{M}\right)N_A = \frac{3}{2}RT$, because $mN_A = M$

4.13 The Maxwell Distribution of Speeds

- **Symbols** → Let v represent a particle's speed, N the total number of particles, $f(v)$ the Maxwell distribution of speeds, Δ a *finite* change, and d an *infinitesimal* change (*not* density) in the following relationships.

- **For a *finite* range of speeds**

$$\Delta N = Nf(v)\Delta v, \quad \text{where } f(v) = 4\pi N\left(\frac{M}{2RT}\right)^{3/2} v^2 e^{-Mv^2/2RT}$$

- **For an *infinitesimal* range of speeds** $\frac{dN}{N} = f(v)dv$

Note: Calculus can be used with the Maxwell distribution of speeds to obtain the following properties that are not mentioned in the text.

***Average* speed** $\langle c\rangle = \int_0^\infty vf(v)dv = \sqrt{\frac{8RT}{\pi M}}$

***Most probable* speed** $c_{mp} = \sqrt{\dfrac{2RT}{M}}$, where $\dfrac{df(v)}{dv} = 0$ (maximum in distribution)

Real Gases (Sections 4.14-4.16)

Key Concepts

real gases, intermolecular forces, compression factor (Z), gas liquefaction, Joule-Thomson effect, van der Waals equation, van der Waals coefficients, virial equation, virial coefficients

4.14 Intermolecular Forces

- **Definition** → Attractions and repulsions between atoms and molecules
- **Evidence** → Gases condense to liquids when cooled or compressed (attraction). Liquids are difficult to compress (repulsion).
- **Compression factor (Z)** → A measure of the effect of intermolecular forces

$$Z \equiv \frac{PV_m}{RT} = \frac{V_m}{V_m^{ideal}}$$

For an ideal gas, $Z = 1$.
For H_2, $Z > 1$ at all P (repulsions dominate).
For NH_3, $Z < 1$ at low P (attractions dominate).

4.15 The Liquefaction of Gases

- **Joule-Thomson effect** → When attractive forces dominate, a real gas *cools* as it expands. (Compression and expansion through a small hole lowers temperature further.)

4.16 Equations of State of Real Gases

- **van der Waals equation** →

$$\left(P + a\frac{n^2}{V^2}\right)(V - nb) = nRT$$

where a and b are the **van der Waals coefficients** (determined experimentally).

- **Rearranged form** →

$$P = \frac{nRT}{V - nb} - a\frac{n^2}{V^2}$$

where a are the attractive effects and b are the repulsive effects.

Example 4.16a Use both the ideal gas law and van der Waals equation to calculate the pressure of CO_2 gas at 298 K and 200 K (5 K above the sublimation temperature). Assume the molar volume is exactly 1 L·mol^{-1}.

Note: The van der Waals coefficients are $a = 3.640$ L^2·atm·mol^{-1} and $b = 4.267 \times 10^{-2}$ L·mol^{-1}.

Solution Use the ideal gas law and the rearranged form of the van der Waals equation.

$$T = 298\text{ K} \Rightarrow P = \frac{nRT}{V} = \frac{(0.082058\text{ L·atm·K}^{-1}\text{·mol}^{-1})(298\text{ K})}{(1\text{ L·mol}^{-1})} = 24.45\text{ atm}$$

$$P = \frac{nRT}{V - nb} - a\frac{n^2}{V^2} = \frac{(0.082058\text{ L·atm·K}^{-1}\text{·mol}^{-1})(298\text{ K})}{(1 - 4.267 \times 10^{-2})(\text{L·mol}^{-1})} - \frac{(3.640\text{ L}^2\text{·atm·mol}^{-1})}{(1\text{ L·mol}^{-1})^2}$$

$$= (25.543 - 3.640)\text{ atm} = 21.90\text{ atm}$$

Change from ideal = (24.45 – 21.90)/(24.45) = (2.55/24.45) = 0.104 or 10.4%

$$T = 200\text{ K} \Rightarrow P = \frac{nRT}{V} = \frac{(0.082058\text{ L·atm·K}^{-1}\text{·mol}^{-1})(200\text{ K})}{(1\text{ L·mol}^{-1})} = 16.41\text{ atm}$$

$$P = \frac{nRT}{V - nb} - a\frac{n^2}{V^2} = \frac{(0.082058\text{ L·atm·K}^{-1}\text{·mol}^{-1})(200\text{ K})}{(1 - 4.267 \times 10^{-2})(\text{L·mol}^{-1})} - \frac{(3.640\text{ L}^2\text{·atm·mol}^{-1})}{(1\text{ L·mol}^{-1})^2}$$

$$= (17.143 - 3.640)\text{ atm} = 13.50\text{ atm}$$

Change from ideal = (16.41 – 13.50)/(16.41) = (2.91/16.41) = 0.177 or 17.7%

Example 4.16b Use both the ideal gas law and van der Waals equation to calculate the pressure of H_2O gas at 298 K and 373 K (boiling temperature of water). Use the molar volumes of 782.3 L·mol^{-1} at 298 K and 30.61 L·mol^{-1} at 373 K, which correspond to the vapor pressure of liquid water.

Note: The van der Waals coefficients are $a = 5.536$ L^2·atm·mol^{-1} and $b = 3.049 \times 10^{-2}$ L·mol^{-1}.

Answer $P = 0.03126$ atm (ideal gas) and $P = 0.03125$ atm (van der Waals) at 298 K
$P = 0.9999$ atm (ideal gas) and $P = 0.9950$ atm (van der Waals) at 373 K
Surprisingly, water vapor behaves as an ideal gas up to its boiling temperature.

- **Virial equation** → $\boxed{PV = nRT\left(1 + \frac{B}{V_m} + \frac{C}{V_m^2} + \cdots\right)}$

B = second virial coefficient
C = third virial coefficient, *etc.*

Note: **Virial coefficients** depend on temperature and are found by fitting experimental data to the virial equation. This equation is more general than the van der Waals equation but is more difficult to use to make predictions.

Chapter 5 Liquids and Solids

Intermolecular Forces (Sections 5.1-5.5)

Key Concepts

phase, condensed phase, hydration, ion-dipole interaction, hydrated salts, anhydrous salts, trends in hydration, dipole-dipole interaction, instantaneous partial charges and dipole, induced dipole moment, London force, polarizability, hydrogen bond, dimers

5.1 The Formation of Condensed Phases

- **Physical states** → Solid, liquid, and gas
- **Phase** → Form of matter uniform in both chemical composition and physical state
- **Condensed phase**
  → Solid or liquid phase
  → Molecules are very close to each other all the time and intermolecular forces are of major importance
- **Coulomb potential energy** V
  → The interaction between two charges, q_1 and q_2, separated by distance r is (see Chapter 1) ⇒ $V \propto \frac{q_1 q_2}{r}$
  → Almost all intermolecular interactions can be traced back to this fundamental expression.

Note: The term *intermolecular* is used in a general way to include atoms and ions.

5.2 Ion-Dipole Forces

- **Hydration**
  → Attachment of water molecules to ions (cations and anions)
  → Water molecules are polar and have an electric dipole moment μ: a small partial positive charge on each H atom (attracts anions) and a small partial negative charge on the O atom (attracts cations).
- **Ion-dipole interaction**
  → The interaction between an ion with charge $|z|$ and a polar molecule with dipole moment μ at a distance of r is ⇒ $V \propto -\frac{|z|\mu}{r^2}$
  → *Attractive* interaction for proper *alignment* of the ion and dipole: cations attract the *partial* negative charges on the polar molecule and anions attract the *partial* positive charges on the polar molecule.
  → Shorter range (r^{-2}) interaction than the Coulomb potential (r^{-1}): polar molecule needs to be *almost* in contact with the ion.

- **Hydrated compounds**
  → Ion-dipole interactions are much *weaker* than ion-ion interactions, yet they are relatively strong for *small, highly charged* cations
  → Accounts for the formation of salt hydrates such as $CuSO_4 \cdot 5H_2O$ and $CrCl_3 \cdot 6H_2O$

- **Size effects** → Li^+ and Na^+ tend to form hydrated compounds, whereas K^+, Rb^+, and Cs^+ do not. NH_4^+ is similar in size to K^+ and forms *anhydrous* compounds.

- **Charge effects** → Ba^{2+} and K^+ are similar in size, yet Ba^{2+} forms *hydrated* compounds.

Example 5.2a Use the ion-dipole interaction energy to predict whether Li^+ or Tl^{3+} has the stronger attraction to water molecules. Assume that the distance r of closest approach between the ion and water is given by the radius of the ion (see Fig. 1.39 in the text).

Solution The relationship $V \propto -\frac{|z|\mu}{r^2}$ leads to the following ratio for the same polar molecule:

$$\frac{V_2(Tl^{3+})}{V_1(Li^+)} = \frac{|z_2| r_1^2}{|z_1| r_2^2} = \frac{3 \times (58\text{ pm})^2}{1 \times (88\text{ pm})^2} = \frac{10\,092\text{ pm}^2}{7\,744\text{ pm}^2} = 1.3$$

Thus, the energy of interaction between an H_2O molecule and a Tl^{3+} ion is about 30% greater than that between an H_2O molecule and a Li^+ ion. Because Li^+ forms hydrated salts, we predict that Tl^{3+} will also form hydrated salts.

Example 5.2b Calculate the radius of a cation with a charge of +2 that has exactly the same interaction energy with water molecules as does Na^+ (r = 102 pm).

Answer r_2 = 144 pm, which is slightly larger than Ba^{2+} (r = 136 pm)

5.3 Dipole-Dipole Forces

- **Dipole alignment** → In solids, molecules with dipole moments tend to *align* with each other. The partial positive charge on one molecule is attracted to the partial negative charge on another.

- **Dipole-dipole interaction**
  → The interaction between a polar molecule with dipole moment μ_1 and another polar molecule with dipole moment μ_2 at a distance of r, when the dipoles are *aligned,* is ⇒ $V \propto -\frac{\mu_1 \mu_2}{r^3}$
  → *Attractive* interaction in solids for *alignment* of dipoles
  → Shorter range (r^{-3}) interaction than ion-dipole (r^{-2}) or Coulomb potential (r^{-1}) Polar molecules need to be *almost* in contact with each other

- **Dipole-dipole interactions in gas-phase molecules**
  → Dipole-dipole interactions are much *weaker* in gases than in solids. Since the molecules are in motion (rotating as well), they experience only a *weak* net attraction because of *occasional* alignment. The interaction is ⇒ $V \propto -\frac{\mu_1^2 \mu_2^2}{r^6}$

- **Dipole-dipole interactions in liquid-phase molecules**
  → Same relationship as in the gas phase, but the interaction is slightly *stronger* because the molecules are closer.
  → The liquid *boiling point* is a measure of the *strength* of the intermolecular forces in a liquid. For molecules with the same numbers and types of atoms, but with a different placement of bonds (isomers), the relative boiling point of the liquid is *often* related to the strength of the dipole-dipole interaction.
  → The *stronger* the dipole moment, the *higher* the boiling point.

Example 5.3 If *cis*-dibromoethene has a nonzero dipole moment (≈2.4 D) and *trans*-dibromoethene has a value of 0 because its bond dipoles cancel, predict which substance has the higher boiling temperature.

Solution The *cis* substance is expected to have the higher boiling temperature because of its nonzero dipole moment. The actual boiling temperatures, however, are not very different: *cis* (112.5°C) and *trans* (108°C).

Note: The boiling-point rule is *not* always successful. The dipole moment of 1,1-dibromoethene (≈1.4 D) is significantly smaller than the value for *cis*-dibromoethene, and the boiling point is lower [1,1 (92°C) and *cis* (112.5°C)]. The *trans*-molecule ($\mu = 0$) is expected to have a *lower* boiling temperature than the 1,1-form, but it actually has a *higher* one. Although there is no dipole moment in *trans*-dibromoethene, *localized* regions of partial charge in one molecule strongly attract regions of partial charge with the opposite sign in another molecule. The ease of molecular packing may also play an important role as it does in solids. This exception holds for the analogous series of dichloroethene isomers (see text examples).

5.4 London Forces

- **Nonpolar molecules** → Condensation of *nonpolar* molecules to form liquids implies the existence of a type of intermolecular interactions other than those described above.
- **London force**
  → London force accounts for the attraction (*always*) between any pair of ground-state molecules (*polar* or *nonpolar*) arising from *instantaneous* partial charges (*instantaneous* dipole moment) in one molecule *inducing* partial charges (*induced* dipole moment) in a neighboring one, where the two molecules are a distance r apart. London force exists between atoms and *rotating* molecules as well.
  → The polarizability α of a molecule is related to the ease of deformation of its electron cloud. Polarizability is proportional to the total number of electrons in the molecule. Since the number of electrons in a molecule is proportional to the molar mass, the polarizability is as well.
  → The interaction between a molecule (polar or nonpolar) with polarizability α_1 and another molecule (polar or nonpolar) with polarizability α_2 at a distance of r is ⇒ $$V \propto -\frac{\alpha_1 \alpha_2}{r^6}$$

  Note: Same (r^{-6}) interaction as dipole-dipole (r^{-6}) with *rotating* molecules, but the London interaction is *usually* of greater strength at normal temperatures.

- **Liquid boiling point**
  - → The boiling point is a measure of the strength of the intermolecular forces in a liquid.
  - → In comparing the interactions discussed in Sections 5.3 and 5.4, the *major* influence on boiling point in both *polar* and *nonpolar* molecules is the London force.

Example 5.4a Predict the relative boiling points of the following *nonpolar* molecules: F_2 (18 electrons), Cl_2 (34 electrons), Br_2 (70 electrons), and I_2 (106 electrons).

Solution The greater the polarizability of the substance, the higher the boiling temperature. The polarizability is proportional to the number of electrons in the molecule. The boiling points are expected to increase in the order of F_2, Cl_2, Br_2, and I_2. The actual boiling temperatures follow this pattern: F_2 (–188°C), Cl_2 (–34°C), Br_2 (59°C), and I_2 (184°C).

Example 5.4b Predict the relative boiling points of the following *polar* molecules: H_2S (18 electrons), H_2Se (36 electrons), and H_2Te (54 electrons).

Solution The greater the polarizability of the substance, the higher the boiling temperature. The polarizability is proportional to the number of electrons in the molecule. The boiling points are expected to increase in the order of H_2S, H_2Se, and H_2Te. The actual boiling temperatures follow this pattern: H_2S (–60°C), H_2Se (–42°C), and H_2Te (–2°C).

Note: Dipole-dipole interactions predict trends opposite to those cited. Also, H_2O (100°C, 10 electrons) does not conform to the general correlation, because it engages in hydrogen bonding (Section 5.5).

- → The dipole-dipole interaction may be decisive, however, for different molecules with the same number of electrons. In this case, the stronger the dipole moment, the higher the boiling point.

Example 5.4c Predict the relative boiling points of propane, C_3H_8, and dimethyl ether, CH_3OCH_3.

Solution Both molecules have 26 electrons and are expected to have nearly the same polarizability. Propane is essentially a nonpolar molecule ($\mu = 0.084$ D), because of the low polarity of C–H bonds. Dimethyl ether has a Lewis structure similar to H_2O, with *two* methyl groups, CH_3, replacing *two* H atoms. The molecular shape is *angular* around the O atom, and the molecule is polar ($\mu = 1.30$ D). In this case, the dipole-dipole interaction is the determining factor, because the London interactions are expected to be similar. The boiling point of dimethyl ether is expected to be greater than that of propane. The actual boiling temperatures are –25°C for dimethyl ether and –42°C for propane.

Example 5.4d Predict the relative boiling points of ethanol, CH_3CH_2OH, and dimethyl ether, CH_3OCH_3.

Solution Both molecules have identical atoms arranged differently (isomers). Ethanol has a Lewis structure similar to H_2O, with the ethyl group, C_2H_5, replacing one H atom. The O–H bond is more polar than the O–C bond, and ethanol is expected to have a greater dipole moment ($\mu = 1.69$ D) and boiling point (78.5°C) than dimethyl ether. The presence of hydrogen bonding (Section 5.5), however, is actually the *major* reason for the higher boiling point of ethanol.

→ For molecules that are nonpolar *and* have the same number of electrons, the *shape* of the molecule may allow for a *greater* London force.

→ Consider the two isomers discussed in the text. Pentane is a *rod*-shaped molecule and 2,2-dimethylpropane is *spherically symmetrical.* The polarizabilities are identical, but the *rod*-shaped molecules approach each other more closely, and the London interactions are strengthened.

5.5 Hydrogen Bonding

- **Hydrogen bonds**
  - → Interaction *specific* to certain types of molecules (strong attractive forces)
  - → Unusually high boiling points in ammonia (NH_3, –33°C), water (H_2O, 100°C), and hydrogen fluoride (HF, 20°C)
  - → An H atom, *covalently* bonded to either an N, O, or F atom in *one* molecule, is strongly attracted to a *lone pair* of electrons on an N, O, or F atom in *another* molecule. The interaction is the *strongest* when the three atoms are in a *straight line*, and the distance between terminal atoms is within a particular range. Common symbol for the H bond is three dots: ···
  - → *Strongest* intermolecular interaction between *neutral* molecules

Example 5.5a Use the three dot notation to show the hydrogen bonds that form in pure ammonia, NH_3; pure water, H_2O; and pure hydrogen fluoride, HF.

Solution
[N–H···:N] between NH_3 and NH_3 molecules in pure NH_3
[O–H···:O] between H_2O and H_2O molecules in pure H_2O
[F–H···:F] between HF and HF molecules in pure HF

Example 5.5b Use the three dot notation to show the *additional* hydrogen bonds that form in a mixture of ammonia, water, and hydrogen fluoride.

Solution
[N–H···:O] and [O–H···:N] between NH_3 and H_2O molecules
[N–H···:F] and [F–H···:N] between NH_3 and HF molecules
[O–H···:F] and [F–H···:O] between H_2O and HF molecules

Note: The lone pair is a *requirement* of hydrogen bond formation.

- **Hydrogen bonding in gas-phase molecules**
  - → Aggregation of some molecules persists in the vapor phase.
  - → In HF, fragments of *zigzag* chains and $(HF)_6$ rings are formed. In CH_3COOH (acetic acid), *dimers* are formed. The abbreviated (no C–H bonds shown) Lewis structure for the dimer $(CH_3COOH)_2$ is ⇒

```
      :O:···H—O:
      //        \
H3C—C            C—CH3
      \         //
      :O—H···:O:
```

- **Importance of hydrogen bonding**
  - → Accounts for the open structure of solid water
  - → Maintains the shape of biological molecules
  - → Binds the two strands of DNA together

Liquid Structure (Sections 5.6-5.7)

Key Concepts

long-range order, short-range order, viscosity, surface tension, capillary action, adhesion, cohesion, meniscus

5.6 Order in Liquids

- **Liquid phase**
  → Between the extremes of gas and solid phases
  → Mobile molecules with restricted motion
- **Long-range order** → Characteristic of a *cystalline solid*
  Atoms or molecules are positioned in an orderly pattern that is repeated over long distances.
- **Short-range order**
  → Characteristic of the *liquid phase*
  Atoms or molecules are positioned in an orderly pattern at nearest-neighbor distances only.
  → Local order is maintained by a continual process of forming and breaking nearest-neighbor interactions.

5.7 Viscosity and Surface Tension

- **Viscosity**
  → Resistance of a substance to flow
  → The *greater* the viscosity, the *slower* the flow
  → Viscous liquids include those with hydrogen bonding between molecules [H_3PO_4 (phosphoric acid) and $C_3H_8O_3$ (glycerol): many H bonds], liquid phases of metals, and long chain molecules that can be entangled [hydrocarbon oil and greases].
- **Viscosity and temperature** → *Usually* viscosity *decreases* with *increasing T*
  Exception: unusual behavior of sulfur (see text)
- **Surface tension**
  → Tendency of molecules at the surface of a liquid to be pulled into the body of the liquid by an imbalance in the intermolecular forces. Also, the *inward pull* that determines the resistance of a liquid to an increase in surface area.
  → A measure of the *force* that must be applied to *surface molecules* so that they experience the same *force* as molecules in the *interior* of the liquid. Also, a measure of the tightness of the surface layer.
  Symbol: γ (gamma) Units: $N{\cdot}m^{-1}$ or $J{\cdot}m^{-2}$
- **Capillary action**
  → Rise of liquids up narrow tubes when the forces of *adhesion* are greater than the forces of *cohesion*

→ **adhesion**: Forces that bind a *substance* to a *surface*.
→ **cohesion:** Forces that bind *molecules* of a substance together to form a *bulk material*.

- **Meniscus**
  → Curved surface that a liquid forms in a tube
  Adhesive forces greater than *cohesive* forces (forms a ◡ shape)
  Cohesive forces greater than *adhesive* forces (forms a ◠ shape)
  → Glass surfaces have exposed O atoms and O–H groups to which hydrogen-bonded liquids like H_2O can bind. In this case, the *adhesive* forces are greater than the *cohesive* ones. Water *wets* glass, forms a ◡ shape at the surface, and undergoes a capillary rise in a glass tube.
  → Mercury liquid does not bind to glass surfaces. The *cohesive* forces are greater than the *adhesive* ones. Mercury does *not* wet glass, forms a ◠ shape at the surface, and undergoes a capillary *lowering* in a glass tube.

- **Capillary rise relation** → $h = \dfrac{2\gamma}{gdr}$ or $\gamma = \dfrac{1}{2}gdrh$

d = density of liquid | r = radius of the tube
g = gravitational constant | h = height of liquid rise in tube

Units of γ: $\left(\dfrac{\text{m}}{\text{s}^2}\right)\left(\dfrac{\text{kg}}{\text{m}^3}\right)(\text{m})(\text{m}) = \left(\dfrac{\text{kg}}{\text{s}^2}\right) = \left(\dfrac{\text{kg}\cdot\text{m}}{\text{s}^2\cdot\text{m}}\right) = \left(\dfrac{\text{N}}{\text{m}}\right)$

Example 5.7a Benzene was found to rise to a height of 3.36 cm in a tube of radius 0.200 mm at 25°C. Given that $g = 9.80665\ \text{m}\cdot\text{s}^{-2}$, estimate the density of liquid benzene at the temperature of the experiment.
Values of γ ($\text{mN}\cdot\text{m}^{-1}$) are given in Table 5.3 in the text.

Solution Rearrange the capillary rise relation to calculate the unknown density. Look up the value of the surface tension in Table 5.3 in the text. If all units are converted into the SI base values, the calculated density will then have units of $\text{kg}\cdot\text{m}^{-3}$. Express density in the more commonly used units of $\text{g}\cdot\text{cm}^{-3}$.
$h = 3.36\ \text{cm} = 3.36 \times 10^{-2}\ \text{m}$ $r = 0.200\ \text{mm} = 2.00 \times 10^{-4}\ \text{m}$
$\gamma = 28.88\ \text{mN}\cdot\text{m}^{-1} = 2.888 \times 10^{-2}\ \text{N}\cdot\text{m}^{-1}$

$$d = \frac{2\gamma}{ghr} = \frac{2(2.888 \times 10^{-2}\ \text{N}\cdot\text{m}^{-1})}{(9.80665\ \text{m}\cdot\text{s}^{-2})(3.36 \times 10^{-2}\ \text{m})\left(2.00 \times 10^{-4}\ \text{m}\right)} = 8.76 \times 10^{2}\ \text{kg}\cdot\text{m}^{-3}$$

$$d = \left(8.76 \times 10^{2}\ \text{kg}\cdot\text{m}^{-3}\right)\left(10^{3}\ \text{g}\cdot\text{kg}^{-1}\right)\left(10^{-2}\ \text{m}\cdot\text{cm}\right)^{3} = 0.876\ \text{g}\cdot\text{cm}^{-3}$$

Example 5.7b Acetic acid was found to rise to a height of 2.70 cm in a tube of radius 0.200 mm. The density of acetic acid at the same temperature is $1.049\ \text{g}\cdot\text{cm}^{-3}$. Calculate the surface tension of acetic acid in units of $\text{mN}\cdot\text{m}^{-1}$ at the temperature of the experiment.

Answer $\gamma = 27.8\ \text{mN}\cdot\text{m}^{-1}$

Solid Structures (Section 5.8-5.10)

Key Concepts

crystalline solid, lattice, amorphous solid, crystal faces, solids (metallic, ionic, network, molecular), close-packed structure (hexagonal hcp and cubic ccp), coordination number, slip planes, malleability, hole (tetrahedral and octahedral), body-centered cubic structure bcc, primitive cubic structure pc, unit cell, face-centered cubic structure fcc (ccp), Bravais lattices

5.8 Classification of Solids

- **Amorphous solid** → Atoms or molecules (neutral or charged) lie in random positions.
- **Lattice** → An *orderly* array of points in three dimensions
- **Crystalline solid** → Atoms, ions, or molecules are associated with points in a lattice.
- **Crystal faces** → Flat, well-defined planar surfaces at definite interplanar angles.
- **Classification of crystalline solids**

  metallic solids: Cations in a sea of electrons
  ionic solids: Mutual attractions of cations and anions
  molecular solids: Discrete molecules held together by the *intermolecular forces* discussed earlier
  network solids: Atoms bonded *covalently* to their neighbors throughout the entire solid

5.9 Metallic Solids

- **Close-packed structures**
  → Atoms occupy smallest total *volume* with the *least* amount of empty space.
  → Metal atoms are treated as *spheres*, where *r* is the radius of each sphere.
  → Type of metal determines whether a close-packed structure is assumed.
- **Close-packed layer**
  → Atoms (*spheres*) are arranged in a layer (planar) with the *least* amount of empty space. The contour of the layer has *dips* or *depressions*.
  → Each atom has *six* nearest neighbors in the layer (hexagonal pattern), and the layers are stacked to lie in the *dips*.
- **Stacked layers** → In the final stacked structure, each atom has *three* nearest neighbors in the layer above, *six* in its original layer, and *three* in the layer below for a total of 12 nearest neighbors.
- **Coordination number** → Number of nearest neighbors of each atom in the solid
- **Two arrangements of stacking close-packed layers**

  Pattern: ABABABAB... hcp ≡ *hexagonal close-packing*
  ABCABCABC... ccp ≡ *cubic close-packing*

- **Occupied space in a close-packed structure** → 74% of space is occupied by the spheres and 26% is *empty*.
- **Tetrahedral hole** → Formed when a *dip* between *three* atoms in one layer is *covered* by *another* atom in an adjacent layer
  Two holes per atom in a close-packed structure (hcp or ccp)
- **Octahedral hole** → Space between *six* atoms, *four* of which are at the corners of a square plane. The square plane is oriented 45° with respect to two close-packed layers. *Two* atoms in the square plane are in layer **A** and the other *two* in layer **B**. There is *one* additional atom in layer **A** that is *above* the square plane and *another* in layer **B** that is *below* the square plane. The result is an octahedral arrangement of 6 atoms with a hole in the center.
  One hole per atom in a close-packed structure (hcp or ccp)

  Figure of an octahedron and an octahedron rotated to show the relation to two close-packed layers (**A** *and* **B**)

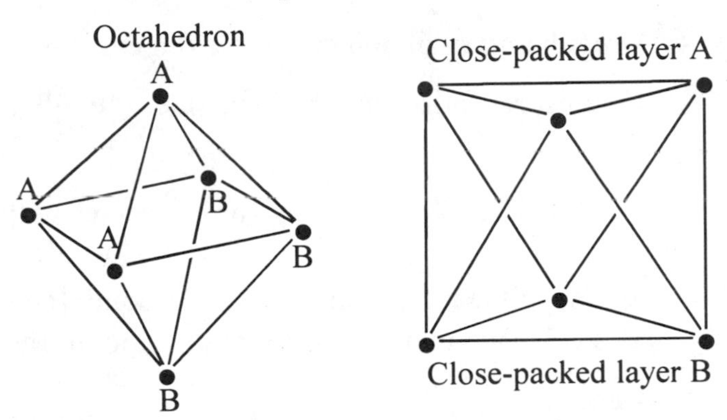

- **Slip plane**
  → Plane of atoms that may slip or slide along an adjacent plane under stress
  → Characteristic of a *cubic close-packed* structure (ccp), which is also called *face-centered cubic* (fcc)
  → Accounts for the *malleability* of ccp metallic crystals (may be bent, flattened, or pounded into different shapes)
  → Metallic crystals with the hcp structure tend to be brittle.
- **Body-centered cubic (bcc) structure**
  → Coordination number of 8 (*not* close-packed)
  → Spheres touch along the *body* diagonal of a cube
  Can be forced into close-packed structures under high pressure
- **Primitive cubic (pc) structure**
  → Coordination number of 6 (*not* close-packed)
  → Spheres touch along the *edge* of the cube
  Only one known example: Po (polonium)

Crystal Structure	Coordination Number	Occupied Space	Examples
ccp (fcc)	12	74%	Ca, Sr, Ni, Pd, Pt, Cu, Ag, Au, Al, Pb [Group 18: noble gases at low temperature, except He]
hcp	12	74%	Be, Mg, Ti, Co, Zn, Cd, Tl
bcc	8	68%	Group 1: alkali metals, Ba, Cr, Mo, W, Fe
pc	6	52%	Po (covalent character)

5.10 Unit Cells

- **Lattice** → A regular array of points in three dimensions
- **Unit cell** → Smallest repeating unit that generates the full array of points
- **Crystal**
  → Constructed by associating atoms, ions, or molecules with each lattice point
  → Metallic crystals: One metal atom per lattice point [ccp (fcc), bcc, pc]
  Two metal atoms per lattice point [hcp]
- **Unit cells for metals**
  → Cubic system: pc, bcc, and fcc (ccp)
  → Hexagonal system: *primitive unit cell* with *two* atoms per lattice point (hcp)
- **Unit cells in general**
  → Edge lengths: a, b, c
  → Angles between two edges: α (between edges b and c)
  β (between edges a and c)
  γ (between edges a and b)
- **Other crystal systems** → Tetragonal, orthorhombic, rhombohedral, monoclinic, and triclinic
- **Bravais lattices**
  → 14 basic patterns of arranging points in three dimensions
  → Each pattern has a different unit cell, as shown below.

The 14 Bravais Lattices

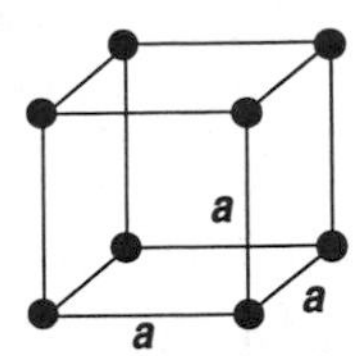

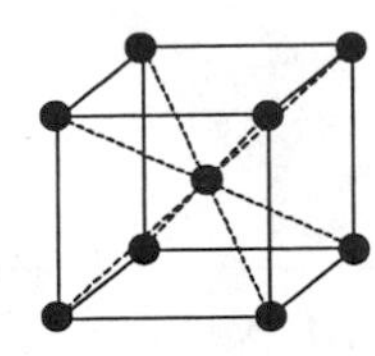

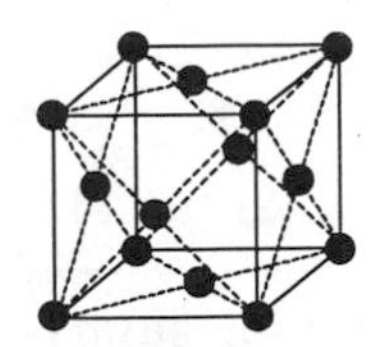

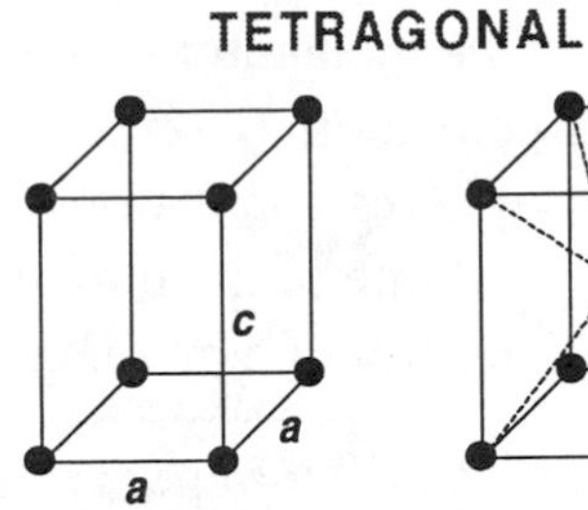

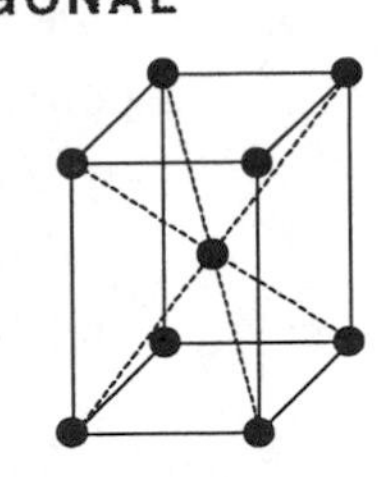

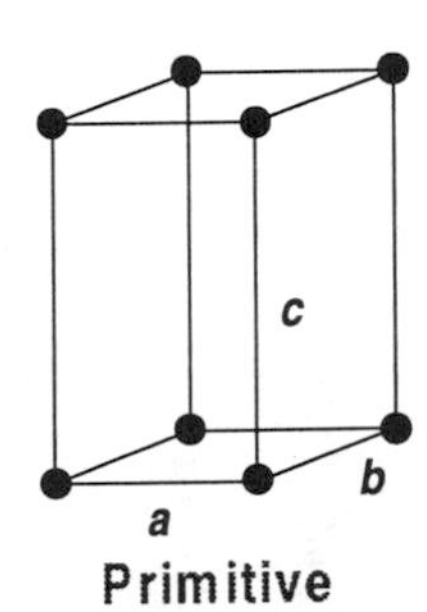

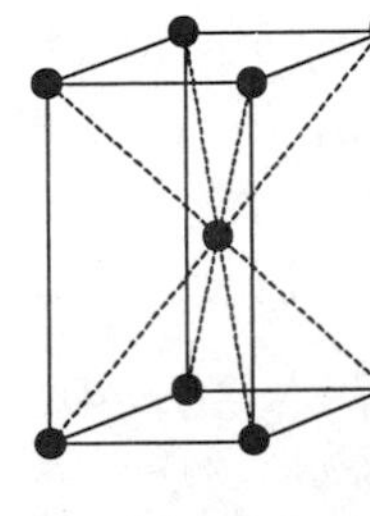
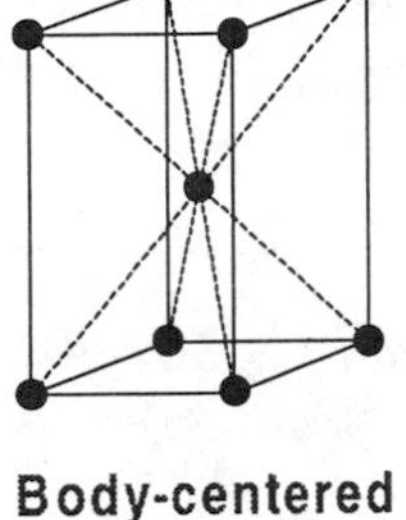

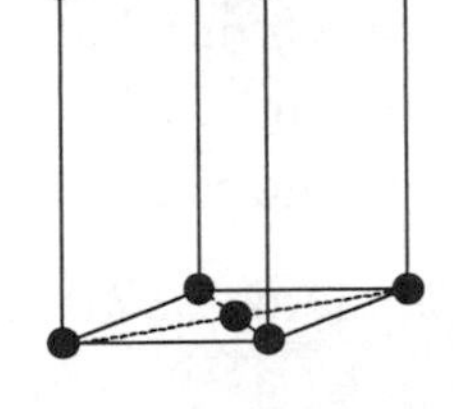

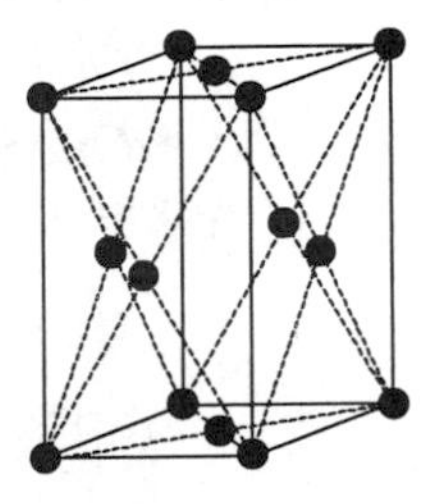

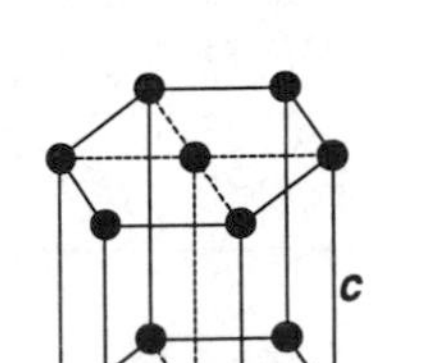

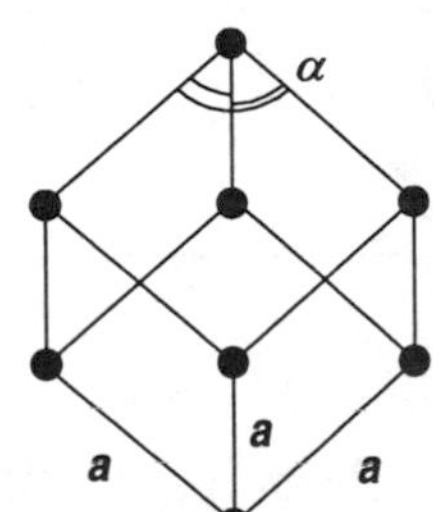

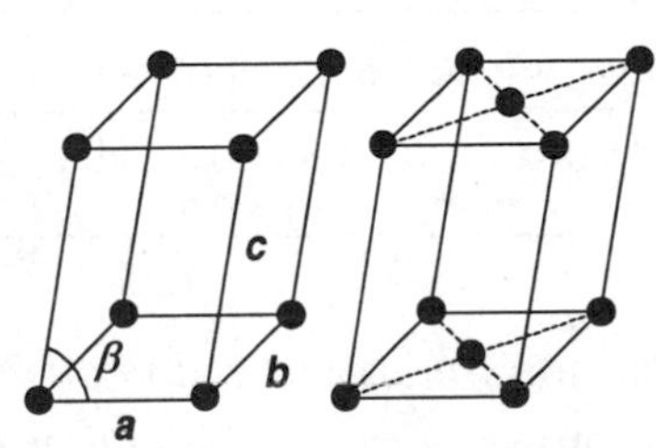

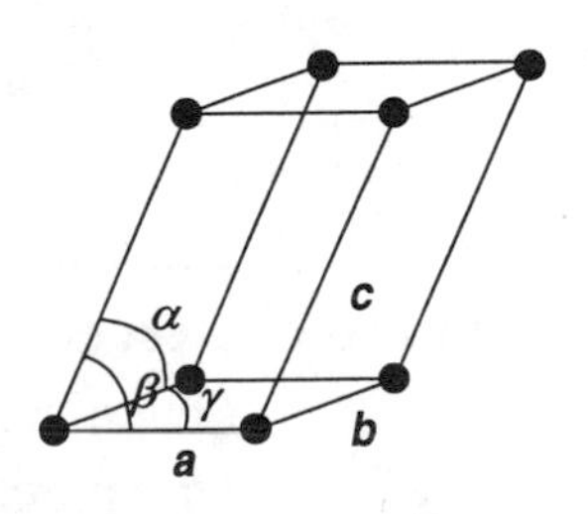

- **Volume of several unit cells**

  Cubic: $V = a \times a \times a = a^3$ [$\alpha = \beta = \gamma = 90°$]
  Tetragonal: $V = a \times a \times c = a^2c$ [$\alpha = \beta = \gamma = 90°$]
  Orthorhombic: $V = a \times b \times c = abc$ [$\alpha = \beta = \gamma = 90°$]

- **Properties of unit cells**

  Each *corner point* shared among 8 cells [1/8 per cell]
  8 corner points × 1/8 per cell = 1 corner point per cell [*all cells*]

  Each *face-centered point* shared between 2 cells [1/2 per cell]
  6 face points × 1/2 per cell = 3 face points per cell [*face-centered cells*]
  2 face points × 1/2 per cell = 1 face point per cell [*end face-centered cells*]

  Each *body-centered point* unshared [1 per cell]
  1 body point × 1 per cell = 1 body point per cell [*body-centered cells*]

- **Properties of *cubic system* unit cells** → *One* metal atom occupies each *lattice point*

Primitive

One metal atom (*sphere*) at each *corner point* in a unit cell
[*one* atom (atomic radius r) per unit cell (edge length a)]
Spheres *touch* along an *edge:* $a = 2r$
Volume of the unit cell: $V = a^3 = (2r)^3 = 8r^3$
Density of a pc metal (M = molar mass):

$$d = \frac{(1 \text{ atom per cell})(\text{mass of one atom})}{(\text{volume of one unit cell})} = \frac{M/N_A}{a^3} = \frac{M/N_A}{8r^3}$$

Body-centered

One metal atom (*sphere*) at each *corner point* in a unit cell
One metal atom (*sphere*) at the *center* of the unit cell
Total atoms in unit cell = 8 (1/8) + 1 (1) = 2
[*two* atoms (atomic radius r) per unit cell (edge length a)]
Spheres *touch* along the *body diagonal:* $\sqrt{3}a = 4r$ or $a = \frac{4}{\sqrt{3}}r$

Volume of the unit cell: $V = a^3 = \frac{4^3}{3^{3/2}}r^3$

Density of a bcc metal:

$$d = \frac{(2 \text{ atoms per cell})(\text{mass of one atom})}{(\text{volume of one unit cell})} = \frac{(2)(M/N_A)}{a^3} = \frac{(2)(M/N_A)}{(4^3/3^{3/2})r^3}$$

Face-centered

One metal atom (*sphere*) at each *corner point* in a unit cell
One metal atom (*sphere*) at the *center* of each *face* in the unit cell
Total atoms in unit cell = 8 (1/8) + 6 (1/2) = 4
[*four* atoms (atomic radius r) per unit cell (edge length a)]
Spheres *touch* along the *face diagonal:* $\sqrt{2}a = 4r$ or $a = \frac{4}{\sqrt{2}}r = \sqrt{8}r$

Volume of the unit cell: $V = a^3 = 8^{3/2}r^3$
Density of a fcc (ccp) metal:

$$d = \frac{(4 \text{ atoms per cell})(\text{mass of one atom})}{(\text{volume of one unit cell})} = \frac{(4)(M/N_A)}{a^3} = \frac{(4)(M/N_A)}{(8^{3/2})r^3}$$

- **Calculations** → Determining the *unit cell* type from the *measured density* and *atomic radius* of metals in the *cubic* system

Note: The *atomic radius* of a metal atom is actually determined from the crystal type and the edge length of a unit cell, which are both obtained by x-ray diffraction (see Major Technique 3 in the text).

Example 5.10a The *density* of lead, Pb, is 11.34 g·cm^{-3} and the *atomic radius* is 175 pm. Assuming the unit cell belongs to the *cubic system*, determine whether it is *primitive, body-centered*, or *face-centered.*

Solution Calculate the density of the metal by using the relations previously developed for each of the three cubic types. The value that best agrees with the measured density is evidence for a particular unit cell type. Because density has units of g·cm^{-3}, convert the atomic radius to units of cm. The ratio of the molar mass to the Avogadro constant gives the mass of a single atom in units of g.

$M_{\text{Pb}} = 207.19\text{ g·mol}^{-1}$; $r = 175\text{ pm} \times (10^{-10}\text{ cm·pm}^{-1}) = 1.75 \times 10^{-8}\text{ cm}$

Primitive: $$d = \frac{M/N_A}{8r^3} = \frac{(207.19/6.022\,14\times10^{23})}{8(1.75\times10^{-8})^3} = 8.02\text{ g·cm}^{-3}$$

Body-centered: $$d = \frac{2\,(M/N_A)}{(4^3/3^{3/2})\,r^3} = \frac{2(207.19/6.022\,14\times10^{23})}{(4^3/3^{3/2})(1.75\times10^{-8})^3} = 10.4\text{ g·cm}^{-3}$$

Face-centered: $$d = \frac{4\,(M/N_A)}{(8^{3/2})\,r^3} = \frac{4\,(207.19/6.022\,14\times10^{23})}{(8^{3/2})(1.75\times10^{-8})^3} = 11.3\text{ g·cm}^{-3}$$

The unit cell type is most likely *face-centered cubic*, because the density for that type is closest to the measured value of 11.34 g·cm^{-3}.

Example 5.10b The *density* of tungsten, W (M = 183.85 g·mol^{-1}), is 19.30 g·cm^{-3} and the *atomic radius* is actually 137 pm. Assuming the unit cell belongs to the *cubic system*, determine whether it is *primitive, body-centered*, or *face-centered.*

Answer The unit cell type is most likely *body-centered cubic*, because the density for that type (19.3 g·cm^{-3}) is closest to the measured value of 19.30 g·cm^{-3}.

The Impact on Materials: Metals (Sections 5.11-5.12)

Key Concepts

properties in terms of electron mobility, alloys (homogeneous and heterogeneous), substitutional alloy, interstices, interstitial alloy

5.11 The Properties of Metals

- **Mobility of electrons**
  - → Responsible for characteristic luster and light reflectivity properties of a metal
  - → Accounts for *malleability, ductility*, and *electrical conductivity* of metals
  - → Produces a relatively strong metallic bond that leads to high melting points for most metals

5.12 Alloys

- **Heterogeneous** → Mixture of crystalline phases, various samples have different compositions; for example, solders and mercury amalgams

- **Homogeneous** → Atoms of different elements are distributed uniformly; for example, brass, bronze, and coinage metals.
- **Substitutional**
  - → Atoms nearly the same size (< 15% difference) can substitute for each other more or less freely (mainly, *d*-block elements)
  - → Lattice is distorted and electron flow is hindered. Alloy has lower thermal and electrical conductivity than pure element, but it is harder and stronger. Homogeneous examples include the copper-zinc alloy, brass (up to 40% zinc in copper), and bronze (metal other than zinc or nickel in copper: casting bronze is 10% Sn and 5% Pb).
- **Interstitial**
  - → Solute atoms that are much smaller in size (> 60% difference) can occupy holes or *interstices* in a lattice.
  - → Interstitial atoms interfere with electrical conductivity and the movement of atoms in the lattice. Restricted motion makes the alloy harder and stronger than the host metal. Examples include low-carbon steel (carbon in iron, which is soft enough to be stamped), high-carbon steel (hard and brittle unless subject to heat treatment), and stainless steel (a mixture of iron with metals such as chromium and nickel, which aid in its resistance to corrosion).

The Impact on Materials: Nonmetallic Solids (Sections 5.13-5.16)

Key Concepts

ionic structures, rock-salt (NaCl) structure, coordination number convention, radius ratio, cesium-chloride structure, zinc-blende structure, molecular solids, network solids, allotropes, liquid crystals, mesophase, anisotropic material, isotropic material, nematic phase, smectic phase, cholesteric phase, thermotropic liquid crystals, lyotropic liquid crystals

5.13 Ionic Structures

- **Model**
  - → Spheres of *different* radii and *opposite* charges representing cations and anions
  - → Larger spheres (*usually* anions) *customarily* occupy the unit cell lattice points. The smaller spheres (*usually* cations) fill *holes* in the unit cell.
- **Radius ratio**
  - → Ratio of radius of the *smaller* sphere to the radius of the *larger* sphere
  - → Defined for ions of the *same* charge number [$z_{cation} = |z_{anion}|$]

Radius Ratio – Crystal Structure Correlations

radius ratio $\leq$ 0.414:	tetrahedral *holes* fill (fcc for *larger* spheres) [*zinc-blende structure* (one form of ZnS)]
0.414 < radius ratio < 0.732:	octahedral *holes* fill (fcc for *larger* spheres) [*rock-salt structure* (NaCl)]
radius ratio $\geq$ 0.732:	cubic *holes* fill (pc for *larger* spheres) [*cesium-chloride structure* (CsCl)]

Note: r(tetrahedral hole) < r(octahedral hole) < r(cubic hole)

- **Coordination number** → Number of nearest-neighbor ions of *opposite* charge
- **Coordination of ionic solid** → Represented as (cation coordination number, anion coordination number)

- **Properties of the *ionic structure* of unit cells**

Zinc blende

Radius ratio (ZnS) = $r(Zn^{2+})/r(S^{2-})$ = (74 pm)/(184 pm) = 0.40
S^{2-} ions at the *corners* and *faces* of a fcc unit cell
[*four* anions (atomic radius r_{anion}) per unit cell (edge length a)]
Zn^{2+} ions at *four* of the *eight* tetrahedral holes in the unit cell
[*four* cations (atomic radius r_{cation}) per unit cell (edge length a)]
Cation and anion spheres *touch* in the tetrahedral locations.
Each cation has *four* nearest-neighbor anions, and each anion has *four* nearest-neighbor cations: (4,4)-coordination.

Rock salt

Radius ratio (NaCl) = $r(Na^{+})/r(Cl^{-})$ = (102 pm)/(181 pm) = 0.564
Cl^{-} ions at the *corners* and *faces* of a fcc unit cell
[*four* anions (atomic radius r_{anion}) per unit cell (edge length a)]
Na^{+} ions at all *four* of the *four* octahedral holes in the unit cell
[*four* cations (atomic radius r_{cation}) per unit cell (edge length a)]
Cation and anion spheres *touch* along the edge of the cell.
Each cation has *six* nearest-neighbor anions, and each anion has *six* nearest-neighbor cations: (6,6)-coordination
Spheres *touch* along the *edge:* $a = 2r_{cation} + 2r_{anion} = 566$ pm
Volume of the unit cell: $V = a^3 = 1.81 \times 10^8\ \text{pm}^3 = 1.81 \times 10^{-22}\ \text{cm}^3$
Density of NaCl: M(58.44 g·mol^{-1}) and d(experimental) = 2.17 g·cm^{-3}

$$d = \frac{(4\ \text{NaCl per cell})(M/N_A)}{a^3} = \frac{4(58.44/N_A)}{(1.81 \times 10^{-22})} = 2.14\ \text{g}\cdot\text{cm}^{-3}$$

Cesium chloride

Radius ratio (CsCl) = $r(\text{Cs}^+)/r(\text{Cl}^-)$ = (170 pm)/(181 pm) = 0.939

Cl^- ions at the *corners* of a pc (*primitive cubic*) unit cell

[*one* anion (atomic radius r_{anion}) per unit cell (edge length a)]

Cs^+ ion at the *center* of the *cube* in the unit cell

[*one* cation (atomic radius r_{cation}) per unit cell (edge length a)]

Cation and anion spheres *touch* along the *body diagonal* of the cell.

Each cation has *eight* nearest-neighbor anions and each anion has *eight* nearest-neighbor cations: (8,8)-coordination.

Spheres *touch* along the *body diagonal:*

$$\sqrt{3}a = 2r_{\text{cation}} + 2r_{\text{anion}} = 702 \text{ pm} \quad \text{and} \quad a = 405 \text{ pm}$$

Volume of the unit cell: $V = a^3 = 6.64 \times 10^7 \text{ pm}^3 = 6.64 \times 10^{-23} \text{ cm}^3$

Density of CsCl: $M(168.36 \text{ g·mol}^{-1})$ and $d(\text{experimental}) = 3.99 \text{ g·cm}^{-3}$

$$d = \frac{(1 \text{ CsCl per cell})(M/N_A)}{a^3} = \frac{1(168.36/N_A)}{(6.64 \times 10^{-23})} = 4.21 \text{ g·cm}^{-3}$$

- **Calculations**

→ Determining the *ionic structure* in the *cubic* system from the *ionic radii* of cations and anions by using the *radius-ratio rule*

→ Determining the *density* of the *ionic solid* from the *density* of the *unit cell*
Note: Usually within 10% of the experimental value

Example 5.13a The *ionic structure* of CsF is the *rock-salt structure*. The *ionic radii* of Cs^+ and F^- are 170 pm and 133 pm, respectively. Show that the *radius-ratio rule* is *not* followed in this case. Describe the properties of the unit cell. Calculate the density of CsF and compare your answer to the experimental value of 4.12 g·cm^{-3}.

Solution Calculate the *radius ratio* and use the guidelines given to show that the *ionic structure* predicted is not correct. Place the larger ions at lattice points within the fcc unit cell. Describe the locations of the smaller ions within the cell. Calculate the density of the unit cell and compare your answer to the experimental value.

Radius ratio (CsF) = $r(\text{F}^-)/r(\text{Cs}^+)$ = (133 pm)/(170 pm) = 0.782 ≥ 0.70
The predicted *ionic structure* is the *cesium-chloride structure*. The rule is not followed for the alkali fluorides in Group 1 except for LiF.

The larger ion is Cs^+, and cations are located at the *corners* and *faces* of the fcc unit cell. The F^- anions are placed in the *octahedral holes*. The unit cell contains *four* cations and *four* anions. The spheres *touch* along the *edge:*
$a = 2r_{\text{cation}} + 2r_{\text{anion}} = 606$ pm and $V = a^3 = 2.23 \times 10^8 \text{ pm}^3 = 2.23 \times 10^{-22} \text{ cm}^3$.
The density of the unit cell is given by

$$d = \frac{(4 \text{ CsF per cell})(M/N_A)}{a^3} = \frac{4(151.91/N_A)}{(2.23 \times 10^{-22})} = 4.52 \text{ g·cm}^{-3}.$$

The calculated value is in good agreement with the experimental value.

Example 5.13b The *ionic structure* of NH_4Cl follows the *radius-ratio rule*. The *ionic radii* of NH_4^+ and Cl^- are 143 pm and 181 pm, respectively. Use the *radius-ratio rule* to predict the *ionic structure*. Describe the properties of the unit cell. Calculate the density of NH_4Cl and compare your answer to the experimental value of 1.527 $g{\cdot}cm^{-3}$.

Answer *Radius ratio* = 0.79, conforming to the *cesium-chloride structure*. The larger ion is Cl^-, and anions are located at the *corners* of the pc unit cell. The NH_4^+ cation is placed in the *cubic hole*. The unit cell contains *one* cation and *one* anion. The spheres *touch* along the *body diagonal:* $\sqrt{3}\,a = 2r_{\text{cation}} + 2r_{\text{anion}} =$ 648 pm. $a = 374$ pm. The volume of the unit cell is $V = a^3 = 5.23 \times 10^7\ \text{pm}^3$ $= 5.23 \times 10^{-23}\ \text{cm}^3$. The density of the unit cell is given by

$$d = \frac{(1\ \text{NH}_4\text{Cl per cell})(\,M/N_\text{A})}{a^3} = \frac{1(53.49/N_\text{A})}{(5.23\times10^{-23})} = 1.70\ \text{g}{\cdot}\text{cm}^{-3}.$$

The calculated value is in good agreement with the experimental value.

Note: There are many exceptions to the *radius-ratio rule*. It should be used with caution.

5.14 Molecular Solids

- **Crystals**
  - → Solid structures reflect the nonspherical nature of molecules and the weak intermolecular forces that hold them together.
  - → Characterized by low melting temperatures and less hardness than ionic solids

Example 5.14 Account for the observation that ice floats on liquid water and that solid benzene sinks in liquid benzene in terms of their solid structure.

Solution In the structure of ice, each H_2O molecule forms hydrogen bonds with *four* neighboring molecules. The resulting structure is an open network of molecules that collapses when the solid melts (Fig. 5.51 and Fig. 5.52 in the text). Other molecular solids (benzene, carbon dioxide, *etc.*) are held together only by *weaker* and *less directional* London forces. The resulting structure is a dense arrangement of molecules that expands when the solid melts (Fig. 5.53 in the text). Solid ice floats on liquid water, whereas solid benzene sinks to the bottom of liquid benzene.

5.15 Network Solids

- **Crystals**
  - → Atoms joined to neighbors by *strong* covalent bonds that form a network extending throughout the solid
  - → Characterized by high melting and boiling temperatures and hard and rigid structures
- **Ceramics** → Usually, they are oxides with a network structure that has great strength and stability, because covalent bonds must break to deform the crystal. Such materials tend to shatter rather than bend under stress. Examples include quartz, silicates, and high-temperature superconductors.

- **Elemental network solids** → *Allotropes* are forms of an element with different solid structures

Example 5.15 Describe the structure and properties of diamond and graphite, the allotropes of carbon, which are both network solids.

Solution Diamond is an insulator consisting of a three-dimensional network of σ-bonds formed by the overlap of sp^3 hybrid orbitals. The tetrahedral network accounts for the hardness and high thermal conductivity of the solid (see Fig. 5.54). The unit cell is fcc with two carbon atoms per lattice point. Carbon atoms occupy the *corners*, *face* positions, and *four* of the *eight* tetrahedral holes in the unit cell.

Graphite consists of flat sheets of sp^2 hybrid σ-bonds and *delocalized* π-bonds formed from *unhybridized p*-orbitals overlapping side by side out of the plane of the sheets. The sheets are held together by weak intermolecular forces, and atom impurities between the sheets allow them to *slide* over one another. Graphite is a lubricant and a good conductor of electricity in the plane of the sheets (see Fig. 5.55).

5.16 Liquid Crystals

- **Properties**
  → Substance flows like a viscous liquid, yet molecules are characterized by a moderately ordered array similar to that in a crystal.
  → Intermediate state of matter with the fluid properties of a liquid and some molecular ordering similar to a crystal is called a *mesophase*.

- **Isotropic material** → Properties independent of the direction of measurement
  Ordinary liquids have viscosity values that are the same in every direction.

- **Anisotropic material** → Properties depend on the direction of measurement
  Rod-shaped molecules form a liquid crystal, in which molecules are free to slide past one another along their axes but resist motion perpendicular to that direction.

- **Classes of liquid crystals** → Differ in the arrangement of the molecules
  See Fig. 5.57 through Fig. 5.59 in the text.

nematic phase: Molecules lie together in the same direction but are *staggered*

smectic phase: Molecules lie together in the same direction in *layers*

cholesteric phase: Molecules form *nematiclike layers*, but the molecules of neighboring layers are *rotated* with respect to each other. The resulting liquid crystal has a *helical* arrangement of molecules.

- **Mode of preparation**
  → *Thermotropic liquid crystals* are made by melting the solid phase. The liquid crystal then becomes a liquid above a characteristic temperature.
  → *Lyotropic liquid crystals* are layered structures that are produced by the action of a solvent on a solid or a liquid.

Chapter 6 Thermodynamics: The First Law

Systems, States, and Energy (Sections 6.1-6.8)

Key Concepts

thermodynamics, statistical thermodynamics, system, surroundings, open system, closed system, isolated system, work, internal energy, translational kinetic energy, rotational kinetic energy, vibrational kinetic energy, degree of freedom, equipartition theorem, internal energy of an ideal gas, heat, adiabatic wall, first law of thermodynamics, state function, expansion work, nonexpansion work, reversible isothermal expansion, reversible process, irreversible process, exothermic process, endothermic process, calorimetry, calorimeter, heat capacity, specific heat capacity, molar heat capacity

Overview

- **Thermodynamics** → Branch of science concerned with the relationship between heat and other forms of energy
- **Laws of thermodynamics**
  → Generalizations based on experience with *bulk* matter
  → Not derivable, but understandable without recourse to knowledge of the behavior of atoms and molecules
- **First law of thermodynamics**
  → Consequence of the **law of conservation of energy**
  → Quantitative description of energy changes in physical (for example, phase changes) and chemical (for example, reactions) processes
  → Energy changes in relation to heat and work:
  A paddlewheel stirrer immersed in a liquid produces heat from mechanical work (temperature rise in the liquid).
  A hot gas expanding in an insulated cylinder coupled to a flywheel produces mechanical work from heat (temperature drop in the gas).
- **Second law of thermodynamics**
  → Criterion for spontaneity (see Chapter 7)
  → Explains why some chemical reactions occur spontaneously and others do not
- **Statistical thermodynamics**
  → Laws of thermodynamics reflected in the behavior of large numbers of atoms or molecules in a sample
  → Link between the atomic level and the behavior of bulk matter
- **Chapter goals** → Gain insight into heat and work, and their relationship to energy changes in physical and chemical processes

6.1 Systems

- **The "universe" or "world"** → Composed of a system and its surroundings
- **System** → Portion of the universe of interest
  Examples include a beaker of ethanol, a frozen pond, 20 g of $CaCl_2(s)$, and the earth itself.
- **Surroundings**
  → Remainder of the universe (everything that is *not* in the system)
  → Where observations and measurements of the system are made
  For example, the observation of heat transferred to or from a system is made in the surroundings.
- **Boundary** → Dividing surface between the system and the surroundings

Example 6.1a Suggest a system, boundary, and surroundings for 10 moles of propane gas in a rigid metal cylinder.

Answer One choice is that the 10 moles of propane gas constitute the system, the metal cylinder and everything outside of it the surroundings, and the inside walls of the cylinder a real boundary. This is an appropriate way to study the properties of the gas alone.

Example 6.1b Suggest a system, boundary, and surroundings for 500 mL of water in an open beaker.

Answer To study the properties of water alone, it should constitute the system. The walls of the beaker constitute a real, physical boundary and the phase boundary between water and air constitutes an imaginary boundary. The surroundings consist of everything outside the boundaries including the beaker, the platform on which it rests, and the atmosphere.

- **Types of systems**

Open system → Both *matter* and *energy* can be exchanged between the system and surroundings.

An example is a half-liter of water in an open beaker (see Example 6.1b). Matter can cross the boundary (water evaporates). Energy can cross the boundary (heat from the surroundings may enter the system).

Closed system → *Energy* can be exchanged between the system and surroundings, but *matter* cannot.

An example is the refrigerant in an air-conditioning system. As the refrigerant expands, energy (heat) is withdrawn from the space to be cooled. As the refrigerant is compressed, energy (heat) is supplied to another space that is warmed. The mass of the refrigerant is unchanged.

Isolated system → Neither *matter* nor *energy* can be exchanged between the system and surroundings.
An example of an *approximately* isolated system is a substance in a well-insulated container such as ice in a Styrofoam ice-chest or coffee in a covered Styrofoam cup.

6.2 Energy and Work

- **Work, *w***
  - → Fundamental thermodynamic property (see Section A.2 of the text)
  - → Movement against an opposing force
  - → Definition: work = force × distance
  - → Units: (force) *newton*, N; $1\ \text{N} = 1\ \text{kg}\cdot\text{m}\cdot\text{s}^{-2}$
    (work) *joule*, J; $1\ \text{J} = 1\ \text{N}\cdot\text{m} = 1\ \text{kg}\cdot\text{m}^2\cdot\text{s}^{-2}$

Example 6.2a Describe three commonplace examples of how work is done on or by a system.

Answer Compression of a spring: If the spring is the system, work is done on the system by the surroundings to compress the spring.

Compression of a gas mixture as in an automobile engine: If the gas mixture is the system, work is done on the system by the surroundings (piston) to compress the gases.

Muscle contraction: As the muscle contracts, work is done on the surroundings. Both mechanical and electrical work are involved as well as chemical processes.

Example 6.2b A plumber of mass 65 kg carries a toolbox of mass 15 kg to a fifth floor walk-up apartment 15 m above ground level. Calculate the work required for this process. Recall that (difference in potential energy) = mgh from Section A.2 of the text.

Answer The difference in potential energy between the fifth and ground floors is equal to the work done by the plumber. The force is mg and the distance is h. Both the plumber's mass and that of the toolbox must be elevated.

$m = (65 + 15)\ \text{kg} = 80\ \text{kg}$
$g = 9.81\ \text{m}\cdot\text{s}^{-2}$
$h = 15\ \text{m}$
$w = (80\ \text{kg})(9.81\ \text{m}\cdot\text{s}^{-2})(15\ \text{m}) = 1.18 \times 10^4\ \text{J} = 11.8\ \text{kJ}$

- **Internal energy, *U***
  - → Internal energy is the *total* capacity of a system to do work.
  - → *Only* a change in U ($\Delta U = U_{\text{final}} - U_{\text{initial}}$) is measurable.
  - → An extensive property (see Section A.1 in the text)
  - → For energy transferred to a system by doing work on the system, $\Delta U = w$.

Example 6.2c Describe the internal energy change and work performed when a spring is compressed or expanded.

Answer Work is done on the system by the surroundings to compress the spring. The compressed spring has a greater capacity to do work than the uncompressed spring. In each case, the internal energy of the system is increased by the amount of work done on it.

$$\Delta U > 0 \text{ and } \Delta U = U_{\text{final}} - U_{\text{initial}} = w$$

Therefore, $w > 0$ when work is done *on* the system.

Work is done on the surroundings by the system to expand the spring. The expanded spring has a lesser capacity to do work than the unexpanded spring. The internal energy of the system is decreased by the amount of work done by it.

$$\Delta U < 0 \text{ and } \Delta U = U_{\text{final}} - U_{\text{initial}} = w$$

Therefore, $w < 0$ when work is done *by* the system.

Example 6.2d Describe the internal energy change and work performed when a battery is recharged.

Answer Electrical work is done on the system by the surroundings in recharging a battery. The charged battery has a greater capacity to do work than the discharged battery. The internal energy of the system is increased by the amount of work done on it. As in the case of mechanical work, $w > 0$ when electrical work is done *on* the system.

6.3 The Molecular Origin of Internal Energy

- **Internal energy, U** → Energy stored as potential and kinetic energy of molecules
- **Potential energy** → Energy an object has by virtue of its position
- **Kinetic energy**
  → Energy an object has by virtue of its motion
  → For atoms, kinetic energy is translational in nature.
  → For molecules, kinetic energy has translational, rotational, and vibrational components.
  Translation is the motion of the molecule as a *whole.*
  (See Section 1.5 in the text for the particle-in-a-box translational energy levels.)
  Rotation and vibration are motions *within* the molecule, and do not change the center of gravity (mass) of the molecule.
  (See Box 2.2 in the text for the rigid-rotor rotational energy levels.)
  (See Major Technique 1 in the text for a description of vibrational motion.)

Translational, rotational, and vibrational motion of F_2

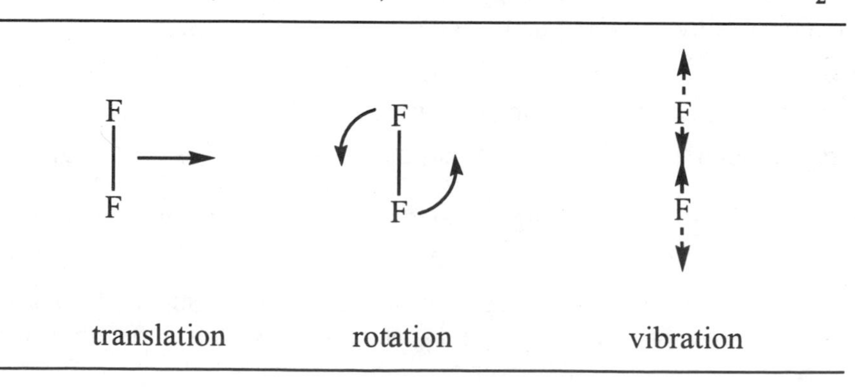

- **Degrees of freedom**
  - → Modes of motion (translational, rotational, and vibrational)
  - → *Mainly* translational and rotational degrees of freedom store internal energy at room temperature
- **Equipartition theorem**
  - → *Average* kinetic energy of each degree of freedom of a molecule in a sample at temperature T is equal to $\frac{1}{2}kT$. The Boltzmann constant $k = 1.380\,658 \times 10^{-23}$ J·K^{-1}.
  - → A molecule moving in *three* dimensions has *three* translational degrees of freedom.
    $U(\text{translation}) = 3 \times \frac{1}{2}kT = \frac{3}{2}kT$
    For 1 mol of molecules, $U_m(\text{translation}) = N_A(\frac{3}{2}kT) = \frac{3}{2}RT$. $(R = N_A k)$
  - → A linear molecule has *two* rotational degrees of freedom.
    $U(\text{rotation, linear}) = 2 \times \frac{1}{2}kT = kT$
    For 1 mol of molecules, $U_m(\text{rotation, linear}) = N_A(kT) = RT$.
  - → A nonlinear molecule has *three* rotational degrees of freedom.
    $U(\text{rotation, nonlinear}) = 3 \times \frac{1}{2}kT = \frac{3}{2}kT$
    For 1 mol of molecules, $U_m(\text{rotation, nonlinear}) = N_A(\frac{3}{2}kT) = \frac{3}{2}RT$.
  - → A molecule of an *ideal gas* does *not* interact with its neighbors and the potential energy is 0. The internal energy is independent of volume and depends *only* on temperature, $U_m(T)$.
  - → A molecule of a liquid, solid, or real gas does interact with its neighbors. The potential energy is an important component of the internal energy, which has a significant dependence on volume. The internal energy depends on both temperature and volume, $U_m(T, V)$.

Example 6.3a Calculate the internal energy of 1 mol of Ar at 300 K. Assume the atoms behave as an ideal gas.

Solution An atom has only three degrees of freedom in translational motion. The internal energy per mole is then

$$U_m = U_m(\text{translation}) = \tfrac{3}{2}RT = (1.5)(8.314\,51\ \text{J·K}^{-1}\text{·mol}^{-1})(300\ \text{K})$$
$$= 3.74 \times 10^3\ \text{J·mol}^{-1}$$
$$= 3.74\ \text{kJ·mol}^{-1}$$

Example 6.3b Calculate the internal energy of 1 mol of H_2 at 300 K. Assume the molecules behave as an ideal gas.

Solution All diatomic molecules are linear. The H_2 molecule has three degrees of freedom in translational motion and two degrees of freedom in rotational motion. The internal energy per mole is then

$$U_m = U_m(\text{translation}) + U_m(\text{rotation, linear}) = \tfrac{3}{2}RT + RT = \tfrac{5}{2}RT$$
$$= (2.5)(8.314\,51\ \text{J·K}^{-1}\text{·mol}^{-1})(300\ \text{K}) = 6.24 \times 10^3\ \text{J·mol}^{-1} = 6.24\ \text{kJ·mol}^{-1}$$

Example 6.3c Calculate the internal energy of 1 mol of H_2O vapor at 300 K. Assume the molecules behave as an ideal gas.

Solution If we use the VSEPR model (Section 3.2), the H_2O molecule is predicted to be nonlinear. Thus, it has three degrees of freedom in translational motion and three degrees of freedom in rotational motion. The internal energy per mole is then

$$U_m = U_m(\text{translation}) + U_m(\text{rotation, nonlinear}) = \tfrac{3}{2}RT + \tfrac{3}{2}RT = 3RT$$
$$= (3)(8.314\,51\ \text{J·K}^{-1}\text{·mol}^{-1})(300\ \text{K}) = 7.48 \times 10^3\ \text{J·mol}^{-1} = 7.48\ \text{kJ·mol}^{-1}$$

Notes: In all three examples, the internal energy is linearly proportional to the temperature of the ideal gas. The vibrational contribution to the internal energy of an ideal gas is small at room temperature and can usually be ignored. It has a temperature dependence, however, that is decidedly nonlinear at normal temperatures. At *very* high temperature, each vibrational degree of freedom of a molecule has a contribution of $\frac{1}{2}kT$ from kinetic energy and $\frac{1}{2}kT$ from potential energy if vibration is treated as a harmonic oscillator. The total contribution to the internal energy is then kT per degree of freedom for vibrational motion.

6.4 Heat

- **Heat, q**
  - → Fundamental thermodynamic property
  - → Energy transferred as a result of a temperature difference, $\Delta T = T_2 - T_1 > 0$
  - → Energy flows as *heat* from a region of high temperature, T_2, to one of low temperature, T_1.
  - → SI unit: *joule*, $1\ \text{J} = 1\ \text{N·m} = 1\ \text{kg·m}^2\text{·s}^{-2}$

    Common unit: The *calorie*, 1 cal ≡ 4.184 J, is widely used in biochemistry, organic chemistry, and related fields.

- **Sign of q defined in terms of the system**
  - → If heat flows from the surroundings into the system, $q > 0$ (system gains heat).
  - → If heat flows from the system into the surroundings, $q < 0$ (system loses heat).
  - → If no heat flows between the system and surroundings, $q = 0$ (*adiabatic* process). For example, a thermally insulating wall (Styrofoam or vacuum flask) prevents the passage of heat.

- **Internal energy, U**
  - → For energy transferred to a system by the flow of heat, $\Delta U = q$ $(w = 0)$.
  - → If the system is initially at temperature T_1 and $\Delta T = T_2 - T_1 > 0$, then $q > 0$ and $\Delta U > 0$.

Example 6.4a Calculate the heat flow and change in internal energy if the temperature of 1 mol of Ar is increased from 300 K to 900 K. Assume the atoms behave as an ideal gas and no work is done.

Solution From Example 6.3a, $U_m = 3.74\ \text{kJ·mol}^{-1}$ at 300 K. Tripling the temperature leads to $U_m = 11.22\ \text{kJ·mol}^{-1}$ at 900 K.

$$q = \Delta U_m = U_m(900\ \text{K}) - U_m(300\ \text{K}) = (11.22 - 3.74)\ \text{kJ·mol}^{-1} = 7.48\ \text{kJ·mol}^{-1}$$

Example 6.4b Calculate the heat flow and change in internal energy if the temperature of 1 mol of H_2 is increased from 300 K to 900 K. Assume the molecules behave as an ideal gas and no work is done.

Solution In Example 6.3b, $U_m = 6.24\ \text{kJ·mol}^{-1}$ at 300 K. Tripling the temperature leads to $U_m = 18.72\ \text{kJ·mol}^{-1}$ at 900 K.

$q = \Delta U_m = U_m(900\ \text{K}) - U_m(300\ \text{K}) = (18.72 - 6.24)\ \text{kJ·mol}^{-1} = 12.48\ \text{kJ·mol}^{-1}$

Example 6.4c Calculate the heat flow and change in internal energy if the temperature of 1 mol of H_2O is increased from 300 K to 900 K. Assume the molecules behave as an ideal gas and no work is done.

Solution In Example 6.3c, $U_m = 7.48\ \text{kJ·mol}^{-1}$ at 300 K. Tripling the temperature leads to $U_m = 22.44\ \text{kJ·mol}^{-1}$ at 900 K.

$q = \Delta U_m = U_m(900\ \text{K}) - U_m(300\ \text{K}) = (22.44 - 7.48)\ \text{kJ·mol}^{-1} = 14.96\ \text{kJ·mol}^{-1}$

Note: In all three examples, q and the change in internal energy are positive values, because heat flows from the surroundings into the system. The amount of heat required to increase the temperature (and internal energy) is smaller for atoms than for molecules, and is smaller for linear molecules than for nonlinear molecules. The greater the number of degrees of freedom, the greater the *capacity* of a species to store energy.

6.5 The First Law

- **Combining heat and work**
  - → $\boxed{\Delta U = q + w}$
  - → Change in internal energy of a system is the *sum* of the heat *added* ($q > 0$) *to* or *removed* ($q < 0$) *from* the system *plus* the work done *on* ($w > 0$) or *by* ($w < 0$) the system.
- **First law of thermodynamics**
  - → The internal energy of an *isolated* system is constant.
  - → It is an extension of the law of conservation of energy.
  - → It is a generalization based on experience and cannot be proven.

  If the first law were false, one could construct a "perpetual motion machine" by starting with an isolated system, removing it from isolation, and letting it do work on the surroundings. It could then be placed in isolation again, and its internal energy allowed to return to the initial value. This feat has never been accomplished despite many attempts.

Example 6.5 An electric stirrer performs 40 kJ of work on a beaker of water and water transfers 1.5×10^3 cal of heat to the surroundings. Calculate the change in internal energy of water in kJ.

Solution Convert calories to the units of kilojoules.

$(1.5 \times 10^3\ \text{cal})(4.184\ \text{J·cal}^{-1})(10^{-3}\ \text{kJ·J}^{-1}) = 6.3\ \text{kJ}$

Work is done on the system (water), so $w = 40\ \text{kJ} > 0$.

Heat is transferred from the system to the surroundings, so $q = -6.3\ \text{kJ} < 0$.

$\Delta U = q + w = (-6.3 + 40)\text{ kJ} = 34\text{ kJ}$
$\Delta U > 0$ and the internal energy of the system has increased.

6.6 State Functions

- **State of a system** → Defined when all of its properties are fixed

Example 6.6a Describe the state of a system composed of 1 mol of Ar(g) at 1 bar and 300 K.

Solution The pressure and temperature are uniform throughout the gas. The ideal gas equation uniquely determines the volume of the gas, $V = nRT/P$. All other properties applicable to a noble gas under these conditions, such as density, refractive index, and heat capacity, are fixed as well.

Example 6.6b Describe the state of the following system: a partition that separates a sample of liquid water at 10°C from a sample of water at 30°C is removed.

Solution The resulting system is not in any given state, because its properties are not fixed. The temperature and density are not uniform throughout. The system will eventually approach a uniform temperature and density as a result of diffusion. The state of the system can then be defined.

- **State function**
  - → Property that depends only on the current state of the system
  - → Independent of the manner in which the state was prepared

  Examples of state functions include temperature, T; pressure, P; volume, V; and internal energy, U. (Heat, q, and work, w, are *not* state functions.)

- **Path**
  - → Sequence of intermediate steps linking an initial and a final state
  - → Consider the transition from state 1 to state 2. The changes in the state functions are independent of path. The quantities ΔU, ΔP, ΔV, and ΔT are uniquely determined. The amount of heat transferred or the work done depends on the sequence of steps followed.

Example 6.6c For spring break, two students decide to fly from Philadelphia (initial state) to Denver (final state). Two of their classmates choose to drive from Philadelphia to Denver by way of New Orleans (intermediate state). Describe one variable whose change is the same for the two groups (state function) and another that is not.

Solution The destination (final state) and origin (initial state) are the same for both groups. The distances traveled depend on the path taken, and, presumably, are not the same for the plane and car.

Example 6.6d A laboratory technician assigned to stir an insulated beaker of water inadvertently leaves the stirrer on overnight. In the morning, the technician notes that the temperature of the water has risen 3°C. The same temperature rise can also be accomplished by heating the water using a heating mantle. Compare the two processes in terms of the change in internal energy, and the heat and work done on the system.

Solution For the two processes, the initial and final states are the same, so ΔU is also the same. In the first process, mechanical work is done on the system, $q = 0$, and $\Delta U = w$. In the second process, $w = 0$, and $\Delta U = q$. Both q and w are clearly path dependent.

Note: Because the changes in state functions are path independent, the most convenient path may be chosen for a given change in state both in the *laboratory* and in *calculations*.

6.7 Expansion Work

- **Nonexpansion work** → Does not involve a change in volume
  Examples include electrical current and change of position.
- **Expansion work** → Change in volume of the system against an external pressure
  An example includes inflating a tire or balloon.
- **Constant volume process, $\Delta V = 0$**
  → No expansion work is done
  → $\Delta U = q$ (no work of any kind is done)
  → Heat absorbed or released by a system equals the change in internal energy for a constant volume process if no other forms of work are done: $\Delta U = q$.
- **Calculating expansion work with a constant external pressure, P_{ex}**
  → Consider a gas confined in a volume, V, by a piston with surface area, A.
  → Work = force × distance = $F \times d = (P_{ex} \times A) \times d = P_{ex}\Delta V$ (see Section 4.2), where d is the distance the piston is displaced.
  → $\boxed{w = -P_{ex}\Delta V}$ By convention, work done *on* the system is positive (+) in sign.

  Units: $1\ \text{Pa·m}^3 = 1\ \text{kg·m}^{-1}\text{·s}^{-2} \times 1\ \text{m}^3 = 1\ \text{kg·m}^2\text{·s}^{-2} = 1\ \text{J}$
  $1\ \text{L·atm} = 10^{-3}\ \text{m}^3 \times 101\,325\ \text{Pa} = 101.325\ \text{Pa·m}^3 = 101.325\ \text{J}$ (exactly)

Example 6.7a The pressure exerted on an ideal gas at 2.00 atm and 300 K is reduced suddenly to 1.00 atm while heat is transferred to maintain the initial temperature of 300 K. Calculate q, w, and ΔU in *joules* for this process.

Solution For an ideal gas undergoing an isothermal process, $\Delta U = 0$. The work done at constant external pressure is $w = -P_{ex}\Delta V$, where $\Delta V = V_{final} - V_{initial}$.

$V_{initial} = (nRT)/P_{initial} = (1\ \text{mol})(0.082\,06\ \text{L·atm·K}^{-1}\text{·mol}^{-1})(300\ \text{K})/(2.00\ \text{atm})$
$= 12.3\ \text{L}$

$V_{final} = (nRT)/P_{final} = (1\ \text{mol})(0.082\,06\ \text{L·atm·K}^{-1}\text{·mol}^{-1})(300\ \text{K})/(1.00\ \text{atm})$
$= 24.6\ \text{L}$

$\Delta V = V_{final} - V_{initial} = 24.6\ \text{L} - 12.3\ \text{L} = 12.3\ \text{L}$ (system expands)

$w = -P_{ex}\Delta V = -(1.00\ \text{atm})(12.3\ \text{L}) = -12.3\ \text{L·atm}$
$= (-12.3\ \text{L·atm})(101.325\ \text{J·L}^{-1}\text{·atm}^{-1}) = -1.25 \times 10^3\ \text{J}$

Work is done by the system on the surroundings.

$q = \Delta U - w = 0 - (-1.25 \times 10^3\ \text{J}) = +1.25 \times 10^3\ \text{J}$

Heat is absorbed by the system from the surroundings.

- **Reversible process**
  → A process that can be reversed by an *infinitesimal* change in a variable
  → For a change in pressure, the external pressure must always be infinitesimally different from the pressure of the system. A net change is effected by a series of infinitesimal changes in the external pressure followed by infinitesimal adjustments of the system.
  → For a change in temperature, the temperature of the system must change by a series of infinitesimal amounts of either heat flow or of work done.
  → Processes in which change occurs by *finite* amounts are *irreversible* in nature.

- **Reversible, isothermal expansion or compression of an ideal gas**
  → $$\boxed{w = -nRT\ln\frac{V_{\text{final}}}{V_{\text{initial}}} = -nRT\ln\frac{P_{\text{initial}}}{P_{\text{final}}}}$$
  → The change in pressure is carried out in infinitesimal steps.
  → See derivation in Section 6.7 in the text.

Example 6.7b Suppose the pressure change in Example 6.7a is carried out *reversibly* as well as *isothermally*. Calculate q, w, and ΔU in *joules* for this process and compare the answers to those in the *irreversible* expansion in one step.

Answer The internal energy of an ideal gas depends only on temperature, and $\Delta U = 0$. Check the two formulas for isothermal, reversible work.

$$w = -nRT\ln\frac{V_{\text{final}}}{V_{\text{initial}}} = -(1.00\text{ mol})(8.314\,51\text{ J}\cdot\text{K}^{-1}\cdot\text{mol}^{-1})(300\text{ K})\ln\left(\frac{24.6\text{ L}}{12.3\text{ L}}\right)$$

$$\text{or } w = -nRT\ln\frac{P_{\text{initial}}}{P_{\text{final}}} = -(1.00\text{ mol})(8.314\,51\text{ J}\cdot\text{K}^{-1}\cdot\text{mol}^{-1})(300\text{ K})\ln\left(\frac{2.00\text{ atm}}{1.00\text{ atm}}\right)$$

$w = -1.73 \times 10^3$ J and $q = 1.73 \times 10^3$ J

Note: The absolute value of the work done by the system in a *reversible* expansion is greater than the value obtained in an *irreversible* expansion. It is possible to show that the *maximum* amount of work is always done *by* the system in a *reversible* expansion.

Example 6.7c Suppose the pressure change in Example 6.7a is carried out in the opposite direction (compression). Calculate q, w, and ΔU in *joules* for this process in both a *reversible* and *irreversible* way.

Answer The labels for *initial* and *final* states are interchanged. The calculation for *reversible* work results simply in a change in sign.

$$w = -nRT\ln\frac{V_{\text{final}}}{V_{\text{initial}}} = -(1.00\text{ mol})(8.314\,51\text{ J}\cdot\text{K}^{-1}\cdot\text{mol}^{-1})(300\text{ K})\ln\left(\frac{12.3\text{ L}}{24.6\text{ L}}\right)$$

$= +1.73 \times 10^3$ J and $q = -1.73 \times 10^3$ J

The calculation for *irreversible* work results in a quite different value.

$$w = -P_{\text{ex}}\Delta V = -(2.00\text{ atm})(-12.3\text{ L}) = +24.6\text{ L}\cdot\text{atm}$$
$$= (24.6\text{ L}\cdot\text{atm})(101.325\text{ J}\cdot\text{L}^{-1}\cdot\text{atm}^{-1}) = +2.50 \times 10^3\text{ J} \quad\text{and}\quad q = -2.50 \times 10^3\text{ J}$$

Note: The value of the work done *on* the system in a *reversible* compression is smaller than the value obtained in an *irreversible* compression. It is possible to show that the *minimum* amount of work is always done *on* the system in a *reversible* compression.

6.8 The Measurement of Heat: Calorimetry

- **Heat change in a process**
  - → An *exothermic* process is one in which heat is released by the system into the surroundings: $q < 0$.
    Examples include the combustion of hydrocarbons (see Section 3.7) and the condensation of water at constant temperature.
  - → An *endothermic* process is one in which heat is absorbed by the system from the surroundings: $q > 0$.
    Examples include breaking a chemical bond (see Section 2.16) and the vaporization of water at constant temperature.

- **Calorimeter**
  - → Device used to monitor or measure heat transfer in a system by the temperature changes that take place in the surroundings
  - → In calorimetry, $q_{cal} = q_{surr}$.
  - → For an exothermic process, $q_{cal} > 0$; the temperature of the calorimeter increases, $\Delta T > 0$.
  - → For an endothermic process, $q_{cal} < 0$; the temperature of the calorimeter decreases, $\Delta T < 0$.
  - → The heat capacity of a calorimeter, C_{cal}, is the amount of heat absorbed *by the calorimeter* per degree Celsius rise in temperature.

$$\boxed{q_{cal} = C_{cal}\Delta T}$$

  A calorimeter has a heat capacity that is determined experimentally by the addition of a known amount of heat supplied electrically and the measurement of the resulting temperature rise in the calorimeter (see Example 6.3 in the text).
  Units of heat capacity: $J{\cdot}(°C)^{-1}$ or $J{\cdot}K^{-1}$

- **Heat capacity *C* of a pure substance**
  - → Heat capacity of a pure substance is an *extensive* property.
  - → Amount of heat absorbed *by a sample of the substance* per degree Celsius rise in temperature

- **Specific heat capacity, C_s, of a pure substance**
  - → Specific heat capacity of a pure substance is an *intensive* property.
  - → Amount of heat absorbed *by one gram* of a substance per degree Celsius rise in temperature (values for common materials are given in Table 6.1)
  - → $C_s = C/m$ Units: $J{\cdot}(°C)^{-1}{\cdot}g^{-1}$ or $J{\cdot}K^{-1}{\cdot}g^{-1}$
  - → $\boxed{q = C\Delta T = mC_s\Delta T}$

- **Molar heat capacity, C_m, of a pure substance**
  → Molar heat capacity of a pure substance is an *intensive* property.
  → Amount of heat absorbed *by one mole* of a substance per degree Celsius rise in temperature (values for some substances are given in Table 6.1)
  → $C_m = C/n$ Units: $J{\cdot}(°C)^{-1}{\cdot}mol^{-1}$ or $J{\cdot}K^{-1}{\cdot}mol^{-1}$
  → $\boxed{q = C\Delta T = nC_m\Delta T}$
  → Values of molar heat capacities increase with increasing molecular complexity. They depend on the temperature and the state of the substance. $C_m(\text{liquid}) > C_m(\text{solid})$
- **Use of calorimetry to measure heat capacities**
  → Add a known mass, *m*, or amount, *n*, of a substance at a given temperature to a calibrated calorimeter, which is initially at a lower temperature. Wait until the substance and calorimeter have a final common temperature.
  → Heat lost by substance = heat gained by the calorimeter
  → mC_s (decrease in temperature of substance) = C_{cal} (increase in temperature of calorimeter) All quantities are known except C_s.
  → nC_m (decrease in temperature of substance) = C_{cal} (increase in temperature of calorimeter) All quantities are known except C_m.
- **Use of calorimetry to determine the heat output from a reaction**
  → Add known amounts of reactants to a calibrated calorimeter. Record the common temperature of the reactants and calorimeter after the equilibration interval, which should be much faster than the reaction time. Initiate the reaction (combustion) or allow the reaction to reach completion. Record the final temperature after the products and calorimeter equilibrate.
  → For an *exothermic* reaction, the heat lost by the reaction system is equal to the heat gained by the calorimeter.
  → $q_{reaction} = -q_{cal} = -C_{cal}$ (increase in temperature of calorimeter) < 0
  → For an *endothermic* reaction, the heat gained by the reaction system is equal to the heat lost by the calorimeter.
  → $q_{reaction} = -q_{cal} = -C_{cal}$ (decrease in temperature of calorimeter) > 0

Note: The heat changes in chemical reactions are normally determined for a constant temperature process. Because the temperature of the system changes, a calculation is made to return the products to the same initial temperature as the reactants. The correction requires knowledge of the heat capacities of all the products and remaining reactants.

Enthalpy (Sections 6.9-6.13)

Key Concepts

enthalpy, heat transfer at constant pressure, heat capacity of gases, heat capacity at constant volume, heat capacity at constant pressure, molecular origin of heat capacity, phase transitions, enthalpy of vaporization, enthalpy of fusion, enthalpy of freezing, enthalpy of sublimation, heating curves, temperature of a substance at its melting or boiling point

Overview

- ***Study Guide*** **notation (*not* used in the text)**
  → q_V is the heat transfer for a system of constant volume.
  → q_P is the heat transfer for a system at constant pressure.
- **Constant volume process**
  → $\boxed{\Delta U = q_V}$ for a system in which only expansion work is possible
  → Combustion reactions are studied in a bomb calorimeter of constant volume.
- **Constant pressure process**
  → $\boxed{\Delta H = q_P}$ a new state function for a system at constant pressure
  → Most chemical reactions occur at constant pressure largely because reaction vessels are usually open to the atmosphere.

6.9 Heat Transfers at Constant Pressure

- **Definition of enthalpy, *H***
  → $\boxed{H = U + PV}$ a state function because U, P, and V are state functions
  → Change in enthalpy of a system, $\Delta H = \Delta U + \Delta(PV)$.
  At constant pressure, $\Delta H = \Delta U + P\Delta V = q_P + w + P\Delta V = q_P - P_{ex}\Delta V + P\Delta V = q_P$ for a system open to the atmosphere, and for which only expansion work may occur.
- **Heat change for a constant pressure process**
  → An *exothermic* process is one in which heat is released by the system into the surroundings, $\Delta H = q_P < 0$.
  → An *endothermic* process is one in which heat is absorbed by the system from the surroundings, $\Delta H = q_P > 0$.

Example 6.9 A gas expands against a constant external pressure and does 25 kJ of expansion work on the surroundings. During the process, 60 kJ of heat is absorbed by the system. Determine the values of ΔH and ΔU.

Solution Because this is a constant pressure process, $\Delta H = q_P = 60$ kJ.
$\Delta U = \Delta H - P\Delta V = \Delta H + w$
$w = -25$ kJ (work done by the system is *negative*)
$\Delta U = 60 \text{ kJ} - 25 \text{ kJ} = 35 \text{ kJ}$

6.10 Heat Capacities of Gases

- **Definition of heat capacity** → $\boxed{C = \dfrac{q}{\Delta T}}$

- **Heat capacity at constant volume** → $\boxed{C_V = \dfrac{q_V}{\Delta T} = \dfrac{\Delta U}{\Delta T}}$

- **Heat capacity at constant pressure** → $\boxed{C_P = \dfrac{q_P}{\Delta T} = \dfrac{\Delta H}{\Delta T}}$

- **Molar heat capacities**
  → $C_{V,\mathrm{m}} = C_V/n$ and $C_{P,\mathrm{m}} = C_P/n$
  → $C_{V,\mathrm{m}} \approx C_{P,\mathrm{m}}$ *liquids and solids*
  → $C_{P,\mathrm{m}} > C_{V,\mathrm{m}}$ *gases*

- **Molar heat capacity relationship for an ideal gas (see derivation in text)**
  → $C_P = C_V + nR$ *any amount of an ideal gas*
  → $\boxed{C_{P,\mathrm{m}} = C_{V,\mathrm{m}} + R}$ *one mole of an ideal gas*

Example 6.10 The heat capacity, C_P, of 1.500 moles of F_2 is 46.95 $\mathrm{J{\cdot}K^{-1}}$ at 25°C. Assuming the gas is ideal, calculate the molar heat capacity at constant volume, $C_{V,\mathrm{m}}$, for F_2 at 25°C.

Solution $C_{P,\mathrm{m}} = C_P/n = (46.95\ \mathrm{J{\cdot}K^{-1}})/(1.500\ \mathrm{mol}) = 31.30\ \mathrm{J{\cdot}K^{-1}{\cdot}mol^{-1}}$

$C_{V,\mathrm{m}} = C_{P,\mathrm{m}} - R = (31.30 - 8.314\,51)\ \mathrm{J{\cdot}K^{-1}{\cdot}mol^{-1}} = 22.99\ \mathrm{J{\cdot}K^{-1}{\cdot}mol^{-1}}$

6.11 The Molecular Origin of the Heat Capacities of Gases

- **Contributions to the molar heat capacity for a monatomic ideal gas**
  → Consider only translational motion.
  → $U_\mathrm{m} = \frac{3}{2}RT$ and $\Delta U_\mathrm{m} = \frac{3}{2}R\Delta T$ (molar internal energy depends only on T)
  → $\boxed{C_{V,\mathrm{m}} = \dfrac{\Delta U_\mathrm{m}}{\Delta T} = \dfrac{\frac{3}{2}R\Delta T}{\Delta T} = \frac{3}{2}R} = 12.471\,77\ \mathrm{J{\cdot}K^{-1}{\cdot}mol^{-1}}$

- **Contributions to the molar heat capacity for a linear-molecule ideal gas**
  → Consider translational motion and 2 degrees of freedom in rotational motion.
  → $U_\mathrm{m} = \frac{5}{2}RT$ and $\Delta U_\mathrm{m} = \frac{5}{2}R\Delta T$ (molar internal energy depends only on T)
  → $\boxed{C_{V,\mathrm{m}} = \dfrac{\Delta U_\mathrm{m}}{\Delta T} = \dfrac{\frac{5}{2}R\Delta T}{\Delta T} = \frac{5}{2}R} = 20.786\,28\ \mathrm{J{\cdot}K^{-1}{\cdot}mol^{-1}}$

- **Contributions to the molar heat capacity for a nonlinear-molecule ideal gas**
  - → Consider translational motion and 3 degrees of freedom in rotational motion.
  - → $U_m = 3RT$ and $\Delta U_m = 3R\Delta T$ (molar internal energy depends only on T)
  - → $\boxed{C_{V,m} = \dfrac{\Delta U_m}{\Delta T} = \dfrac{3R\Delta T}{\Delta T} = 3R} = 24.943\,53\ \text{J·K}^{-1}\text{·mol}^{-1}$

Note: The contribution from molecular rotation is for room temperature. At lower temperatures, this contribution diminishes, and the heat capacity approaches the value for a monatomic gas. At room temperature, a small contribution to the heat capacity from molecular vibrational motion leads to a value slightly larger than the ones given above.

Example 6.11a Calculate the molar heat capacity at constant volume, $C_{V,m}$, for F_2 at 25°C. Compare the calculated value to the experimental one determined in Example 6.10.

Solution Diatomic molecules are linear.
$C_{V,m} = \frac{5}{2}R = 20.786\,28\ \text{J·K}^{-1}\text{·mol}^{-1}$
The experimental value of $22.99\ \text{J·K}^{-1}\text{·mol}^{-1}$ is slightly larger than the calculated one. The difference is associated with vibrational motion of the F_2 molecule.

Example 6.11b A 100-L vessel containing 6.00 mol of $H_2(g)$ at 2.00 atm is cooled to 203.0 K. Assuming the gas behaves ideally, calculate ΔU and ΔH for this process.

Solution Calculate the initial temperature, T_{initial}.
$T_{\text{initial}} = PV/nR = [(2.00\ \text{atm})(100\ \text{L})]/[(6.00\ \text{mol})(0.082\,06\ \text{L·atm·K}^{-1}\text{·mol}^{-1})]$
$T_{\text{initial}} = 406.2\ \text{K}$
$\Delta T = T_{\text{final}} - T_{\text{initial}} = (203.0 - 406.2)\ \text{K} = -203.2\ \text{K}$
$C_{V,m} = \frac{5}{2}R = 20.786\,28\ \text{J·K}^{-1}\text{·mol}^{-1} = 0.020\,79\ \text{kJ·K}^{-1}\text{·mol}^{-1}$
$C_{P,m} = C_{V,m} + R = 29.100\,79\ \text{J·K}^{-1}\text{·mol}^{-1} = 0.029\,10\ \text{kJ·K}^{-1}\text{·mol}^{-1}$
From Sections 6.8 and 6.10
$\Delta U = nC_{V,m}\Delta T = (6.00\ \text{mol})(\,0.020\,79\ \text{kJ·K}^{-1}\text{·mol}^{-1})(-203.2\ \text{K}) = -25.3\ \text{kJ}$
$\Delta H = nC_{P,m}\Delta T = (6.00\ \text{mol})(\,0.029\,10\ \text{kJ·K}^{-1}\text{·mol}^{-1})(-203.2\ \text{K}) = -35.5\ \text{kJ}$

Note: This equation applies even though pressure is not constant. For an ideal gas, enthalpy depends only on temperature.

6.12 Enthalpies of Phase Changes

- **Phase transitions at constant temperature and pressure**

→ Vaporization:	liquid → vapor	$\Delta H_{\text{vap}} = H_{\text{vapor,m}} - H_{\text{liquid,m}} > 0$ (*endothermic*)
Condensation:	vapor → liquid	$\Delta H_{\text{cond}} = H_{\text{liquid,m}} - H_{\text{vapor,m}} < 0$ (*exothermic*)
→ Fusion (melting):	solid → liquid	$\Delta H_{\text{fus}} = H_{\text{liquid,m}} - H_{\text{solid,m}} > 0$ (*endothermic*)
Freezing:	liquid → solid	$\Delta H_{\text{freez}} = H_{\text{solid,m}} - H_{\text{liquid,m}} < 0$ (*exothermic*)
→ Sublimation:	solid → vapor	$\Delta H_{\text{sub}} = H_{\text{vapor,m}} - H_{\text{solid,m}} > 0$ (*endothermic*)
Deposition:	vapor → solid	$\Delta H_{\text{dep}} = H_{\text{solid,m}} - H_{\text{vapor,m}} < 0$ (*exothermic*)

→ $\Delta H_{vap} = -\Delta H_{cond}$ $\Delta H_{fus} = -\Delta H_{freez}$ $\Delta H_{sub} = -\Delta H_{dep}$

→ $\Delta H_{sub} = \Delta H_{fus} + \Delta H_{vap}$ (for the same T and P)

Values of $\Delta H_{fus}°$ and $\Delta H_{vap}°$ for selected substances are given in Table 6.2.

Liquids may evaporate at any temperature, but they boil only at the temperature at which the vapor pressure of the liquid is equal to the external pressure of the atmosphere.

Example 6.12 Considering the relative strength of intermolecular forces in the liquid phase, arrange the following substances in the order of increasing enthalpy of vaporization at the boiling point of the liquid: CH_3OH, C_6H_6, Hg, CH_4.

Solution The intermolecular interactions of significance for the substances listed are London forces for C_6H_6 and CH_4, metallic bonding for Hg, and hydrogen bonding for CH_3OH. The relative strength of intermolecular forces increases in the order London forces, hydrogen bonding, and metallic interactions. The strength of London interactions increases with the number of electrons in the molecule (polarizability); thus, C_6H_6 has stronger interactions than CH_4.

The expected order of increasing enthalpy of vaporization is CH_4, C_6H_6, CH_3OH, and Hg. The values of standard enthalpies of vaporization in Table 6.2 are 8.2, 30.8, 35.3, and 59.3 $kJ{\cdot}mol^{-1}$, respectively.

6.13 Heating Curves

- **Heating (cooling) curve**

→ The heating (cooling) curve is a graph showing the variation of the temperature, T, of a sample as heat is added (removed) at a constant rate.

→ Two types of behavior are seen in a heating curve. The temperature increase of a single phase to which heat is supplied has a positive slope that depends on the value of the heat capacity. The larger the heat capacity, the more gentle the slope. At the temperature of a phase transition, two phases are present and the slope is 0 (no temperature change). The *greater* the heat associated with the phase change, the *longer* the length of the flat line. As more heat is added to the substance, the relative amounts of the two phases change. The temperature does not rise again until only one phase remains.

In this schematic heating curve, line V with positive slope represents the heating of a solid phase. The flat line W shows the phase transition between solid and liquid (fusion). Line X represents the heating of the liquid after the solid has completely melted. Line Y shows the phase transition between liquid and vapor (boiling at constant external pressure). Finally, line Z represents the heating of the vapor after the liquid disappears. Because the enthalpy of vaporization is greater than the enthalpy of fusion, the length of line Y is longer than line W.

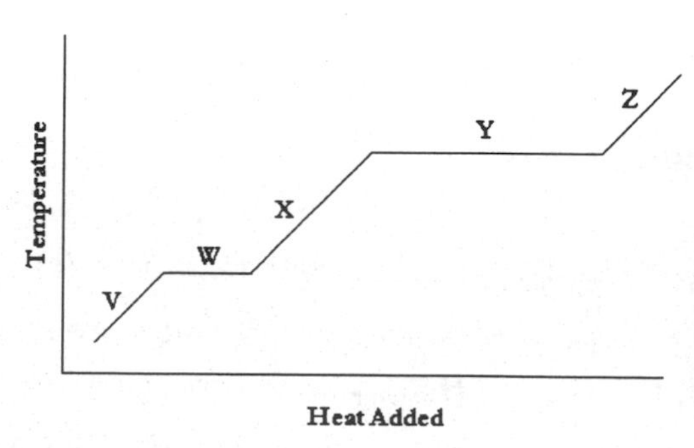

The Enthalpy of Chemical Change (Sections 6.14-6.22)

Key Concepts

reaction enthalpy, thermochemical equation, combustion, standard state, standard reaction enthalpy, Hess's law, reaction sequence, overall reaction, standard enthalpy of combustion, standard enthalpy of formation, lattice enthalpy, Born-Haber cycle, bond enthalpy, average bond enthalpy, temperature dependence of reaction enthalpy, Kirchhoff's law

Overview

- **Reaction enthalpy, ΔH_r**
  - → Applying the principles of thermodynamics to chemical change
  - → A *thermochemical equation* is a chemical equation with the corresponding enthalpy change, ΔH (in kJ).
  - → The *reaction enthalpy*, ΔH_r, has units of $kJ{\cdot}mol^{-1}$, where "mol" refers separately to each reactant and product.
  - → Determining the relation between ΔH and ΔU for chemical reactions
  - → Variation of reaction enthalpy with temperature (Kirchhoff's law)
- **Standard reaction enthalpy, $\Delta H_r°$**
  - → Standard state of a substance is its pure form at an external pressure of *exactly* 1 bar.
  - → Reaction enthalpies of chemical reactions may be combined by using Hess's law. The overall reaction enthalpy is the sum of the values for the individual steps into which a reaction is divided.
  - → Combustion is reaction with oxygen, O_2(g). *Standard enthalpy of combustion*, $\Delta H_c°$, is the standard reaction enthalpy of a combustion reaction occurring under standard conditions. For hydrocarbons, $\Delta H_c° < 0$ (*exothermic*).
  - → A *formation reaction* is one that produces one mole of a substance (product) from its elements (reactants) in their *most stable form*.
  - → *Lattice enthalpy*, ΔH_L, of an ionic compound is the heat *released* when a solid forms from gaseous ions. A Born-Haber cycle includes lattice formation in a multi-step process, in which all other steps have known reaction enthalpies.
  - → The strength of a chemical bond is measured by its bond enthalpy, ΔH_B.

6.14 Reaction Enthalpies

- **Thermochemical equation** → A chemical equation with a corresponding enthalpy change, ΔH, expressed in kJ for the *stoichiometric* number of moles of each reactant and product

- **Enthalpy of reaction, ΔH_r** → The enthalpy change in a thermochemical reaction expressed in $kJ{\cdot}mol^{-1}$, where "mol" refers to the number of moles of each substance as determined by the stoichiometric coefficient in the balanced equation

Example 6.14a Reaction with oxygen is a combustion reaction. Write the thermochemical equation for the combustion of hydrogen gas at 25°C. For the reaction of 1 mol $H_2(g)$, $q_P = -285.83$ kJ. Determine the reaction enthalpy, ΔH_r.

Solution The chemical equation $H_2(g) + O_2(g) \rightarrow H_2O(l)$ is balanced by inspection. Note that water at 25°C is a liquid. The balanced equation with the smallest whole integer coefficients is

$$2\,H_2(g) + O_2(g) \rightarrow 2\,H_2O(l)$$

For a constant pressure reaction, $q_P = \Delta H$. The heat change is given for 1 mol $H_2(g)$, but the balanced equation requires 2 mol. Thus, $\Delta H = 2(-285.83 \text{ kJ}) = -571.66$ kJ. The thermochemical equation is

$$2\,H_2(g) + O_2(g) \rightarrow 2\,H_2O(l) \qquad \Delta H = -571.66 \text{ kJ}$$

The reaction enthalpy is $\Delta H_r = -571.66 \text{ kJ·mol}^{-1}$. Here "mol" refers to the enthalpy change for 2 mol $H_2(g)$, 1 mol $O_2(g)$, and 2 mol $H_2O(l)$. More explicitly, $\Delta H_r = -571.66 \text{ kJ·(2 mol } H_2)^{-1}$, $\Delta H_r = -571.66 \text{ kJ·(1 mol } O_2)^{-1}$, and $\Delta H_r = -571.66 \text{ kJ·(2 mol } H_2O)^{-1}$.

Example 6.14b Use the thermochemical equation given below to determine how much heat is *released*, q_{surr}, when 350 g of propane, $C_3H_8(g)$, is burned completely in a backyard barbecue at 25°C.

$$C_3H_8(g) + 5\,O_2(g) \rightarrow 3\,CO_2(g) + 4\,H_2O(l) \qquad \Delta H = -2220 \text{ kJ}$$

Solution The reaction enthalpy is $\Delta H_r = -2220 \text{ kJ·mol}^{-1}$. Here "mol" refers to the enthalpy change for 1 mol $C_3H_8(g)$, 5 mol $O_2(g)$, 3 mol $CO_2(g)$, and 4 mol $H_2O(l)$. More explicitly, $\Delta H_r = -2220 \text{ kJ·(1 mol } C_3H_8)^{-1}$. The molar mass of propane is $3(12.01) + 8(1.0079) = 44.09 \text{ g·mol}^{-1}$. The number of moles of propane is then $(350 \text{ g})/(44.09 \text{ g·mol}^{-1}) = 7.938$ mol. The heat change for the system is $q_P = n\Delta H_r = (7.938 \text{ mol})(-2220 \text{ kJ·mol}^{-1}) = -1.76 \times 10^4$ kJ. The heat released to the surroundings is $q_{surr} = -q_P = 1.76 \times 10^4 \text{ kJ} = 17.6$ MJ.

6.15 The Relation Between ΔH and ΔU

- **Reactions with liquids and solids only**
  - → Recall that $\Delta H = \Delta U + \Delta(PV)$
  - → $\Delta H \approx \Delta U \Rightarrow q_P \approx q_V$
  - → Only a small change in the volume of reactants compared to products at constant P
- **Reactions with ideal gases only**
  - → $\Delta H = \Delta U + \Delta(PV) = \Delta U + \Delta(nRT) = \Delta U + (\Delta n)RT$ *constant temperature*
  - → The equation is exact for ideal gases only.
- **Reactions with liquids, solids, and gases**
  - → $\Delta H \approx \Delta U + (\Delta n_{gas})RT$ *constant temperature*
  - → $\Delta n_{gas} = n_{final} - n_{initial}$ = change in the number of moles of gas between products and reactants
  - → The equation is a very good approximation. The value of $(\Delta n_{gas})RT$ is usually much smaller than ΔU and ΔH.

Example 6.15a Calculate the change in internal energy, ΔU, for the combustion reaction in Example 6.14b.

Solution

$C_3H_8(g) + 5\,O_2(g) \rightarrow 3\,CO_2(g) + 4\,H_2O(l) \qquad \Delta H = -2220$ kJ

$\Delta H \approx \Delta U + (\Delta n_{gas})RT$ and $\Delta U \approx \Delta H - (\Delta n_{gas})RT$

$\Delta n_{gas} = n(CO_2) - [n(C_3H_8) + n(O_2)] = 3 - (1 + 5) = 3 - 6 = -3$

Be careful to exclude liquid water.

$(\Delta n_{gas})RT = (-3\text{ mol})(8.314\,51\text{ J}\cdot\text{K}^{-1}\cdot\text{mol}^{-1})(298.15\text{ K}) = -7437\text{ J} = -7.437\text{ kJ}$

$\Delta U \approx \Delta H - (\Delta n_{gas})RT = -2220\text{ kJ} - (-7.437\text{ kJ}) = -2213\text{ kJ}$

Note: Be careful to use consistent energy units, either J or kJ. The internal energy change is a more positive value than the enthalpy change for a reaction that *consumes* gas. The difference between ΔH and ΔU is small, only about 0.3%.

Example 6.15b Calculate the change in internal energy, ΔU, for the dissociation of hydrogen gas into atoms at 25°C, given the following thermochemical equation:

$$H_2(g) \rightarrow 2\,H(g) \qquad \Delta H = +435.94\text{ kJ}$$

Solution

$\Delta n_{gas} = n(H) - n(H_2) = 2 - 1 = +1$

$(\Delta n_{gas})RT = (+1\text{ mol})(8.314\,51\text{ J}\cdot\text{K}^{-1}\cdot\text{mol}^{-1})(298.15\text{ K}) = +2478.97\text{ J} = +2.479\text{ kJ}$

$\Delta U \approx \Delta H - (\Delta n_{gas})RT = +435.94\text{ kJ} - (+2.479\text{ kJ}) = +433.46\text{ kJ}$

Note: The internal energy change is a more negative value than the enthalpy change for a reaction that *produces* gas. The difference between ΔH and ΔU is small, only about 0.6%.

6.16 Standard Reaction Enthalpies

- **Standard conditions**
  - → Values of reaction enthalpy, ΔH_r, depend on pressure, temperature, and the physical state of each reactant and product.
  - → A standard set of conditions is used to report reaction enthalpies.
- **Standard state**
  - → Standard state of a substance is its pure form at a pressure of *exactly* 1 bar.
  - → For a solute in a liquid solution, the standard state is a solute concentration of 1 mol·L^{-1} at a pressure of *exactly* 1 bar.
  - → Standard state of liquid water is pure water at 1 bar.
    Standard state of ice is pure ice at 1 bar.
    Standard state of water vapor is pure water vapor at 1 bar.
- **Standard reaction enthalpy, $\Delta H_r°$**
  - → Reaction enthalpy when reactants in their standard states form products in their standard states
  - → Degree symbol, °, is added to the reaction enthalpy.
    Reaction enthalpy has a *weak* pressure dependence, so it is a very good approximation to use standard enthalpy values even when the pressure is *not* exactly 1 bar.

- **Temperature convention**
  - → Most thermochemical data are reported for 298.15 K (25°C).
  - → Temperature convention is *not* part of the definition of the standard state.

Note: Compare the definition of standard state in thermodynamics to the definition of standard temperature and pressure, STP, for gases (Section 4.8). For gases, standard temperature is 0°C (273.15 K) and standard pressure is 1 atm (1.013 25 bar).

Example 6.16 Compare the following two thermochemical reactions at 25°C. Comment on the nature of the difference in standard reaction enthalpy between reaction (2) and reaction (1).

(1) $C_3H_8(g) + 5\,O_2(g) \rightarrow 3\,CO_2(g) + 4\,H_2O(l)$ $\Delta H° = -2220$ kJ

(2) $C_3H_8(g) + 5\,O_2(g) \rightarrow 3\,CO_2(g) + 4\,H_2O(g)$ $\Delta H° = -2044$ kJ

Solution Reaction (2) is less exothermic than reaction (1). Reaction (2) produces water vapor and reaction (1) produces liquid water. For reaction (1), $\Delta H_r° = -2220\ \text{kJ·mol}^{-1}$. For reaction (2), $\Delta H_r° = -2044\ \text{kJ·mol}^{-1}$. Recall that "mol" refers to 1 mol C_3H_8, 5 mol O_2, 3 mol CO_2, and 4 mol H_2O. The enthalpy of vaporization of water is 44.01 kJ·mol^{-1} at 25°C. The difference in standard reaction enthalpies between reaction (2) and reaction (1) is 176 kJ·(4 mol $H_2O)^{-1}$ or 44 kJ·mol^{-1}, which is the enthalpy of vaporization of water at 25°C. Thus, the difference corresponds to the vaporization of 4 mol of liquid water at 25°C. The thermochemical equations may be combined by subtracting reaction (1) from reaction (2). The result is

$$4\,H_2O(l) \rightarrow 4\,H_2O(g) \qquad \Delta H° = +176\ \text{kJ}$$

Changing the stoichiometric coefficients to the simplest whole integers yields

$$H_2O(l) \rightarrow H_2O(g) \qquad \Delta H_r° = +44\ \text{kJ·mol}^{-1} = \Delta H_{vap}$$

Combining thermochemical equations is an application of Hess's law, which is discussed in the next section.

6.17 Combining Reaction Enthalpies: Hess's Law

- **Hess's law**
  - → Overall reaction enthalpy is the sum of the reaction enthalpies of the steps into which a reaction can be divided.
  - → Consequence of the fact that the reaction enthalpy is path independent and depends only on the initial reactants and final products
  - → Used to calculate enthalpy changes for reactions that are difficult to carry out in the laboratory
  - → Used to calculate reaction enthalpies for unknown reactions
- **Using Hess's law**
  - → Write the thermochemical equation for the reaction whose reaction enthalpy is unknown.

→ Write thermochemical equations for intermediate reaction steps whose reaction enthalpies are known. These steps must sum up to the overall reaction whose enthalpy is unknown.

→ Sum the reaction enthalpies of the intermediate reaction steps to obtain the unknown reaction enthalpy (see Toolbox 6.1 in the text).

Example 6.17 Determine the standard reaction enthalpy for

(3) $2\,SO_2(g) + O_2(g) \rightarrow 2\,SO_3(g)$ $\Delta H° = ?$

Use Hess's law to combine combustion reactions (1) and (2) given below.

(1) $2\,S(s) + 3\,O_2(g) \rightarrow 2\,SO_3(g)$ $\Delta H° = -792$ kJ

(2) $S(g) + O_2(g) \rightarrow SO_2(g)$ $\Delta H° = -297$ kJ

Solution In reaction (1), the proper number of moles of product SO_3 is present, and the reaction may be combined as written. In reaction (2), the number of moles of SO_2 needs to be doubled and the reactants and products interchanged. The thermochemical equations that combine to give reaction (3) are then

(1)	$2\,S(s) + 3\,O_2(g) \rightarrow 2\,SO_3(g)$	$\Delta H° = -792$ kJ
−2×(2)	$2\,SO_2(g) \rightarrow 2\,S(s) + 2\,O_2(g)$	$\Delta H° = +594$ kJ
(3)	$2\,SO_2(g) + O_2(g) \rightarrow 2\,SO_3(g)$	$\Delta H° = -198$ kJ

6.18 The Heat Output of Reactions

- **Heat output in a reaction**
  - → Heat is treated as a reactant or product in a stoichiometric relation.
  - → In an endothermic reaction, heat is treated as a reactant. In an exothermic reaction, heat is treated as a product.
- **Standard enthalpy of combustion, $\Delta H_c°$**
  - → Change in enthalpy per mole of a substance that is burned in a combustion reaction under standard conditions (see values in Table 6.3 in the text)
  - → Combustion of hydrocarbons is exothermic, and heat is treated as a product.

Example 6.18a Write a balanced thermochemical equation for the combustion of one mole of octane (use Table 6.3). Suppose an average automobile travels 11 000 miles in a year and averages 21.0 miles per gallon of gasoline. Approximating gasoline as octane and assuming a gallon of gasoline has a mass of 2.66 kg, calculate the enthalpy produced from the combustion of fossil fuel and used by an average automobile per year. Assume the complete combustion of octane in the engine.

Solution The thermochemical equation for one mole of octane is (Table 6.3)

$$C_8H_{18}(l) + \tfrac{25}{2}O_2(g) \rightarrow 8\,CO_2(g) + 9\,H_2O(l) \qquad \Delta H_c° = -5471\ \text{kJ}\cdot\text{mol}^{-1}$$

(11 000 miles / 21.0 miles·gallon^{-1}) = 524 gallons of gasoline or octane

(524 gallons × 2.66 kg·gallon^{-1}) = 1.39×10^3 kg = 1.39×10^6 g

molar mass of octane = [8(12.01) + 18(1.0079)] = 114.22 g·mol^{-1}

(1.39×10^6 g / 114.22 g·mol^{-1}) = 1.22×10^4 mol octane per year

(5471 kJ·mol^{-1} × 1.22×10^4 mol) = 6.67×10^7 kJ = 66.7 GJ per year

Example 6.18b How many grams of $CO_2(g)$ are added to the atmosphere by the average automobile in one year? Use the results of Example 6.18a.

Answer 4.29×10^6 g = 4.29 Mg of $CO_2(g)$

6.19 Standard Enthalpies of Formation

- **Treatment of chemical reactions**
  - → Possible thermochemical equations are nearly innumerable.
  - → Method to handle this problem is the use of a common reference for each substance. The reference is the *most stable* form of the elements.
  - → Standard enthalpies of formation of substances from their elements are tabulated and used to calculate standard reaction enthalpies for other types of reactions.

- **Standard enthalpy of formation, $\Delta H_f°$**
  - → Enthalpy change for the formation of one mole of a substance from the most stable form of its elements under standard conditions
  - → Elements (most stable form) → substance (one mole) $\Delta H° = \Delta H_f°$
    For example, $C(s) + 2H_2(g) + \frac{1}{2}O_2(g) \rightarrow CH_3OH(l)$ $\Delta H_f° = -238.86$ kJ.mol^{-1}
  - → Values of $\Delta H_f°$ at 298.15 K for many substances are listed in Appendix 2A.
  - → Most stable form of the elements at 1 bar (standard state) and 298.15 K (thermodynamic temperature convention):
    Metals and semimetals are solids with atoms arranged in a lattice (see Chapter 5). The exception is Hg(l).
    Nonmetal solids are B(s), C(s, graphite), P(s, white), S(s, rhombic), Se(s, black), and $I_2(s)$.
    Monatomic gases are He(g), Ne(g), Ar(g), Kr(g), Xe(g), and Rn(g). *Group 18*
    Diatomic gases are $H_2(g)$, $N_2(g)$, $O_2(g)$, $F_2(g)$, and $Cl_2(g)$.
    Nonmetal liquid is $Br_2(l)$.

  Note: P(s, red) is actually more stable than P(s, white), but the white form is chosen because it is more readily available. Also, some of the nonmetal solids have molecular forms, for example, P_4 and S_8.

- **Standard enthalpy of reaction, $\Delta H_r°$**
  - → Use the most stable form of the elements as a reference state to obtain enthalpy changes for any reaction, if $\Delta H_f°$ values are known for each reactant and product.
  - → $$\boxed{\Delta H_r° = \sum n\Delta H_f°(\text{products}) - \sum n\Delta H_f°(\text{reactants})}$$
    In this expression, values of n are the stoichiometric coefficients and the symbol Σ (sigma) means a summation. The procedure is to convert the reactants into the most stable form of their elements [$-\Sigma n\Delta H_f°$(reactants)] and then recombine the elements into products [$+\Sigma n\Delta H_f°$(products)]. This is an application of Hess's law.
  - → Use $\Delta H_f°$ values in Appendix 2A to obtain $\Delta H_r°$ (see Toolbox 6.2 in the text).

Example 6.19a Using enthalpy of formation values in Appendix 2A, calculate the enthalpy of combustion of two moles of octane.

Solution The thermochemical equation for the combustion of two moles of octane is

$$2\,C_8H_{18}(l) + 25\,O_2(g) \rightarrow 16\,CO_2(g) + 18\,H_2O(l) \qquad \Delta H_c° = ?$$

The standard enthalpy of formation of an element in its most stable form is 0. Therefore, $\Delta H_f°[O_2(g)] = 0$.

$$\Delta H_r° = \sum n\Delta H_f°(\text{products}) - \sum n\Delta H_f°(\text{reactants}) = \Delta H_c°$$
$$= 16\Delta H_f°[CO_2(g)] + 18\Delta H_f°[H_2O(l)] - 2\Delta H_f°[C_8H_{18}(l)]$$
$$= 16(-393.51\ \text{kJ·mol}^{-1}) + 18(-285.83\ \text{kJ·mol}^{-1}) - 2(-249.9\ \text{kJ·mol}^{-1})$$
$$= -6296.15 - 5144.94 + 499.8\ \text{kJ·mol}^{-1}$$
$$= -10\,941.3\ \text{kJ·mol}^{-1} \ \text{at } 298.15\ \text{K}$$

Recall that "mol" refers to 2 mol of C_8H_{18}, 25 mol O_2, 16 mol CO_2, and 18 mol H_2O.

Example 6.19b Using enthalpy of formation values in Appendix 2A, calculate the enthalpy of dissociation of one mole of hydrogen molecules into hydrogen atoms.

Answer

$$H_2(g) \rightarrow 2\,H(g) \qquad \Delta H_r° = ?$$
$$\Delta H_r° = 2\Delta H_f°[H(g)] = 2(217.97\ \text{kJ·mol}^{-1}) = 435.94\ \text{kJ·mol}^{-1} \ \text{at } 298.15\ \text{K}$$

This enthalpy change is also known as the bond enthalpy, ΔH_B(H–H), defined in Section 6.21.

Note: It is possible to obtain an unknown enthalpy of formation if a reaction enthalpy and the other enthalpies of formation in that reaction are known. See Example 6.13 in the text.

6.20 The Born-Haber Cycle

- **Ionic solids**
  - → Cations and anions are arranged in a three-dimensional lattice (see Section 5.13) and held in place mainly by coulombic interactions (see Section 2.2).
  - → Enthalpy required to separate the ions in the solid into gas phase ions is the lattice enthalpy, $\Delta H_L = H_m(\text{ions, g}) - H_m(\text{ions, s}) > 0$. See values in Table 6.5 in the text.
- **Born-Haber cycle**
  - → Thermodynamic cycle constructed to evaluate the lattice enthalpy from other reactions, in which the enthalpy changes are known
  - → Steps in the Born-Haber cycle that start and end with the most stable form of the elements in amounts appropriate to form one mole of the ionic compound

    As an example, consider $CaF_2(s)$. Start with one mole of Ca(s) and one mole of $F_2(g)$. Two moles of F atoms are required.

    (1) *Atomize:* Atomization of the metal element that will form the cation and of the nonmetal element that will form the anion $\Delta H(\text{atomization}) > 0$

    $Ca(s) \rightarrow Ca(g) \qquad \Delta H_r° = \Delta H_f°[Ca(g)] = 178\ \text{kJ·mol}^{-1}$

    $F_2(g) \rightarrow 2\,F(g) \qquad \Delta H_r° = 2\Delta H_f°[F(g)] = 2(79\ \text{kJ·mol}^{-1}) = 158\ \text{kJ·mol}^{-1}$

    $\Delta H(\text{atomization}) = 178 + 158\ \text{kJ·mol}^{-1} = +336\ \text{kJ·mol}^{-1}$

(2) *Ionize (cation):* Ionize the gaseous metal atom to form the gaseous cation. Several ionization steps may be required. Because electrons created in the formation of cations in the ionization steps are consumed in the formation of anions in the electron-gain steps (3), energy and enthalpy are considered to be interchangeable when treating ionization and electron gain.
ΔH(ionization, cation) > 0

$Ca(g) \rightarrow Ca^{+}(g) + e^{-}(g)$ $\Delta H_r° = I_1[Ca] = 590\ kJ{\cdot}mol^{-1}$
$Ca^{+}(g) \rightarrow Ca^{2+}(g) + e^{-}(g)$ $\Delta H_r° = I_2[Ca] = 1145\ kJ{\cdot}mol^{-1}$
$Ca(g) \rightarrow Ca^{2+}(g) + 2\,e^{-}(g)$ $\Delta H_r° = I_1 + I_2 = 590 + 1145\ kJ{\cdot}mol^{-1} = 1735\ kJ{\cdot}mol^{-1}$
ΔH(ionization, cation) = $+1735\ kJ{\cdot}mol^{-1}$

(3) *Ionize (anion):* Attach electron(s) to the gaseous nonmetal atom to form the gaseous anion. Several electron affinity values may be required. Recall that for electron gain, $\Delta H = -E_{ea}$ for each electron added.
ΔH(ionization, anion) < or > 0

$F(g) + e^{-}(g) \rightarrow F^{-}(g)$ $\Delta H_r° = -E_{ea}[F] = -328\ kJ{\cdot}mol^{-1}$
$2\,F(g) + 2\,e^{-}(g) \rightarrow 2\,F^{-}(g)$ $\Delta H_r° = -2E_{ea} = -656\ kJ{\cdot}mol^{-1}$
ΔH(ionization, anion) = $-656\ kJ{\cdot}mol^{-1}$

(4) *Latticize:* Form the lattice of ions in the solid from the gaseous ions. The enthalpy of lattice formation is the negative of the lattice enthalpy, ΔH_L.
ΔH(lattice formation) < 0

$Ca^{2+}(g) + 2\,F^{-}(g) \rightarrow CaF_2(s)$ $\Delta H_r° = -\Delta H_L$ *value unknown*
ΔH(lattice formation) = $-\Delta H_L$

(5) *Elementize:* Form the most stable form of the elements from the ionic solid.
ΔH(element formation) < 0

$CaF_2(s) \rightarrow Ca(s) + F_2(g)$ $\Delta H_r° = -\Delta H_f°[CaF_2(s)] = 1220\ kJ{\cdot}mol^{-1}$
ΔH(element formation) = $+1220\ kJ{\cdot}mol^{-1}$

(6) *Lattice enthalpy:* The sum of all the enthalpy changes for the complete cycle is 0.
0 = ΔH(atomization) + ΔH(ionization, cation) + ΔH(ionization, anion) + ΔH(lattice formation) + ΔH(element formation)

$0 = 336 + 1735 + (-656) + (-\Delta H_L) + 1220\ kJ{\cdot}mol^{-1} = 2635\ kJ{\cdot}mol^{-1} - \Delta H_L$
$\Delta H_L = +2635\ kJ{\cdot}mol^{-1}$ for $CaF_2(s) \rightarrow Ca^{2+}(g) + 2\,F^{-}(g)$

- **Summary** → Strength of interactions between ions in a solid is determined by the lattice enthalpy, which is obtained from a Born-Haber cycle.

6.21 Bond Enthalpies

- **Bond enthalpy, ΔH_B**
  - → Enthalpy change accompanying the breaking of a chemical bond in the gas phase
  - → Difference between the standard molar enthalpy of the fragments of a molecule and the molecule itself in the gas phase
  - → Reactants and products in their standard states (pure substance at 1 bar) at 298.15 K
  - → Bond enthalpies for diatomic molecules are given in Table 6.6. See Example 6.19b.

Example 6.21a Write a thermochemical equation for breaking one mol of C–H bonds in methane, CH_4. The enthalpy of formation of $CH_3(g)$ is 138.9 kJ·mol^{-1}. Use Appendix 2A for other data.

Solution

$CH_4(g) \rightarrow CH_3(g) + H(g) \qquad \Delta H_r° = \Delta H_B(\text{C–H})$

$\Delta H_r° = \Delta H_f°[CH_3(g)] + \Delta H_f°[H(g)] - \Delta H_f°[CH_4(g)]$

$= 138.9 + 217.97 - (-74.81)\ \text{kJ·mol}^{-1} = 431.7\ \text{kJ·mol}^{-1} = \Delta H_B(\text{C–H})$

The C–H bond has an *average* bond enthalpy of 412 kJ·mol^{-1} in polyatomic molecules (see Table 6.7). Note that "mol" refers to the amount of C–H bonds. Breaking the first C–H bond in methane requires more heat than subsequent ones.

Example 6.21b Write a thermochemical equation for breaking one mol of carbonyl C–H bonds in ethanal (acetaldehyde), CH_3COH. The enthalpy of formation of $CH_3CO(g)$ is –18.8 kJ·mol^{-1}. Use Appendix 2A for other data.

Solution

$CH_3COH(g) \rightarrow CH_3CO(g) + H(g) \qquad \Delta H_r° = \Delta H_B(\text{C–H})$

$\Delta H_r° = \Delta H_f°[CH_3CO(g)] + \Delta H_f°[H(g)] - \Delta H_f°[CH_3COH(g)]$

$= -18.8 + 217.97 - (-192.30)\ \text{kJ·mol}^{-1} = 391.5\ \text{kJ·mol}^{-1} = \Delta H_B(\text{C–H})$

The carbonyl C–H bond in ethanal is considerably weaker than the *average* C–H bond enthalpy in polyatomic molecules.

- **Average bond enthalpies**
  - → In polyatomic molecules, the bond strength between a pair of atoms varies from molecule to molecule (see Examples 6.21a and 6.21b above).
  - → Variations are not very large, and the average values of bond enthalpies are a guide to the strength of a bond in any molecule containing the bond (see Table 6.7).

Example 6.21c Calculate the average C–H bond enthalpy in methane, CH_4. Use the data in Appendix 2A.

Solution

$CH_4(g) \rightarrow C(g) + 4H(g) \qquad \Delta H_r° = 4\Delta H_B(\text{C–H})$

$\Delta H_r° = \Delta H_f°[C(g)] + 4\Delta H_f°[H(g)] - \Delta H_f°[CH_4(g)]$

$= 716.68 + 4(217.97) - (-74.81)\ \text{kJ·mol}^{-1} = 1663.37\ \text{kJ·mol}^{-1}$

$\Delta H_B(\text{C–H}) = \Delta H_r°/4 = 415.84\ \text{kJ·mol}^{-1}$

The average C–H bond enthalpy in methane is slightly larger than the *average* C–H bond enthalpy in polyatomic molecules.

Example 6.21d Calculate the average C=O bond enthalpy in carbon dioxide, CO_2. The enthalpy of formation of O(g) is 249.17 kJ·mol^{-1}. Use Appendix 2A for other data.

Solution

$CO_2(g) \rightarrow C(g) + 2\,O(g) \qquad \Delta H_r° = 2\Delta H_B(C{=}O)$

$\Delta H_r° = \Delta H_f°[C(g)] + 2\Delta H_f°[O(g)] - \Delta H_f°[CO_2(g)]$

$= 716.68 + 2(249.17) - (-393.51)$ kJ·mol^{-1} = 1608.53 kJ·mol^{-1}

$\Delta H_B(C{=}O) = \Delta H_r°/2 = 804.27$ kJ·mol^{-1}

The C=O double bond in carbon dioxide is considerably stronger than the *average* C=O bond enthalpy of 743 kJ·mol^{-1} in polyatomic molecules.

- **Using average bond enthalpies**
  - → Atoms in the gas phase are the reference states used to estimate enthalpy changes for any gaseous reaction. A value of ΔH_B is required for each bond in the reactant and product molecules.
  - → $$\Delta H_r° \approx \sum_{\text{reactants}} n\Delta H_B(\text{bonds broken}) - \sum_{\text{products}} n\,\Delta H_B(\text{bonds formed})$$

    In this expression, values of n are the stoichiometric coefficients and the symbol Σ (sigma) means a summation. The procedure is to convert the reactants into gaseous atoms [$+\Sigma n\Delta H_B$(bonds broken)] and then recombine the atoms into products [$-\Sigma n\Delta H_B$(bonds formed)]. This is an application of Hess's law.
  - → Compare bond enthalpy to dissociation energy in Sections 2.16 and 2.17. In examples 2.17a and 2.17b in this Study Guide, values of average bond enthalpies, not dissociation energies, are used. The procedure is identical to the one shown in the above equation. These two examples are good illustrations of how bond enthalpies are used to estimate reaction enthalpies for gas-phase reactions.

6.22 The Variation of Reaction Enthalpy with Temperature

- **Temperature dependence of reaction enthalpy**
  - → $\Delta H_r°$ needs to be measured at the temperature of interest.
  - → Approximation method is possible from the value of $\Delta H_r°$ measured at one temperature, if heat capacity data on the reactants and products are available.

- **Kirchhoff's law**
  - → The difference in molar heat capacities of the products and reactants in a chemical reaction is

    $$\Delta C_P = \sum nC_{P,\text{m}}(\text{products}) - \sum nC_{P,\text{m}}(\text{reactants})$$
  - → An estimation of the reaction enthalpy at a temperature of T_2, if the value at a temperature of T_1 and the difference in molar heat capacities are known, is

    $$\Delta H_{r,2}° = \Delta H_{r,1}° + \Delta C_P(T_2 - T_1)$$
  - → A major assumption is that the molar heat capacities are independent of temperature.

Example 6.22a Calculate a value for the enthalpy of vaporization of liquid water at 298.15 K. Use Kirchhoff's law to estimate the enthalpy of vaporization of liquid water at the normal boiling point of 373.15 K. Compare the value to the one in Table 6.2 in the text. Use the data in Appendix 2A.

Solution $H_2O(l) \rightarrow H_2O(g) \qquad \Delta H_r° = \Delta H_{vap}°$

$$\Delta H_r° = \Delta H_f°[H_2O(g)] - \Delta H_f°[H_2O(l)]$$
$$= -241.82 - (-285.83)\ \text{kJ·mol}^{-1} = +44.01\ \text{kJ·mol}^{-1} = \Delta H_{vap}° \text{ at } 298.15\ \text{K}$$

$$\Delta C_P = C_{P,m}[H_2O(g)] - C_{P,m}[H_2O(l)] = 33.58 - 75.29\ \text{J·K}^{-1}\text{·mol}^{-1}$$
$$= -41.71\ \text{J·K}^{-1}\text{·mol}^{-1} = -0.041\,71\ \text{kJ·K}^{-1}\text{·mol}^{-1}$$

$$\Delta H_{vap}°(373.15\ \text{K}) = \Delta H_{vap}°(298.15\ \text{K}) + \Delta C_P\,(373.15 - 298.15\ \text{K})$$
$$= 44.01 + (-0.041\,71)(75.00) = 44.01 - 3.13\ \text{kJ·mol}^{-1}$$
$$= 40.88\ \text{kJ·mol}^{-1}$$

The value in Table 6.2 is 40.7 kJ·mol^{-1}. The approximation is an excellent one in this case.

Example 6.22b The bond enthalpy of H–H is 436 kJ·mol^{-1} at 298.15 K. Use Kirchhoff's law to estimate the bond enthalpy at 0 K. Use the data in Appendix 2A.

Solution $H_2(g) \rightarrow 2\,H(g) \qquad \Delta H_r° = \Delta H_B(\text{H–H}) = 436\ \text{kJ·mol}^{-1} \text{ at } 298.15\ \text{K}$

$$\Delta C_P = 2C_{P,m}[H(g)] - C_{P,m}[H_2(g)] = 2(20.78) - 28.82\ \text{J·K}^{-1}\text{·mol}^{-1}$$
$$= +12.74\ \text{J·K}^{-1}\text{·mol}^{-1} = +0.012\,74\ \text{kJ·K}^{-1}\text{·mol}^{-1}$$

$$\Delta H_B(0\ \text{K}) = \Delta H_B(298.15\ \text{K}) + \Delta C_P\,(0 - 298.15\ \text{K})$$
$$= 436 + (0.012\,74)(-298.15) = 436 - 3.8\ \text{kJ·mol}^{-1}$$
$$= 432\ \text{kJ·mol}^{-1}$$

The spectroscopic dissociation energy of $H_2(g)$ is 432.07 kJ·mol^{-1}, which corresponds to a thermodynamic temperature of absolute zero.

Chapter 7 Thermodynamics: The Second and Third Laws

Entropy (Sections 7.1-7.7)

Key Concepts

natural direction of change, spontaneous change, entropy S, second law of thermodynamics, entropy (function of T, V, and P), entropy of vaporization, standard entropy of vaporization, Trouton's rule, entropy of fusion, standard entropy of fusion, third law of thermodynamics, Boltzmann formula for entropy, microstates, residual entropy, entropy and disorder, standard molar entropy, standard reaction entropy

Overview

- **Second law of thermodynamics**
  - → Summarizes the *origin* of *all* chemical change
  - → Predicts spontaneous direction of chemical reaction (criterion for spontaneity)
  - → Allows prediction of equilibrium state of chemical reaction
  - → Does not predict *reaction rates* or *rate of approach* to the equilibrium state (Thermodynamics *is not* concerned with time.)
- **Entropy (S) and free energy (G)** → Essential concepts for understanding the second law of thermodynamics
- **Third law of thermodynamics** → Basis for numerical scales for entropy and free energy of substances
- **The big question:** What is the *cause* of *spontaneous change?*

7.1 Spontaneous Change

- **Spontaneous (or *natural*) change** → Occurs *without* an external influence

Example 7.1a List a few examples of processes that occur spontaneously. Comment on the timescale of each process.

Answer Spontaneous processes proceed at different rates. Water flows downhill (*fast change*). A hot body cools to reach the temperature of its surroundings (*slow change*). Iron rusts or oxidizes in air to form iron (III) oxide (*very slow, but noticeable change*). A mixture of hydrogen and oxygen gases react to form liquid water (*stable mixture over long periods of time; however, a spark can cause the mixture to explode*). Diamond converts to graphite (*infinitesimally slow rate*). Thus, spontaneous changes need not be fast.

- **Nonspontaneous change** → Can be effected by using an external influence (energy from the surroundings)

Example 7.1b List a few examples of processes that are nonspontaneous in nature. Comment on how each nonspontaneous change can be effected.

Answer Water does not flow spontaneously uphill, but it can be forced to do so by pumping. A cold body can be warmed using heat from the surroundings. Liquid water can be *electrolyzed* to form hydrogen and oxygen gases. Graphite can be converted to diamond under extremely high pressures.

7.2 Entropy and Disorder

- **Spontaneous changes** → Any spontaneous change is accompanied by an *increase* in the *disorder* of the *system* plus the *surroundings*.
- **Entropy (S)** → A measure of disorder; *increase* in disorder leads to an *increase* in S
- **Second law**

  → For a spontaneous change, entropy of an *isolated* system *increases*.

  → For a spontaneous change, entropy of the *universe increases*
  [The universe is considered to be a (somewhat large) isolated system.]

- **Macroscopic definition of entropy**

$$dS = \frac{dq_{rev}}{T}$$

(See Section 7.3)

- **Finite changes in heat**

$$\Delta S = \frac{q_{rev}}{T}, \text{ for constant } T$$

  → Heat transfer processes are carried out *reversibly* (temperature of surroundings *equals* that of the system) to evaluate ΔS.

Example 7.2 Suppose 1.00 kJ of heat is withdrawn ($q < 0$) very slowly from a 10 000-kg bronze statue at 25°C. Calculate ΔS for the bronze statue (system).

Solution The mass is so large that removal of 1.00 kJ does not effectively change the temperature of the statue. The heat is removed so slowly that the process is *reversible* for all practical purposes. Use the above equation for the *finite* change in heat.

$$\Delta S = \frac{q_{rev}}{T} = \frac{-1.00 \text{ kJ}}{(273.15+25)\text{K}} = -3.35\times10^{-3} \text{ kJ·K}^{-1} = -3.35 \text{ J·K}^{-1}$$

Notes: $\Delta S < 0$ implies that the final state is more ordered than the initial state, as expected. The units for entropy are J·K^{-1} or kJ·K^{-1} (energy per unit of absolute temperature).

- **Entropy is an extensive property** → Proportional to the amount of sample
- **Entropy is a state function** → Changing the path does not change ΔS.

Note: According to the definition of ΔS, we must calculate ΔS by using a *reversible path*. The result applies to *any* path, because ΔS is *independent of path*.

7.3 Changes in Entropy

- **Temperature dependence of entropy** (*any substance*)
  - → Let n = number of moles of a substance, C_P = heat capacity (J·K^{-1}) at constant P, $C_{P,\text{m}}$ = molar heat capacity (J·K^{-1}·mol^{-1}) at constant P, C_V = heat capacity at constant V, $C_{V,\text{m}}$ = molar heat capacity at constant V, T_1 = initial temperature, and T_2 = final temperature.
  - → Isobaric heating of a substance (constant P)
    $$\Delta S = C_P \ln\frac{T_2}{T_1} = nC_{P,\text{m}} \ln\frac{T_2}{T_1}$$
  - → Isochoric heating of a substance (constant V)
    $$\Delta S = C_V \ln\frac{T_2}{T_1} = nC_{V,\text{m}} \ln\frac{T_2}{T_1}$$
  - → Above relationships assume that the heat capacity is *constant* over the temperature range of T_1 to T_2.

Example 7.3a The specific heat capacity C_s of Fe is 0.449 J·K^{-1}·g^{-1}. Calculate the entropy change when *exactly* two moles of Fe are cooled from 300 K to 200 K at constant pressure.

Solution We expect the final, cooler state to be more ordered than the initial state; hence ΔS should be negative. Convert the specific heat capacity to the molar heat capacity by the relationship: $mC_s = nC_{P,\text{m}}$ and $M = m/n$. Use the above equation for calculating ΔS at constant pressure.

$C_{P,\text{m}} = MC_s = (55.847\ \text{g·mol}^{-1})(0.449\ \text{J·K}^{-1}\text{·g}^{-1}) = 25.08\ \text{J·K}^{-1}\text{·mol}^{-1}$

$n = 2$ (exactly), $T_1 = 300$ K, and $T_2 = 200$ K

$$\Delta S = nC_{P,\text{m}} \ln\frac{T_2}{T_1} = (2\ \text{mol})(25.08\ \text{J·K}^{-1}\text{·mol}^{-1}) \ln\frac{200\ \text{K}}{300\ \text{K}} = -20.3\ \text{J·K}^{-1}$$

Note: $\Delta S < 0$ as expected.

- **Volume dependence of entropy** (*ideal gas*)
  - → Let n = number of moles of an *ideal gas*, R = universal gas constant (8.314 51 J·K^{-1}·mol^{-1}), V_1 = initial volume, and V_2 = final volume.
  - → Isothermal volume change of an *ideal gas* (constant T)
    $$\Delta S = \frac{q_\text{rev}}{T} = -\frac{w_\text{rev}}{T} = nR\ln\frac{V_2}{V_1}$$
  - → Recall: $\Delta U = 0$; see Section 6.3.
  - → Recall that we must choose a *reversible* path to evaluate ΔS, but the result applies to *any* path, because entropy is a state function.

- **Pressure dependence of entropy** (*ideal gas*)
  → Let n = number of moles of an *ideal gas*, P_1 = initial pressure, and P_2 = final pressure.
  → Isothermal pressure change of an *ideal gas* (constant T), $\Delta U = 0$

$$\frac{V_2}{V_1} = \frac{P_1}{P_2} \quad \text{and} \quad \boxed{\Delta S = nR\ln\frac{V_2}{V_1} = nR\ln\frac{P_1}{P_2}}$$

Example 7.3b *Exactly* three moles of an ideal gas undergo an expansion from 20.0 L to 80.0 L at constant T. Calculate ΔS for this change in state.

Solution The final state is more disordered than the initial state, so we expect $\Delta S > 0$. Use the equation that gives the isothermal volume dependence of S for an *ideal gas.*

$$\Delta S = nR\ln\frac{V_2}{V_1} = (3\text{ mol})(8.314\,51\text{ J}\cdot\text{K}^{-1}\cdot\text{mol}^{-1})\ln\frac{80.0\text{ L}}{20.0\text{ L}} = 34.6\text{ J}\cdot\text{K}^{-1}$$

Example 7.3c *Exactly* four moles of an ideal gas undergo a pressure increase from 0.500 atm to 1.75 atm at constant T. Calculate ΔS for this change in state.

Answer

$$\Delta S = nR\ln\frac{P_1}{P_2} = (4\text{ mol})(8.314\,51\text{ J}\cdot\text{K}^{-1}\cdot\text{mol}^{-1})\ln\frac{0.500\text{ atm}}{1.75\text{ atm}} = -41.7\text{ J}\cdot\text{K}^{-1}$$

Note: The two changes in entropy computed for an ideal gas do *not* depend on the temperature of the gas.

7.4 Entropy Changes Accompanying Changes of Physical State

- **Changes of physical state**
  → fusion ≡ fus (solid to liquid)
  → vaporization ≡ vap (liquid to vapor)
  → sublimation ≡ sub (solid to vapor)
  → solid-to-solid phase changes; for example, Sn(gray) → Sn(white)

- **Transition temperature**
  → Temperature of substance remains *constant*
  → Transfer of heat is *reversible*
  → Heat supplied is identified with the *enthalpy change*, because the pressure is constant ($q_{rev} = \Delta H$)

- **Entropy of vaporization**
  → $\boxed{\Delta S_{vap} = \dfrac{\Delta H_{vap}}{T_b}}$, where T_b is the *boiling temperature*
  → ΔS_{vap} is the *entropy of vaporization.* The units are $\text{J}\cdot\text{K}^{-1}\cdot\text{mol}^{-1}$ or $\text{kJ}\cdot\text{K}^{-1}\cdot\text{mol}^{-1}$.

→ $\Delta S_{vap}°$ is the *standard entropy of vaporization.* The symbol ° is added to state properties for the liquid and vapor in their *standard states* (that is, both pure and both at 1 bar).

Example 7.4a Liquid methane, $CH_4(l)$, vaporizes *reversibly* at 1 bar and −161.5°C. Under these conditions, the enthalpy of vaporization is +8.18 kJ·mol^{-1}. Calculate $\Delta S_{vap}°$ for the vaporization of *exactly* 1 mol of $CH_4(l)$.

Solution The phase change is $CH_4(l) \rightarrow CH_4(g)$. Convert the *boiling* temperature to the absolute scale and substitute the information into the equations given above.

$T = 273.15 - 161.5 = 111.65$ K and $\Delta H_{vap}° = 8.18$ kJ·mol^{-1} $= 8.18 \times 10^3$ J·mol^{-1}.

$$\Delta S_{vap}° = \frac{\Delta H_{vap}°}{T_b} = \frac{8.18 \times 10^3 \text{ J·mol}^{-1}}{111.65 \text{ K}} = 73.3 \text{ J·K}^{-1}\text{·mol}^{-1} \quad (73.3 \text{ J·K}^{-1} \text{ for 1 mol})$$

Because $\Delta S > 0$, the final state is more disordered than the initial state, as expected.

Note: This equation can be used only for *reversible processes.* The spontaneous boiling of a *superheated* liquid occurs *irreversibly.*

- **Trouton's rule**
  - → Standard entropies of vaporization of liquids ≈ 85 J·K^{-1}·mol^{-1} or $\Delta S_{vap}°$ *(liq)* ≈ 85 J·K^{-1}·mol^{-1}
  - → Approximately the same increase in *disorder* occurs when *any* liquid is vaporized. *Disorder* is greatest in the gas phase in which the molecules are far apart and moving rapidly. Exceptions occur for *hydrogen-bonded* liquids, for example, water, H_2O (109 J·K^{-1}·mol^{-1}), and methanol, CH_3OH (105 J·K^{-1}·mol^{-1}).

- **Entropy of fusion (melting)**
  - → $\boxed{\Delta S_{fus}° = \dfrac{\Delta H_{fus}°}{T_f}}$, where T_f is the *melting temperature*
  - → $\Delta S_{fus}°$ is the *standard entropy of fusion.*
  - → $\Delta S_{fus}°$ is much smaller than $\Delta S_{vap}°$, because a liquid is only slightly more *disordered* than a solid.

Example 7.4b Solid methane, $CH_4(s)$, melts *reversibly* at 1 bar and −182.5°C. Under these conditions, the enthalpy of fusion is +0.936 kJ·mol^{-1}. Calculate $\Delta S_{fus}°$ for the melting of *exactly* 1 mol of $CH_4(s)$.

Solution The phase change is $CH_4(s) \rightarrow CH_4(l)$. Convert the *melting* temperature (identical to the *freezing* temperature) to the absolute scale and substitute the information into the equations given above.

$T = 273.15 - 182.5 = 90.65$ K and $\Delta H_{vap}° = 0.936$ kJ·mol^{-1} $= 936$ J·mol^{-1}

$$\Delta S_{fus}° = \frac{\Delta H_{fus}°}{T_f} = \frac{936 \text{ J·mol}^{-1}}{90.65 \text{ K}} = 10.3 \text{ J·K}^{-1}\text{·mol}^{-1} \text{ or } 10.3 \text{ J·K}^{-1} \text{ for 1 mol}$$

Because $\Delta S > 0$ for *melting*, the final state is more disordered than the initial state, as expected.

Notes: This equation can be used only for *reversible processes*. The spontaneous *freezing* of a *supercooled* liquid occurs *irreversibly*. For the *reversible* processes of *freezing*, *condensation*, and *deposition*, $\Delta S < 0$.

7.5 A Molecular Interpretation of Entropy

- **Absolute value of the entropy** → If entropy S is a measure of disorder, then a perfectly ordered state of matter (perfect crystal) *should* have *zero* entropy.

- **Third law of thermodynamics** → Entropies of all perfect crystals are the same at the absolute *zero* of temperature.

  (Thermal motion almost ceases at $T = 0$ K, and by convention $S = 0$ for perfect crystals at the absolute *zero* of temperature.)

- **Boltzmann approach** → Entropy increases as the number of ways that molecules or atoms can be arranged in a sample (at the same total energy) increases.

- **Boltzmann formula** → $\boxed{S = k \ln W}$, where k is the Boltzmann constant

  $k = 1.380\,658 \times 10^{-23}\ \text{J·K}^{-1}$ [**Note:** $R = N_A k$]

  W = number of *microstates* available to the system at a certain energy

- **Microstate**
  - → *One* permissible arrangement of atoms or molecules in a sample with a certain total energy.
  - → W = total number of permissible arrangements corresponding to the same total energy.
  - → If only one arrangement is possible, $W = 1$ and $S = 0$.

- **Scaling S**
  - → For N molecules or atoms, the number of *microstates* is related to the number of molecular *orientations* permissible: $W = (orientations)^N$ and

$$\boxed{S = k \ln W = k \ln(orientations)^N = N k \ln(orientations)}$$

  - → For N_A (Avogadro number) molecules or atoms, the number of *microstates* $W = (orientations)^{N_A}$ and

$$\boxed{S = k \ln(orientations)^{N_A} = N_A k \ln(orientations) = R \ln(orientations)}$$

Units are $\text{J·K}^{-1}\text{·mol}^{-1}$

- **Residual entropy** → For some solids, $S > 0$ at $T = 0$ (These solids retain some *disorder*.)

Example 7.5a In solid carbon monoxide, CO(s), the entropy at 0 K is found to be 4.6 $J{\cdot}K^{-1}{\cdot}mol^{-1}$. When a substance has a *nonzero* entropy at 0 K, the effect is called *residual entropy*. Calculate the number of *orientations* of CO molecules possible in the crystal and compare it to the value of 2, which is the value expected for a random orientation (*disorder*) of molecules at 0 K. (Each carbon monoxide molecule can be arranged in one of two ways, see Figs. 7.5 and 7.6 in the text.)

Solution Use the relation for S as a function of *orientations* given above. Solve for the number of *orientations* and compare your answer to the expected value of 2.

$$S = R\ln(orientations);\ \ (orientations) = e^{S/R} = e^{(4.6/8.314\,51)} = e^{(0.553\,25)} = 1.7 < 2$$

The value is less than 2 and indicates that there is some ordering caused by the alignment of the small permanent dipoles of neighboring CO molecules at 0 K.

Example 7.5b In solid H_2O, the *residual* entropy at 0 K is found to be 3.4 $J{\cdot}K^{-1}{\cdot}mol^{-1}$. Calculate the number of *orientations* of H_2O molecules possible in the crystal and compare your answer to the value of 1.5, which is expected for a random orientation (*disorder*) of molecules at 0 K. [The arrangement of H_2O molecules is calculated from a tetrahedral arrangement of possible hydrogen bonds (4 possible ways). When considering neighboring molecules, not all of these are permissible. A detailed analysis shows that of the 16 ways of arranging four H atoms around one O atom, only 6 have two short and two long OH distances. Thus, the final number of orientations is $4 \times (6/16) = 3/2$.]

Answer $(orientations) = e^{S/R} = e^{(3.4/8.314\,51)} = e^{(0.408\,92)} = 1.5 = 3/2$

The values are in excellent agreement.

- **Compare Boltzmann's *molecular interpretation* with the *thermodynamic approach***

Volume increase in an ideal gas

macroscopic approach: $\Delta S = nR\ln\dfrac{V_2}{V_1}$; If $V_2 > V_1$, then $\Delta S > 0$.

microscopic approach: Consider a gas container to be a box. From particle-in-a-box theory, as V increases, energy levels accessible to gas molecules pack more closely together and W increases (see Fig. 7.8). So $\Delta S = S_{V_2} - S_{V_1} = k\ln\dfrac{W_2}{W_1} > 0$.

Temperature increase in an ideal gas

macroscopic approach: $\Delta S = nC_{(V\text{ or }P,\text{m})}\ln\dfrac{T_2}{T_1}$; If $T_2 > T_1$, then $\Delta S > 0$.

microscopic approach: Use particle-in-a-box theory again. At low T, gas molecules occupy a small number of energy levels (W = small); at higher T, more energy levels are available (W = larger) (see Fig. 7.9). So, $\Delta S = S_{T_2} - S_{T_1} = k\ln\dfrac{W_2}{W_1} > 0$.

Note: For liquids and solids, similar reasoning for the T dependence of S applies.
Summary → Macroscopic and microscopic approaches give the same predictions for ΔS. The Boltzmann approach provides deep insight into entropy on the molecular level in terms of the energy states available to a system.

7.6 Standard Molar Entropies, S_{m}° (T)

- **Standard (P = 1 bar) molar entropies of *pure* substances**

  S_{m}° (T) is determined experimentally from *heat capacity* data ($C_{P,\text{m}}$), *enthalpy* data (ΔH°) for phase changes, and the third law of thermodynamics.

- **Quantitatively** → For solid to liquid to vapor at final temperature T (The existence of only one solid phase is assumed.)

$$S_{\text{m}}^{\circ}(T) = S_{\text{m}}^{\circ}(0) + \int_{0}^{T_{\text{f}}} \frac{C_{P,\text{m}}(\text{solid})\text{d}T}{T} + \frac{\Delta H_{\text{fus}}}{T_{\text{f}}} + \int_{T_{\text{f}}}^{T_{\text{b}}} \frac{C_{P,\text{m}}(\text{liquid})\text{d}T}{T} + \frac{\Delta H_{\text{vap}}}{T_{\text{b}}} + \int_{T_{\text{b}}}^{T} \frac{C_{P,\text{m}}(\text{gas})\text{d}T}{T}$$

→ According to the third law, S_{m}° (0) = 0 and

$$\boxed{S_{\text{m}}^{\circ}(T) = \int_{0}^{T_{\text{f}}} \frac{C_{P,\text{m}}(\text{solid})\text{d}T}{T} + \frac{\Delta H_{\text{fus}}}{T_{\text{f}}} + \int_{T_{\text{f}}}^{T_{\text{b}}} \frac{C_{P,\text{m}}(\text{liquid})\text{d}T}{T} + \frac{\Delta H_{\text{vap}}}{T_{\text{b}}} + \int_{T_{\text{b}}}^{T} \frac{C_{P,\text{m}}(\text{gas})\text{d}T}{T}}$$

Note: For solids, only the first term in the equation is used if *only* one solid phase exists. For liquids, the first three terms are used. For gases, all five terms are required.

Example Plots of $C_{P,\text{m}}$ as a function of T and $C_{P,\text{m}}/T$ as a function of T for solid copper Cu are shown below. In the second graph, the area under the curve (T = 0 to T = 298.15 K) yields the experimental value of the entropy for Cu: [S_{m}° (298.15 K) = 33.15 J·K^{-1}·mol^{-1}].

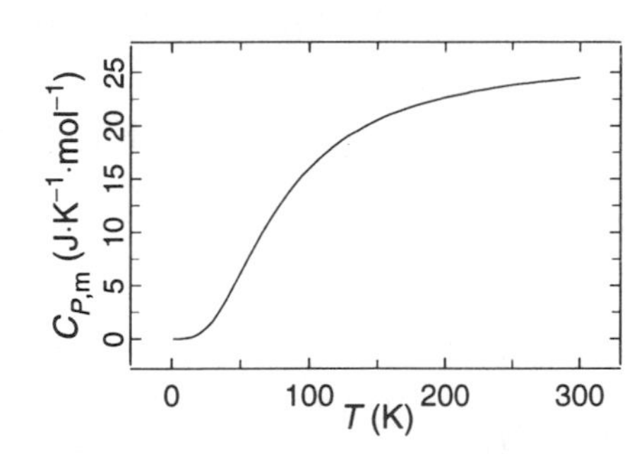

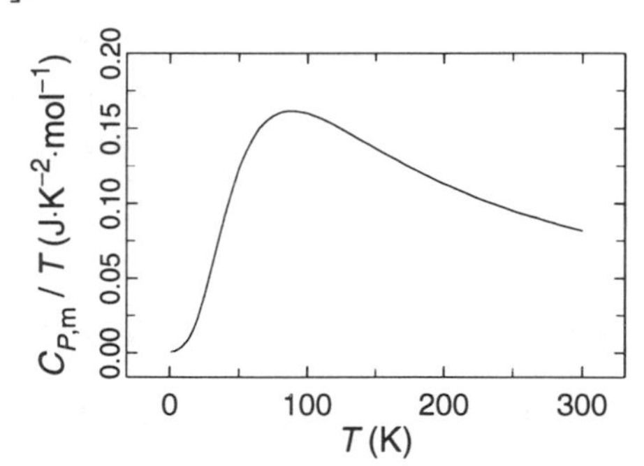

- **Appendix 2A**

  → List of values of *experimental* standard molar entropies at 25°C (298.15 K) for many substances. Abbreviated list in Table 7.3.
  [Recall *standard* ≡ *pure* substance at 1 bar.]

→ In general, entropies of gases are larger than those of solids or liquids ($S_{gas} > S_{liquid}$ *or* S_{solid}), as expected from the association of entropy with *randomness* and *disorder*.

Example 7.6 Use your general understanding of entropy and the properties of substances to arrange the following substances at 25°C in order of increasing standard molar entropy: carbon, C(diamond); methane, $CH_4(g)$; water, $H_2O(l)$; and lead, Pb(s).

Solution At the same *T* and *P*, we expect the gas (methane) to be the most disordered, followed by the liquid (water), and then the solids (diamond and lead). Which solid has the higher standard entropy, diamond or lead? Diamond has an extremely rigid lattice with atoms that are fixed in place and vibrate only slightly. The metal Pb is somewhat malleable. We therefore expect Pb to have the higher standard molar entropy; so the answer is C(diamond) < Pb(s) < $H_2O(l)$ < $CH_4(g)$.

Notes: Some solids have *higher* standard entropy values than certain liquids [compare KI(s), $S_m°(298.15\text{ K}) = 106.32\text{ J·K}^{-1}\text{·mol}^{-1}$ and H_2O, $S_m°(298.15\text{ K}) = 69.91\text{ J·K}^{-1}\text{·mol}^{-1}$], but virtually all condensed phases (solids and liquids) have $S_m°$ values lower than those for gases.

Also, *heavier* elements or species typically have *larger* molar entropies than *lighter* ones. Heavier species have more energy levels available to them [a better reason to place Pb(s) after C(diamond) in the list above].

Examine the $S_m°$ values in Appendix 2A to get a good qualitative feel for standard molar entropies.

7.7 Standard Reaction Entropies, $\Delta S_r°$

- **Standard reaction entropies, $\Delta S_r°$** → Calculated for any chemical reaction in a manner similar to the one used to calculate the standard enthalpy of reaction, $\Delta H_r°$

- **For any chemical reaction** → $\boxed{\Delta S_r° = \sum nS_m°(\text{products}) - \sum nS_m°(\text{reactants})}$

  n represents values of the stoichiometric coefficients in the balanced chemical equation

Example 7.7a Calculate $\Delta S_r°$ for the reaction $2\,Li(s) + Cl_2(g) \rightarrow 2\,LiCl(s)$. Use $S_m°$ values in Appendix 2A.

Solution The molar entropy of the gas Cl_2 is much larger than the value of either Li or LiCl, both of which are solids. Qualitatively then, we expect a net *decrease* in entropy as the reaction proceeds from left to right ($\Delta S_r° < 0$).

Quantitatively, $\Delta S_r° = \sum nS_m°(\text{products}) - \sum nS_m°(\text{reactants})$

$= 2S_m°(LiCl) - [2S_m°(Li) + S_m°(Cl_2)] = 2(59.33) - [2(29.12) + 222.96]$

$= 118.66 - 281.20 = -162.54\text{ J·K}^{-1}\text{·mol}^{-1}$

Example 7.7b Calculate $\Delta S_r°$ for the reaction $CaCO_3(s) \rightarrow CaO(s) + CO_2(g)$. Use $S_m°$ values in Table 7.3.

Answer ($\Delta S_r° > 0$) $\Delta S_r° = 160.6\text{ J·K}^{-1}\text{·mol}^{-1}$

Note: The entropy of the gas in the two preceding examples dominates the change in entropy for the reaction. The following generalizations are useful:

$\Delta S_r°$ is *positive* if there is a net *production* of gas in a reaction.

$\Delta S_r°$ is *negative* if there is a net *consumption* of gas in a reaction.

- **Properties of S and ΔS**

→ S is an *extensive* property of state, like enthalpy.

→ For any process or change in state:

Reverse the process, *change* the sign of ΔS.

Change the amounts of all materials in a process; make a *proportional change* in the value of ΔS.

Add *two* reactions together to get a *third* one; *add* the ΔS values for the first *two* reactions together to get ΔS for the *third* reaction.

Global Changes in Entropy (Sections 7.8-7.10)

Key Concepts

isolated system (system + surroundings), total entropy change (system + surroundings), criterion for spontaneity, equilibrium (thermal, mechanical, physical, chemical), criterion for equilibrium

Overview

- **For a *spontaneous* process, the entropy change of the *system* may be *negative*.**

Example Calculate ΔS when one mole of *supercooled* water freezes *spontaneously* to form ice at −10°C. Assume that the molar heat capacity of ice (36.6 $J{\cdot}K^{-1}{\cdot}mol^{-1}$) and water (76.0 $J{\cdot}K^{-1}{\cdot}mol^{-1}$) are both *independent* of T.
[$\Delta H_{fus}°(H_2O) = 6.007\ kJ{\cdot}mol^{-1}$]

Solution Calculate $\Delta S°$ for the process carried out in the *three reversible* steps shown below. Use some of the equations described earlier in this chapter.

$H_2O(l, -10°C) \rightarrow H_2O(l, 0°C) \rightarrow H_2O(s, 0°C) \rightarrow H_2O(s, -10°C)$

$$\Delta S° = C_{P,m}(\text{water}) \ln\left(\frac{273.15\ K}{263.15\ K}\right) + \frac{-\Delta H_{fus}°(H_2O)}{273.15\ K} + C_{P,m}(\text{ice}) \ln\left(\frac{263.15\ K}{273.15\ K}\right)$$

The first term in the above expression corresponds to heating one mole of water *reversibly* (*slowly*) from −10°C (263.15 K) to 0°C (273.15 K). The second term corresponds to the *reversible* freezing of one mole of water at its normal melting point. The third term gives $\Delta S°$ for the reversible cooling of one mole of ice from 0°C to −10°C.

The *sum* of all three terms gives $\Delta S°$ for the process:

$$H_2O(l, -10°C) \rightarrow H_2O(s, -10°C)$$

Inserting the values for the molar heat capacities and the enthalpy of fusion, we obtain $\Delta S° = 2.83 - 21.99 - 1.37 = -20.53\ J{\cdot}K^{-1}{\cdot}mol^{-1}$, a *negative* value that appears to violate the second law.

Note: The entropy change for the overall process is dominated by the second term, the crystallization of water. The first and third terms are small and tend to cancel.

- **Comments**
  - → Higher life forms also *seem* to violate the second law in that they are able to take smaller molecules and create large, highly organized assemblies of cells.
  - → The preceding examples are not puzzling if we recall that the *second law* holds only for *isolated systems,* to which neither of these examples corresponds.

7.8 The Surroundings

- **Second law** → For *spontaneous* change, the entropy of an *isolated* system *increases.*
- ***Any* spontaneous change**
  - → Entropy of system *plus* surroundings *increases.*
  - → Criterion for spontaneity *includes* the *surroundings.*
  - → $\Delta S_{tot} = \Delta S + \Delta S_{surr}$
- **Criterion**
  - → A process is *spontaneous* as written if $\Delta S_{tot} > 0$.
  - → A process is *nonspontaneous* as written if $\Delta S_{tot} < 0$. The *reverse* process is spontaneous.
  - → For a system at *equilibrium,* $\Delta S_{tot} = 0$ (Section 7.10).
- **Calculating ΔS_{surr} for a process at constant T and P** → $\boxed{\Delta S_{surr} = \dfrac{q_{surr}}{T} = -\dfrac{\Delta H}{T}}$

Note: In this equation, ΔH is the enthalpy change for the *system.* At constant pressure, the *heat* associated with the process is *reversibly* transferred to the surroundings. The heat capacity of the surroundings is vast, so its temperature remains constant. The process itself may be *reversible* or *irreversible.*

Example 7.8 Calculate ΔS_{tot} for the crystallization of one mole of water to form ice at −10°C: $H_2O(l, -10°C) \rightarrow H_2O(s, -10°C)$. See the previous example.

Solution In the previous example, $\Delta S = -20.53\ J{\cdot}K^{-1}$ for one mole of water. The heat change, ΔH, for the system is given by q_P of the *three steps* in the previous example.

$$q_P = \Delta H = (76.0\ J{\cdot}K^{-1})(10\ K) + (-6007\ J) + (36.6\ J{\cdot}K^{-1})(-10\ K) = -5613\ J$$

$$\Delta S_{\text{surr}} = -\frac{\Delta H}{T} = -\frac{(-5613\text{ J})}{(263.15\text{ K})} = 21.33\text{ J·K}^{-1}$$

$\Delta S_{\text{tot}} = \Delta S + \Delta S_{\text{surr}} = -20.53 + 21.33 = 0.80\text{ J·K}^{-1}$
ΔS_{tot} is > 0, as expected for this spontaneous process.

7.9 The Overall Change in Entropy

- **Spontaneity depends on ΔS *and* ΔS_{surr}** → Four **cases** arise because *both* ΔS and ΔS_{surr} can be either positive (+) or negative (–).

Case	[ΔS_{tot}	=	ΔS	+	ΔS_{surr}]	Spontaneity
1	+		+		+	*always* spontaneous
2	?		–		+	spontaneous, if $\lvert\Delta S_{\text{surr}}\rvert > \lvert\Delta S\rvert$
3	?		+		–	spontaneous, if $\lvert\Delta S\rvert > \lvert\Delta S_{\text{surr}}\rvert$
4	–		–		–	*never* spontaneous

- **Application to chemical reactions**
  → *Exothermic* reactions ($\Delta S_{\text{surr}} > 0$) correspond to Cases 1 and 2, that is, *spontaneous* if $\Delta S > 0$ or if $\lvert\Delta S_{\text{surr}}\rvert > \lvert\Delta S\rvert$ when $\Delta S < 0$.
  → *Endothermic* reactions ($\Delta S_{\text{surr}} < 0$) correspond to Cases 3 and 4, that is, *spontaneous* only if $\Delta S > 0$ *and* $\lvert\Delta S\rvert > \lvert\Delta S_{\text{surr}}\rvert$.

Example 7.9 Predict whether ethyne gas, $C_2H_2(g)$, can form spontaneously from its elements under standard conditions (1 bar) at 25°C (298.15 K).

Solution Evaluate $\Delta S_{\text{tot}} = \Delta S + \Delta S_{\text{surr}}$ for the process

$$2\text{C(graphite)} + H_2(g) \rightarrow C_2H_2(g)$$

Recall, $\Delta S = \Delta S_r° = \sum nS_m°(\text{products}) - \sum nS_m°(\text{reactants})$ and $\Delta S_{\text{surr}} = -(\Delta H/T)$

$\Delta S_r° = S_m°[C_2H_2(g)] - \{2S_m°[\text{C(graphite)}] + S_m°[H_2(g)]\}$
$= 200.83 - [2(5.74) + (130.57)] = 58.78\text{ J·K}^{-1}$

To calculate ΔS_{surr}, assume that $\Delta H = \Delta H_r°$. The surroundings gain the heat lost by the system ($\Delta H = q_P$) reversibly as the reaction occurs.

$\Delta H_r° = \sum n\Delta H_f°(\text{products}) - \sum n\Delta H_f°(\text{reactants})$
$= \Delta H_f°[C_2H_2(g)] - \{2\Delta H_f°[\text{C(graphite)}] + \Delta H_f°[H_2(g)]\}$
$= \Delta H_f°[C_2H_2(g)] = 226.73\text{ kJ·mol}^{-1} = 226.73\text{ kJ·mol}^{-1} = 2.2673 \times 10^5\text{ J·mol}^{-1}$,

because the heat of formation of the most stable form of an element is 0 under standard conditions.

Thus, $\Delta S_{\text{surr}} = -\frac{\Delta H}{T} = -\frac{\Delta H_r°}{T} = -\frac{2.2673\times10^5\text{ J·mol}^{-1}}{298.15\text{ K}} = -760.46\text{ J·K}^{-1}\text{·mol}^{-1}$.

Then, $\Delta S_{\text{tot}} = \Delta S + \Delta S_{\text{surr}} = 58.78 - 760.46 = -701.68\text{ J·K}^{-1}\text{·mol}^{-1}$

Because $\Delta S_{tot} < 0$, ethyne (acetylene) will *not* form spontaneously from its elements.
The reverse reaction, decomposition of acetylene into its elements is spontaneous, because ΔS_{tot} (reverse reaction) = $+701.68\ J{\cdot}K^{-1}{\cdot}mol^{-1} > 0$.

- **For a given change in state (same ΔS), the *path* can affect ΔS_{surr}, ΔS_{tot}, and *spontaneity*** (see Example 7.11 in the text).

  → Comparison of a *reversible* and an *irreversible* isothermal expansion of an ideal gas for the same initial and final states:

  (1) ΔS (*irreversible*) = ΔS (*reversible*) (entropy is a state function)

  (2) $\Delta U = q + w = 0$ and $\Delta H = 0$, because energy and enthalpy of an ideal gas depend only on T; and $\Delta T = 0$ for this process.

  (3) *q and w* are different for the two processes (q and w are *path* functions). Heat given off to the surroundings ($-q = q_{surr}$) is different for the two processes.

  (4) From Chapter 6, we know that a *reversible* process does the *maximum* work w. Because $q + w = 0$, a *reversible* process delivers *less* heat to the surroundings.

  (5) Because $\Delta S_{surr} = q_{surr}/T$, ΔS_{surr} (*irreversible*) > ΔS_{surr} (*reversible*).

  (6) Thus, ΔS_{tot} (*irreversible*) > ΔS_{tot} (*reversible*) = 0.

  (7) The *reversible* process corresponds to one in which the system is only *infinitesimally* removed from equilibrium as the process proceeds, whereas the *irreversible* process [ΔS_{tot}(*irreversible*) > 0] is *spontaneous*.

7.10 Equilibrium

- **System at equilibrium** → No tendency to change in either the forward or the reverse direction without an external influence

- **Types of equilibrium**

thermal: No tendency for heat to flow in or out of the system; for example, an aluminum rod at room temperature

mechanical: No tendency for any part of a system to move; for example, an undeformed spring has no tendency to stretch or compress

physical: Two phases of a substance at the transition temperature with no tendency for either phase to increase in mass; for example, steam and water at the normal boiling point (100°C)

chemical: A mixture of reactants and products with no *net* tendency for either to form; for example, a saturated solution of sucrose in water in contact with solid sucrose

- **Universal thermodynamic criterion for equilibrium**
  → $\Delta S_{tot} = 0$ for any system at equilibrium
  If this were not true, ΔS_{tot} would be greater than 0 in either the forward or the reverse direction and *spontaneous change* would occur until $\Delta S_{tot} = 0$.
  → Total entropy change may be calculated to determine whether a system is at equilibrium (see Example 7.12 in the text).

Free Energy (Sections 7.11-7.15)

Key Concepts

(Gibbs) free energy G, reaction free energy ΔG_r, standard reaction free energy ΔG_r°, standard free energy of formation ΔG_f°, thermodynamically stable compound, thermodynamically unstable compound, labile, nonlabile, inert, nonexpansion work, coupled processes

Overview

- **Gibbs free energy (G)**
  → State property of the *system:* $G \equiv H - TS$
  → For any process or reaction at *constant* T and P (typical laboratory conditions), the change in free energy of the system (ΔG) determines *spontaneity.*
  → A criterion for *spontaneity* that depends *only* on the *system*

- **ΔG and spontaneity** → For a process or reaction at constant T and P:
  if $\Delta G < 0$, the process or reaction is *spontaneous* as written
  if $\Delta G = 0$, the process or reaction is at *equilibrium*
  if $\Delta G > 0$, the process or reaction is *nonspontaneous* as written

- **Properties of G and ΔG**
  → G is an *extensive* state property, like enthalpy and entropy.
  → Reversing a process or reaction changes the *sign* of ΔG.
  → Changing the amount of material changes the magnitude of ΔG proportionately.
  → Combining several reactions to obtain a new one requires adding the values of ΔG for each reaction to obtain the value for the new reaction.

7.11 Focusing on the System

- **Alternative criterion for *spontaneity***

  (1) $\Delta S_{tot} = \Delta S + \Delta S_{surr}$ (always true)

  (2) If P and T are constant, $\Delta S_{surr} = -\Delta H / T$

(3) If P and T are constant, $\Delta S_{tot} = \Delta S - (\Delta H/T)$ (function of *system only*)

(4) Definition of Gibbs free energy, $G \equiv H - TS$

(5) If T is constant, $\Delta G = \Delta H - \Delta(TS) = \Delta H - T\Delta S$

(6) According to (5) and (3), $\Delta G = -T\Delta S_{tot}$

- **Summary**
  → If a process at constant T and P is *spontaneous* ($\Delta S_{tot} > 0$), then $\Delta G < 0$.
  → If a process at constant T and P is *nonspontaneous* ($\Delta S_{tot} < 0$), then $\Delta G > 0$.
  → If a process at constant T and P is at *equilibrium* ($\Delta S_{tot} = 0$), then $\Delta G = 0$ as well.

- **Dependence of spontaneity on ΔS and ΔH of a system at constant T and P**
  → Four cases arise because *both* ΔS and ΔH can be *either* positive (+) *or* negative (–).

Case	[ΔG	= ΔH	– $T\Delta S$]	Spontaneity
1	–	–	+	*always* spontaneous
2	?	–	–	spontaneous, if $\lvert\Delta H\rvert > T\lvert\Delta S\rvert$
3	?	+	+	spontaneous, if $T\lvert\Delta S\rvert > \lvert\Delta H\rvert$
4	–	+	–	*never* spontaneous

- **Details of the four cases**

Case 1

→ *Exothermic* process or reaction ($\Delta H < 0$)
→ *Enthalpy* and *entropy* favor *spontaneity*
Enthalpy- and *entropy*-driven process or reaction
→ *Temperature* dependence: if ΔH and ΔS are T independent, then the reaction is *spontaneous* at all T.

Case 2

→ *Exothermic* process or reaction ($\Delta H < 0$)
→ *Only enthalpy* favors *spontaneity*
Enthalpy-driven process or reaction
→ *Temperature* dependence: if ΔH and ΔS are T independent, then the reaction is *spontaneous* at low T (*enthalpy* "wins") but *nonspontaneous* at high T.
A crossover temperature exists at which $\Delta H = T\Delta S$ and the system is at equilibrium.

Case 3

→ *Endothermic* process or reaction ($\Delta H > 0$)
→ *Only entropy* favors *spontaneity*
Entropy-driven process or reaction

→ *Temperature* dependence: if ΔH and ΔS are T independent, then the reaction is *nonspontaneous* at low T and *spontaneous* at high T (*entropy* "wins").
A crossover temperature exists at which $\Delta H = T\Delta S$ and the system is at equilibrium.

Case 4

→ *Endothermic* process or reaction ($\Delta H > 0$)
→ *Nonspontaneous* process or reaction ($\Delta G > 0$)
→ *Temperature* dependence: if ΔH and ΔS are T independent, then the reaction is *nonspontaneous* at all T.

Example 7.11 For the gas-phase reaction $2\,SO_3(g) \rightarrow 2\,SO_2(g) + O_2(g)$, $\Delta H_r° = 197.8\ kJ{\cdot}mol^{-1}$ and $\Delta S_r° = 0.188\ kJ{\cdot}K^{-1}{\cdot}mol^{-1}$ at 25°C (calculated from values of *standard* formation enthalpies and *standard* entropies in Appendix 2A). Assume that both $\Delta H_r°$ and $\Delta S_r°$ are *independent* of temperature, and calculate the crossover temperature at which the reactants and products are at *equilibrium*. Describe the behavior of the reaction above and below the crossover temperature.

Solution The reaction is endothermic and $\Delta S_r° > 0$, which corresponds to Case 3. At 25°C, calculate $\Delta G_r°$ from the definition of the free energy to see whether the reaction is spontaneous. Assuming that $\Delta H_r°$ and $\Delta S_r°$ are temperature independent, solve for the crossover temperature at the *equilibrium* condition ($\Delta G = 0$).

$$\Delta G_r° = \Delta H_r° - T\Delta S_r° = 197.8\ kJ{\cdot}mol^{-1} - (298.15\ K)(0.188\ kJ\ K^{-1}{\cdot}mol^{-1})$$
$$= 141.7\ kJ{\cdot}mol^{-1}$$

Because $\Delta G_r° > 0$, the reaction is *nonspontaneous* at 25°C. The reaction will become spontaneous at temperatures above the crossover value defined by $\Delta G_r° = 0 = \Delta H_r° - T\Delta S_r°$.
Solving for T, one obtains $T = \Delta H_r°/\Delta S_r°$.

$$T = \frac{\Delta H_r°}{\Delta S_r°} = \frac{197.8\ kJ{\cdot}mol^{-1}}{0.188\ kJ{\cdot}K^{-1}{\cdot}mol^{-1}} = 1050\ K\ \text{(3 signifcant figures)}$$

Above 1050 K, $\Delta G_r° < 0$ and the reaction is *spontaneous*.

Note: This reaction is *entropy* driven above 1050 K. As the temperature increases above 1050, ΔG becomes *more* negative. As this occurs, we say that the *driving force* for the reaction to occur increases. The *more negative* $\Delta G_r°$ becomes, the *larger* the driving force.

7.12 Reaction Free Energy

- **Chemical reactions** → Reaction free energy, ΔG_r, and standard reaction free energy, $\Delta G_r°$, are defined in a manner similar to the reaction enthalpy, ΔH_r, and standard reaction enthalpy, $\Delta H_r°$.

- **Definitions**

→ If G_m is the molar free energy of a reactant or product with a stoichiometric coefficient of n, then the reaction free energy is

$$\boxed{\Delta G_r = \sum nG_m(\text{products}) - \sum nG_m(\text{reactants})}$$

→ If $G_m°$ is the standard molar free energy of a reactant or product with a stoichiometric coefficient of n, then the standard reaction free energy is

$$\Delta G_r° = \sum nG_m°(\text{products}) - \sum nG_m°(\text{reactants})$$

→ Values of G_m or $G_m°$ cannot be determined directly. Thus, the equations given above *cannot* be used to determine ΔG_r and $\Delta G_r°$.

→ Values of ΔG_r and $\Delta G_r°$ are determined from the free energies of *formation*, ΔG_f and $\Delta G_f°$.

- **Standard free energy of formation, $\Delta G_f°$**
  - → Standard free energy of *formation* of a compound or element
  - → Free energy change for the formation of one mole of a compound from the most stable form of its elements under standard conditions (1 bar)
  - → $\Delta G_f° \equiv 0$ for *all elements* in their most stable form (same convention as enthalpy).
- **Calculating $\Delta G_f°$ for a given compound**

1. *Write and balance the formation reaction*
   One mole of compound on the product side and the most stable form of the elements with appropriate coefficients on the reactant side
2. *Calculate $\Delta H_f°$ and $\Delta S_f°$ for the reaction by using the data in* Appendix 2A
   The value of $\Delta S_f°$ is obtained from the standard molar entropy values as follows: $\Delta S_f° = S_m°(\text{compound}) - \sum nS_m°(\text{reactants})$
3. *Solve for the formation of one mole of compound, using the expression*

$$\begin{aligned}\Delta G_f° &= \Delta H_f° - T\Delta S_f° \\ &= \Delta H_f°(\text{compound}) - T\left[S_m°(\text{compound}) - \sum nS_m°(\text{reactants})\right]\end{aligned}$$

Example 7.12a Calculate the standard free energy of *formation* of aluminum oxide, Al_2O_3, at 25°C $\{\Delta G_f°[Al_2O_3(s)]\}$.

Solution Write the formation reaction for $Al_2O_3(s)$. Use Appendix 2A to find the values of the standard formation enthalpy of $Al_2O_3(s)$ and the standard molar entropies of all reactants and products in the formation reaction.

Use $\Delta G_f° = \Delta H_f° - T\{S_m°[Al_2O_3(s)] - \sum nS_m°[\text{reactants}]\}$.

Because $\Delta G_f°$ has the units of $kJ\cdot mol^{-1}$, convert the standard entropies of reactants and products into the units of $kJ\cdot K^{-1}\cdot mol^{-1}$ before using the above relation.

Standard formation reaction at 25°C: $2\,Al(s) + \frac{3}{2}O_2(g) \rightarrow Al_2O_3(s)$

$$\begin{aligned}&\Delta G_f°[Al_2O_3(s)] \\ &= \Delta H_f°[Al_2O_3(s)] - T\{S_m°[Al_2O_3(s)] - 2\,S_m°[Al(s)] - (3/2)S_m°[O_2(g)]\} \\ &= -1675.7 - (298.15)\{0.050\,92 - 2(0.028\,33) - (3/2)(0.205\,14)] \\ &= -1675.7 + 93.46\ kJ\cdot mol^{-1} = -1582.2\ kJ\cdot mol^{-1}\end{aligned}$$

Thus, $Al_2O_3(s)$ will form *spontaneously* from its elements under standard conditions at 25°C.

Example 7.12b Calculate the standard free energy of *formation* of liquid octane at 25°C $\{\Delta G_f^\circ [C_8H_{18}(l)]\}$.

Answer $\Delta G_f^\circ [C_8H_{18}(l)] = +\,7.7\ \text{kJ·mol}^{-1}$, so liquid octane will *not* form *spontaneously* from its elements under standard conditions at 25°C.

- **Thermodynamically *stable* compound**
  → *Negative* standard free energy of formation ($\Delta G_f^\circ < 0$)
  → Thermodynamic tendency to *form* from its elements

- **Thermodynamically *unstable* compound**
  → *Positive* standard free energy of formation ($\Delta G_f^\circ > 0$)
  → Thermodynamic tendency to *decompose* into its elements

Examples The compound $Al_2O_3(s)$ in Example 7.12a is thermodynamically stable, whereas liquid octane in Example 7.12b is thermodynamically unstable.

Many metal oxides with *very negative* ΔG_f° values are the major components of ores such as bauxite, Al_2O_3, and hematite, Fe_2O_3. These oxides are quite stable and are said to be in a "thermodynamic pit" from which the (useful) metals are liberated with great difficulty.

In contrast to metal oxides, petroleum products (also occurring naturally) have *less negative* ΔG_f° values; some are even positive (liquid octane). These compounds react readily to form compounds with lower free energy; hence their use as fuels.

Observe the ΔG_f° values for some of the compounds in Appendix 2A. You should develop a good feeling for which types of compounds are most and least thermodynamically stable. Compare the property of thermodynamic stability to the natural abundance of the compounds.

- **Properties of thermodynamically *unstable* compounds**
  → *Many* decompose into their elements over a *long* time span.
  → *Thermodynamically unstable* substances (like liquid octane and diamond) that decompose into their elements *slowly* are said to be *kinetically stable.*

- **Classification of thermodynamically *unstable* compounds**

  labile: Decompose *or* react readily, for example, TNT and NO

  nonlabile: Decompose *or* react slowly, for example, liquid octane

  inert: Show virtually no reactivity *or* decomposition, for example, diamond

- **Calculating the standard reaction free energy ΔG_r° for a given chemical reaction**

  1. *Write the balanced chemical equation for the reaction.*
  2. *Calculate ΔH_r° and ΔS_r° for the reaction by using the data in* Appendix 2A.
  3. *Solve for ΔG_r°, using the expression*

$$\Delta G_r° = \Delta H_r° - T\Delta S_r°$$
$$= \left[\sum n\Delta H_f°(\text{products}) - \sum n\Delta H_f°(\text{reactants})\right]$$
$$- T\left[\sum nS_m°(\text{products}) - \sum nS_m°(\text{reactants})\right]$$

4. *An alternative procedure is to calculate $\Delta G_r°$ from the $\Delta G_f°$ values that are also listed in* Appendix 2A.

$$\Delta G_r° = \sum n\Delta G_f°(\text{products}) - \sum n\Delta G_f°(\text{reactants})$$

This equation yields a faster result, but it does *not* allow for an estimation of the *temperature* dependence of $\Delta G_r°$.

Example 7.12c Calculate $\Delta G_r°$ at 25°C for the reduction of Al_2O_3(s) with C(graphite) [that is, charcoal] to form CO_2(g) and Al(s) metal. Is the process *spontaneous* under standard conditions? If not, is there a crossover temperature above which it becomes *spontaneous*?

Solution The balanced equation is $2\,Al_2O_3(s) + 3\,C(\text{graphite}) \rightarrow 3\,CO_2(g) + 4\,Al(s)$. Use the last equation given above and Appendix 2A. Recall that $\Delta G_f°$ and $\Delta H_f°$ for the most stable form of the elements ≡ 0, which is the case for graphite and aluminum metal. If the process is not spontaneous, determine the crossover temperature from $\Delta G_r° = \Delta H_r° - T\Delta S_r° = 0$, assuming both $\Delta H_r°$ and $\Delta S_r°$ are temperature independent. We expect $\Delta S_r° > 0$, since there is one gas-phase product.

$$\Delta G_r° = \sum n\Delta G_f°(\text{products}) - \sum n\Delta G_f°(\text{reactants})$$
$$= 3\Delta G_f°[CO_2(g)] + 4G_f°[Al(s)] - \{2\Delta G_f°[Al_2O_3(s)] + 3\Delta G_f°[C(\text{graphite})]\}$$
$$= 3(-394.36\ \text{kJ·mol}^{-1}) - 2(-1582.3\ \text{kJ·mol}^{-1}) = 1981.5\ \text{kJ·mol}^{-1}$$

Since $\Delta G_r°$ is positive, the reaction is not *spontaneous* under standard conditions.

$$\Delta H_r° = \sum n\Delta H_f°(\text{products}) - \sum n\Delta H_f°(\text{reactants})$$
$$= 3\Delta H_f°[CO_2(g)] + 4\Delta H_f°[Al(s)] - \{2\Delta H_f°[Al_2O_3(s)] + 3\Delta H_f°[C(\text{graphite})]\}$$
$$= 3(-393.51\ \text{kJ·mol}^{-1}) - 2(-1675.7\ \text{kJ·mol}^{-1}) = 2170.87\ \text{kJ·mol}^{-1}$$

$$\Delta S_r° = \sum nS_m°(\text{products}) - \sum nS_m°(\text{reactants})$$
$$= 3S_m°[CO_2(g)] + 4S_m°[Al(s)] - \{2S_m°[Al_2O_3(s)] + 3S_m°[C(\text{graphite})]\}$$
$$= \{3(0.213\,74\ \text{kJ·K}^{-1}\text{·mol}^{-1}) + 4(0.028\,33\ \text{kJ·K}^{-1}\text{·mol}^{-1})\}$$
$$- \{2(0.050\,92\ \text{kJ·K}^{-1}\text{·mol}^{-1}) + 3(0.005\,740\ \text{kJ·K}^{-1}\text{·mol}^{-1})\}$$
$$= 0.635\,48\ \text{kJ·K}^{-1}\text{·mol}^{-1}$$

$$T = \frac{\Delta H_r°}{\Delta S_r°} = \frac{2170.87\ \text{kJ·mol}^{-1}}{0.635\,48\ \text{kJ·K}^{-1}\text{·mol}^{-1}} = 3416\ \text{K} = 3143°\text{C}$$

= crossover temperature

Clearly, this temperature is extremely high and cannot be attained under normal circumstances. We therefore expect that Al metal will not be produced efficiently by this reaction.

Example 7.12d Calculate the crossover temperature above which the reduction of magnetite, Fe_3O_4(s), with C(graphite) to form CO_2(g) and Fe(s) metal becomes *spontaneous*?

Answer $T = 943$ K = 670°C. This temperature is easily reached in a furnace, so we expect Fe(s) to be much easier to obtain in this way than Al(s). Indeed, Al was an expensive, exotic material until the development of the efficient Hall process for its production in 1886.

Note: In both of the preceding examples, we have ignored the complication that solids melt at sufficiently high temperature. Our results are only qualitative.

7.13 Free Energy and Nonexpansion Work

- **Processes at constant T and P**

Expansion work

→ $dw = -PdV$ (*infinitesimal* change in volume, see Chapter 6)

→ $w = -P\Delta V$ (*finite* change in volume, constant opposing P)

→ Expansion work is performed against an opposing pressure.
Recall that the sign of w is defined in terms of the *system*.
If work is done *on* the system *by* the surroundings, $w > 0$;
if work is done *by* the system *on* the surroundings, $w < 0$.

Nonexpansion work

→ dw_e (*infinitesimal* change)

→ w_e (*finite* change)

→ Any other type of work, including electrical work, mechanical work, work of muscular contraction, work involved in neuronal signaling, and that of chemical synthesis (making chemical bonds)

- **Relationship between nonexpansion work and (finite) free energy changes**

For a reversible process, $w_e = \Delta G$ (constant P and T)

Notes: For any process at constant P and T, w_e is equal to the *maximum* nonexpansion work that can be obtained from the process. A *reversible* process is one in which the maximum amount of work is done by the system on the surroundings (negative sign of w). Real systems are always *irreversible* in nature and have inefficiencies such that the maximum nonexpansion work will *not* be obtained.

Example 7.13 What is the maximum amount of electrical work that can be obtained from burning 10.0 g of propane C_3H_8(g) in a fuel cell that is operating at 1 bar and 298.15 K and also contains *excess* oxygen?

Solution The balanced equation (fuel cell reaction) for the complete oxidation of propane is

$$C_3H_8(g) + 5\,O_2(g) \rightarrow 3\,CO_2(g) + 4\,H_2O(l)$$

To solve the problem, calculate $\Delta G_r°$ and the maximum nonexpansion work, w_e, obtained from the oxidation of 1 mole of propane. Finally, convert to the free energy released for 10.0 g. Note that *liquid* water is formed because the temperature is 298.15 K.

$\Delta G_r° = \sum n\Delta G_f°(\text{products}) - \sum n\Delta G_f°(\text{reactants})$
$= 3\Delta G_f°[CO_2(g)] + 4G_f°[H_2O(l)] - \{\Delta G_f°[C_3H_8(g)] + 5\Delta G_f°[O_2(g)]\}$
$= 3(-394.36\ \text{kJ·mol}^{-1}) + 4(-237.13\ \text{kJ·mol}^{-1}) - (-23.49\ \text{kJ·mol}^{-1})$
$= -2108.11$ kJ for 1 mol $C_3H_8(g)$

$w_e = \Delta G_r° = -2108.11\ \text{kJ·mol}^{-1}\ [C_3H_8(g)]$

For 10.0 g of propane, (10.0 g)/(M = 44.094 g·mol^{-1}) = 0.2268 mol.
Finally, (0.2268 mol)(−2108.11 kJ·mol^{-1}) = −478 kJ

Because the free energy of the system decreases ($\Delta G_r° < 0$) the nonexpansion work ($w_e < 0$) will be done *by* the system *on* the surroundings. Thus, the *maximum* amount of electrical work obtainable is 478 kJ.

7.14 The Effect of Temperature

- **For any chemical reaction at constant *P* and *T***

$$\begin{aligned}\Delta G_r° &= \Delta H_r° - T\Delta S_r° \\ &= \left[\sum n\Delta H_f°(\text{products}) - \sum n\Delta H_f°(\text{reactants})\right] \\ &\quad - T\left[\sum nS_m°(\text{products}) - \sum nS_m°(\text{reactants})\right]\end{aligned}$$

→ Consider the discussion in Section 7.11 of this *Study Guide*, where the temperature independence of $\Delta H_r°$ and $\Delta S_r°$ is assumed. Cases 1–4 apply to any chemical reaction. The temperature dependence of $\Delta G_r°$ arises *mainly* from the ($-T\Delta S_r°$) term.

Example 7.14 Over what temperature range is each of the following reactions *spontaneous* at 1 bar?

(1) $H_2(g) + \frac{1}{2}O_2(g) \rightarrow H_2O(g)$

(2) $2NO_2(g) \rightarrow 2NO(g) + O_2(g)$

Solution Use $\Delta G_r° = \Delta H_r° - T\Delta S_r°$. Calculate $\Delta H_r°$ and $\Delta S_r°$. Calculate $\Delta G_r°$ at 25°C to determine the reaction *spontaneity* under normal temperatures. Set $\Delta G_r° = 0$ and solve for the temperature at which the crossover from *spontaneity* to *nonspontaneity* (or *vice versa*) occurs.

(1) $\Delta H_r° = \Delta H_f°[H_2O(g)] = -241.82\ \text{kJ·mol}^{-1}$
$\Delta S_r° = S_m°[H_2O(g)] - \{S_m°[H_2(g)] + (1/2)S_m°[O_2(g)]\}$
$= 0.18883\ \text{kJ·K}^{-1}\text{·mol}^{-1} - \{0.13068 + (1/2)(0.20514\ \text{kJ·K}^{-1}\text{·mol}^{-1})\}$
$= -0.04442\ \text{kJ·K}^{-1}\text{·mol}^{-1}$

Enthalpy favors *spontaneity*, entropy does not.

$$T = \frac{\Delta H_r°}{\Delta S_r°} = \frac{-241.82\ \text{kJ·mol}^{-1}}{-0.04442\ \text{kJ·K}^{-1}\text{·mol}^{-1}} = 5444\ \text{K} = \text{crossover } T$$

At 298.15 K, $\Delta G_r° = \Delta H_r° - T\Delta S_r° = -241.82 - (298.15)(-0.04442)$ $= -228.58$ kJ·mol^{-1}. The reaction is *spontaneous* at room temperature and is driven by *enthalpy*. As T increases, $\Delta G_r°$ becomes more positive, until at 5444 K, a crossover occurs ($\Delta G_r° = 0$). Above this temperature, $H_2O(g)$ should decompose *spontaneously* into its elements. This *reverse* reaction will be *entropy* driven. The reaction as written above is *spontaneous* for $T < 5444$ K, a maximum temperature that is inaccessible under normal conditions.

(2) $\Delta H_r° = 2\,\Delta H_f°[\text{NO(g)}] - 2\,\Delta H_f°[\text{NO}_2\text{(g)}] = 2(90.25\text{ kJ}) - 2(33.18\text{ kJ})$
$= 114.14$ kJ·mol^{-1}

$\Delta S_r° = 2\,S_m°[\text{NO(g)}] + S_m°[\text{O}_2\text{(g)}] - 2\,S_m°[\text{NO}_2\text{(g)}]$
$= 2(0.21076\text{ kJ·K}^{-1}\text{·mol}^{-1}) + (0.20514\text{ kJ·K}^{-1}\text{·mol}^{-1})$
$- 2(0.24006\text{ kJ·K}^{-1}\text{·mol}^{-1}) = 0.14654\text{ kJ·K}^{-1}\text{·mol}^{-1}$

Entropy favors *spontaneity*, enthalpy does not.

$$T = \frac{\Delta H_r°}{\Delta S_r°} = \frac{114.14\text{ kJ·mol}^{-1}}{0.146\,54\text{ kJ·K}^{-1}\text{·mol}^{-1}} = 778.90\text{ K} \approx 779\text{ K} = \text{crossover } T$$

At 298.15 K, $\Delta G_r° = \Delta H_r° - T\Delta S_r° = (114.14) - (298.15)(0.14654)$ $= 70.45$ kJ·mol^{-1}.

The reaction as written is *nonspontaneous* at 25°C. It becomes *spontaneous* and *entropy* driven when T > 779 K.

7.15 Free Energy Changes in Biological Systems

- **Metabolism**

  Cells use energy from the surroundings (sunlight for plants, food for animals) to build useful molecules *via* chemical reactions. In metabolic processes, some reaction steps have a *positive* (unfavorable) reaction free energy. How does nature solve this problem for biological systems? Most reactions take place *inside* cells, so changing the temperature to make a *nonspontaneous* reaction *spontaneous* is not possible. The solution is the use of *coupled reactions.*

- **Coupled reactions**

  → A way to make *nonspontaneous* reactions occur without changing the temperature or pressure

  → A prominent process in biological systems, also found in nonbiological ones

- **Concept of coupled chemical reactions**

  → If reaction (1) has a *positive* reaction free energy [$\Delta G_r°(1) > 0$] (*nonspontaneous*), it can be coupled to reaction (2) with a *more negative* reaction free energy [$\Delta G_r°(2) < 0$], such that reaction (3), the sum of the two reactions, has a *negative* reaction free energy [$\Delta G_r°(3) < 0$]:

  $\Delta G_r°(3) = \Delta G_r°(1) + \Delta G_r°(2) < 0$ (overall reaction is *spontaneous*)

Example The sugar glucose (a food) is converted to pyruvate in a series of steps. The overall process may be represented as

glucose + other reactants → 2 pyruvate + other products,

$$\Delta G_r° = -80.6\ \text{kJ·mol}^{-1}$$

The *biochemical standard state* (not described here) corresponds more nearly to typical conditions in a cellular environment and *replaces* the previous definition. The relatively large *negative* value of $\Delta G_r°$ indicates that the overall reaction occurs spontaneously under *biochemical* standard conditions. The first step in the metabolic process is the conversion of glucose into glucose-6-phosphate:

glucose + phosphate → glucose-6-phosphate + H_2O(l),

$$\Delta G_r° (1) = +14.3\ \text{kJ·mol}^{-1}$$

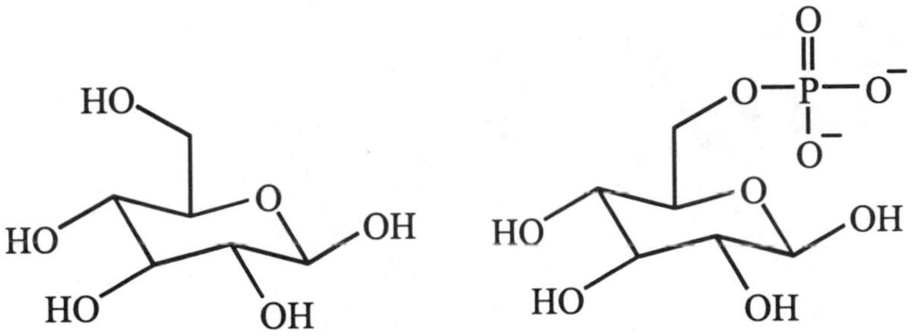

The large *positive* standard free energy implies *nonspontaneity* for reaction (1). To generate glucose-6-phosphate, reaction (1) is coupled with reaction (2), the hydrolysis of adenosine triphosphate (ATP) to yield adenosine diphosphate (ADP). Reaction (2) has a favorable $\Delta G_r° < 0$ value (*spontaneous*):

ATP + H_2O → ADP + phosphate, $\Delta G_r° (2) = -31.0\ \text{kJ·mol}^{-1}$

Coupling reactions (1) and (2) gives a net reaction (3) that is *spontaneous:*

glucose + ATP → glucose-6-phosphate + ADP,
$\Delta G_r° = \Delta G_r° (1) + \Delta G_r° (2) = -16.7\ \text{kJ·mol}^{-1}$

Note: Reactions in metabolic pathways are *catalyzed* by *enzymes*. Step (1) is catalyzed by the enzyme hexokinase. Biological catalysts are discussed in Section 13.15 and sugars in Section 19.14 and 19.15.

Chapter 8 Physical Equilibria

Phases and Phase Transitions (Sections 8.1-8.8)

Key Concepts

phase, phase transition, vapor pressure, volatile, sublimation, physical equilibrium, dynamic equilibrium, Clausius-Clapeyron equation, boiling, standard boiling point, normal boiling point T_b, freezing, melting, normal freezing point T_f, supercooling, anomalous behavior of water, phase diagram, component, phase boundary, definition of the absolute temperature scale, degrees of freedom, phase rule, triple point, critical point, critical temperature T_c, critical pressure P_c, supercritical fluid

Overview

- **Phase**
  - → Sample of matter uniform in chemical composition and physical state
  - → Gas (g), liquid (l), solid (s), solution (gas, liquid or solid in a liquid), solid solution (typically one solid dissolved in another)

  Examples: *Gases* oxygen [$O_2(g)$], hydrogen [$H_2(g)$], carbon dioxide [$CO_2(g)$], water vapor [$H_2O(g)$]
  Liquids water [$H_2O(l)$], methanol [$CH_3OH(l)$], benzene [$C_6H_6(l)$]
  Solids carbon [C(graphite)], carbon [C(diamond)], ice [$H_2O(s)$]
  Solutions gasoline, seawater, carbonated water
  Solid solution solder composed of Sn and Pb

- **Phase transition** → Conversion of one phase into another
  Melting of ice [$H_2O(s) \rightarrow H_2O(l)$]
  Sublimation of dry ice [$CO_2(s) \rightarrow CO_2(g)$]

- **Physical equilibrium** → State in which two or more phases of a substance coexist without a tendency to change
  Partial melting of ice [$H_2O(s) \rightleftharpoons H_2O(l)$]

- **Goals** → Developing insight into phase equilibria, phase transitions, and the properties and behavior of solutions

8.1 Vapor Pressure

- **Liquids** → Evaporate to form a gas or vapor
  Puddle of rainwater evaporates. The reaction $H_2O(l) \rightarrow H_2O(g)$ goes to completion and the puddle disappears. Phase equilibrium is *not* attained.

- **Gases** → Condense to form a liquid or solid at sufficiently low temperature
  Water vapor condenses to form rain. Steam condenses on cold surfaces. In a closed system, both liquid and gas phases exist in *equilibrium*.
  $H_2O(l) \rightleftharpoons H_2O(g)$

- **Liquid-gas phase equilibrium** → *Dynamic equilibrium*
  Rate of evaporation equals rate of condensation

  Example: In a *covered* jar half-filled with water at room temperature, the liquid and gas phases are in equilibrium, $H_2O(l) \rightleftharpoons H_2O(g)$. Water vapor exhibits a characteristic equilibrium pressure, its *vapor pressure.*

- **Vapor pressure, *P***
  → Characteristic pressure of a vapor above a *confined* liquid or solid when they are in dynamic equilibrium (*closed system*)

  Examples: Several solids with measurable vapor pressures are camphor, naphthalene, paradichlorobenzene (mothballs), and dry ice [$CO_2(s)$].

  → Depends on temperature, increasing rapidly with increasing temperature
  → Different vapor pressure for different liquids and solids
- **Sublimation** → Evaporation of solid to form a gas [$CO_2(s) \rightarrow CO_2(g)$]
- **Volatility**
  → Related to the ability to evaporate
  → Liquids with high vapor pressures are *volatile* substances.

Example 8.1a Use Table 8.2 to explain why ethanol in an open beaker at 25°C evaporates faster than water in an open beaker at 25°C.

Solution According to Table 8.2, the vapor pressure of ethanol is greater than that of water at 25°C. Thus, ethanol is more volatile than water at 25°C and evaporates more quickly.

Example 8.1b Use Table 8.3 to explain why a beaker of water evaporates more quickly at 60°C than at 25°C.

Solution The vapor pressure of $H_2O(l)$ is greater at 60°C than at 25°C. Water is more volatile at the higher temperature, and therefore evaporates more quickly.

- **Characteristics of equilibrium between phases**
  → *Dynamic equilibrium* occurs when molecules enter and leave the individual phases at the same rate, and the total amount in each phase remains unchanged.
  → Molar free energies of the individual phases are equal.
  substance (phase 1) $\rightleftharpoons$ substance (phase 2), $\Delta G_m = 0$.

8.2 Volatility and Molecular Properties

- ***Increasing* strength of intermolecular forces**
  → *Decrease* in volatility
  → *Decrease* in vapor pressure at a given temperature
  → *Increase* in the normal boiling point, T_b (defined in Section 8.4)

- **London forces**
  - → More massive molecules are less volatile.
  - → More electrons in a molecule produce stronger forces.
- **Dipolar forces**
  - → For molecules with the same number of electrons, dipolar molecules are associated with less volatile substances than are molecules with only London forces.
  - → A *greater* dipole moment yields a *lower* volatility.
- **Hydrogen bonding** → Molecules that form H-bonds may produce even less volatile substances than dipolar molecules.
- **Ionic forces** → Salts are essentially nonvolatile.

Note: As with many rules in chemistry, the above qualitative guidelines must be applied cautiously.

Example 8.2a Arrange the following liquids in order of *increasing* volatility at room temperature: pentane [C_5H_{12}], hexane [C_6H_{14}], water [H_2O], chloroacetic acid [ClH_2CCOOH].

Solution Draw the Lewis structures to determine whether the species are nonpolar, polar, or require hydrogen bonding. Pentane and hexane are both nonpolar, whereas both chloroacetic acid and water form hydrogen bonds (neat chloroacetic acid exists primarily as the hydrogen-bonded dimer). The abbreviated Lewis structure for the chloroacetic acid dimer is similar to that of acetic acid.

```
            :O:···H—O:
            //        \
ClH2C—C              C—CH2Cl
            \         //
            :O—H···:O:
```

Volatility increases with decreasing strength of intermolecular forces. Volatility of H-bonded substances $<$ dipolar species $<$ substances with London forces only. So, we expect [$ClCH_2COOH$, H_2O] $<$ [C_5H_{12}, C_6H_{14}]. Both H_2O and $ClCH_2COOH$ form H-bonds of the type [O–H···:O], but $ClCH_2COOH$ has a higher molar mass, and is therefore expected to be less volatile. Of the remaining hydrocarbons, C_6H_{14} has the higher molar mass and should be less volatile. So, in order of increasing volatility, we expect $ClCH_2COOH < H_2O < C_6H_{14} < C_5H_{12}$.

Some degree of educated guessing was used to obtain this order. For example, in comparing chloroacetic acid and water, the increased molar mass of chloroacetic acid is assumed to tilt the balance between acetic acid and water, because both substances have H-bonding capability. This insight leads to the correct answer.

Example 8.2b The *shapes* of the planar molecules of *ortho*- and *para*-dichlorobenzene are shown in Margin Figures 3 and 4 in Section 5.3. Which substance is expected to be more volatile?

Solution The *p*-dichlorobenzene molecule is symmetrical and has no dipole moment. It is a nonpolar molecule in contrast to polar *o*-dichlorobenzene. The molar masses of these isomers are the same, so the nonpolar molecule *p*-dichlorobenzene is expected to be more volatile.

- **Quantitative approach to vapor pressure** (see derivation in text)

  → $\ln P = \dfrac{-\Delta G_{\text{vap}}°}{RT}$ *and* $\Delta G_{\text{vap}}° = \Delta H_{\text{vap}}° - T\Delta S_{\text{vap}}°$ *lead to*

$$\boxed{\ln P = \frac{-\Delta H_{\text{vap}}°}{RT} + \frac{\Delta S_{\text{vap}}°}{R}}$$

- **Insight from the above equation**
  - → Because $\Delta S_{\text{vap}}°$ is about the same for *all* liquids (Trouton's rule), the vapor pressure of a liquid depends mainly on $\Delta H_{\text{vap}}°$, which is always a positive quantity.
  - → *Stronger* intermolecular forces result in a *larger* $\Delta H_{\text{vap}}°$ and a *decrease* in vapor pressure.
  - → For a given liquid, an *increase* in temperature results in an *increase* in vapor pressure.

8.3 The Variation of Vapor Pressure with Temperature

- **Solve the vapor pressure equation for two temperatures** (P_2 at T_2 and P_1 at T_1)

  → $\boxed{\ln \dfrac{P_2}{P_1} = \dfrac{\Delta H_{\text{vap}}°}{R}\left(\dfrac{1}{T_1} - \dfrac{1}{T_2}\right)}$ Clausius-Clapeyron equation

- **Uses of the Clausius-Clapeyron equation**
  - → Measure the vapor pressure of a liquid at two temperatures (P_2 at T_2 and P_1 at T_1) to *estimate* the standard enthalpy of vaporization $\Delta H_{\text{vap}}°$.
  - → Measure $\Delta H_{\text{vap}}°$ and the vapor pressure of a liquid at one temperature (P_1 at T_1) to *estimate* the vapor pressure at a different temperature (P_2 at T_2).
  - → Measure $\Delta H_{\text{vap}}°$ and the vapor pressure at one temperature (P_1 at T_1). *Estimate* the normal boiling point of the liquid ($T_b = T_2$, when $P_2 = 1$ atm) or the standard boiling point of the liquid (T_2, when $P_2 = 1$ bar).
  - → Determine $\Delta H_{\text{vap}}°$ and the vapor pressure at the normal boiling temperature (P_1 at T_b). *Estimate* the vapor pressure at a different temperature (P_2 at T_2).

  Note: In all cases, the term *estimate* is used because the assumption that $\Delta H_{\text{vap}}°$ is independent of temperature is only an approximation (see Example 6.22a).

Example 8.3a The vapor pressure of octane [$C_8H_{18}(l)$] was found to be 10.0 mmHg at 19.2°C and 100 mmHg at 65.7°C. Estimate $\Delta H_{\text{vap}}°$ for octane.

Solution Rearrange the terms in the Clausius-Clapeyron equation to solve for $\Delta H_{\text{vap}}°$. Be sure to convert temperature to kelvin.

Answer

$$\Delta H_{\text{vap}}° = \left(\frac{R}{\left(T_1^{-1} - T_2^{-1}\right)}\right)\ln\frac{P_2}{P_1} = \left(\frac{8.314\,51\ \text{J·K}^{-1}\text{·mol}^{-1}}{\left((292.35\ \text{K})^{-1} - (338.85\ \text{K})^{-1}\right)}\right)\ln\frac{100\ \text{mmHg}}{10.0\ \text{mmHg}}$$

$= 40\,786\ \text{J·mol}^{-1} = 40.8\ \text{kJ·mol}^{-1}$

Note: The units of pressure in the logarithmic term cancel.

Example 8.3b Use the results of the previous example to estimate the vapor pressure of octane in units of mmHg and kPa at 30°C.

Answer $P = 18.2$ mmHg = 2.43 kPa

8.4 Boiling

- **Liquid in an open container** → Rate of vaporization is greater than the rate of condensation, so the liquid evaporates.
- **Boiling** → Vapor pressure of liquid equals atmospheric pressure. Rapid vaporization occurs throughout the bulk of liquid.
- **Boiling point** → Temperature at which the liquid begins to boil
- **Normal boiling point** → Boiling point at 1 atm, T_b
  Within experimental error, $T_b = 99.974°C \approx 100°C$ for water.
- **Standard boiling point** → Boiling point at 1 bar (see Example 8.4c)

Example 8.4a Arrange the following liquids in order of *increasing* boiling point at 1 atm: pentane [C_5H_{12}], hexane [C_6H_{14}], water [H_2O], and chloroacetic acid [ClH_2CCOOH].

Solution See the results of Example 8.2a. As the strength of the intermolecular force *increases*, the volatility *decreases* and the normal boiling point *increases*. The answer is then

	increasing normal boiling point T_b						
	C_5H_{12}	<	C_6H_{14}	<	H_2O	<	ClH_2CCOOH
T_b (°C)	36	<	69	<	100	<	188

As an example of the care that must be taken in making these comparisons, note that the normal boiling point of nonane [C_9H_{20}], a liquid with only London forces, is 151°C, well above that of water with its strong H-bonding.

- **Effect of pressure on the boiling point**
  → An *increase* of pressure on a liquid leads to an *increase* in the boiling point. The pressure cooker makes use of this fact.
  → A *decrease* of pressure on a liquid leads to a *decrease* in the boiling point. This property explains why water boils at a lower temperature on a mountaintop. It also allows for vacuum distillation at low temperatures to purify liquids that decompose at higher temperatures.

Example 8.4b At the top of a mountain, water is found to boil at 95.0°C. Estimate the pressure of the atmosphere on the mountaintop. The enthalpy of vaporization of water at its normal boiling point (100°C) is 40.656 kJ·mol^{-1}.

Solution Convert the two temperatures to kelvin. Use the Clausius-Clapeyron equation with $P_1 = 1$ atm, $T_1 = 373.15$ K, and $T_2 = 368.15$ K.

$$\ln\frac{P_2}{P_1} = \frac{\Delta H_{vap}^{\circ}}{R}\left(\frac{1}{T_1} - \frac{1}{T_2}\right) = \frac{40\,656\ \text{J·mol}^{-1}}{8.314\,51\ \text{J·K}^{-1}\text{·mol}^{-1}}\left(\frac{1}{373.15\ \text{K}} - \frac{1}{368.15\ \text{K}}\right)$$

$$\ln\frac{P_2}{(1\ \text{atm})} = -0.178 \quad \text{and} \quad P_2 = 0.84\ \text{atm}$$

The pressure on the mountaintop is about 16% lower than at sea level.

Note: The enthalpy of vaporization of water at 100°C (40.656 kJ·mol^{-1}) is about 8% lower than the value at 25°C (44.0 kJ·mol^{-1}). A relatively small error is introduced by ignoring the temperature variation of the enthalpy of vaporization. Smaller changes in temperature lead to smaller errors in predicting P_2.

Example 8.4c Estimate the *standard* boiling point of water at 1 bar. Assume the enthalpy of vaporization is a constant value of 40.656 kJ·mol^{-1}.

Solution Convert pressure in bar into the value in atm. Use the *normal* boiling point of water as the reference: P_1 = 1 atm and T_1 = 373.15 K. Rearrange the Clausius-Clapeyron equation and solve for T_2.

P_2 = (1 bar)/(1.013 25 bar·atm^{-1}) = 0.986 92 atm

$$T_2 = \frac{1}{\left(\frac{1}{T_1} - \frac{R}{\Delta H_{vap}^{\circ}}\ln\frac{P_2}{P_1}\right)} = \frac{1}{\left(\frac{1}{373.15\ \text{K}} - \frac{(8.314\,51\ \text{J·K}^{-1}\text{·mol}^{-1})}{(40\,656\ \text{J·mol}^{-1})}\ln(0.986\,92)\right)}$$

$$= \frac{1}{(2.679\,9 + 0.002\,69)\times 10^{-3}\ \text{K}^{-1}} = 372.77\ \text{K} \quad \text{or} \quad 99.6°\text{C}$$

Note: The standard boiling point of water is only about 0.4 K lower than the normal boiling point.

8.5 Freezing and Melting

- **Freezing and melting** → Two common phase transitions, one the reverse of the other
- **Freezing (melting)** → Solidification of a liquid (liquefaction of a solid)
- **Freezing (melting) point** → Temperature at which liquid freezes (solid melts)
  Freezing and melting points are identical.
- **Normal freezing point, T_f** → Temperature at which solid begins to freeze at 1 atm
  Within experimental error, T_f = 0°C for water.
- **Standard freezing point** → Temperature at which solid begins to freeze at 1 bar
  Difference between T_f and the standard freezing point is insignificant (see Example 8.4c).

- **Supercooled liquid**
  → A very pure liquid may exist below its freezing point if it is cooled extremely slowly.
  → Thermodynamically unstable with respect to formation of solid
  → If heat is withdrawn slowly from very pure water, the temperature may drop below 0°C, indicating supercooling. At some point, a tiny crystal (nucleus) of ice forms, and the entire sample crystallizes rapidly to form ice whose temperature rises to 0°C from the heat released by sudden freezing.

- **Pressure dependence of the freezing point:**

  Most substances → Solid phase is more dense than the liquid phase (smaller molar volume), so the solid *sinks* as it forms. Increasing pressure favors the phase with the smallest density; thus, the freezing point *increases* with *increasing* pressure.

  Few substances → Solid phase is less dense than the liquid phase (larger molar volume) and the solid *floats* as it forms. An increase in pressure favors the phase with the greatest density; thus, the freezing point *decreases* with *increasing* pressure. Examples include water and bismuth.

  Note: The *anomalous* behavior of water is critical for the survival of certain species. If ice sank, bodies of water would freeze solid in winter, killing aquatic life.

8.6 Phase Diagrams

- **Component** → A single substance, for example, aluminum, octane, water, or oxygen
- **Single-component phase diagram** → Map showing the most stable phase of a substance at different pressures and temperatures
- **Phase boundary**
  → Lines separating regions on a phase diagram
  → Represents a set of P and T values for which *two* phases coexist in dynamic equilibrium
- **Triple point**
  → Point where *three* phase boundaries intersect
  → Corresponds to a single value of P and T for which *three* phases coexist in dynamic equilibrium
- **Critical point** → Terminus of the liquid-vapor phase boundary at high temperature
- **Critical temperature, T_c** → The temperature above which a gas cannot condense into a liquid. Only one phase is observed above T_c.

  Note: Critical point and critical temperature are introduced in Section 8.8 in the text. The definitions given above are repeated in this Study Guide on p. 156.

Phase Diagram of Water

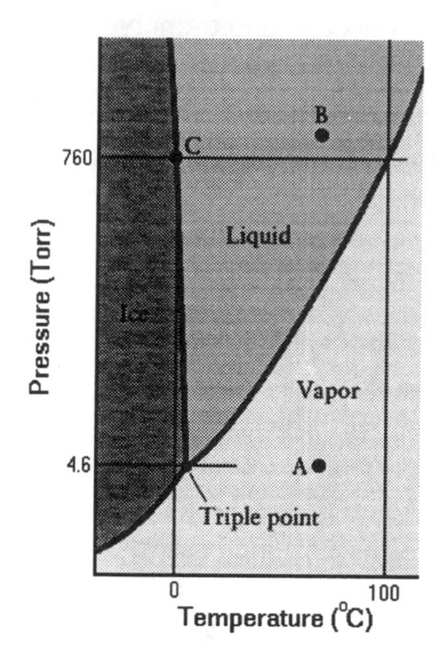

Note the regions (bounded areas) labeled ice (solid), liquid, and vapor (gas). In each area, only a single phase (ice, water or water vapor) is stable. Within each region, pressure and/or temperature may be varied independently without an accompanying phase transition. Two independent variables, or *two* degrees of freedom, characterize a region. The three lines that separate the regions define the phase boundaries of solid-liquid, liquid-vapor and solid-vapor. At a boundary, only *P* or *T* may be varied if the two phases are to remain in equilibrium. Only one independent variable, or *one* degree of freedom, exists for water in states corresponding to the boundary. All three phases coexist at the *triple point* with its specific *P* and *T* or *zero* degrees of freedom.

Note: See Section 8.7 for a definition of degrees of freedom.

Example 8.6a A sample of water is kept at 20°C and 760 Torr. Which phase is the most stable? A second sample of water is kept at 150°C and 750 Torr. Which phase is the most stable?

Solution At 20°C and 760 Torr to the right of point C, the stable phase is liquid. At 150°C and 750 Torr to the right of the liquid-vapor boundary line, the stable phase is vapor.

Example 8.6b Use the water phase diagram to describe the physical states and phase changes of water as the pressure is increased from 5 Torr to 800 Torr at 70°C, that is, for the transition from point A to point B on the diagram.

Solution Although the phase diagram is not to scale, point A is approximately 5 Torr at 70°C (vapor) and point B is near 800 Torr at 70°C (liquid). The initial state is vapor. Increasing pressure causes a transition to the liquid-vapor boundary, at which point liquid begins to form. At this pressure, liquid and vapor are in equilibrium. Following an increase in pressure, the vapor condenses. An additional increase to 800 Torr causes no change in physical state. The sample is a liquid in this region because the applied pressure is greater than the vapor pressure.

Example 8.6c Describe the properties of water at the triple point.

Solution The following three equilibria are satisfied at the triple point located on the diagram at a pressure of 4.6 Torr and a temperature of 0.01°C (273.16 K exactly).

$H_2O(s) \rightleftharpoons H_2O(l)$ $\qquad$ $H_2O(s) \rightleftharpoons H_2O(g)$ $\qquad$ $H_2O(l) \rightleftharpoons H_2O(g)$

The system has zero degrees of freedom at the triple point. Any change in *P* or *T* results in the disappearance of at least one phase.

- **Slope of the solid-liquid phase boundary line**
  → Positive for most substances:
  Solid sinks because it is more dense than liquid.
  At constant temperature, a pressure increase yields no phase change for the solid.
  At constant temperature, a pressure increase may cause the liquid to solidify.
  → Negative for H_2O:
  Because the solid is less dense than the liquid, ice floats.
  At constant temperature, a pressure increase may cause ice to melt.
  At constant temperature, a pressure increase yields no phase change for the liquid.

8.7 The Phase Rule

- **Degree of freedom f** → An intensive variable, such as P or T, that can be varied independently without changing the number of phases in equilibrium

- **One phase (α) by itself**
  → An area on a phase diagram corresponding to ice, for example
  → Molar free energy = $G_m(\alpha; P, T) = G_m(\text{ice}; P, T)$
  → Two degrees of freedom, P and T, exist; that is, P and/or T may be varied without a phase change.

- **Two phases (α, β) in equilibrium**
  → Represented by a line on a phase diagram, for example, the ice-water boundary
  → $G_m(\alpha; P, T) = G_m(\beta; P, T)$
  → One degree of freedom (P or T) exists.
  → Only P or T may be varied if equilibrium of phases is to be maintained.

- **Three phases (α, β, γ) in equilibrium**
  → Shown by a point on a phase diagram, for example, the triple point of water
  → $G_m(\alpha; P, T) = G_m(\beta; P, T) = G_m(\gamma; P, T)$
  → Zero degrees of freedom exist; only one value each of P and of T will allow three phases to remain in equilibrium.

- **Equilibrium of four or more phases is impossible**

- **Gibbs phase rule**
  → Summarizes and generalizes the above observations
  → $\boxed{f = c - p + 2}$

  f = number of degrees of freedom
  c = number of components (individual substances) in system
  p = number of phases in equilibrium

- **Increasing the pressure on graphite to make diamond and liquid carbon**
  → Consider what occurs when the pressure on graphite at 3000 K is increased from 1 bar to 500 kbar. Use the phase rule and the phase diagram of carbon given below as the basis for an explanation.

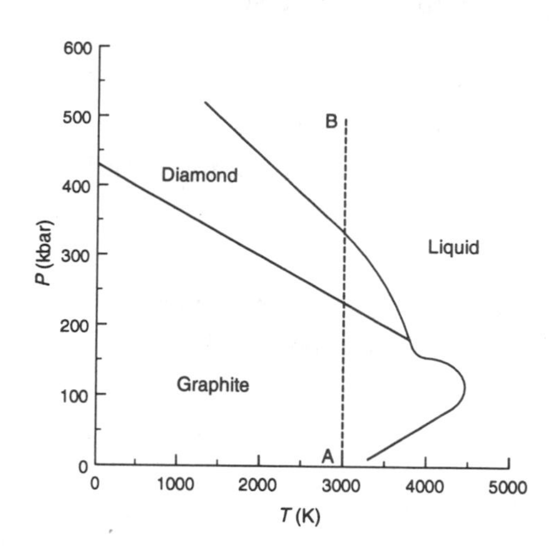

In the initial state of graphite (point A), $f = 2$. Temperature is held constant (3000 K), and the pressure is allowed to vary freely. The C(graphite)-C(diamond) phase boundary is reached at approximately 230 kbar. At this point, C(diamond) forms and exists in equilibrium with C(graphite). Here, $f = 1$. All the C(graphite) is eventually converted to C(diamond). This process may be quite slow. In the diamond region, $f = 2$ again. Additional increases in the pressure on C(diamond) result in the equilibrium of C(diamond) and liquid carbon ($f = 1$) at about 400 kbar. When all the diamond has melted, $f = 2$, and the pressure on C(l) is free to increase to attain the final state (point B).

Note: There are three allotropes of carbon: graphite, diamond and buckminsterfullerene (C_{60}); the latter, discovered in 1985, is composed of soccer-ball-shaped molecules. The thermodynamic stability of buckminsterfullerene has not yet been determined. The validity of its inclusion on the C phase diagram is, therefore, uncertain. (Metastable phases, such as supercooled water, do not appear on phase diagrams.) The crystal structure is face-centered cubic with C_{60} molecules at the corners and faces of a cubic unit cell. The unit cell is shown below.

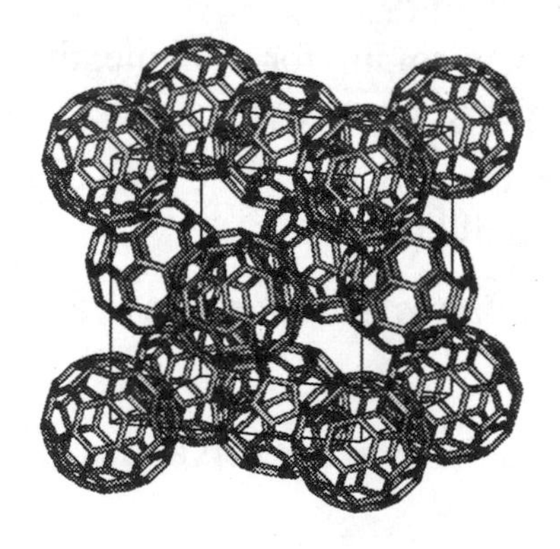

8.8 Critical Properties

- **Critical point** → Terminus of the liquid-gas phase boundary at high temperature
- **Critical temperature** → The temperature above which a vapor (gas) cannot condense into a liquid
  Only one phase is observed above T_c.
- **Supercritical fluid** → Substance above its T_c

Phase Diagram of Water

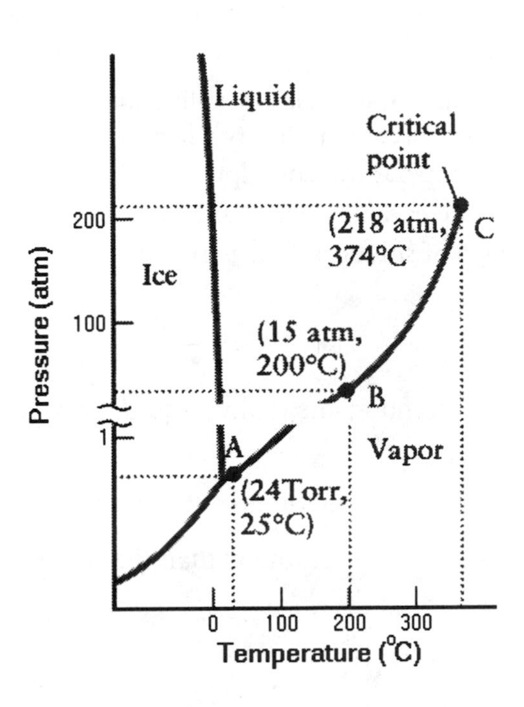

This discontinuous diagram shows both low and high pressure regions. Point A represents typical "room temperature" conditions. Liquid water has a vapor pressure of 23.76 Torr (0.031 26 atm) at 25°C (298.15 K). At 200°C (473.15 K), the vapor pressure is 15 atm (point B). As the temperature is further increased, the densities of the two phases (liquid and vapor) in equilibrium approach one another and become nearly equal at the critical point (point C). Above the critical temperature, only one phase with the properties of a very dense vapor remains. The *critical pressure*, P_c, is the vapor pressure measured at the critical temperature. The temperature, pressure, and density values for water at the critical point are $T_c = 374.1°C$ (647 K), $P_c = 218.3$ atm, and $d_c = 0.32$ g·cm^{-3}, respectively.

- **Intermolecular forces** → Both T_c and T_b show a tendency to *increase* with the *increasing* strength of intermolecular forces.

Substance	T_c (K)	T_b (K)
He (helium)	5.2	4.3
Ar (argon)	150	88
Xe (xenon)	290	166
NH_3 (ammonia)	405	240
H_2O (water)	647	373

Note: The strength of the London forces *increases* with increasing molar mass (number of electrons) of the noble gas atom. With *stronger* intermolecular forces, the more massive atoms have *higher* critical and normal boiling temperatures. Ammonia and water have strong hydrogen-bonding interactions. Water forms more hydrogen bonds per molecule than does ammonia. Therefore, water has critical and normal boiling temperatures *higher* than those of ammonia.

Solubility (Sections 8.9-8.15)

Key Concepts

types of solutions, saturated solution, molar solubility s, like-dissolves-like rule, hydrophilic, hydrophobic, micelle, cell membrane, surfactant, Henry's law, Henry's constant k_H, solubility as a function of temperature, enthalpy of solution ΔH_{sol}, limiting enthalpy of solution, lattice enthalpy ΔH_L, enthalpy of hydration ΔH_{hyd}, ion hydration enthalpy, entropy of solution, free energy of solution

Overview

- **Solution** → Homogeneous mixture of two or more components (solid, liquid, or gas) See Fundamentals section G in text.
- **Solvent** → Usually, the component of the solution present in the *larger* amount
- **Solutes** → Dissolved substances usually present in *smaller* amounts than the solvent
- **Solid solution**
  → Alloys such as solder (mixture of Sn and Pb)
  → Doped semiconductors (Si doped with P)
  See Sections 3.13 and 5.12.
- **Liquid solution** → Glucose in water, benzene in octane, sodium chloride in water
- **Gaseous solution**
  → Any mixture of gases
  → Air (N_2, O_2, Ar, water vapor, CO_2, and other gaseous constituents) from which particulate matter and aerosols have been removed

8.9 The Molecular Nature of Dissolving

- **Two component solution** → Contains one solute and one solvent species
- **Interactions** → Solute-solute, solvent-solvent, and solute-solvent

Example 8.9a Describe the intermolecular interactions that exist when solid sodium chloride NaCl (solute) dissolves in liquid water H_2O (solvent) to form a liquid solution.

Answer Sodium chloride is an ionic solid (Section 5.13) for which the coulomb interaction (solute-solute) between ions of opposite charge prevails. In liquid water, hydrogen-bond interactions (Section 5.5) account for liquid cohesion (solvent-solvent). As sodium chloride dissolves, water molecules *solvate* the Na^+ and Cl^- ions. The interactions are of the ion-dipole type (solute-solvent).

Example 8.9b Describe the enthalpy and entropy changes that occur when a solute dissolves in a solvent to form a solution.

Answer An enthalpy change (*enthalpy of solution*) is associated with the replacement of some solute-solute and solvent-solvent interactions with solute-solvent interactions.

An entropy change (*entropy of solution*) is associated with the disappearance of the highly-ordered solid phase and the appearance of a homogeneous solution phase.

- **Unsaturated solution**
  → All solute added to solvent dissolves.
  → Amount of solute dissolved in the solvent is *less* than the maximum possible amount.

- **Saturated solution**
  → Solubility limit of solute has been reached, with any additional solute present as a precipitate.
  → *Maximum* amount of solute has been dissolved in solvent.
  → Dissolved and undissolved solute molecules are in *dynamic* equilibrium with each other; in other words, the rate of dissolution equal the rate of precipitation.

- **Supersaturated solution**
  → Under certain conditions, an amount of solute greater than the solubility limit can be dissolved in the solvent.
  → Because *more* than the maximum or equilibrium amount of solute has been dissolved in the solvent, the system is thermodynamically unstable.
  A slight disturbance (seed crystal) causes precipitation and rapid return to equilibrium.

- **Solubility limit** → Depends on the nature of both the solute and the solvent molecules

Example 8.9c Write an equation that describes the dynamic equilibrium occurring in a saturated solution of sodium chloride in water prepared by adding an amount of sodium chloride to water in excess of the solubility limit.

Answer $NaCl(s) \rightleftharpoons Na^{+}(aq) + Cl^{-}(aq)$
The symbol $\rightleftharpoons$ is used to show the equilibrium of NaCl in solid and hydrated ionic forms, that is, the equality of dissolution and precipitation rates. The symbol (aq) is used to indicate that both sodium and chloride ions are *hydrated* by the solvent water (aqueous solution).

- **Molar solubility** → Molar concentration of a saturated solution of a substance
  Units: (moles of solute)/(liter of solution) or $mol{\cdot}L^{-1} \equiv M$
  For example, a saturated aqueous solution of AgCl has a molar concentration of 1.33×10^{-5} $mol{\cdot}L^{-1}$ or 1.33×10^{-5} M at 25°C.

- **Gram solubility** → Mass concentration of a saturated solution of a substance
  Units: (grams of solute)/(liter of solution) or $g{\cdot}L^{-1}$
  For example, a saturated aqueous solution of AgCl has a mass concentration of 1.91×10^{-3} $g{\cdot}L^{-1}$ at 25°C.

- **Molal solubility** → Molality of a saturated solution of a substance
  Units: (moles of solute)/(kg of solvent) or $mol{\cdot}kg^{-1}$
  For example, a saturated aqueous solution of AgCl has a molality of 1.33×10^{-5} $mol{\cdot}kg^{-1}$ at 25°C.

8.10 The Like-Dissolves-Like Rule

- **Rule** → If the solute-solute and solvent-solvent intermolecular forces (London, dipole, hydrogen-bonding, or ionic) are similar, larger solubilities are predicted. *Lesser* solubilities are expected if the forces are dissimilar.
- **Use** → A *qualitative* guide to predict and understand the solubility of various solute species in different solvents

Example 8.10 Suggest appropriate solvents for dissolving oil, glucose, and potassium iodide.

Answer Oil is composed of long chain hydrocarbons. The failure of oil and water to mix is an observation of everyday life. Oil molecules are bound together by London forces, whereas water associates primarily by hydrogen bonds. The like-dissolves-like rule suggests a solvent with only cohesive London forces is needed. Gasoline, a mixture of shorter chain hydrocarbons, is a possible solvent.

Glucose, $C_6H_{12}O_6$, is a sugar with five –OH groups capable of forming hydrogen bonds (see the example in Section 7.15 of this Study Guide). It is expected to be soluble in hydrogen-bonding solvents such as water. Conversely, it should be insoluble in nonpolar solvents such as hexane.

Potassium iodide, KI, is an ionic compound. Because both water and ammonia are highly polar molecules, they are expected to be effective in solvating both K^+ and I^- ions. Ethyl alcohol molecules, C_2H_5OH, are less polar than water, so KI is expected to be less soluble in ethanol than in water. A similar argument can be used to explain the *slight* solubility of KI in acetone, $(C_2H_5)_2CO$.

- **Hydrophilic** → Water-attracting
- **Hydrophobic** → Water-repelling
- **Soaps** → Long chain molecules with hydrophobic and hydrophilic ends that are soluble in both polar and nonpolar solvents
- **Surfactant** → Surface-active molecule with a hydrophilic head and a hydrophobic tail
- **Micelle** → Spherical aggregation of surfactant molecules with hydrophobic ends in the interior and hydrophilic ends on the surface

Note: The term hydrophobic is somewhat of a misnomer. In actuality, an *attraction* between solute and solvent molecules *always* exists. In the case of water, the solute-solvent interaction disrupts the local structure of water determined mainly by hydrogen bonding. The stronger the solute-H_2O interaction, the greater the likelihood that the local solvent structure will be disrupted by hydrated solute molecules.

8.11 Pressure and Gas Solubility: Henry's Law

- **Pressure dependence of gas solubility:**

Qualitative features

→ For a gas and liquid present in a container, an increase in pressure of gas leads to an increase in its solubility in the liquid.

→ Gas molecules strike the liquid surface and some dissolve. An increase in gas pressure leads to an increase in the number of impacts per unit time, thereby increasing solubility.

→ In a gas mixture, the solubility of each component depends on its partial pressure because molecules strike the surface independently of one another.

Quantitative features

→ Henry's law $\boxed{s = k_H P}$

→ This equation means the solubility, s, of a gas in a liquid is directly proportional to the partial pressure, P, of the gas above the liquid.

→ $s \equiv$ molar solubility (see Overview)

→ $k_H \equiv$ Henry's law constant, a function of T (see Section 8.12), and gas and solvent species
Units: s ($mol{\cdot}L^{-1}$), k_H ($mol{\cdot}L^{-1}{\cdot}atm^{-1}$), and P (atm)

Example 8.11 Calculate the molar solubility and gram solubility of O_2, N_2, Ar, and CO_2 in air dissolved in a liter of water open to the atmosphere at 20°C. Assume that air is 78.09% N_2, 20.95% O_2, 0.93% Ar, and 0.03% CO_2 by volume. The pressure of the atmosphere is 1 atm.

Solution Henry's law applies to each component separately. The partial pressure of each gas is given by Dalton's law. Values of Henry's law constant are given in Table 8.5 of the text. The gram solubility is equal to the product of the molar solubility and the molar mass of the solute.

At one atmosphere total pressure, Dalton's law gives partial pressures of 0.7809 atm N_2, 0.2095 atm O_2, 0.0093 atm Ar, and 0.0003 atm CO_2.

$s = k_H P$ for each component.

$s(N_2) = (7.0 \times 10^{-4}\ mol{\cdot}L^{-1}{\cdot}atm^{-1})(0.7809\ atm) = 5.5 \times 10^{-4}\ mol{\cdot}L^{-1}$
$s \times M = (5.5 \times 10^{-4}\ mol{\cdot}L^{-1})(28.0\ g{\cdot}mol^{-1}) = 1.5 \times 10^{-2}\ g{\cdot}L^{-1}$

$s(O_2) = (1.3 \times 10^{-3}\ mol{\cdot}L^{-1}{\cdot}atm^{-1})(0.2095\ atm) = 2.7 \times 10^{-4}\ mol{\cdot}L^{-1}$
$s \times M = (2.7 \times 10^{-4}\ mol{\cdot}L^{-1})(32.0\ g{\cdot}mol^{-1}) = 8.6 \times 10^{-3}\ g{\cdot}L^{-1}$

$s(Ar) = (1.5 \times 10^{-3}\ mol{\cdot}L^{-1}{\cdot}atm^{-1})(0.0093\ atm) = 1.4 \times 10^{-5}\ mol{\cdot}L^{-1}$
$s \times M = (1.4 \times 10^{-5}\ mol{\cdot}L^{-1})(39.95\ g{\cdot}mol^{-1}) = 5.6 \times 10^{-4}\ g{\cdot}L^{-1}$

$s(CO_2) = (2.3 \times 10^{-2}\ mol{\cdot}L^{-1}{\cdot}atm^{-1})(0.0003\ atm) = 7 \times 10^{-6}\ mol{\cdot}L^{-1}$
$s \times M = (7 \times 10^{-6}\ mol{\cdot}L^{-1})(44.0\ g{\cdot}mol^{-1}) = 3 \times 10^{-4}\ g{\cdot}L^{-1}$

8.12 Temperature and Solubility

- **Dependence of molar solubility on temperature**
  → Many solute species display simple behavior.
  → For solid and liquid solutes, solubility *usually* increases with increasing temperature.
  → For gaseous solutes, solubility *usually* decreases with increasing temperature.
  → Despite these trends, solubility behavior can be complex.
  → The hydrated form of a solid solute that dissolves is often different from the form that arises in precipitation from a saturated solution. Also, different hydrated forms may precipitate at different temperatures.

Examples: The nature of a solid solute precipitated from a saturated solution is often different from that of the solid initially dissolved. In the precipitate, the solute may be hydrated. At room temperature, anhydrous potassium hydroxide, KOH(s), readily dissolves in water. Precipitation of solid from a saturated solution of KOH produces the dihydrate salt, $KOH{\cdot}2H_2O$. Therefore, the chemical equilibrium is

$$KOH{\cdot}2H_2O(s) \rightleftharpoons K^+(aq,\ saturated) + OH^-(aq,\ saturated).$$

As a function of temperature, the solubility of LiCl between –50 to +160°C displays three discontinuities. These correspond to temperatures at which one solid hydrated form is transformed into another. The four regions between the three discontinuities correspond, from low to high temperature, to precipitation of $LiCl{\cdot}3H_2O$, $LiCl{\cdot}2H_2O$, $LiCl{\cdot}H_2O$, and LiCl, respectively. Note also that saturated aqueous solutions freeze at much lower temperatures than pure water (see Section 8.17), accounting for the low temperatures (< 0°C) of some of these saturated solutions.

8.13 The Enthalpy of Solution

- **Solutions of ionic substances** → $A_mB_n(s) \rightarrow mA^{n+}(aq) + nB^{m-}(aq)$
- **Enthalpy of solution, ΔH_{sol}**
  → Enthalpy change per mole of substance dissolved
  → Depends on *concentration* of solute
- **Limiting enthalpy of solution**
  → Refers to the formation of a very dilute solution
  Values are given in Table 8.6 of the text.
  → Use of the limiting enthalpy of solution avoids serious complications arising from the interionic interactions that occur in more concentrated solutions, because ions are far apart in very dilute solutions.
  Note that all the topics that follow refer to the limiting enthalpy condition.

- **Nature of the formation of solutions of ionic substances**
  - → Conceptualized as a two step process: sublimation of an ionic solid to form gaseous ions followed by the solvation of the gaseous ions to form an ionic solution
  - → The lattice enthalpy, ΔH_L, which is associated with the formation of gas phase ions from the ionic solid, is required in first step (see Table 6.5 in the text).
  - → The enthalpy of hydration, ΔH_{hyd}, which is associated with the formation of the ionic solution (hydrated ions) from the gas phase ions, is required in the second step (see Table 8.7 in the text).
  - → ΔH_L *always* has a positive value, whereas ΔH_{hyd} *always* has a negative one.
  - → $\Delta H_{sol} = \Delta H_L + \Delta H_{hyd}$ is, then, the difference of two large numbers.
- **Endothermic process:** $\Delta H_L > |\Delta H_{hyd}|$
- **Exothermic process:** $\Delta H_L < |\Delta H_{hyd}|$

Note: For small, highly charged ions, both ΔH_L and $|\Delta H_{hyd}|$ have large values.

Example 8.13a Use the values in Tables 6.5 and 8.7 to estimate the limiting enthalpy of solution for potassium fluoride, KF. Compare your answer to the value in Table 8.6.

Solution The values of ΔH_L and ΔH_{hyd} are 826 and –844 $kJ \cdot mol^{-1}$, respectively. The value of the limiting enthalpy of solution is –17.7 $kJ \cdot mol^{-1}$.

$$\begin{array}{rll} KF(s) \rightarrow & K^+(g) + F^-(g) & \Delta H_L = +826\ kJ\cdot mol^{-1} \\ K^+(g) + F^-(g) \rightarrow & K^+(aq) + F^-(aq) & \Delta H_{hyd} = -844\ kJ\cdot mol^{-1} \\ \hline KF(s) \rightarrow & K^+(aq) + F^-(aq) & \Delta H_{sol} = \Delta H_L + \Delta H_{hyd} = -18\ kJ\cdot mol^{-1} \end{array}$$

The two values of ΔH_{sol} are in excellent agreement.

Example 8.13b Use the values in Tables 6.5 and 8.7 to estimate the limiting enthalpy of solution for potassium chloride, KCl. Compare your answer to the value in Table 8.6.

Solution The values of ΔH_L and ΔH_{hyd} are 717 and –701 $kJ \cdot mol^{-1}$, respectively. The value of the limiting enthalpy of solution is +17.2 $kJ \cdot mol^{-1}$.

$$\begin{array}{rll} KCl(s) \rightarrow & K^+(g) + Cl^-(g) & \Delta H_L = +717\ kJ\cdot mol^{-1} \\ K^+(g) + Cl^-(g) \rightarrow & K^+(aq) + Cl^-(aq) & \Delta H_{hyd} = -701\ kJ\cdot mol^{-1} \\ \hline KCl(s) \rightarrow & K^+(aq) + Cl^-(aq) & \Delta H_{sol} = \Delta H_L + \Delta H_{hyd} = +16\ kJ\cdot mol^{-1} \end{array}$$

Agreement between the two values of ΔH_{sol} is not as good as that found in the above example, but the sign is correct.

Notes: In both examples, ΔH_{sol} is a small number arising from the difference of two large numbers. For KF, $\Delta H_{sol} < 0$ and the solution process is *exothermic*. For KCl, $\Delta H_{sol} > 0$ and the process is *endothermic*. Both salts, however, are very soluble in water (see Section 8.15). The difference between KF and KCl is the replacement of the small F^- ion with the larger Cl^-. This change reduces both the lattice enthalpy and enthalpy of hydration for KCl in comparison to KF. The two properties decrease by different amounts resulting in different signs of the enthalpy of solution. In general, predictions of enthalpies of solution by this method are inherently unreliable. A small inaccuracy in either large number may lead to a value with the wrong sign.

8.14 Individual Ion Hydration Enthalpies

- **Enthalpy of hydration of H^+** → Value of -1130 kJ·mol^{-1} obtained by a special technique for very dilute solutions
- **Values for other ions** → The enthalpy of hydration of H^+ can be used in conjunction with cation-anion values like those in Table 8.7 to calculate individual ion hydration enthalpies. (See Table 8.8 in the text.)
- **Trends in individual ion hydration enthalpies** → Small, highly charged ions have large negative values.

Example 8.14 Use the values of ion hydration enthalpies in Table 8.8 to estimate the enthalpy of hydration of calcium bromide, $CaBr_2$.

Answer The ion hydration enthalpy values are -1657 kJ·mol^{-1} for Ca^{2+} and -309 kJ·mol^{-1} for Br^-. One mole of $CaBr_2$ contains two moles of Br^- ions. The estimated hydration enthalpy for $CaBr_2$ is then
$Ca^{2+} + 2\,Br^- = -1657 + 2(-309) = -2275$ kJ·mol^{-1}.
The value for $CaCl_2$ in Table 8.7 is -2337 kJ·mol^{-1}, larger than the $CaBr_2$ value because Cl^- is smaller than Br^-.

8.15 The Free Energy of Solution

- **Solutions of ionic substances** → $A_mB_n(s) \rightarrow mA^{n+}(aq) + nB^{m-}(aq)$
- **Free energy of solution, ΔG_{sol}**
  → Free energy change per mole of substance dissolved
  → Depends on *concentration* of solute
  → $\Delta G_{sol} = \Delta H_{sol} - T\Delta S_{sol}$ at constant T.
- **Nature of solubility**
  → A substance will dissolve if $\Delta G_{sol} < 0$, and will continue to dissolve until a saturated solution, for which $\Delta G_{sol} = 0$, is created. Both ΔH_{sol} and ΔS_{sol} change with increasing concentration to make ΔG_{sol} more positive. If the solute is consumed before ΔG_{sol} reaches 0, then an unsaturated solution results with the potential to dissolve more solute. With excess solute present, ΔG_{sol} reaches 0 and a saturated solution results.
  → For endothermic enthalpies of solution, the increase in entropy of solution drives the solubility process. A substance with a *large* endothermic enthalpy of solution is *usually* insoluble.

Colligative Properties (Sections 8.16-8.18)

Key Concepts

colligative properties, vapor-pressure lowering, Raoult's law, ideal solution, real solution, nonideal solution, boiling-point elevation, boiling-point constant k_b, freezing-point depression, freezing-point constant k_f, van't Hoff i factor, cryoscopy, osmosis, semipermeable membrane, osmotic pressure, van't Hoff equation, osmometry, reverse osmosis

Overview

- **Nonvolatile solute** → Affects the physical properties of a solvent
- **Solution**
  → *Vapor-pressure lowering:* A solution has a *lower* vapor pressure than pure solvent.
  → *Boiling-point elevation:* A solution has a *higher* boiling point than pure solvent.
  → *Freezing-point depression:* A solution has a *lower* freezing point than pure solvent.
  → *Osmosis:* The tendency of a solvent to flow through a membrane into a more concentrated solution is called *osmosis*.
  → *Osmotic pressure:* A *semipermeable membrane* allows solvent molecules to pass through, whereas solute species cannot. For a solution separated by a semipermeable membrane from pure solvent, the solvent tends to flow through the membrane into the solution. The pressure needed to stop of the flow of solvent is called the *osmotic pressure*.
- **Colligative properties**
  → Depend on the relative numbers of solute and solvent molecules in solution
  → Do *not* depend on the chemical identity of the solute
  → Can be used to determine an unknown molar mass
- **Four colligative properties** → Vapor-pressure lowering, freezing-point depression, boiling-point elevation, and osmotic pressure

8.16 Vapor-Pressure Lowering

- **Qualitative features**
  → The vapor pressure of a solvent in a solution containing a *nonvolatile solute* (such as sucrose or sodium chloride in water) is lower than that of the pure solvent.
  → In general, the vapor pressure of any volatile solution component is proportional to its mole fraction in solution.

- **Quantitative features**

→ Raoult's law $\boxed{P = x_{\text{solvent}} P_{\text{pure}}}$

→ The vapor pressure P of a solvent is equal to the product of its mole fraction in solution, x_{solvent}, and its vapor pressure when pure, P_{pure}. The quantities P and P_{pure} are measured at the same temperature.

Example 8.16a The vapor pressure of water at 25°C is 23.76 mmHg. Determine the vapor pressure of a solution containing 20.00 g of glucose, $C_6H_{12}O_6$, in 400.0 g of water at 25°C.

Solution Use Raoult's law. Convert mass quantities into mole values, and obtain the mole fraction of water. The molar mass of glucose is 180.155 g·mol^{-1}, and that of water is 18.016 g·mol^{-1}.

$$n_{\text{solute}} = \frac{m_{\text{solute}}}{M_{\text{solute}}} = \frac{20.00\ \text{g}}{180.155\ \text{g·mol}^{-1}} = 0.1110\ \text{mol}$$

$$n_{\text{solvent}} = \frac{m_{\text{solvent}}}{M_{\text{solvent}}} = \frac{400.0\ \text{g}}{18.016\ \text{g·mol}^{-1}} = 22.20\ \text{mol}$$

$$x_{\text{solvent}} = \frac{n_{\text{solvent}}}{n_{\text{solute}} + n_{\text{solvent}}} = \frac{22.20\ \text{mol}}{0.1110\ \text{mol} + 22.20\ \text{mol}} = 0.9950$$

$$P = x_{\text{solvent}} P_{\text{pure}} = (0.9950)(23.76\ \text{mmHg}) = 23.64\ \text{mmHg}$$

The effect is not very large because the amount of solute is much smaller than the amount of solvent.

Example 8.16b The vapor pressure of water at 20°C is 0.023 08 atm. When 12.00 g of urea is dissolved in 500.0 g of water, the vapor pressure is observed to be 0.022 92 atm at 25°C. Determine the molar mass of urea.

Solution Use Raoult's law to obtain x_{solvent}. Convert the mass of water into mole units. The first quantity to determine is the amount of solute. Once this is obtained, the molar mass can be calculated.

$$n_{\text{solvent}} = \frac{m_{\text{solvent}}}{M_{\text{solvent}}} = \frac{500.0\ \text{g}}{18.016\ \text{g·mol}^{-1}} = 27.75\ \text{mol}$$

$$x_{\text{solvent}} = \frac{P}{P_{\text{pure}}} = \frac{0.022\,92\ \text{atm}}{0.023\,08\ \text{atm}} = 0.9931 = \frac{n_{\text{solvent}}}{n_{\text{solute}} + n_{\text{solvent}}} = \frac{27.75\ \text{mol}}{n_{\text{solute}} + 27.75\ \text{mol}}$$

$$n_{\text{solute}} = \frac{27.75\ \text{mol}}{0.9931} - 27.75\ \text{mol} = 27.94\ \text{mol} - 27.75\ \text{mol} = 0.19\ \text{mol}$$

$$M_{\text{solute}} = \frac{m_{\text{solute}}}{n_{\text{solute}}} = \frac{12.00\ \text{g}}{0.19\ \text{mol}} = 63\ \text{g·mol}^{-1}$$

The actual molecular formula and molar mass of urea are $CO(NH_2)_2$ and 60.06 g·mol^{-1}, respectively. Because vapor pressure changes are so small, vapor-pressure lowering is not normally used to measure an unknown molar mass.

- **Ideal solution**
  - → An *ideal* solution is a hypothetical solution that obeys Raoult's law exactly, at all concentrations of solute.
  - → Solute-solvent interactions are the same as solvent-solvent interactions, therefore the enthalpy of solution, ΔH_{sol}, is zero.
  - → Entropy of solution, $\Delta S_{sol} > 0$.
  - → Free energy of solution, $\Delta G_{sol} < 0$ leads to a lowering of vapor pressure of the solvent (see Section 8.2).
  - → The process of solution formation is entropy driven.
  - → An ideal solution is typically formed by a similar solute and solvent species, such as hexane and heptane.
  - → Real solutions do not obey Raoult's law at all concentrations, but they follow Raoult's law at the limit of low solute concentration.

- **Nonideal solution**
  - → Solution that does not obey Raoult's law at a certain concentration
  - → Solute-solvent interactions differ from solvent-solvent
  - → ΔH_{sol} is not equal to zero.
  - → Real solutions approximate ideal behavior at concentrations below 10^{-1} mol·kg^{-1} for *nonelectrolyte* solutions and below 10^{-2} mol·kg^{-1} for *electrolyte* solutions.
  - → Real solutions behave ideally as the solute concentration approaches zero.

8.17 Boiling-Point Elevation and Freezing-Point Depression

- **Boiling-point elevation**
  - → $\boxed{\Delta T_b = k_b \times \text{molality}}$ k_b is the boiling-point constant

    The value of k_b depends on the nature of the solvent and has units of K·kg·mol^{-1}. This equation is *not* given in the text, but values of k_b for different solvents are given in Table 8.9.
  - → This effect arises from the influence of the solute on the entropy of the solvent.
  - → The boiling-point elevation equation holds for the addition of a nonvolatile solute to a solvent forming a dilute solution that is approximately ideal.
  - → Because solute lowers the solvent vapor pressure, the solution will not boil at the normal boiling point of the pure solvent. The temperature must be increased above that value in order to increase the vapor pressure of the solution to atmospheric pressure. Boiling is, therefore, achieved at a higher temperature ($\Delta T_b > 0$).

- **Freezing-point depression**
  - → $\boxed{-\Delta T_f = k_f \times \text{molality}}$ k_f is the freezing-point constant

    The value of k_f depends on the nature of the solvent and has units of K·kg·mol^{-1}.
  - → This effect arises from the influence of the solute on the entropy of the solvent.

→ The freezing-point depression equation holds for the addition of a nonvolatile solute to a solvent forming a dilute solution that is approximately ideal.

→ Vapor-pressure lowering of the solution produces a decrease in the triple-point temperature, which is the intersection of the liquid-vapor (vapor pressure) and solid-vapor phase boundaries. The solid-liquid phase boundary that originates at the triple point is moved slightly to the left on the phase diagram in Section 8.6 of this Study Guide (the solid-vapor boundary is unchanged). The freezing temperature is thereby lowered ($\Delta T_f < 0$).

Example 8.17a Calculate the boiling-point elevation and freezing-point depression for the sucrose solution of Example 8.16a. The constants k_b and k_f for water are 0.51 and 1.86 K·kg·mol^{-1}, respectively (see Table 8.9).

Solution Calculate the molality of the solution and use the two equations given above. Note that the new boiling point is *higher* and the new freezing point is *lower*.

$$\text{molality} = \frac{n_{\text{solute}}}{m_{\text{solvent}}(\text{kg})} = \frac{0.1110 \text{ mol}}{0.4000 \text{ kg}} = 0.2775 \text{ mol·kg}^{-1}$$

$\Delta T_b = (0.51 \text{ K·kg·mol}^{-1})(0.2775 \text{ mol·kg}^{-1}) = 0.14 \text{ K}$

The new boiling point is 373.15 K + 0.14 K = 373.29 K or 100.29°C.

$-\Delta T_f = (1.86 \text{ K·kg·mol}^{-1})(0.2775 \text{ mol·kg}^{-1}) = 0.516 \text{ K}$

The new freezing point is 273.15 K – 0.516 K = 272.634 K or –0.516°C.

Note: The freezing-point effect is larger and easier to measure.

Example 8.17b Camphor, $C_{10}H_{16}O$, is a solvent with a large freezing-point constant ($k_f = 39.7$ K·kg·mol^{-1}); it was once commonly used to determine the molar masses of unknown soluble substances. The general technique is called *cryoscopy*. Estimate the molar mass of phenanthrene, $C_{14}H_{10}$, if a solution of 0.0113 g of phenanthrene dissolved in 0.0961 g of camphor has a freezing-point depression of 27.0°C. Compare your answer to the molar mass calculated from the molecular formula.

Solution The unknown molality is first obtained from the freezing-point depression relation. The change in temperature is the same in units of °C or K. The mass of solvent yields the number of moles of solute. The ratio of the mass of solute to the amount in moles yields the molar mass.

$$\text{molality} = \frac{\text{freezing-point depression}}{k_f} = \frac{27.0 \text{ K}}{39.7 \text{ K·kg·mol}^{-1}} = 0.680 \text{ mol·kg}^{-1}$$

$$\text{molality} = \frac{\text{moles of solute}}{\text{kg of solvent}} = \frac{(0.0113 \text{ g}/M)}{9.61\times10^{-5} \text{ kg}} = 0.680 \text{ mol·kg}^{-1}$$

$$M = \frac{(0.0113 \text{ g})}{(9.61\times10^{-5} \text{ kg})(0.680 \text{ mol·kg}^{-1})} = 173 \text{ g·mol}^{-1}$$

Using atomic molar masses, $M = 178.22$ g·mol^{-1}. The two values are in qualitative agreement.

- **Freezing-point depression corrected for ionization or aggregation of the solute**
  - → $- \Delta T_f = i k_f \times \text{molality}$
  - → *i*, called the van't Hoff factor, is determined experimentally. The "apparent" molality is calculated from the amount of unreacted solute.
  - → *i* is an adjustment used to treat *electrolytes* (for example, *ionic solids,* strong *acids* and *bases*), and molecular solutes that *aggregate* (for example, acetic acid dimers).
  - → For electrolytes, *i* is the number of moles of ions formed by each mole of solute dissolved in 1 kg of solvent. If ionization is incomplete, *i* is used to determine the percent ionization of the electrolyte (see Example 8.17c below).
  - → For molecular aggregates, *i* is used to determine the average size of the aggregates.

Example 8.17c The freezing-point depression of an aqueous solution of K_2SO_4 (174.26 g·mol^{-1}) prepared by dissolving 4.00 g of solute in 96.0 g of water is 0.950 K. Calculate the van't Hoff factor and determine the percent dissociation (ionization) of the solute. Recall that k_f for water is 1.86 K·kg·mol^{-1}.

Solution

$$\text{molality} = \frac{n_{\text{solute}}}{m_{\text{solvent}}\,(\text{kg})} = \frac{(4.00\text{ g})/(174.26\text{ g·mol}^{-1})}{0.0960\text{ kg}} = 0.239\text{ mol·kg}^{-1}$$

$$i = \frac{\text{freezing-point depression}}{k_f \times \text{molality}} = \frac{0.950\text{ K}}{(1.86\text{ K·kg·mol}^{-1})(0.239\text{ mol·kg}^{-1})} = 2.14$$

The actual molality is 2.14 times larger than the apparent value or 0.511 mol·kg^{-1}. Let x be equal to the number of moles of K_2SO_4 that dissociate per kg of water.

$$K_2SO_4(aq) \rightleftharpoons 2\,K^+(aq) + SO_4^{2-}(aq)$$

equilibrium: $(0.239 - x)$ $\quad 2x \quad$ x

For x moles of solute that dissociate, 3x moles of ions are formed per kg of water.
actual molality = 0.511 = $(0.239 - x) + 3x$, and $x = 0.136$ mol·kg^{-1}.

Percent ionization (dissociation) = (0.136/0.239) × 100% = 56.9%
For dilute solutions of K_2SO_4, the percent ionization approaches 100% and $i \approx 3$.

Example 8.17d The freezing-point depression of a 0.077 34 mol·kg^{-1} solution of tetrakis(isoamyl)ammonium thiocyanate, $[(CH_3)_2CH(CH_2)_2]_4N(SCN)$, in benzene ($k_f$ = 5.12 K·kg·mol^{-1}) is 0.01573 K. Calculate the van't Hoff factor and estimate the extent of aggregation of the solute.

Solution

$$i = \frac{\text{freezing-point depression}}{k_f \times \text{molality}} = \frac{0.015\,73\text{ K}}{(5.12\text{ K·kg·mol}^{-1})(0.077\,34\text{ mol·kg}^{-1})} = 0.0397$$

The actual molality is 0.0397 times smaller than the apparent value or 0.003 07 mol·kg^{-1}. Assume that only one "average" aggregate species is formed. Let x be equal to the number of moles of the aggregate formed per kg of water.

$$\text{n solute} \rightarrow (\text{solute})_n$$

$$(0.07734 - nx) \qquad x$$

For x moles of the aggregate formed, nx moles of solute react per kg of water.
If aggregation is complete, $(0.07734 - nx) = 0$ and $x = 0.0307$.
$[0.07734 - n(0.0307)] = 0$ and n = 25.2 ("average" aggregate species)

Note: If aggregation is complete, n = (1/*i*). The large aggregate species is possibly a micelle with the hydrophobic isoamyl (hydrocarbon) groups facing outward towards the hydrophobic benzene solvent molecules and the polar groups clustered in the interior.

8.18 Osmosis

- **Qualitative features**
  - → *Osmosis* is the flow of solvent molecules into a solution from which it is separated by a semipermeable membrane. It is used primarily to determine an unknown molar mass, particularly for large molecules such as polymers and proteins.
  - → *Osmometry* is the technique used to determine an unknown molar mass of a solute, if the mass concentration is measured.
  - → A *semipermeable membrane* allows only certain types of molecules to pass through. Typical membranes allow passage of water and small molecules, but not large molecules or ions.
  - → In osmosis, the free energy of the solution is lower than that of the solvent. Dilution is, therefore, a *spontaneous process*. The free energy of the solution can be increased by applying pressure on the solution. The increased pressure at equilibrium is called the *osmotic pressure*, Π.
  - → In a static apparatus open to the atmosphere, pressure is applied by the increased height of the raised column of solution (flow of solvent through the membrane into the solution). In a dynamic apparatus in a closed system, the pressure is applied by the increased force of a piston confining the solution (preventing solvent flow).
  - → If the external pressure $P < \Pi$, *osmosis* (dilution) is *spontaneous*.
    If $P = \Pi$, the system is at equilibrium (no net flow).
    If $P > \Pi$, *reverse osmosis* occurs (flow of solvent in solution to the pure solvent). Used to purify seawater.

- **Quantitative features**
  - → van't Hoff equation $\boxed{\Pi = iRT\mathcal{M}}$
  - → $\mathcal{M}$ = molarity of the solution = $\dfrac{\text{moles of solute}}{\text{liters of solution}}$

    i = van't Hoff factor, R = gas constant, T = temperature in kelvin
    Use units of Π in atm and of R in $\text{L·atm·K}^{-1}\text{·mol}^{-1}$.

Example 8.18a Estimate the *maximum* osmotic pressure in units of atmospheres expected from an aqueous solution of 0.0050 M magnesium nitrate hexahydrate, $Mg(NO_3)_2\cdot 6\,H_2O$, at 298 K.

Solution In order to use the van't Hoff equation, we need to estimate i. The six waters of hydration in the solid are incorporated into the solvent on addition of the solute. If the salt dissolves completely, the reaction is

$Mg(NO_3)_2(s) \rightarrow Mg^{2+}(aq) + 2\,NO_3^-(aq)$.

Thus, the maximum number of ions formed is $i = 3$. Strong interionic interactions would prevent the completion of the reaction and thereby reduce the value of i. Because the solution is fairly dilute, this effect is negligible.

$$\Pi = iRT\mathcal{M} = (3)(0.082\,058\ \text{L·atm·K}^{-1}\text{·mol}^{-1})(298\ \text{K})(0.0050\ \text{mol·L}^{-1}) = 0.37\ \text{atm}$$

Example 8.18b At 20°C, 0.200 g of a protein dissolved in 150.0 mL of water exerts an osmotic pressure equivalent to a column of water 1.80 cm tall. Estimate the approximate molar mass of the protein.

Solution Use the van't Hoff equation to obtain the molarity of the solution. A protein is a molecular solute, so i is likely to be 1. The osmotic pressure is calculated from the density of the liquid ($\approx$1 g·cm^{-3}), the gravitational constant, and the height of the column (see Section 4.2). The temperature must be converted into kelvin. Once the molarity is obtained, the molar mass follows as in previous examples.

$$\Pi = dhg = (1\ \text{g·cm}^{-3})(1.80\ \text{cm})(9.806\,65\ \text{m·s}^{-2})\times(10^{-3}\ \text{kg·g}^{-1})(10^{2}\ \text{cm·m}^{-1})^{2} = 177\ \text{Pa}$$

$$\Pi = \frac{177\ \text{Pa}}{101\,325\ \text{Pa·atm}^{-1}} = 1.75\times10^{-3}\ \text{atm}$$

$$T = 20 + 273.15 = 293.15\ \text{K}$$

$$\mathcal{M} = \frac{\Pi}{iRT} = \frac{1.75\times10^{-3}\ \text{atm}}{(1)(0.082\,058\ \text{L·atm·K}^{-1}\text{·mol}^{-1})(293.15\ \text{K})} = 7.27\times10^{-5}\ \text{mol·L}^{-1}$$

Assuming there is no volume change upon addition of the solute to the solvent, the volume of the solution is 150.0 mL and it contains

$(0.1500\ \text{L})(7.27\times10^{-5}\ \text{mol·L}^{-1}) = 1.09\times10^{-5}$ moles of solute

$$M = \frac{m}{n} = \frac{0.200\ \text{g}}{1.09\times10^{-5}\ \text{mol}} = 1.83\times10^{4}\ \text{g·mol}^{-1}$$

Binary Liquid Mixtures (Sections 8.19-8.21)

Key Concepts

vapor pressure of ideal binary mixture, distillation, temperature-composition diagram, distillate, fractional distillation, fractions, deviations from Raoult's law, enthalpy of mixing ΔH_{mix}, azeotrope, minimum-boiling azeotrope, maximum-boiling azeotrope

Overview

- **Binary mixture**

  According to Dalton's law, the total vapor pressure of an ideal solution of two volatile substances is equal to the sum of each component's partial pressures. The partial pressure of each component is given by Raoult's law.

- **Distillation**

  The vapor phase of a binary mixture is enriched in the more volatile component, which preferentially condenses on a cold surface. The process of condensing a vapor above a liquid mixture is called *distillation*. The vapor composition changes with position in a distillation column and the fraction at the top corresponds to many distillation-condensation enrichments. In this case, the process is known as *fractional distillation*, which is a method for purifying liquids.

- **Azeotropes**

  Positive or negative deviations from ideal behavior correspond to positive or negative *enthalpies of mixing*, respectively. For positive deviation, the lowest boiling point of possible mixtures is often below that of either component, and a mixture is said to be a *minimum-boiling azeotrope*. In this case, a mixture cannot be separated by distillation. For negative deviation, the highest boiling point of possible mixtures is often above that of either component, and a mixture is said to be a *maximum-boiling azeotrope*. In this case, one component of a mixture can be separated by distillation.

8.19 The Vapor Pressure of a Binary Mixture

- **Ideal solution with two volatile components A and B**

  → Let $P_{A,pure}$ = vapor pressure of pure A and $P_{B,pure}$ = vapor pressure of pure B.

  → Each component obeys Raoult's law in an ideal solution at all concentrations.

  → $P_A = x_{A,liquid} P_{A,pure}$, where $x_{A,liquid}$ = mole fraction of A in the liquid mixture.

  → $P_B = x_{B,liquid} P_{B,pure}$, where $x_{B,liquid}$ = mole fraction of B in the liquid mixture.

  → For a binary mixture, $x_{A,liquid} + x_{B,liquid} = 1$.

  → P_A and P_B are now the partial pressures of A and B, respectively, in the vapor above the solution. P_{total} is the total pressure above the solution. Dalton's law of partial pressures for ideal gases is $P_{total} = P_A + P_B$ (see Section 4.10).

  → Combination of Raoult's law and Dalton's law produces the expression

$$\boxed{P_{total} = P_A + P_B = x_{A,liquid} P_{A,pure} + x_{B,liquid} P_{B,pure}}$$

  → Ideal solutions with two volatile components are formed when the components are very similar (hexane/octane and benzene/toluene), and both components are completely soluble in each other. The composition may range from $0 < x < 1$; the labels "solute" and "solvent" are not usually used.

- **Composition of the vapor above a binary ideal solution**

  → Each vapor component obeys Dalton's law.

  → $P_A = x_{A,vapor} P$, where $x_{A,vapor}$ = mole fraction of A in the vapor mixture and $P = P_{total}$.

  → $P_B = x_{B,vapor} P$, where $x_{B,vapor}$ = mole fraction of B in the vapor mixture.

  → For a binary mixture, $x_{A,vapor} + x_{B,vapor} = 1$.

→ Combination of Raoult's law and Dalton's law yields an expression for $x_{A,vapor}$

$$x_{A,vapor} = \frac{P_A}{P} = \frac{P_A}{P_A + P_B} = \frac{x_{A,liquid}P_{A,pure}}{x_{A,liquid}P_{A,pure} + x_{B,liquid}P_{B,pure}}$$

→ If $P_{A,pure} \neq P_{B,pure}$, then $x_{A,vapor} \neq x_{A,liquid}$ and $x_{B,vapor} \neq x_{B,liquid}$. The vapor is always *richer* in the more volatile component.

Example 8.19a Benzene, C_6H_6, and *n*-hexane, C_6H_{14}, form nearly ideal solutions. At 300°C, the vapor pressure of pure benzene is 0.1355 atm and that of pure hexane is 0.2128 atm. Determine $x_{hexane,vapor}$ and $x_{benzene,vapor}$ for a solution containing 1.000 mol each of hexane and benzene at equilibrium with its vapor phase.

Solution The liquid composition is $x_{H,liquid} = x_{B,liquid} = 0.5000$. Because the vapor pressures of the pure liquids differ, the vapor is expected to be richer in the more volatile component, and $x_{hexane,vapor}$ should be greater than 0.5000. Use Raoult's law and Dalton's law to determine the vapor composition.

$$P_{hexane} = x_{hexane,liquid}P_{hexane,pure} = (0.5000)(0.2128 \text{ atm}) = 0.1064 \text{ atm}$$

$$P_{benzene} = x_{benzene,liquid}P_{benzene,pure} = (0.5000)(0.1355 \text{ atm}) = 0.067\,75 \text{ atm}$$

$$x_{hexane,vapor} = \frac{P_{hexane}}{P_{hexane} + P_{benzene}} = \frac{0.1064 \text{ atm}}{0.1064 \text{ atm} + 0.06775 \text{ atm}} = 0.6110$$

$$x_{benzene,vapor} = 1 - x_{hexane,vapor} = 0.3890$$

The vapor is indeed richer in the more volatile component. If the vapor were removed from the liquid and condensed, the resulting solution is enriched in hexane. Further evaporation and condensation (fractional distillation) leads to a richer and richer solution (separation) of hexane.

Example 8.19b A solution of hexane and benzene at 300°C is found to have a total vapor pressure of 0.1750 atm. Determine $x_{hexane,liquid}$ and $x_{hexane,vapor}$.

Hint: Recall that the mole fractions of the two components sum to unity in both the liquid and vapor phases.

Answer $x_{hexane,liquid} = 0.551$ and $x_{hexane,vapor} = 0.623$

8.20 Distillation

- **Temperature-composition phase diagram (ideal solution)**
  - → Plot of the temperature of an equilibrium mixture *vs.* composition (mole fraction of one component)
  - → Two curves are plotted on the same diagram, T *vs.* x_{liquid} and T *vs.* x_{vapor}.
  - → At a given temperature, the composition of each phase at equilibrium is given.
  - → See Fig. 8.40 in the text for a graphical illustration of a temperature-composition diagram for a mixture of benzene and toluene.

- **Distillation** → Purification of a liquid by evaporation and condensation
- **Distillate** → Vapor produced during distillation that is condensed and collected in the final stage
- **Fractional distillation**
  → Continuous separation (purification) of two or more liquids by repeated evaporation and condensation steps on a vertical column
  → Each step takes place on a fractionating column in small increments.
  → At the top of the column, condensed vapor (distillate) is collected in a series of samples or *fractions*; the most volatile fraction is collected first.
  → Progress may be displayed on a temperature-composition phase diagram.

8.21 Azeotropes

- **Nonideal solutions with volatile components**
  → Most mixtures of liquids are not ideal.
  → For *nonideal solutions* with *volatile* components: Raoult's law is not obeyed, the enthalpy of mixing, $\Delta H_{mix} \neq 0$, and solute-solvent interactions are different from solvent-solvent interactions.

- **Azeotrope**
  → A solution that, like a pure liquid, distills at a constant temperature without a change in composition
  → At the azeotrope temperature, $x_{liquid} = x_{vapor}$ for each component.

- **Positive deviation from Raoult's law**
  → Vapor pressure of the mixture is *greater* than the value predicted by Raoult's law.
  → Enthalpy of mixing is endothermic, $\Delta H_{mix} > 0$.
  → Relatively *weak* solute-solvent interactions
  → See Fig. 8.43a for a graphical illustration of the vapor-pressure behavior of a mixture of ethanol and benzene liquids. In the ethanol-benzene solution, the strong hydrogen bonds in ethanol are broken and replaced by weaker ethanol-benzene London interactions. Weaker interactions *increase* the vapor pressure and *decrease* the boiling temperature.

- **Minimum-boiling azeotrope**
  → Forms if the minimum boiling temperature is less than either of the pure liquids
  → Mixture cannot be separated by distillation (see Fig. 8.44 in text).

- **Negative deviation from Raoult's law**
  → Vapor pressure of the mixture is *smaller* than the one predicted by Raoult's law.
  → Enthalpy of mixing is exothermic, $\Delta H_{\text{mix}} < 0$.
  → Relatively *strong* solute-solvent interactions
  → See Fig. 8.43b for a graphical illustration of the vapor-pressure behavior of a mixture of acetone and chloroform liquids. In the acetone-chloroform solution, strong interactions occur between a lone pair of electrons on the O atom of acetone and the H atom of chloroform, which is similar to hydrogen bonding. Stronger interactions *decrease* the vapor pressure and *increase* the boiling temperature.

- **Maximum-boiling azeotrope**
  → Forms if the maximum boiling temperature is greater than either of the pure liquids
  → Mixture can be separated by distillation (see Fig. 8.45 in text).

Chapter 9 Chemical Equilibria

Reactions at Equilibrium (Sections 9.1-9.4)

Key Concepts

chemical equilibrium, dynamic equilibrium, reaction free energy, activity, activity coefficient, reaction quotient, equilibrium constant, law of mass-action, equilibrium constant expression, equilibrium constant using partial pressures *K*, equilibrium constant using molar concentrations K_c, homogeneous equilibria, heterogeneous equilibria, activities of pure liquids and solids, activities of solvent in a dilute solution

Overview

- **Chemical equilibrium**
  - → Stage in a chemical reaction when the reaction composition undergoes no further change
  - → Dynamic equilibrium with the forward and reverse reactions occurring at the same rate
  - → Application of the methods of Chapter 8 to chemical change
- **Quantitative aspects**
  - → Free energy changes with composition.
  - → Equilibrium constants and composition
  - → Dynamic equilibrium responds to changes in conditions.
- **Goals** → Developing insight into chemical equilibrium, determining the composition of a system at equilibrium, and understanding homogeneous and heterogeneous processes

9.1 The Reversibility of Reactions

- **Chemical reaction**
  - → Approaching a state of dynamic equilibrium
  - → Rates of the forward and reverse reactions become equal.

Example 9.1 Consider the reaction between gaseous hydrogen and iodine to form hydrogen iodide product (forward reaction). Then consider the reverse reaction of hydrogen iodide as a reactant that forms hydrogen and iodine products. Write chemical reactions for both directions and show a representation of the overall chemical reaction in dynamic equilibrium.

Solution Forward reaction: $H_2(g) + I_2(g) \rightarrow 2\,HI(g)$

Reverse reaction: $2\,HI(g) \rightarrow H_2(g) + I_2(g)$

Dynamic equilibrium: $H_2(g) + I_2(g) \rightleftharpoons 2\,HI(g)$ *or* $2\,HI(g) \rightleftharpoons H_2(g) + I_2(g)$

At chemical equilibrium, the reaction may be represented either way.

9.2 Thermodynamics and Chemical Equilibrium

- **ΔG_r as a function of the composition of the reaction mixture**

  → $\boxed{G_m(J) = G_m°(J) + RT \ln P_J}$ for ideal gases from Section 8.2

  → $\boxed{G_m(J) = G_m°(J) + RT \ln a_J}$ for solutes and pure substances

- ***Activity* of substance J, a_J**

  → Pure number that is unitless

  → Idealized systems

  $a_J = P_J$ for an ideal gas

  $P_J =$ partial pressure of substance J divided by the units of pressure (bar)

  $a_J = [J]$ for a solute in a dilute solution

  $[J] =$ molarity of substance J divided by the units of molarity ($mol{\cdot}L^{-1}$)

  $a_J = 1$ for a pure solid or liquid

- ***Activity coefficient*, γ_J**

  → Pure number that is unitless

  → Real systems

  $a_J = \gamma_J P_J$ for a real gas

  $a_J = \gamma_J[J]$ for any concentration of solute in a solution

  $a_J = 1$ for a pure solid or liquid

  → *Activity coefficient*, γ_J, is equal to 1 for ideal gases and solutes in dilute solutions.

- **ΔG_m for changes in pressure or concentration in idealized systems**

  → Standard state of a substance is the pure form at 1 bar

  → Standard state of a solute is the concentration of 1 $mol{\cdot}L^{-1}$

Example 9.2a Calculate the change in molar free energy if the partial pressure of oxygen in a gas mixture is decreased from $P_1 = 0.200$ bar to $P_2 = 0.100$ bar at 298 K.

Solution At P_1, $G_m(1) = G_m° + RT \ln P_1$ and at P_2, $G_m(2) = G_m° + RT \ln P_2$.

$\Delta G_m = G_m(2) - G_m(1) \equiv G_m° + RT \ln P_2 - G_m° - RT \ln P_1 = RT \ln (P_2/P_1)$

$= RT \ln (0.500) = (2.48\ kJ{\cdot}mol^{-1})(-0.693) = -1.72\ kJ{\cdot}mol^{-1}$

Note: The units of pressure are implicitly divided by 1 bar in each case.

Example 9.2b Calculate the change in molar free energy if the molar concentration of glucose in water is increased from $[\text{glucose}]_1 = 1.00 \times 10^{-3}\ mol{\cdot}L^{-1}$ to $[\text{glucose}]_2 = 3.00 \times 10^{-3}\ mol{\cdot}L^{-1}$ at 298 K.

Answer $\Delta G_m = 2.72\ kJ{\cdot}mol^{-1}$

Note: The units of molarity are implicitly divided by 1 $mol{\cdot}L^{-1}$ in each case.

- **ΔG_r for chemical reactions (see Section 7.12)**

  → Generalized chemical reaction: $aA + bB \rightarrow cC + dD$

→ $\boxed{\Delta G_r{}^\circ = \sum nG_m{}^\circ(\text{products}) - \sum nG_m{}^\circ(\text{reactants})}$

→ $\Delta G_r{}^\circ$ is the difference in molar free energies of the products and reactants in their standard states. The stoichiometric coefficients, n, are used as pure numbers.

→ Evaluated by $\boxed{\Delta G_r{}^\circ = \sum n\Delta G_f{}^\circ(\text{products}) - \sum n\Delta G_f{}^\circ(\text{reactants})}$ Appendix 2A

→ $\boxed{\Delta G_r = \sum nG_m(\text{products}) - \sum nG_m(\text{reactants})}$

→ ΔG_r is the reaction free energy at any definite, fixed composition of the reaction mixture.

→ Evaluated by $\boxed{\Delta G_r = \Delta G_r{}^\circ + RT\ln Q}$

→ Q is called the **reaction quotient**

→ Real systems: $\boxed{Q = \dfrac{a_C^c a_D^d}{a_A^a a_B^b}}$ a has two different meanings here

→ Ideal gases: $\boxed{Q = \dfrac{P_C^c P_D^d}{P_A^a P_B^b}}$ Q is unitless

→ Dilute solutions: $\boxed{Q = \dfrac{[C]^c[D]^d}{[A]^a[B]^b}}$ Q is unitless

- **Chemical Equilibrium**
  → $\Delta G_r = 0$ and $Q = K$ (equilibrium constant)
  → K has the same form as Q, but it is uniquely determined by the equilibrium composition of the reaction system.
  → If $Q < K$, reactants are in excess and reaction proceeds towards products (→).
  → If $Q > K$, products are in excess and reaction proceeds towards reactants (←).
  → If $Q = K$, reactants and products are at chemical equilibrium (⇌).

→ Real systems: $\boxed{K = \left(\dfrac{a_C^c a_D^d}{a_A^a a_B^b}\right)_{\text{equilibrium}}}$ law of mass-action, K is unitless

- **Evaluation of K** → $\boxed{\Delta G_r{}^\circ = -RT\ln K}$

Example 9.2c Calculate the equilibrium constant for the gas-phase reaction at 298.15 K:

$$H_2(g) + I_2(g) \rightleftharpoons 2HI(g)$$

Solution Use the free energies of formation in Appendix 2A to obtain the standard-state free energy change for the reaction. Then modify the above relation to obtain K.

$$\Delta G_r° = 2\,\Delta G_f°\,[\text{HI(g)}] - \Delta G_f°\,[\text{I}_2\text{(g)}] = (2 \times 1.70) - 19.33 = -15.93\ \text{kJ·mol}^{-1}$$
$$= -15\,930\ \text{J·mol}^{-1}$$

$$K = e^{-\frac{\Delta G_r°}{RT}} = e^{-\frac{(-15\,930\ \text{J·mol}^{-1})}{(8.314\,51\ \text{J·K}^{-1}\text{mol}^{-1})(298.15\ \text{K})}} = e^{6.426} = 6.18 \times 10^{2}$$

Note: In Example 9.3 in the text, $I_2(s)$ is used in place of $I_2(g)$ and the balancing coefficients differ. The values of $\Delta G_r°$ and K depend on the stoichiometric coefficients used to balance the chemical equation and the physical state of each reactant and product. In the above example, $\Delta G_r°$ refers to one mole of H_2 gas reacting with one mole of I_2 gas to form two moles of HI gas. In the text example, $\Delta G_r°$ refers to one-half mole of H_2 gas reacting with one-half mole of I_2 solid to form one mole of HI gas.

Example 9.2d Consider the reaction system in Example 9.2c at 298.15 K. Use the equilibrium constant obtained therein and assume the gases behave ideally to determine the condition of the reaction system for (a) $P_{H_2} = P_{I_2} = 10^{-6}$ bar and $P_{HI} = 10^{-3}$ bar and (b) $P_{H_2} = P_{I_2} = P_{HI} = 10^{-6}$ bar.

Solution Calculate the reaction quotient in each case, and compare Q to the value of the equilibrium constant, $K = 6.18 \times 10^2$.

(a) $Q = \dfrac{P_{HI}^{\,2}}{P_{H_2}P_{I_2}} = \dfrac{(10^{-3})^2}{(10^{-6})(10^{-6})} = \dfrac{10^{-6}}{10^{-12}} = 10^6$ and $Q > K$ (reaction: reactants ←)

(b) $Q = \dfrac{P_{HI}^{\,2}}{P_{H_2}P_{I_2}} = \dfrac{(10^{-6})^2}{(10^{-6})(10^{-6})} = \dfrac{10^{-12}}{10^{-12}} = 1$ and $Q < K$ (reaction: → products)

Note: Although the values of Q and K depend on the balanced set of stoichiometric coefficients chosen, the condition of the reaction system relating to chemical equilibrium is independent of how the equation is balanced. In the reaction quotient expression, each unit of pressure is implicitly divided by the standard pressure of 1 bar, and Q is unitless.

9.3 Equilibrium Constants

- **Gas-phase equilibria (ideal gas reaction mixture)**

→ $a\text{A(g)} + b\text{B(g)} \rightleftharpoons c\text{C(g)} + d\text{D(g)}$ (general gas phase reaction at equilibrium)

→ $$\boxed{K = \left(\frac{P_C^{\,c}P_D^{\,d}}{P_A^{\,a}P_B^{\,b}}\right)_{\text{equilibrium}}}$$ ideal gases (activity = partial pressure)

→ $$\boxed{K_c = \left(\frac{[C]^c[D]^d}{[A]^a[B]^b}\right)_{\text{equilibrium}}}$$ ideal gases (activity = molar concentration)

→ Both K and K_c are unitless.

→ In general, $P_J = \frac{n_J RT}{V} = RT\frac{n_J}{V} = RT[J]$

→ Relationship between K and K_c as derived in the text:

→ $\boxed{K = (RT)^{\Delta n} K_c}$

$\Delta n = (c + d) - (a + b)$

$= \Sigma$(coefficients of products) − Σ(coefficients of reactants)

Example 9.3a The equilibrium constant K for the gas phase reaction

$$2NOCl(g) \rightleftharpoons 2NO(g) + Cl_2(g)$$

is 1.8×10^{-2} at 227°C. Calculate the value of K_c at the same temperature.

Solution Evaluate Δn, convert °C to K, and calculate K_c. Be certain to use the correct units for the gas constant ($R = 8.314\,51 \times 10^{-2}$ L·bar·K^{-1}·mol^{-1}).

$\Delta n = (2 + 1) - (2) = 1$

$T = 227 + 273 = 500$ K

$K_c = (RT)^{-\Delta n} K = (0.0831451 \times 500)^{-1}(1.8 \times 10^{-2}) = 4.3 \times 10^{-4}$

Example 9.3b Calculate the value of K_c for the chemical equilibrium in Example 9.2c.

Answer $\Delta n = (2) - (1 + 1) = 0$ and $K_c = K = 6.18 \times 10^2$

- **Dependence of K on the direction of reaction and the balancing coefficients**

→ $a\text{A(g)} + b\text{B(g)} \rightleftharpoons c\text{C(g)} + d\text{D(g)} \quad\Rightarrow\quad K_1 = \left(\frac{a_C^c a_D^d}{a_A^a a_B^b}\right)_{\text{equilibrium}}$

→ $c\text{C(g)} + d\text{D(g)} \rightleftharpoons a\text{A(g)} + b\text{B(g)} \quad\Rightarrow\quad K_2 = \left(\frac{a_A^a a_B^b}{a_C^c a_D^d}\right)_{\text{equilibrium}} = \frac{1}{K_1} = K_1^{-1}$

→ $na\text{A(g)} + nb\text{B(g)} \rightleftharpoons nc\text{C(g)} + nd\text{D(g)} \quad\Rightarrow\quad K_3 = \left(\frac{a_C^{nc} a_D^{nd}}{a_A^{na} a_B^{nb}}\right)_{\text{equilibrium}} = K_1^{\,n}$

Example 9.3c If $K_1 = 1.8 \times 10^{-2}$ is the value of the equilibrium constant for the gas phase reaction in Example 9.3a, then calculate the value of the equilibrium constant K_2 for the reaction

$$2NO(g) + Cl_2(g) \rightleftharpoons 2NOCl(g)$$

Solution An interchange of reactants and products leads to a new equilibrium constant that is the reciprocal of the original value.

$$K_2 = \frac{1}{K_1} = \frac{1}{1.8 \times 10^{-2}} = 56 \text{ or } 5.6 \times 10^1$$

Example 9.3d If $K_1 = 1.8 \times 10^{-2}$ is the value of the equilibrium constant for the gas phase reaction in Example 9.3a, then calculate the value of the equilibrium constant K_3 for the reaction

$$\mathrm{NOCl(g)} \rightleftharpoons \mathrm{NO(g)} + \tfrac{1}{2}\mathrm{Cl_2(g)}$$

Answer $K_3 = K_1^{1/2} = \sqrt{K_1} = \sqrt{1.8 \times 10^{-2}} = 1.3 \times 10^{-1}$

- **Combining chemical reactions**
  - → Adding reaction 1 with equilibrium constant K_1 to reaction 2 with equilibrium constant K_2 yields a combined reaction with equilibrium constant K that is the *product* of K_1 and K_2.

    $(\Delta G_1° + \Delta G_2°)$ leads to $K = K_1 \times K_2$
  - → Subtracting reaction 2 with individual equilibrium constant K_2 from reaction 1 with equilibrium constant K_1 yields a combined reaction with equilibrium constant K that is the *quotient* of K_1 and K_2.

    $(\Delta G_1° - \Delta G_2°)$ leads to $K = K_1 \div K_2 = K_1 \times (K_2)^{-1}$

Example 9.3e Consider the following two reactions and equilibrium constants at 25°C:

$$\mathrm{H_2(g) + I_2(g) \rightleftharpoons 2HI(g)} \qquad K_1 = 6.18 \times 10^{2}$$

$$\mathrm{I_2(s) \rightleftharpoons I_2(g)} \qquad K_2 = 4.11 \times 10^{-4}$$

Calculate the equilibrium constant for the reaction obtained by addition of the two given reactions at the same temperature.

Solution For the addition of two reactions, the equilibrium constant for the combined reaction is the *product* of the equilibrium constants of the individual reactions.

Note: In the second heterogeneous reaction, $K_2 = P_{\mathrm{I_2}}$ because the activity of $\mathrm{I_2(s)} = 1$. The vapor pressure of solid iodine at equilibrium is then 4.11×10^{-4} bar.

$$\mathrm{H_2(g) + I_2(s) \rightleftharpoons 2HI(g)} \qquad K = K_1 \times K_2 = 0.254$$

Example 9.3f Consider the following reactions and equilibrium constants at 25°C:

$$\mathrm{H_2(g) + Br_2(l) \rightleftharpoons 2HBr(g)} \qquad K_1 = 5.35 \times 10^{18}$$

$$\mathrm{Br_2(l) \rightleftharpoons Br_2(g)} \qquad K_2 = 0.285$$

Calculate the equilibrium constant for the reaction obtained by subtracting the second reaction from the first at the same temperature.

Solution For the subtraction of reaction 2 from reaction 1, the equilibrium constant for the combined reaction is the *quotient* of the equilibrium constants of the individual reactions.

Note: In the second heterogeneous reaction, $K_2 = P_{\mathrm{Br_2}}$ because the activity of $\mathrm{Br_2(l)} = 1$. The vapor pressure of liquid bromine at equilibrium is then 0.285 bar.

$$\mathrm{H_2(g) + Br_2(g) \rightleftharpoons 2HBr(g)} \qquad K = K_1 \div K_2 = 1.88 \times 10^{19}$$

9.4 Heterogeneous Equilibria

- **Homogeneous equilibria** → Reactants and products all in the same phase
- **Heterogeneous equilibria** → Reactants and products with different phases
  Activities of pure (or nearly pure) solids and liquids are all equal to 1.

Example 9.4 Write an equilibrium constant expression for each of the following heterogeneous reactions.

(a) $CaCO_3(s) \rightleftharpoons CaO(s) + CO_2(g)$
(b) $MgCl_2(s) \rightleftharpoons Mg^{2+}(aq) + 2Cl^-(aq)$
(c) $NH_3(g) + H_2O(l) \rightleftharpoons NH_4^+(aq) + OH^-(aq)$

Solution Express the equilibrium constant in terms of the activity quotient with equilibrium values specified. Replace each activity in the expression for K with a partial pressure for each gas and a molar concentration for each solute in solution. The activity of a pure or nearly pure solid or liquid is 1. Let eq ≡ equilibrium.

Answer

(a) $$K = \left(\frac{a_{CaO(s)}a_{CO_2(g)}}{a_{CaCO_3(s)}}\right)_{eq} = \left(\frac{(1)P_{CO_2}}{(1)}\right)_{eq} = \left(P_{CO_2}\right)_{eq}$$

(b) $$K = \left(\frac{a_{Mg^{2+}(aq)}a_{Cl^-(aq)}^{\ 2}}{a_{MgCl_2(s)}}\right)_{eq} = \left(\frac{[Mg^{2+}][Cl^-]^2}{(1)}\right)_{eq} = \left([Mg^{2+}][Cl^-]^2\right)_{eq}$$

(c) $$K = \left(\frac{a_{NH_4^+(aq)}a_{OH^-(aq)}}{a_{NH_3(g)}a_{H_2O(l)}}\right)_{eq} = \left(\frac{[NH_4^+][OH^-]}{P_{NH_3}(1)}\right)_{eq} = \left(\frac{[NH_4^+][OH^-]}{P_{NH_3}}\right)_{eq}$$

Using Equilibrium Constants (Sections 9.5-9.7)

Key Concepts

extent of reaction, direction of reaction, equilibrium table, calculating equilibrium constants, calculating equilibrium composition (exact and approximate solutions)

Overview

- **Equilibrium constant**
  → Determines the composition of the reaction mixture at equilibrium
  → Predicts the amount of product (high or low) expected
  → Predicts direction of reaction at arbitrary concentrations of reactants and products
  → Rate of reaction approaching equilibrium is undetermined, however.

- **Equilibrium composition**
  → Exact solution for any adjustment to equilibrium
  → Method of approximation for very small adjustment to equilibrium

9.5 The Extent of Reaction

- **Guidelines for attaining equilibrium**
  - → If $K > 10^3$, products are favored.
  - → If $10^{-3} < K < 10^3$, neither reactants nor products are strongly favored.
  - → If $K < 10^{-3}$, reactants are favored.

Example 9.5a Suppose that the equilibrium pressures of ICl and Cl_2 in the reaction $2\,ICl(g) \rightleftharpoons Cl_2(g) + I_2(g)$ are 0.50 bar and 0.063 bar, respectively. What is the equilibrium pressure of I_2 if $K = 0.10$?

Solution According to the guidelines, the value of K indicates that neither reactants nor products are strongly favored. Rearrange the equilibrium constant expression to solve for the unknown pressure of I_2.

$$K = \left(\frac{P_{Cl_2}P_{I_2}}{P_{ICl}^2}\right)_{eq} \quad \text{and} \quad P_{I_2} = \frac{KP_{ICl}^2}{P_{Cl_2}} = \frac{(0.10)(0.50)^2}{(0.063)} = 0.40 \text{ bar}$$

Example 9.5b Suppose that the equilibrium concentrations of N_2 and NH_3 in the reaction $N_2(g) + 3H_2(s) \rightleftharpoons 2NH_3(g)$ are 0.11 M and 1.5 M, respectively. What is the equilibrium concentration of H_2 if $K_c = 4.4 \times 10^4$?

Solution The value of K_c is in the range that strongly favors products. Solving the equilibrium constant expression requires the evaluation of the cube root of the concentration of H_2. Use the y^x key on your calculator.

$$K_c = \left(\frac{[NH_3]^2}{[N_2][H_2]^3}\right)_{eq} \quad \text{and} \quad [H_2]^3 = \frac{[NH_3]^2}{K_c[N_2]}$$

$$[H_2] = \sqrt[3]{\frac{[NH_3]^2}{K_c[N_2]}} = \left(\frac{(1.5)^2}{(4.4\times10^4)(0.11)}\right)^{1/3} = \left(4.65\times10^{-4}\right)^{1/3} = 0.077 \text{ M}$$

Note: Qualitative predictions require careful analysis. In this example, the product of concentration terms in the denominator must be small with respect to the concentration term in the numerator. The constraint is $[N_2][H_2]^3 << [NH_3]^2$ at equilibrium. One of the reactants may have, however, an equilibrium concentration the same or greater than NH_3 as long as the other one is much smaller. For example, if $[NH_3] = [N_2] = 1.5$ M at equilibrium, then $[H_2] = 0.032$ M.

9.6 The Direction of Reaction

- **Guidelines for approach to equilibrium** (see Example 9.2d in the Study Guide)
  - → If $Q > K$, the concentration of products is too high or the concentration of reactants is too low. Reaction proceeds in the reverse direction, toward reactants.
  - → If $Q < K$, the concentration of reactants is too high or the concentration of products is too low. Reaction proceeds to form products in the forward direction.
  - → If $Q = K$, the mixture has its equilibrium composition and no change occurs.

9.7 Equilibrium Tables

- **Format of the equilibrium table**
  - → Reactant and product species taking part in the reaction are identified.
  - → The initial composition of the system is listed.
  - → Next, the changes needed to reach equilibrium are listed.
  - → The final equilibrium composition is entered on the last line.
  - → The composition is entered into the equilibrium constant expression and the unknown quantity (equilibrium constant or reaction species concentration) is determined. The solution may allow for an approximate method.

- **Calculating equilibrium constants from pressures or concentrations**

Example 9.7a A sample of HI(g) is placed in a flask at a pressure of 0.100 bar. After equilibrium is attained, the partial pressure of HI(g) is 0.050 bar. Evaluate K for the reaction $2\,HI(g) \rightleftharpoons H_2(g) + I_2(g)$.

Solution Set up an equilibrium table and let x be the unknown equilibrium pressure of each product, H_2 and I_2. The value of x is determined from the change in pressure of HI ($-2x$) from the initial to the equilibrium value. The final pressure of each reactant and product is entered into the equilibrium constant expression to obtain K.

	Species		
	HI	H_2	I_2
1. Initial pressure	0.100	0	0
2. Change in pressure	$-2x$	$+x$	$+x$
3. Equilibrium pressure	$0.100 - 2x$	x	x

$P_{HI} = 0.100 - 2x = 0.050$ and $x = 0.025 = P_{H_2} = P_{I_2}$

$$K = \frac{P_{H_2} P_{I_2}}{P_{HI}^2} = \frac{(0.025)^2}{(0.050)^2} = 0.25$$

Example 9.7b The equilibrium concentrations of reactants and products for the reaction $SO_2(g) + Cl_2(g) \rightleftharpoons SO_2Cl_2(g)$ are found to be $[SO_2] = 0.025$ M, $[Cl_2] =$ 0.025 M, and $[SO_2Cl_2] = 4.1 \times 10^{-4}$ M. Determine the value of K_c.

Answer

$$K_c = \frac{[SO_2Cl_2]}{[SO_2][Cl_2]} = \frac{(4.1 \times 10^{-4})}{(0.025)^2} = 0.66$$

- **Calculating the equilibrium composition**

Example 9.7c A sample of $C_2H_6(g)$ is placed in a flask at a pressure of 2.00 bar at 900 K. The equilibrium constant K for the reaction $C_2H_6(g) \rightleftharpoons C_2H_4(g) + H_2(g)$ is 0.050 at 900 K. Determine the composition of the equilibrium mixture and the percent decomposition of C_2H_6.

Solution Set up an equilibrium table and let x be the unknown equilibrium pressure of each product C_2H_4 and H_2. The equilibrium pressure of C_2H_6 is then $2.00 - x$. Solve for x by rearranging the equilibrium constant expression. The new form is a quadratic equation (see Toolbox 9.1 in the text), which is solved exactly by the quadratic formula. The equilibrium pressures can then be obtained. The percent decomposition is the loss of reactant divided by the initial pressure times 100% or $x/2.00 \times 100\%$.

	Species		
	C_2H_6	C_2H_4	H_2
1. Initial pressure	2.00	0	0
2. Change in pressure	$-x$	$+x$	$+x$
3. Equilibrium pressure	$2.00 - x$	x	x

$$K = \frac{P_{C_2H_4}P_{H_2}}{P_{C_2H_6}} = \frac{(x)^2}{(2.00 - x)} = 0.050$$

Rearranging yields a quadratic equation: $x^2 + (0.050)x - 0.10 = 0$

Recall: $ax^2 + bx + c = 0 \Rightarrow x = \dfrac{-b \pm \sqrt{b^2 - 4ac}}{2a}$ (exact solution)

Then, $a = 1$, $b = 0.050$, and $c = -0.10$.

$$x = \frac{-(0.050) \pm \sqrt{(0.050)^2 - 4(1)(-0.10)}}{2(1)} = \frac{-(0.050) \pm \sqrt{0.4025}}{2}$$

$$= \frac{-(0.050) \pm (0.6344)}{2} = \frac{-(0.050) + (0.6344)}{2} = \frac{0.5844}{2} = 0.29$$

The negative root is rejected because it is unphysical.

$P_{C_2H_4} = P_{H_2} = x = 0.29$ bar and $P_{C_2H_6} = 2.00 - x = 1.71$ bar

$$\text{Percent decomposition} = \frac{\text{change in pressure}}{\text{initial pressure}} \times 100\% = \frac{0.29}{2.00} \times 100\% = 14.5\%$$

Note: If K is less than 10^{-3} in a problem of this type, the adjustment to equilibrium is small. The pressure of reactant then changes by only a small amount. In this case, an approximate solution that avoids the quadratic equation is possible.

Example 9.7d Samples of HBr(g) and H_2(g) are placed in a flask at initial concentrations of 0.20 M and 0.15 M, respectively. The equilibrium constant K_c for the reaction $2\,HBr(g) \rightleftharpoons H_2(g) + Br_2(g)$ is 7.7×10^{-11}. Determine the composition of the equilibrium mixture, assuming constant volume and temperature.

Solution Set up an equilibrium table and let x be the unknown equilibrium concentration of Br_2. The equilibrium concentration of HBr is $0.20 - 2x$. The equilibrium concentration of H_2 is $0.15 + x$. Solve for x by rearranging the equilibrium constant expression. The rearranged form is a quadratic equation that need not be solved exactly in this case. Because the equilibrium constant is very small, an approximation method is appropriate. The approximations are $[HBr] \approx 0.20$ and $[H_2] \approx 0.15$ and the resulting equation is easy to solve. The last steps are to check the validity of the two approximations and to recalculate K_c.

	HBr	H_2	Br_2
	Species		
1. Initial concentration	0.20	0.15	0
2. Change in concentration	$-2x$	$+x$	$+x$
3. Equilibrium concentration	$0.20 - 2x$	$0.15 + x$	x

$$K_c = \frac{[H_2][Br_2]}{[HI]^2} = \frac{(0.15 + x)(x)}{(0.20 - 2x)^2} = 7.7 \times 10^{-11}$$

Approximations: $[HBr] = 0.20 - 2x \approx 0.20$ and $[H_2] = 0.15 + x \approx 0.15$

$$K_c = \frac{(0.15)(x)}{(0.20)^2} = 7.7 \times 10^{-11} \text{ and } x = \frac{(7.7 \times 10^{-11})(0.20)^2}{(0.15)} = 2.1 \times 10^{-11} = [Br_2]$$

Check the approximations and evaluate the equilibrium constant using the equilibrium concentrations. The approximation method is generally valid if the change from initial to equilibrium concentrations is less than 5%.

$$\frac{x}{0.15} \times 100\% = \frac{2.1 \times 10^{-11}}{0.15} \times 100\% = 1.4 \times 10^{-8}\% << 5\%$$

$$\frac{2x}{0.20} \times 100\% = \frac{4.2 \times 10^{-11}}{0.20} \times 100\% = 2.1 \times 10^{-8}\% << 5\%$$

$$K_c = \frac{[H_2][Br_2]}{[HI]^2} = \frac{(0.15)(2.1 \times 10^{-11})}{(0.20)^2} = 7.9 \times 10^{-11} \approx 7.7 \times 10^{-11}$$

The Response of Equilibria to Changes in Conditions (Sections 9.8-9.11)

Key Concepts

Le Chatelier's principle, qualitative predictions, quantitative predictions, compressing a reaction mixture, temperature and equilibrium, catalysts

Overview

- **Le Chatelier's principle** → Stress applied to a system in dynamic equilibrium results in an adjustment to a new equilibrium composition that minimizes the stress.
- **Quantitative approach** → Equilibrium table calculation
- **Compressing a gas mixture** → Change in pressure leads to new equilibrium composition if there is a change in the total amount of gas phase reactants and products.

- **Temperature change** → Change in the value of the equilibrium constant
- **Catalyst** → Increases rate of reaction without being consumed and without changing the equilibrium constant

9.8 Adding and Removing Reagents

- **Qualitative features**
  → Use Le Chatelier's principle to predict the qualitative adjustments to obtain a new equilibrium composition that relieves the stress.
  → Assessing the effects of the applied stress on Q relative to K is an alternative approach.

Example 9.8a Use Le Chatelier's principle and reaction quotient arguments to describe the effect on the equilibrium $2\,SO_2(g) + O_2(g) \rightleftharpoons 2\,SO_3(g)$ of (a) addition of SO_2, (b) removal of O_2, (c) addition of SO_3, and (d) removal of SO_3.

Solution Both reactant and product pressures appear in the equilibrium constant expression. Assuming no change in volume or temperature, adding reactant (or removing product) shifts the equilibrium composition to produce more product molecules, thereby relieving the stress. Adding product (or removing reactant) shifts the equilibrium composition to produce more reactant molecules, thereby relieving the stress.

(a) Adding reactant SO_2 results in $Q < K$. The system adjusts by producing more product SO_3 (consuming SO_2 and O_2) until $Q = K$.

(b) Removing reactant O_2 results in $Q > K$. The system adjusts by consuming product SO_3 (producing more SO_2 and O_2) until $Q = K$.

(c) Adding SO_3 results in $Q > K$. The system adjusts by producing more SO_2 and O_2 (consuming SO_3) until $Q = K$.

(d) Removing SO_3 results in $Q < K$. The system adjusts by consuming SO_2 and O_2 (producing more SO_3) until $Q = K$.

Note: In this example, the stress changes only the value of Q. The composition adjusts to return the value of Q to the original value of K. The equilibrium constant does not change. The equilibrium constant has, however, a very weak pressure dependence that is usually safe to ignore.

Example 9.8b Describe the effect of the addition of C(s) on the equilibrium $C(s) + O_2(g) \rightleftharpoons CO_2(g)$.

Answer Addition of a pure liquid or pure solid to an equilibrium mixture has no effect on the equilibrium. The activity of pure liquids and solids is 1.

- **Quantitative features** → Use the equilibrium table method to calculate the new composition of an equilibrium system after a stress is applied.

Example 9.8c Assume that the reaction $2HI(g) \rightleftharpoons H_2(g) + I_2(g)$ has an equilibrium composition as determined in Example 9.7a. If 0.008 bar of HI, 0.008 bar of H_2, and 0.008 bar of I_2 are added to the system, predict the direction of change and calculate the new equilibrium composition.

Solution From Example 9.7a, $P_{HI} = 0.050$ bar, $P_{H_2} = P_{I_2} = 0.025$ bar, and $K = 0.25$. After adding 0.008 bar to each reactant and product, the *nonequlibrium* pressures are $P_{HI} = 0.058$ bar, $P_{H_2} = 0.033$ bar, and $P_{I_2} = 0.033$ bar.

$Q = \dfrac{(0.033)(0.033)}{(0.058)^2} = 0.324 > K$ and the adjustment to the new equilibrium is in the direction towards the reactants. Set up an equilibrium table and let $-x$ be the unknown adjustment to the initial *nonequilibrium* pressures of each product H_2 and I_2. The value of $+2x$ is then the adjustment to the initial *nonequilibrium* pressure of HI. The equilibrium constant expression is relatively easy to solve in this case.

	Species		
	HI	H_2	I_2
1. Initial pressure	0.058	0.033	0.033
2. Change in pressure	$+2x$	$-x$	$-x$
3. Equilibrium pressure	$0.058 + 2x$	$0.033 - x$	$0.033 - x$

$$K = \frac{P_{H_2}P_{I_2}}{P_{HI}^2} = \frac{(0.033-x)^2}{(0.058+2x)^2} = 0.25 \quad \text{and} \quad \sqrt{K} = \frac{(0.033-x)}{(0.058+2x)} = \sqrt{0.25} = 0.50$$

$0.033 - x = 0.029 + x \qquad 2x = 0.004 \qquad x = 0.002$

$P_{H_2} = P_{I_2} = 0.033 - 0.002 = 0.031$ bar

$P_{HI} = 0.058 + 0.004 = 0.062$ bar

Equilibrium expression check: $K = \dfrac{(0.031)^2}{(0.062)^2} = 0.25$

9.9 Compressing a Reaction Mixture

- **Decrease in volume of a reaction mixture at constant temperature**
  - → The partial pressure of each member J of the reaction mixture increases; recall $P_J = (n_J/V)RT$ (see Chapter 4).
  - → To minimize the increase in pressure, the equilibrium shifts to form fewer gas molecules, if possible.

Example 9.9a Use Le Chatelier's principle and reaction quotient arguments to describe the effect on the equilibrium $COCl_2(g) \rightleftharpoons CO(g) + Cl_2(g)$ of decreasing the volume of the reaction mixture.

Solution A decrease in volume at constant temperature leads to an increase in pressure. The system adjusts by decreasing the total number of gas molecules, if possible. A decrease of gas molecules occurs if the shift is toward the reactant side.

$$K = \frac{P_{CO}P_{Cl_2}}{P_{COCl_2}} = \frac{(n_{CO}/V)(n_{Cl_2}/V)}{(n_{COCl_2}/V)}(RT) = \frac{n_{CO}n_{Cl_2}}{n_{COCl_2}}\left(\frac{RT}{V}\right) \quad \text{and } Q > K$$

A decrease in volume leads to a decrease in gas molecules. The amount of product molecules decreases while the amount of reactant molecules increases. The net result is a decrease in the total number of gas phase molecules.

Example 9.9b Use Le Chatelier's principle and reaction quotient arguments to describe the effect on the equilibrium $H_2(g) + CO_2(g) \rightleftharpoons H_2O(g) + CO(g)$ of decreasing the volume of the reaction mixture.

Solution A decrease in volume at constant temperature leads to an increase in pressure. The system adjusts by decreasing the total number of gas molecules, if possible. In this case, any decrease in gas phase molecules on one side of the reaction leads to an equivalent increase on the other side. There is no net change in the number of gas phase molecules. The amounts of all gases do not change. The partial pressure of each reactant and product gas increases by the same factor.

$$K = \frac{P_{H_2O}P_{CO}}{P_{H_2}P_{CO_2}} = \frac{(n_{H_2O}/V)(n_{CO}/V)}{(n_{H_2}/V)(n_{CO_2}/V)} = \frac{n_{H_2O}n_{CO}}{n_{H_2}n_{CO_2}} \quad \text{and } Q = K$$

The volume terms cancel and a decrease in volume has no effect on the equilibrium amounts of reactant and product molecules.

Example 9.9c Use Le Chatelier's principle and reaction quotient arguments to describe the effect on the equilibrium $CaCO_3(s) \rightleftharpoons CaO(s) + CO_2(g)$ of decreasing the volume of the reaction mixture.

Solution A decrease in volume at constant temperature leads to an increase in pressure. The system adjusts by decreasing the total number of gas molecules. A decrease of gas molecules occurs if the shift is toward the reactant side.

$$K = P_{CO_2} = (n_{CO_2}/V)(RT) \quad \text{and } Q > K$$

A decrease in volume leads to a corresponding decrease in CO_2 gas molecules. Carbon dioxide reacts with calcium oxide to produce more calcium carbonate.

- **Increase in volume of a reaction mixture at constant temperature**
  - → The partial pressure of each member J of the reaction mixture decreases; recall $P_J = (n_J/V)RT$.
  - → To compensate for the decrease in pressure, the equilibrium shifts to form more gas molecules, if possible.

Example 9.9d Predict the changes that occur in the equilibrium composition of the systems in Examples 9.9a through 9.9c by increasing the volume of the reaction mixture.

Solution If one increases the volume of a gas mixture, the result is generally the opposite of that obtained by a decrease in volume. The exception is Example 9.9b, in which there is no change in the number of gas phase molecules in the reaction.

- **Increase in pressure of a reaction mixture by introducing an inert gas**
  - → The partial pressure of each member J of the reaction mixture does not change; recall $P_J = (n_J/V)RT$.
  - → The equilibrium composition is unaffected. The equilibrium constant has a weak pressure dependence that is safe to ignore unless the pressure change is large.

9.10 Temperature and Equilibrium

- **Qualitative features**
  - → Use Le Chatelier's principle to predict the qualitative adjustments needed to obtain a new equilibrium composition that relieves the stress.
  - → For endothermic reactions, heat is treated as a reactant. Increasing the temperature supplies heat to the reaction mixture and the composition of the mixture adjusts to form more product molecules. The value of the equilibrium constant K increases. Decreasing the temperature removes heat from the reaction mixture and the composition of the mixture adjusts to form more reactant molecules. The value of the equilibrium constant K decreases.
  - → For exothermic reactions, heat is treated as a product. Increasing the temperature supplies heat to the reaction mixture and the composition of the mixture adjusts to form more reactant molecules. The value of the equilibrium constant K decreases. Decreasing the temperature removes heat from the reaction mixture and the composition of the mixture adjusts to form more product molecules. The value of the equilibrium constant K increases.

Example 9.10a What is the effect of increasing the temperature on the equilibrium composition of the reaction $2H_2(g) + O_2(g) \rightleftharpoons 2H_2O(l)$ $\Delta H_r° = -571.66\ kJ{\cdot}mol^{-1}$

Solution For exothermic reactions, heat is treated as a product. Increasing the temperature supplies heat to the reaction mixture and the composition of the mixture adjusts to form more reactant molecules. The value of the equilibrium constant K decreases.

Note: The value of $\Delta G_r° = -474.26\ kJ{\cdot}mol^{-1}$ and $K = 1.22 \times 10^{83}$ at 25°C.

Example 9.10b What is the effect of increasing the temperature on the equilibrium composition of the reaction $NH_4Cl(s) \rightleftharpoons NH_3(g) + HCl(g)$ $\Delta H_r° = 176.01\ kJ{\cdot}mol^{-1}$?

Solution For endothermic reactions, heat is treated as a reactant. Increasing the temperature supplies heat to the reaction mixture and the composition of the mixture adjusts to form more product molecules. The value of the equilibrium constant K increases.

Note: The value of $\Delta G_r° = 91.12\ kJ{\cdot}mol^{-1}$ and $K = 1.09 \times 10^{-16}$ at 25°C.

Example 9.10c What is the effect of decreasing the temperature on the equilibrium composition of the two reactions in Examples 9.10a and 9.10b?

Solution The effect is the opposite of the trends given above for increasing temperature.

- **Quantitative features**
  - → van't Hoff equation $$\ln\frac{K_2}{K_1} = \frac{\Delta H_r°}{R}\left(\frac{1}{T_1} - \frac{1}{T_2}\right)$$ (see derivation in the text)
  - → The equilibrium constant is K, not K_c.
  - → A major assumption in the derivation is that both $\Delta H_r°$ and $\Delta S_r°$ are independent of temperature between T_1 and T_2.

Example 9.10d Use the van't Hoff equation to calculate the value of the equilibrium constant of the reaction in Example 9.10a at 500 K.

Solution For the reaction as balanced in Example 9.10a, $\Delta H_r° = -571.66\ \text{kJ·mol}^{-1}$ and $K_1 = 1.22 \times 10^{83}$ at 298.15 K. The temperature range is $T_1 = 298.15$ K to $T_2 = 500$ K.

Answer

$$\ln\frac{K_2}{K_1} = \frac{\Delta H_r°}{R}\left(\frac{1}{T_1} - \frac{1}{T_2}\right) = \frac{(-571\,660\ \text{J·mol}^{-1})}{(8.314\,51\ \text{J·K}^{-1}\text{·mol}^{-1})}\left(\frac{1}{298.15\ \text{K}} - \frac{1}{500\ \text{K}}\right)$$

$$= -93.0947$$

$K_2 = K_1\,e^{-93.0947} = (1.22 \times 10^{83})(3.711 \times 10^{-41}) = 4.53 \times 10^{42}$

As expected, the new value of the equilibrium constant is smaller (shifting toward reactants).

Example 9.10e Use the van't Hoff equation to calculate the value of the equilibrium constant of the reaction in Example 9.10b at 800 K.

Solution For the reaction as balanced in Example 9.10b, $\Delta H_r° = 176.01\ \text{kJ·mol}^{-1}$ and $K_1 = 1.09 \times 10^{-16}$ at 298.15 K. The temperature range is $T_1 = 298.15$ K to $T_2 = 800$ K.

Answer

$$\ln\frac{K_2}{K_1} = \frac{\Delta H_r°}{R}\left(\frac{1}{T_1} - \frac{1}{T_2}\right) = \frac{(176\,010\ \text{J·mol}^{-1})}{(8.314\,51\ \text{J·K}^{-1}\text{·mol}^{-1})}\left(\frac{1}{298.15\ \text{K}} - \frac{1}{800\ \text{K}}\right)$$

$$= 44.5400$$

$K_2 = K_1\,e^{44.5400} = (1.09 \times 10^{-16})(2.205 \times 10^{19}) = 2.40 \times 10^{3}$

As expected, the new value of the equilibrium constant is larger (shifting toward products). In this case, the value of the equilibrium constant changes from a small value that favors reactants at 298.15 K to a large value that favors products at 800 K.

9.11 Catalysts and Haber's Achievement

- **Catalyst increases the rate at which a reaction approaches equilibrium**
  - → Rate of forward reaction is increased.
  - → Rate of reverse reaction is increased.
  - → Dynamic equilibrium is unaffected.
  - → The identity of the catalyst is unchanged. The catalyst is *neither* a reactant *nor* a product. It does *not* appear in the chemical equation.
- **Haber's achievement**
  - → Production of ammonia: $3H_2(g) + N_2(g) \rightleftharpoons 2NH_3(g)$
    At 298.15 K, $\Delta H_r° = -92.22\ \text{kJ·mol}^{-1}$, $\Delta G_r° = -32.90\ \text{kJ·mol}^{-1}$, and $K = 5.80 \times 10^5$.
  - → Compression of the gas mixture favors product ammonia.
  - → Removal of ammonia as it is formed encourages more to be formed.
  - → Increasing the temperature is required to increase the slow rate of the reaction at 298.15 K, but it adversely affects the position of the equilibrium. An appropriate catalyst was discovered to circumvent the problem.

Chapter 10 Acids and Bases

Properties of Acids and Bases (Sections 10.1-10.4)

Key Concepts

acid, proton donor, base, proton acceptor, Brønsted-Lowry theory, deprotonation, hydronium ion, proton transfer equilibrium, protonation, weak acids and bases, conjugate base, conjugate acid, amphiprotic, autoprotolysis, autoprotolysis constant K_w, pH, pure water (pH = 7), acidic solution (pH < 7), basic solution (pH > 7), universal indicator paper, pH meter, pOH, pK_w = 14.00, pH + pOH = pK_w

Definitions of Acids and Bases

- **Arrhenius (~1884)** (see Section J in text)
  - → An acid is a *compound* that contains hydrogen and reacts with water to form hydrogen ions. For example, HCl, H_2SO_4, and CH_3COOH are Arrhenius acids.
  - → A base is a *compound* that produces hydroxide ions in water. For example, NaOH, NH_3, and $CO(NH_2)_2$ are Arrhenius bases.
  - → Specific for only one solvent, water
- **Brønsted-Lowry (1923)**
  - → Acid is a proton donor
  - → Base is a proton acceptor
  - → Conjugate acid-base pair
  - → General theory for *any* solvent (for example, H_2O and NH_3) or *no* solvent at all
- **Lewis (1923)** (see Sections 2.12 and 2.13)
  - → Acid is an electron pair acceptor
  - → Base is an electron pair donor
  - → Lewis acid-base complex
  - → Most general theory of all (includes metal cations)

10.1 Proton Transfer Equilibrium

- **Brønsted-Lowry definitions**
  - → Acid (*proton donor*) is *deprotonated* when reacting with water, H_2O.
  - → Water, acting as a base, is *protonated*, thereby forming the *hydronium ion*, H_3O^+: $HBr(aq) + H_2O(l) \rightarrow H_3O^+(aq) + Br^-(aq)$ (*strong* acid reaction).
  - → Base (*proton acceptor*) is *protonated* when reacting with H_2O.
  - → Water, acting as an acid, is *deprotonated*, thereby forming the *hydroxide ion*, OH^-: $NH_2^-(aq) + H_2O(l) \rightarrow NH_3(aq) + OH^-(aq)$ (*strong* base reaction).
  - → For *weak* acids or bases, the reaction is written as a *chemical equilibrium:* $NH_3(aq) + H_2O(l) \rightleftharpoons NH_4^+(aq) + OH^-(aq)$ (*weak* base reaction)

- **Conjugate base** → Formed from acid after the proton is donated:
  Cl^- is the conjugate base of HCl; PO_4^{3-} of HPO_4^{2-}.
- **Conjugate acid** → Formed from base after the proton is accepted:
  NH_4^+ is the conjugate acid of NH_3; H_3PO_4 of $H_2PO_4^-$.
- **Conjugate acid-base pair** → Differ only by the presence of an H^+ species:
  HCN, CN^-; H_3O^+, H_2O; H_2O, OH^-; H_2S, HS^-
- **Solvent**
  → Need not be water; other solvents are HF, H_2SO_4, CH_3COOH, C_2H_5OH, $NH_3 \equiv$ (am).
  → Or *no* solvent at all, as in the reaction $HCl(g) + NH_3(g) \rightarrow NH_4Cl(s)$.
- **Lewis definitions** → Acid is an electron pair *acceptor*, base is an electron pair *donor*. In the preceding example, H^+ (electron pair *acceptor*) in the complex HCl is a Lewis acid, and NH_3 (electron pair *donor*) is a Lewis base. HCl (proton *donor*) is a Brønsted acid, and NH_3 (proton *acceptor*) is a Brønsted base.

Example 10.1a Identify the Brønsted conjugate acid-base pairs in the equilibrium

$$H_2S(aq) + NH_3(aq) \rightleftharpoons HS^-(aq) + NH_4^+(aq)$$

Which reactant acts as a Lewis acid and which as a Lewis base?

Solution Members of a conjugate acid-base pair differ only in the presence of a proton. An acid reactant will have a conjugate base as a product. A base reactant will have a conjugate acid as a product. A Lewis acid is an electron pair acceptor; a Lewis base, an electron pair donor. Once a Lewis acid binds to a Lewis base, a complex is formed.

In this reaction, H_2S donates a proton and becomes HS^-. The species H_2S and HS^- differ only in the presence of an H^+ and constitute a conjugate acid-base pair. The proton is accepted by NH_3, which then forms NH_4^+. The species NH_4^+ and NH_3 constitute the other conjugate acid-base pair, differing only in the presence of a proton.

The species H_2S is a Lewis acid-base complex, because it is made up of the Lewis acid H^+ and the Lewis base HS^-. Therefore, the Lewis acid in this reaction is H^+ in H_2S, which accepts an electron pair from the Lewis base NH_3.

Example 10.1b Identify the Brønsted conjugate acid-base pairs in the equilibrium

$$CN^-(aq) + H_2O(l) \rightleftharpoons HCN(aq) + OH^-(aq)$$

Which reactant acts as a Lewis acid and which as a Lewis base?

Answer The conjugate acid-base pairs are HCN, CN^- and H_2O, OH^-. The Lewis acid is H^+ in the H_2O complex. The Lewis base is CN^-.

10.2 Proton Exchange Between Water Molecules

- **Amphiprotic species** → A molecule or ion that can act as either a Brønsted acid (H^+ *donor*) or a Brønsted base (H^+ *acceptor*).
  H_2O: H^+ *donor* ($\rightarrow OH^-$) or H^+ *acceptor* ($\rightarrow H_3O^+$)
  NH_3: H^+ *donor* ($\rightarrow NH_2^-$) or H^+ *acceptor* ($\rightarrow NH_4^+$)

- **Autoprotolysis** → A type of reaction in which one molecule transfers a proton to another molecule of the same kind
  Autoprotolysis of water: $2H_2O(l) \rightleftharpoons H_3O^+(aq) + OH^-(aq)$
  Autoprotolysis of ammonia: $2NH_3(l) \rightleftharpoons NH_4^+(am) + NH_2^-(am)$

- **Equilibrium constant** → Autoprotolysis of water: $2H_2O(l) \rightleftharpoons H_3O^+(aq) + OH^-(aq)$

Example 10.2a Use Eq. 10 in Chapter 9 of the text to calculate a value of the equilibrium constant for the autoprotolysis of water at 25°C. Use the values of $\Delta G_f°$ in Appendix 2A for $H_2O(l)$ and $OH^-(aq)$. The value of $\Delta G_f°[H_3O^+(aq)]$ is -237.13 kJ·mol^{-1}. Reformulate the equilibrium constant expression in terms of the dilute solution values of activities.

Solution Evaluate $\Delta G_r°$ for the autoprotolysis reaction at 25°C. Use this value in Eq. 10 of Chapter 9.

$$\Delta G_r° = \Delta G_f°[H_3O^+(aq)] + \Delta G_f°[OH^-(aq)] - 2\Delta G_f°[H_2O(l)]$$
$$= (-237.13\ \text{kJ·mol}^{-1}) + (-157.24\ \text{kJ·mol}^{-1}) - 2(-237.13\ \text{kJ·mol}^{-1})$$
$$= 79.89\ \text{kJ·mol}^{-1} = 7.989 \times 10^4\ \text{J·mol}^{-1}$$

$$\ln K = -\frac{\Delta G_r°}{RT} = -\frac{7.989\times10^4\ \text{J·mol}^{-1}}{(8.314\,51\ \text{J·K}^{-1}\text{·mol}^{-1})(298.15\ \text{K})} = -32.227$$

$$K = e^{-32.227} = 1.009\times10^{-14} = 1.0\times10^{-14} = \frac{a_{H_3O^+}\cdot a_{OH^-}}{a_{H_2O}^{\ 2}}$$

$a_{H_2O} = 1$ (nearly pure solvent) and

$a_{H_3O^+} \approx [H_3O^+]$ and $a_{OH^-} \approx [OH^-]$ (for a dilute solution)

$K = [H_3O^+][OH^-] = 1.0 \times 10^{-14} \equiv K_w$ (autoprotolysis constant of water)

- **Pure water at 25°C** → $[H_3O^+] = [OH^-] = 1.0 \times 10^{-7}$ mol·L^{-1}
- **Acidic solution** → $[H_3O^+] > 1.0 \times 10^{-7}$ mol·L^{-1} and $[OH^-] < 1.0 \times 10^{-7}$ mol·L^{-1}
- **Basic solution** → $[H_3O^+] < 1.0 \times 10^{-7}$ mol·L^{-1} and $[OH^-] > 1.0 \times 10^{-7}$ mol·L^{-1}

Example 10.2b Estimate the molarities of H_3O^+ and OH^- in 5.4×10^{-4} M $HClO_4$.

Solution Perchloric acid, $HClO_4$, is a strong acid (see Table J.1), and the reaction is

$$HClO_4(aq) + H_2O(l) \rightarrow H_3O^+(aq) + ClO_4^-(aq)$$

Because the reaction goes to completion, the concentration of H_3O^+ will equal the initial molarity of the monoprotic acid. The OH^- concentration follows from the autoprotolysis equilibrium expression. See Section 10.13 for exceptions.
$[H_3O^+] = 5.4 \times 10^{-4}$ mol·L^{-1}

$$[OH^-] = \frac{K_W}{[H_3O^+]} = \frac{1.0 \times 10^{-14}}{0.000\,54} = 1.852 \times 10^{-11}\ mol{\cdot}L^{-1} = 1.9 \times 10^{-11}\ mol{\cdot}L^{-1}$$

Example 10.2c Estimate the molarities of H_3O^+ and OH^- when 0.030 g of lithium oxide, Li_2O, dissolves in 1.00 L of water.

Solution The oxide anion is a strong base (see Table J.1) and the reaction is

$Li_2O(s) + H_2O(l) \rightarrow 2\,Li^+(aq) + 2\,OH^-(aq)$ *or* $O^{2-}(aq) + H_2O(l) \rightarrow 2\,OH^-(aq)$

Because the reaction goes to completion, the concentration of OH^- can be calculated from *twice* the number of moles of solid that dissolve in 1.00 L of water. The molar mass of Li_2O is 29.88 $g{\cdot}mol^{-1}$. The H_3O^+ concentration follows from the autoprotolysis equilibrium expression. See Section 10.13 for exceptions.

(0.030 g Li_2O)(1/29.88 $g{\cdot}mol^{-1}$) = 1.0×10^{-3} mol Li_2O $\Rightarrow$ 1.0×10^{-3} M Li_2O
1.0×10^{-3} M Li_2O $\Rightarrow$ 2.0×10^{-3} M OH^-

$$[H_3O^+] = \frac{K_W}{[OH^-]} = \frac{1.0 \times 10^{-14}}{0.0020} = 5.0 \times 10^{-12}\ mol{\cdot}L^{-1}$$

Note: A 2.0×10^{-3} M LiOH solution has the same values for $[OH^-]$ and $[H_3O^+]$.

10.3 The pH Scale

- **Concentration variation of H_3O^+** → Values higher than 1 $mol{\cdot}L^{-1}$ and less than 10^{-14} $mol{\cdot}L^{-1}$ are commonly encountered.

- **Logarithmic scale**
  → $pH = -\log [H_3O^+]$ *and* $[H_3O^+] = 10^{-pH}$
  pH values range between 1 and 14 in most cases
  → Pure water at 25°C: pH = 7.00
  → Acidic solution: $pH < 7.00$
  → Basic solution: $pH > 7.00$
  Note: pH has *no* units, but $[H_3O^+]$ has units of molarity.

- **Measurement of pH**
  → *Universal indicator paper* turns different colors at different pH values.
  → *pH meter* measures a difference in electrical potential (proportional to pH) across electrodes that dip into the solution (see Section 12.11).

Example 10.3a Calculate the pH of the solutions in Examples 10.2b and 10.2c.

Solution The appropriate relation is $pH = -\log [H_3O^+]$. Be careful to use the correct log key on your calculator (log *base* 10 ≡ $\log_{10}$).

$[H_3O^+] = 5.4 \times 10^{-4}$ $mol{\cdot}L^{-1}$ in Example 10.2b (5.4×10^{-4} M $HClO_4$)
$pH = -\log [H_3O^+] = -(-3.27) = 3.27$ (*acidic* solution)
$[H_3O^+] = 5.0 \times 10^{-12}$ $mol{\cdot}L^{-1}$ in Example 10.2c ("2.0×10^{-3} M LiOH")
$pH = -\log [H_3O^+] = -(-11.30) = 11.30$ (*basic* solution)

Example 10.3b Calculate the molar concentration of H_3O^+ for one solution that has a pH value of 1.70 and for another that has a pH of 9.80. The appropriate relation is $[H_3O^+] = 10^{-pH}$. Use the 10^x key on your calculator.

Answer $10^{-pH} = 10^{-1.70} = 2.0 \times 10^{-2}\ mol{\cdot}L^{-1}$

$10^{-pH} = 10^{-9.80} = 1.6 \times 10^{-10}\ mol{\cdot}L^{-1}$

10.4 The pOH of Solutions

- **pX as a generalization of pH**
  - → $pX = -\log X$
  - → $pOH = -\log[OH^-]$ *and* $[OH^-] = 10^{-pOH}$
  - → $pK_w = -\log K_w = -\log(1.0 \times 10^{-14}) = 14.00$
- **Relationship between pH and pOH of a solution**
  - → $[H_3O^+][OH^-] = K_w = 1.0 \times 10^{-14}$ at 25°C
  - → $\log([H_3O^+][OH^-]) = \log[H_3O^+] + \log[OH^-] = \log K_w$
  - → $-\log[H_3O^+] - \log[OH^-] = -\log K_w$
  - → $pH + pOH = pK_w = -\log(1.0 \times 10^{-14}) = 14.00$ at 25°C

Example 10.4 Calculate the pOH of the *two* solutions in Example 10.3a.

Solution The appropriate relation is $pOH = pK_w - pH = 14.00 - pH$.

For 5.4×10^{-4} M $HClO_4$, pH = 3.27 *and* pOH = 10.73.

For "2.0×10^{-3} M LiOH", pH = 11.30 *and* pOH = 2.70.

Weak Acids and Bases (Sections 10.5-10.9)

Key Concepts

weak acid, weak base, dynamic equilibrium, acidity constant K_a, proton donating strength, basicity constant K_b, proton accepting strength, conjugate seesaw, quantitative expression: $pK_a + pK_b = pK_w$, solvent leveling, trends in strengths of acids and bases: polarity and strength of chemical bonds for binary acids, comparison of oxidation state of the central atom and strength of oxoacids

Overview

- **Weak acid** → pH is *larger* than that of a strong acid with the same molarity. Incomplete deprotonation
- **Weak base** → pH is *smaller* than that of a strong base with the same molarity. Incomplete protonation
- **Dynamic equilibrium** → A process in which the rates of forward and reverse changes are the same

10.5 Acidity and Basicity Constants

- **General weak acid HA**

  → $HA(aq) + H_2O(l) \rightleftharpoons H_3O^+(aq) + A^-(aq)$

  → $K = \dfrac{a_{H_3O^+}a_{A^-}}{a_{HA}a_{H_2O}}$, but $a_{H_2O} = 1$,

  $a_{HA} \approx [HA]$, $a_{H_3O^+} \approx [H_3O^+]$, and $a_{A^-} \approx [A^-]$.

  → Thus, $K_a = \dfrac{[H_3O^+][A^-]}{[HA]}$. K_a is the *acidity constant.*

  → $pK_a = -\log K_a$ *commonly used*

  Note: See Tables 10.1, 10.7, and 10.9 in the text.

Acids and Their Constants

Species	Formula	K_a	pK_a
Neutral acids			
phosphoric acid	H_3PO_4	7.6×10^{-3}	2.12
hydrofluoric acid	HF	3.5×10^{-4}	3.45
acetic acid	CH_3COOH	1.8×10^{-5}	4.75
hydrocyanic acid	HCN	4.9×10^{-10}	9.31
Cation acids			
pyridinium	$C_5H_5^+$	5.6×10^{-6}	5.24
ammonium	NH_4^+	5.6×10^{-10}	9.25
ethylammonium	$C_2H_5NH_3^+$	1.5×10^{-11}	10.82
Anion acids			
hydrogen sulfate	HSO_4^-	1.2×10^{-2}	1.92
dihydrogen phosphate	$H_2PO_4^-$	6.2×10^{-8}	7.21

- **General weak base B**

  → $B(aq) + H_2O(l) \rightleftharpoons BH^+(aq) + OH^-(aq)$

  → $K = \dfrac{a_{BH^+}a_{OH^-}}{a_B a_{H_2O}}$, but $a_{H_2O} = 1$,

  $a_B \approx [B]$, $a_{BH^+} \approx [BH^+]$, and $a_{OH^-} \approx [OH^-]$.

  → Thus, $K_b = \dfrac{[BH^+][OH^-]}{[B]}$. K_b is the *basicity constant.*

  → $pK_b = -\log K_b$ *commonly used*

  Note: See Table 10.2 in the text for *neutral* bases.

Bases and Their Constants

Species	Formula	K_b	pK_b
Neutral bases			
ethylamine	$C_2H_5NH_2$	6.5×10^{-4}	3.19
methylamine	CH_3NH_2	3.6×10^{-4}	3.44
ammonia	NH_3	1.8×10^{-5}	4.75
hydroxylamine	NH_2OH	1.1×10^{-8}	7.97
Anion bases			
phosphate	PO_4^{3-}	4.8×10^{-2}	1.32
fluoride	F^-	2.9×10^{-11}	10.54

- **Acid-base strength**
  → The *larger* the value of K, the *stronger* the acid or base.
  → The *smaller* the value of pK, the *stronger* the acid or base.

10.6 The Conjugate Seesaw

- **Reciprocity of conjugate acid-base pairs**
  → The *stronger* the acid, the *weaker* the conjugate base.
  → The *stronger* the base, the *weaker* the conjugate acid.

- **Conjugate acid-base pairs of weak acids**
  → The conjugate base of a *weak* acid is classified as a *weak* base and the conjugate acid of a *weak* base is classified as a *weak* acid. "*Weak*" is defined by pK values between 1 and 14.
  → Consider the general *weak* acid HA and its conjugate base A^-.

$$HA(aq) + H_2O(l) \rightleftharpoons H_3O^+(aq) + A^-(aq) \qquad K_a = \frac{[H_3O^+][A^-]}{[HA]}$$

$$A^-(aq) + H_2O(l) \rightleftharpoons HA(aq) + OH^-(aq) \qquad K_b = \frac{[HA][OH^-]}{[A^-]}$$

$$K_a \times K_b = \frac{[H_3O^+][A^-]}{[HA]} \times \frac{[HA][OH^-]}{[A^-]} = [H_3O^+][OH^-] = K_w$$

It also follows that $pK_a + pK_b = pK_w$. See derivation in the text.
If $K_a > 1 \times 10^{-7}$ ($pK_a < 7$), then the acid is *stronger* than its conjugate base.
If $K_b > 1 \times 10^{-7}$ ($pK_b < 7$), then the base is *stronger* than its conjugate acid.
See Table 10.3 in the text.

Example 10.6a Sulfurous acid, H_2SO_3, is a stronger acid than nitrous acid, HNO_2. Which is the stronger base, HSO_3^- or NO_2^-? Do *not* refer to the tables in the text.

Solution Note that both acids are classified as *weak*, as are both bases. The terms *stronger* and *weaker* refer to the *relative* strengths of these *weak* acids and bases. The *stronger* acid has the *weaker* conjugate base. If you memorize the list of the *strong* acids and *strong* bases (see Table J.1 in the text), you can assume all other acids and bases are *weak*. It is very important to remember that the conjugate base of a *strong* acid is a *very* weak base, and the conjugate acid of a *strong* base is a *very* weak acid. The *very* weak acids and bases are *not* classified as weak. "Strong" is defined by $pK < 1$, "very weak" by $pK > 14$.

Because H_2SO_3 is a *stronger* acid than HNO_2, NO_2^- is a *stronger* base than HSO_3^-.

Example 10.6b The trimethylammonium ion, $(CH_3)_3NH^+$, is a stronger acid than the dimethylammonium ion, $(CH_3)_2NH_2^+$. Write the formula of the weaker conjugate base of these two acids. Do *not* refer to the tables in the text.

Answer $(CH_3)_3N$

10.7 The Role of the Solvent in Acid Strength

- **Strong acid in water**
  - → *Stronger* proton donor than H_3O^+ ($pK_a < 1$)
  - → Strong acids are *leveled* to the strength of the acid H_3O^+.
- **Strong base in water**
  - → *Stronger* proton acceptor than OH^- ($pK_b < 1$)
  - → Strong bases are *leveled* to the strength of the base OH^-.
- **Weak acid in water** → *Weaker* proton donor than H_3O^+ ($1 < pK_a < 14$)
- **Weak base in water** → *Weaker* proton acceptor than OH^- ($1 < pK_b < 14$)
- **Strong acid in ammonia** → *Stronger* proton donor than NH_4^+; for example, acetic acid, CH_3COOH, is *weak* in water, but *strong* in ammonia.

10.8 Molecular Structure and Acid Strength

- **Binary acids**
  - → For elements in the *same* period, the more *polar* the H–A bond, the *stronger* the acid. For example, HF is a stronger acid than H_2O, which is stronger than NH_3.
  - → For elements in the *same* group, the *weaker* the H–A bond, the *stronger* the acid. For example, HBr is a stronger acid than HCl, which is stronger than HF.

10.9 The Strengths of Oxoacids (also called oxyacids)

- **Oxoacids**
  → The *greater* the number of oxygen atoms attached to the *same* central atom, the *stronger* the acid. This trend also corresponds to the *increasing* oxidation number of the *same* central atom.
  For example, $HClO_3$ is a stronger acid than $HClO_2$, which is stronger than HClO.
  In HClO (H–Ö–Cl:), consider Cl the central atom, because it is classified as such in the other two acids.]
  → For the *same* number of O atoms attached to the central atom, the *greater* the electronegativity of the central atom, the *stronger* the acid.
  For example, HClO is a stronger acid than HBrO, which is stronger than HIO. The halogen is considered the central atom for comparison to other oxoacids.
- **Carboxylic acids** → The *greater* the electronegativities of the groups attached to the carboxyl (COOH) group, the *stronger* the acid.
  For example, $CHBr_2COOH$ is a stronger acid than $CH_2BrCOOH$, which is stronger than CH_3COOH.

Note: The preceding rules are summarized in Tables 10.4, 10.5, and 10.6 in the text.

The pH of Solutions of Weak Acids and Bases (Sections 10.10-10.12)

Key Concepts

solutions of weak acids or solutions of weak bases, initial concentration, percentage deprotonation (acids) or protonation (bases), consideration of autoprotolysis equilibrium, general solution of main equilibrium, 5% approximation, salt solutions, acidic cation, neutral cation, basic anion, evaluate K_a from K_b of the conjugate base, evaluate K_b from K_a of the conjugate acid

Overview

- **Dilute solution** → First solve the *main* equilibrium expression, then the *secondary* autoprotolysis equilibrium. Use the methods developed in Chapter 9.
- **Main equilibrium**
  → $HA(aq) + H_2O(l) \rightleftharpoons H_3O^+(aq) + A^-(aq)$ for a *weak* acid
  → $B(aq) + H_2O(l) \rightleftharpoons BH^+(aq) + OH^-(aq)$ for a *weak* base
  → $BH^+(aq) + H_2O(l) \rightleftharpoons H_3O^+(aq) + B(aq)$ for an *acidic* cation
  → $A^-(aq) + H_2O(l) \rightleftharpoons HA(aq) + OH^-(aq)$ for a *basic* anion

- **Method of solution** → Assume the starting concentration of each product is 0. Use the *exact* quadratic or 5% approximation method. The final solution should yield $[H_3O^+] \geq 10^{-6}$ (*weak* acid or *acidic* cation) or $\leq 10^{-8}$ (*weak* base or *basic* anion).

- **Corrections**
  → Concentrated solution: Use *activity* in place of *molarity*.
  → *Very* dilute solution: Solve two equilibrium expressions *simultaneously* (see Section 10.14).

10.10 Solutions of Weak Acids

- **General result for a *weak* acid HA at *dilute* concentrations**

Main equilibrium	HA(aq)	+ $H_2O(l)$ ⇌	$H_3O^+(aq)$	+ $A^-(aq)$
Initial molarity	$[HA]_{initial}$	—	0	0
Change in molarity	$-x$		$+x$	$+x$
Equilibrium molarity	$[HA]_{initial} - x$		x	x

Acidity constant expression: $K_a = \dfrac{[H_3O^+][A^-]}{[HA]} = \dfrac{x^2}{[HA]_{initial} - x}$

Quadratic equation: $x^2 + (K_a)x - (K_a[HA]_{initial}) = 0$ *and* $x > 0$ (positive root)

Positive root: $x = \dfrac{-K_a + \sqrt{K_a^2 + 4K_a[HA]_{initial}}}{2} = [H_3O^+] = [A^-] \geq 10^{-6}$ M

$[HA] = [HA]_{initial} - x$

$[OH^-] = K_w/[H_3O^+]$ secondary equilibrium

Percentage deprotonated $= \dfrac{x}{[HA]_{initial}} \times 100\%$

- **Approximation to the general result**

Assume $[HA]_{initial} - x \approx [HA]_{initial}$, then $K_a = \dfrac{x^2}{[HA]_{initial}}$ and $x = \sqrt{K_a[HA]_{initial}}$.

Assumption is justified if percentage deprotonated < 5%.

Example 10.10a Check the answers to Example 10.5 in the text by using the general results given above. The weak acid is 0.10 M CH_3COOH, with $K_a = 1.8 \times 10^{-5}$. The answers from the approximation method are $x = [H_3O^+] = [CH_3COO^-] = 1.3 \times 10^{-3}$ M, $[CH_3COOH] \approx 0.10$ M, and $[OH^-] = 7.69 \times 10^{-12}$ M.
Percentage deprotonated = 1.3%. Also, pH = 2.89 and pOH = 11.11.

Solution Use the quadratic formula to calculate the "exact" value of x. If x differs from the approximate value, complete the calculation of all equilibrium concentrations.

$$x = \frac{-K_a + \sqrt{K_a^2 + 4K_a[\text{HA}]_{\text{initial}}}}{2}$$

$$x = \frac{-(1.8\times10^{-5}) + \sqrt{(1.8\times10^{-5})^2 + 4(1.8\times10^{-5})(0.10)}}{2}$$

$$x = \frac{-(1.8\times10^{-5}) + \sqrt{(3.2\times10^{-10}) + (7.2\times10^{-6})}}{2} = \frac{-(1.8\times10^{-5}) + \sqrt{7.2\times10^{-6}}}{2}$$

$$x = \frac{-(1.8\times10^{-5}) + (2.68\times10^{-3})}{2} = \frac{2.67\times10^{-3}}{2} = 1.33\times10^{-3} \approx 1.3\times10^{-3}\ \text{M}$$

The approximation method is excellent in this case.

Example 10.10b Compare the "exact" answer to Self-Test 10.9A in the text to the one given by the approximate method. The weak acid is 0.20 M lactic acid, with K_a = 8.4×10^{-4}. The "exact" answer is pH = 1.90.

Answer $x = \sqrt{K_a[\text{HA}]_{\text{initial}}} = \sqrt{(8.4\times10^{-4})(0.20)} = \sqrt{1.68\times10^{-4}} = 1.3 \times 10^{-2}\ \text{M} = [H_3O^+]$

pH = $-\log(1.3 \times 10^{-2}) = 1.89$

"Exact" solution: $x = 1.25 \times 10^{-2}$ M and pH = 1.90

Percentage deprotonated = 6.25%, a value slightly above the 5% approximation cutoff.

10.11 Solutions of Weak Bases

- **General result for a *weak* base B at *dilute* concentrations**

Main equilibrium	B(aq) +	H_2O(l) ⇌	BH^+(aq) +	OH^-(aq)
Initial molarity	$[B]_{\text{initial}}$	—	0	0
Change in molarity	$-x$		$+x$	$+x$
Equilibrium molarity	$[B]_{\text{initial}} - x$		x	x

Basicity constant expression: $K_b = \dfrac{[BH^+][OH^-]}{[B]} = \dfrac{x^2}{[B]_{\text{initial}} - x}$

Quadratic equation: $x^2 + (K_b)x - (K_b[B]_{\text{initial}}) = 0$ *and* $x > 0$ (positive root)

Positive root: $x = \dfrac{-K_b + \sqrt{K_b^2 + 4K_b[B]_{\text{initial}}}}{2} = [BH^+] = [OH^-] \geq 10^{-6}$ M

$[B] = [B]_{\text{initial}} - x$

$[H_3O^+] = K_w/[OH^-]$ secondary equilibrium

Percentage protonated = $\dfrac{x}{[B]_{\text{initial}}} \times 100\%$

- **Approximation to the general result**

Assume $[B]_{\text{initial}} - x \approx [B]_{\text{initial}}$, then $K_b = \dfrac{x^2}{[B]_{\text{initial}}}$ and $x = \sqrt{K_b[B]_{\text{initial}}}$.

Assumption is justified if percentage protonated < 5%.

Example 10.11 Check the answers to Example 10.6 in the text by using the general results given above. The weak base is 0.20 M CH_3NH_2 with $K_b = 3.6 \times 10^{-4}$. The answers from the approximation method are $x = [CH_3NH_3^+] = [OH^-] = 8.5 \times 10^{-3}$ M, $[CH_3NH_2] \approx 0.20$ M, and $[H_3O^+] = 1.2 \times 10^{-12}$ M. Percentage protonated = 4.2%. Also, pH = 11.9 and pOH = 2.1.

Solution Use the quadratic formula to calculate the "exact" value of x. If x differs from the approximate value, complete the calculation of all equilibrium concentrations.

$$x = \frac{-K_b + \sqrt{K_b^2 + 4K_b[B]_{initial}}}{2}$$

$$x = \frac{-(3.6\times10^{-4}) + \sqrt{(3.6\times10^{-4})^2 + 4(3.6\times10^{-4})(0.20)}}{2}$$

$$x = \frac{-(3.6\times10^{-4}) + \sqrt{(1.29\times10^{-7}) + (2.88\times10^{-4})}}{2} = \frac{-(3.6\times10^{-4}) + \sqrt{2.88\times10^{-4}}}{2}$$

$$x = \frac{-(3.6\times10^{-4}) + (1.70\times10^{-2})}{2} = \frac{1.66\times10^{-2}}{2} = 8.31\times10^{-3} \approx 8.3\times10^{-3}$$

The approximation holds in this case.

10.12 The pH of Salt Solutions

- **Acidic solutions**
  - → Salts that contain the conjugate acids of *weak* bases
  - → Salts that contain small, highly charged metal cations
- **Basic solutions** → Salts that contain the conjugate bases of *weak* acids
- **Method of solution (main equilibrium)**
  - → For acidic salts, use the same method used for *weak* acids, but replace HA with BH^+. (Note that some are *anions* and some *metal cations.*)
  - → For basic salts, use the same method used for *weak* bases, but replace B with A^-. (All are *anions.*)

Autoprotolysis and pH (Sections 10.13-10.14)

Key Concepts

initial concentration, very dilute solution, charge balance equation, material balance equation, strong acids and bases (*three* equations in *three* unknowns), weak acids (*four* equations in *four* unknowns)

Overview

- **Autoprotolysis** → Important source of H_3O^+ for *very* dilute solutions of acids and bases
- **Charge balance** → Because the overall solution is electrically neutral, the concentration of total positive charges on cations must equal the concentration of total negative charges on anions.
- **Material balance** → All added solute must be accounted for even though it is present as ions in solution.
- **Strong acids and bases** → Solve autoprotolysis equilibrium by using *charge* and *material* balance as constraints; that is, solve *three* equations in *three* unknowns.
- **Weak acids and bases** → Simultaneously solve weak acid *or* base equilibrium and autoprotolysis equilibrium by using *charge* and *material* balance as constraints; that is, solve *four* equations in *four* unknowns.

10.13 Very Dilute Solutions of Strong Acids and Bases

- **Very dilute solution** → Concentration of a strong acid or base is less than 10^{-6} M
- **General result for a *strong* acid HA at *very dilute* concentrations**

Three equations in *three* unknowns

Autoprotolysis equilibrium: $K_w = [H_3O^+][OH^-] = 1.0 \times 10^{-14}$

Charge balance: $[H_3O^+] = [OH^-] + [A^-]$

Material balance: $[A^-] = [HA]_{initial}$ (complete deprotonation)

Solution: $[OH^-] = [H_3O^+] - [HA]_{initial}$

$K_w = [H_3O^+]([H_3O^+] - [HA]_{initial})$

$K_w = [H_3O^+]^2 - [HA]_{initial}[H_3O^+]$, let $x = [H_3O^+]$

Quadratic equation: $x^2 - [HA]_{initial}\, x - K_w = 0$ *and* $x > 0$ (positive root)

Positive root: $$x = \frac{[HA]_{initial} + \sqrt{[HA]^2_{initial} + 4K_w}}{2} = [H_3O^+] \quad \textit{and}$$

$[OH^-] = K_w/[H_3O^+]$ (autoprotolysis equilibrium)

Notes: If $[HA]^2_{initial} >> 4K_w$, then the result is the same as that for a *dilute* solution. The procedure is general for all strong acids *except* H_2SO_4, because the hydrogen sulfate ion, HSO_4^-, is a *weak* acid.

- **General result for a *strong* base MOH at *very dilute* concentrations**

Three equations in *three* unknowns

Autoprotolysis equilibrium: $K_w = [H_3O^+][OH^-] = 1.0 \times 10^{-14}$

Charge balance: $[M^+] + [H_3O^+] = [OH^-]$

Material balance: $[M^+] = [MOH]_{initial}$ (complete dissociation)

Solution: $[OH^-] = [H_3O^+] + [MOH]_{initial}$

$K_w = [H_3O^+]\{[H_3O^+] + [MOH]_{initial}\}$

$K_w = [H_3O^+]^2 + [MOH]_{initial}[H_3O^+]$, let $x = [H_3O^+]$

Quadratic equation: $x^2 + [MOH]_{initial}\,x - K_w = 0$ *and* $x > 0$ (positive root)

Positive root: $$x = \frac{-[MOH]_{initial} + \sqrt{[MOH]^2_{initial} + 4K_w}}{2} = [H_3O^+] \quad \textit{and}$$

$[OH^-] = K_w/[H_3O^+]$ (autoprotolysis equilibrium)

Notes: If $[MOH]^2_{initial} >> 4K_w$, then the result is the same as that for a *dilute* solution. The procedure is general for all strong bases *except* those like $Ba(OH)_2$, because there are two moles of OH^- for every mole of salt. Other strong bases (for example, O^{2-} and CH_3^-) produce aqueous solutions that are equivalent to MOH. Some strong bases (for example, NH_2^-) form conjugate acids that behave as *weak* bases (for example, NH_3) in water.

Example 10.13a Calculate the pH of a 1.5×10^{-8} M solution of the strong acid $HClO_3$.

Solution Use the appropriate quadratic equation to calculate $[H_3O^+]$ for a *very dilute* solution of a strong acid. Then evaluate the pH from pH = $-\log[H_3O^+]$.

$$x = \frac{[HA]_{initial} + \sqrt{[HA]^2_{initial} + 4K_w}}{2}$$

$$x = \frac{(1.5\times10^{-8}) + \sqrt{(1.5\times10^{-8})^2 + 4(1.0\times10^{-14})}}{2} = \frac{1.5\times10^{-8} + \sqrt{4.0225\times10^{-14}}}{2}$$

$$x = \frac{1.5\times10^{-8} + 2.01\times10^{-7}}{2} = \frac{2.16\times10^{-7}}{2} = 1.08\times10^{-7}\ \text{M} = [H_3O^+]$$

pH = $-\log[H_3O^+] = -\log(1.08 \times 10^{-7}) = 6.97$ (slightly *acidic* solution)

The methods of Sections 10.2 and 10.3 yield an *incorrect* answer of pH = 7.82.

Example 10.13b Calculate the pH of a 1.5×10^{-8} M solution of the strong base KOH.

Solution Use the appropriate quadratic equation to calculate $[H_3O^+]$ for a *very dilute* solution of a strong base. Then evaluate the pH.

$$x = \frac{-[\text{MOH}]_{\text{initial}} + \sqrt{[\text{MOH}]^2_{\text{initial}} + 4K_w}}{2} = 9.28 \times 10^{-8}\,\text{M} = [H_3O^+]$$

$\text{pH} = -\log[H_3O^+] = -\log(9.28 \times 10^{-8}) = 7.03$ (slightly *basic* solution)

The methods of Sections 10.2 and 10.3 yield an *incorrect* answer of pH = 6.18.

10.14 Very Dilute Solutions of Weak Acids

- **Very dilute solution** → Concentration of a weak acid or base is less than 10^{-3} M
- **General result for a *weak* acid HA at *very dilute* concentrations**
  Four equations in *four* unknowns

Weak acid equilibrium: $K_a = \dfrac{[H_3O^+][A^-]}{[HA]}$

Autoprotolysis equilibrium: $K_w = [H_3O^+][OH^-] = 1.0 \times 10^{-14}$

Charge balance: $[H_3O^+] = [OH^-] + [A^-]$

Material balance: $[HA]_{\text{initial}} = [HA] + [A^-]$ (incomplete deprotonation)

Solution: $[A^-] = [H_3O^+] - [OH^-] = [H_3O^+] - \dfrac{K_w}{[H_3O^+]}$

$[HA] = [HA]_{\text{initial}} - [A^-]$

$[HA] = [HA]_{\text{initial}} - [H_3O^+] + \dfrac{K_w}{[H_3O^+]}$

Substitute expressions for $[HA]$ and $[A^-]$ into K_a.

$$K_a = \frac{[H_3O^+]\left([H_3O^+] - \dfrac{K_w}{[H_3O^+]}\right)}{[HA]_{\text{initial}} - [H_3O^+] + \dfrac{K_w}{[H_3O^+]}}, \quad \text{let } x = [H_3O^+]$$

$$K_a = \frac{x\left(x - \dfrac{K_w}{x}\right)}{[HA]_{\text{initial}} - x + \dfrac{K_w}{x}}.$$ Rearrange to a cubic form.

Cubic equation: $x^3 + K_a x^2 - (K_w + K_a[HA]_{\text{initial}})x - K_a K_w = 0$

Solve by using mathematical software present on the CD that comes with the text. Only one of the three roots will be physically meaningful. Alternatively, try the approximations discussed in the text.

Example 10.14 Calculate the pH of a 2.0×10^{-4} M solution of the weak acid phenol, C_6H_5OH.

Solution In Table 10.1, $K_a = 1.3 \times 10^{-10}$. Calculate the terms in the cubic equation. Solve with the appropriate software on the CD.

The cubic equation is $x^3 + (1.3 \times 10^{-10})x^2 - (3.6 \times 10^{-14})x - (1.3 \times 10^{-24}) = 0$.
The solution is $x = 1.9 \times 10^{-7}$ M $= [H_3O^+]$ and pH = 6.72.
Compare with $x = 1.6 \times 10^{-7}$ M and pH = 6.80, derived with the method of Section 10.10. The percent deprotonation is 0.08%.

Polyprotic Acids and Bases (Sections 10.15-10.18)

Key Concepts

polyprotic acid, polyprotic base, succession of deprotonations or protonations, pH of a polyprotic acid (use K_{a1} as the main equilibrium; H_2SO_4 is the only exception), salts of polyprotic acids: pH = $\frac{1}{2}(pK_{a1} + pK_{a2})$, concentration of solute species, composition and pH

Overview

- **Successive deprotonations of a polyprotic acid**
  → A polyprotic acid is a species that can donate more than one proton.
  → Stepwise equilibria must be considered if successive K_a values differ by a factor of $\geq 10^3$, which is normally the case. Values are given in Table 10.9 in the text. Note that polyprotic bases also exist. A polyprotic base is a species that can accept more than one proton.
- **Dilute solution** → Identify and solve the *main* equilibrium expression. Use the appropriate *secondary* equilibria to calculate the concentrations of other solute species.
  Extension of methods developed in Sections 10.2 and 10.10.
- **Examples of main equilibria**
  → For H_2SO_4, $HSO_4^-(aq) + H_2O(l) \rightleftharpoons H_3O^+(aq) + SO_4^{-}(aq)$ K_{a2} (HSO_4^-)
  → For H_3PO_4, $H_3PO_4(aq) + H_2O(l) \rightleftharpoons H_3O^+(aq) + H_2PO_4^-(aq)$ K_{a1} (H_3PO_4)
- **Salt solutions** → For a solution of a salt HA^-, pH = $\frac{1}{2}(pK_{a1} + pK_{a2})$.
  The acidity constant K_{a2} refers to the base, HA^-, and K_{a1} to its conjugate acid, H_2A.
- **Composition and pH** → Solution to the stepwise equilibrium expressions for a polyprotic acid as a function of pH

10.15 The pH of a Polyprotic Acid Solution

Example 10.15 Calculate the pH of a 0.025 M solution of the polyprotic acid, H_2CO_3.

Solution Identify and solve the main equilibrium expression, which is K_{a1}. Use the 5% approximation method and check its validity.

$$x = \sqrt{K_{a1}[H_2A]_{initial}} = \sqrt{(4.3 \times 10^{-7})(0.025)} = 1.0 \times 10^{-4}\,M = [H_3O^+]$$

$$\text{Percentage deprotonated} = \frac{1.0 \times 10^{-4}}{0.025} \times 100\% = 0.4\% \text{ (approximation justified)}$$

$$pH = 4.00$$

10.16 Solutions of Salts of Polyprotic Acids

Example 10.16 Calculate the pH of a solution of a salt, HCO_3^-, of the polyprotic acid, H_2CO_3.

Solution Use the relationship given in the text for the pH in terms of pK_{a1} and pK_{a2}. Note that the answer is independent of the concentration of salt.

$$pH = \tfrac{1}{2}(pK_{a1} + pK_{a2}) = (0.5)(6.37 + 10.25) = 8.31$$

10.17 The Concentrations of Solute Species

Example 10.17 Calculate the concentration of all the solute species present in Example 10.15.

Solution In Example 10.15, $[H_3O^+] = 1.0 \times 10^{-4}$ M and $[H_2CO_3] = 0.025$ M. Use the main equilibrium expression to find the concentration of HCO_3^-. Use secondary equilibrium expressions to find the concentrations of CO_3^{2-} and OH^-.

$$x = [H_3O^+] = [HCO_3^-] = 1.0 \times 10^{-4}\,M$$

$$K_{a2} = \frac{[H_3O^+][CO_3^{2-}]}{[HCO_3^-]} = [CO_3^{2-}] = 5.6 \times 10^{-11}\,M$$

$$[OH^-] = \frac{K_w}{[H_3O^+]} = 1.0 \times 10^{-10}\,M$$

Note: The significant species present at equilibrium are H_2CO_3, H_3O^+, and HCO_3^-.

10.18 Composition and pH

Example 10.18 Use the equations in Section 10.18 of the text to calculate the fraction $\alpha(X)$ of three species (H_2CO_3, HCO_3^-, and CO_3^{2-}) present at a pH of 2.0, 7.0, and 12.0.

Solution Use the equations for $\alpha(X)$ given in the text: $K_{a1} = 4.3 \times 10^{-7}$ and $K_{a2} = 5.6 \times 10^{-11}$

$pH = 2.0 \Rightarrow [H_3O^+] = 1.0 \times 10^{-2}$ M

$f = [H_3O^+]^2 + [H_3O^+]K_{a1} + K_{a1}K_{a2} = 1.0 \times 10^{-4}$

$$\alpha(H_2CO_3) = \frac{[H_3O^+]^2}{f} = 1.0$$

$$\alpha(HCO_3^-) = \frac{[H_3O^+]K_{a1}}{f} = 4.3\times 10^{-5}$$

$$\alpha(CO_3^{2-}) = \frac{K_{a1}K_{a2}}{f} = 2.4 \times 10^{-13}$$

$pH = 7.0 \Rightarrow [H_3O^+] = 1.0 \times 10^{-7}$ M

$f = 5.3 \times 10^{-14}$, $\alpha(H_2CO_3) = 0.19$, $\alpha(HCO_3^-) = 0.81$, $\alpha(CO_3^{2-}) = 4.5 \times 10^{-4}$

$pH = 12.0 \Rightarrow [H_3O^+] = 1.0 \times 10^{-12}$ M

$f = 2.45 \times 10^{-17}$, $\alpha(H_2CO_3) = 4.1 \times 10^{-8}$, $\alpha(HCO_3^-) = 0.02$, $\alpha(CO_3^{2-}) = 0.98$

Note: If $\alpha(HCO_3^-) \approx 1$, then $pH = \frac{1}{2}(pK_{a1} + pK_{a2}) = (0.5)(6.37 + 10.25) = 8.31$. As a check of the expression for pH, evaluate f and $\alpha(HCO_3^-)$ by the above method for pH = 8.3. Keep extra significant figures to facilitate the comparison of fractions.

$pH = 8.3 \Rightarrow [H_3O^+] = 5.0 \times 10^{-9}$ M

$f = 2.2 \times 10^{-15}$, $\alpha(H_2CO_3) = 0.011$, $\alpha(HCO_3^-) = 0.977$, $\alpha(CO_3^{2-}) = 0.011$

This confirms an unstated, important result in the derivation of the equation, $pH = \frac{1}{2}(pK_{a1} + pK_{a2})$, for a salt solution of HCO_3^- that $[H_2CO_3] = [CO_3^{2-}]$.

Chapter 11 Aqueous Equilibria

Mixed Solutions and Buffers (Sections 11.1-11.4)

Key Concepts

mixed solutions, buffer, acid buffer, base buffer, designing a buffer, Henderson-Hasselbalch equation, equimolar, buffer capacity

Overview

- **Mixed solutions**
  - → A solution of an acid and a salt (conjugate base) of the acid
  - → A solution of a base and a salt (conjugate acid) of the base
- **Buffer**
  - → A mixed solution that resists a change in pH when strong acids or bases are added
  - → An acid buffer consists of a weak acid and its conjugate base provided as a salt.
  - → A base buffer consists of a weak base and its conjugate acid provided as a salt.
- **Henderson-Hasselbalch equation** → $\mathrm{pH} \approx \mathrm{p}K_a + \log\left(\frac{[\text{base}]_{\text{initial}}}{[\text{acid}]_{\text{initial}}}\right)$
- **Buffer capacity**
  - → Amount of acid or base that can be added to a buffer solution before it loses its ability to resist a change in pH
  - → Buffer solution is effective in the range $\mathrm{pH} = \mathrm{p}K_a \pm 1$.

11.1 Mixed Solutions

- **Mixed solution**
  - → A solution of a weak acid or a weak base that also contains a salt of the acid or base
  - → See Table 10.3 in the text for a listing of conjugate acid-base pairs.
- **pH of a mixed solution**
  - → Requires a standard equilibrium calculation with a table of concentrations
  - → Because salt is present, one of the ions formed has a nonzero concentration before the equilibrium occurs.
- **pH changes**
  - → The pH of a weak acid *increases* when a salt of its conjugate base is added to the solution.
  - → The pH of a weak base *decreases* when a salt of its conjugate acid is added to the solution.

Example 11.1a What is the pH of a solution that is 0.25 M in HCN and 0.15 M in KCN? pK_a(HCN) is 9.31.

Solution Potassium cyanide, like any other potassium salt, ionizes completely in aqueous solution. A solution that is 0.15 M in KCN actually contains 0.15 M K^+(aq) and 0.15 M CN^-(aq). The chemical equilibrium is

$$HCN(aq) + H_2O(l) \rightleftharpoons H_3O^+(aq) + CN^-(aq)$$

The table of concentrations is

	Species		
	HCN	H_3O^+	CN^-
1. Initial concentration	0.25 M	0	0.15 M
2. Change	$-x$	$+x$	$+x$
3. Equilibrium	0.25 M $- x$	x	0.15 M $+ x$

Because $pK_a = 9.31$, $K_a = 10^{-9.31} = 4.9 \times 10^{-10}$. The equilibrium expression is

$$\frac{[H_3O^+][CN^-]}{[HCN]} = K_a$$

Substituting the equilibrium values from the table gives

$$\frac{(x)(0.15+x)}{(0.25-x)} = 4.9\times10^{-10}$$

The equilibrium constant and adjustment to equilibrium are small. The approximations $x \ll 0.15$ and $x \ll 0.25$ lead to $0.15 + x = 0.15$ and $0.25 + x = 0.25$, respectively. Thus,

$$\frac{x(0.15)}{(0.25)} = 4.9\times10^{-10}$$

$$x = \frac{(0.25)(4.9\times10^{-10})}{(0.15)} = 8.2\times10^{-10}$$

Due to the small size of x, the approximations are valid. Because x equals the hydronium ion concentration at equilibrium,

$$[H_3O^+] = 8.2 \times 10^{-10}\ \text{M}$$

$$\text{pH} = -\log [H_3O^+] = -\log (8.2 \times 10^{-10}) = -(-9.09) = 9.09 \text{ (basic solution)}$$

Example 11.1b What is the pH of a solution that is 0.55 M in $C_6H_5NH_2$ and 0.75 M in $C_6H_5NH_3Cl$? $pK_b(C_6H_5NH_2)$ is 9.37.

Answer Follow the procedure in Example 11.1a, with $x = [OH^-]$.
Chemical equilibrium is $C_6H_5NH_2(aq) + H_2O(l) \rightleftharpoons C_6H_5NH_3^+(aq) + OH^-(aq)$
$K_b = 4.3 \times 10^{-10}$; $[OH^-] = 3.2 \times 10^{-10}$ M; pOH = 9.50; pH = 4.50
(acidic solution)

11.2 Buffer Action

- **Buffer**
  → A mixture of weak conjugate acid-base pairs that stabilizes the pH of a solution
  → Provides a source or sink for protons

- **Acid buffer**
  → Consists of a weak acid and its conjugate base provided as a salt
  → Buffers a solution on the acid side of neutrality (pH < 7), if $pK_a < 7$
- **Base buffer**
  → Consists of a weak base and its conjugate acid provided as a salt
  → Buffers a solution on the base side of neutrality (pH > 7), if $pK_b < 7$

Example 11.2a Will a solution of nitrous acid, HNO_2, and sodium nitrite, $NaNO_2$, act as a buffer? If it does, what reaction will occur when a small amount of a strong acid such as HCl(aq) is added?

Solution A solution of a weak acid and a salt of the acid will act as a buffer. Because HNO_2 is a weak acid and $NaNO_2$ is a salt of HNO_2, the mixture will act as a buffer. A strong acid such as HCl(aq) produces hydronium ions in solution. If a small amount of strong acid is added to the buffer solution, the additional hydronium ions will react with NO_2^- and the pH of the buffer solution will change by only a small amount. The reaction of the *additional* hydronium ions is

$$H_3O^+(aq) + NO_2^-(aq) \rightarrow HNO_2(aq) + H_2O(l)$$

Example 11.2b What reaction occurs in the buffer solution of Example 11.2a when a strong base is added?

Answer A strong base produces hydroxide ions in solution. The reaction of the *additional* hydroxide ions in the buffer solution is

$$OH^-(aq) + HNO_2(aq) \rightarrow H_2O(l) + NO_2^-(aq)$$

11.3 Designing a Buffer

- **pH of a buffer**
  → The pH of a buffer solution is given by the Henderson-Hasselbalch equation:

$$\text{pH} \approx pK_a + \log\left(\frac{[\text{base}]_{\text{initial}}}{[\text{acid}]_{\text{initial}}}\right)$$

  → The adjustment from initial concentrations to the equilibrium ones of the conjugate acid-base pair is *usually* negligible.
  → Use of this equation is equivalent to setting up a table of concentrations and solving an equilibrium problem.

Example 11.3a Calculate the pH of a buffer prepared by making up a solution that is 0.50 M nitrous acid, HNO_2, and 0.25 M sodium nitrite, $NaNO_2$. pK_a of HNO_2 is 3.37.

Solution Use the Henderson-Hasselbalch equation with the initial *weak* acid concentration given by the molarity of HNO_2 and the initial *weak* base concentration given by the molarity of NO_2^-. Recall that soluble ionic salts are completely dissociated into ions in aqueous solution.

$$pH = pK_a + \log\left(\frac{[\text{base}]_{\text{initial}}}{[\text{acid}]_{\text{initial}}}\right) = 3.37 + \log\frac{0.25\text{ M}}{0.50\text{ M}} = 3.37 - \log 2 = 3.07$$

Example 11.3b Calculate the pH of a buffer prepared by making up a solution that is 0.55 M urea, $CO(NH_2)_2$, and 0.15 M urea hydrochloride, $CO(NH_2)(NH_3)Cl$. pK_b of $CO(NH_2)_2$ is 13.90.

Answer $pK_a = 14.00 - pK_b = 14.00 - 13.90 = 0.10$

The base is urea and the acid is urea hydrochloride, which dissociates completely into $CO(NH_2)(NH_3)^+$ and Cl^-.

$pH = 0.10 + \log(0.55/0.15) = 0.10 + 0.56 = 0.66$

Example 11.3c Calculate the ratio of the molarities of fluoride ions and hydrofluoric acid, HF, needed to buffer a solution at pH = 3.76. pK_a of HF is 3.45.

Answer

$$\log\left(\frac{[F^-]}{[HF]}\right) = pH - pK_a = 3.76 - 3.45 = 0.31$$

$([F^-]/[HF]) = 10^{0.31} = 2.04$

Note: To make a buffer solution with a $pH > pK_a$, an excess of the conjugate base is required. Conversely, to make a buffer solution with $pH < pK_a$, an excess of the conjugate acid is required.

11.4 Buffer Capacity

- **Exceeding the capacity of a buffer**
  - → Buffers function by providing a weak acid to react with added base and a weak base to react with added acid.
  - → Adding sufficient base to react with *all* of the weak acid or adding sufficient acid to react with *all* of the weak base completely destroys the buffer system.
- **Limits of buffer action**
  - → A buffer has a limit to its buffering ability. As a rule of thumb, a buffer acts most effectively in the range

$$10 > \frac{[\text{base}]_{\text{initial}}}{[\text{acid}]_{\text{initial}}} > 0.1$$

  - → The preceding ratio of molarities translates into the following range of pH values:

$$pH = pK_a \pm 1$$

Example 11.4a Use Tables 10.1 and 10.2 in the text and the rule of thumb given above to select the best conjugate acid-base system(s) to buffer a solution with a pH of 4.50 and the best system(s) to buffer a solution with a pH of 9.50.

Solution Select acid-base systems with a pK_a value within 1 unit of the desired pH value. From the list of acids in Table 10.1, there are two choices for the solution with pH = 4.50; benzoic acid ($pK_a = 4.19$) and its conjugate base, and acetic acid ($pK_a = 4.75$) and its conjugate base. From the list of bases in Table 10.2, the best choice for the pH = 4.50 solution is aniline and its conjugate acid (pK_a = 14.00 – 9.37 = 4.63).

From the list of acids in Table 10.1, there are two choices for the solution with pH = 9.50: hydrocyanic acid (pK_a = 9.31) and its conjugate base, and phenol (pK_a = 9.89) and its conjugate base. From the list of bases in Table 10.2, there are two choices for the solution with pH = 9.50: ammonia and its conjugate acid (pK_a = 14.00 – 4.75 = 9.25), and trimethylamine and its conjugate acid (pK_a = 14.00 – 4.19 = 9.81).

- **Buffer capacity**
  - → The amount of acid or base that can be added before the buffer loses its ability to resist a change in pH
  - → The *greater* the concentrations of the conjugate acid-base pair in the buffer system, the *greater* the buffer capacity

Example 11.4b An acetic acid-potassium acetate buffer solution is 0.100 M in acetic acid and 0.100 M in potassium acetate. How many milliliters of 1.00 M HCl can be added to 500 mL of the buffer solution before its buffer capacity is exceeded?

Solution The reaction that occurs when HCl(aq) is added to this buffer is

$$H_3O^+(aq) + CH_3COO^-(aq) \rightarrow CH_3COOH(aq) + H_2O(l)$$

As HCl is added, the CH_3COO^- reacts and its concentration *decreases*. The concentration of CH_3COOH *increases*. The ratio of the concentration of the conjugate base to the concentration of the conjugate acid of the buffer *decreases*. The buffer capacity will be exceeded when the ratio gets small enough that

$$\frac{[CH_3COO^-]}{[CH_3COOH]} = 0.1$$

To determine when this ratio is reached, we first calculate the amount of acetic acid and acetate ion in the 500 mL (0.500 L) of buffer.

$$\text{amount of } CH_3COO^- = 0.500 \text{ L} \times 0.100 \text{ mol}\cdot\text{L}^{-1} = 0.0500 \text{ mol}$$

$$\text{amount of } CH_3COOH = 0.500 \text{ L} \times 0.100 \text{ mol}\cdot\text{L}^{-1} = 0.0500 \text{ mol}$$

When x mol HCl is added, x mol CH_3COO^- reacts and x mol CH_3COOH is formed. After addition of x mol HCl,

$$\text{amount of } CH_3COO^- = 0.0500 \text{ mol} - x$$

$$\text{amount of } CH_3COOH = 0.0500 \text{ mol} + x$$

Because both acetic acid and the acetate ion are in the same solution, the volumes cancel in the ratio of concentrations. The buffer capacity is exceeded when

$$\frac{[0.0500 \text{ mol} - x]}{[0.0500 \text{ mol} + x]} = 0.1$$

$$0.0500 \text{ mol} - x = 0.1(0.0500 \text{ mol} + x) = 0.00500 \text{ mol} + 0.1x$$

$$0.0500 \text{ mol} - 0.00500 \text{ mol} = 0.0450 \text{ mol} = x + 0.1x = 1.1x$$

$$x = (0.0450 \text{ mol}/1.1) = 0.0409 \text{ mol} = \text{amount of HCl required}$$

The solution of HCl is 1.00 M, and the volume added is $x/(1.00 \text{ mol}\cdot\text{L}^{-1})$ = 0.0409 L = 40.9 mL.   pH = 4.75 – 1.00 = 3.75

Example 11.4c An acetic acid-potassium acetate buffer solution is 0.200 M in acetic acid and 0.200 M in potassium acetate. How many milliliters of 1.00 M HCl can be added to 500 mL of the buffer solution before its buffer capacity is exceeded?

Answer The initial amount of acid and base is now 0.100 mol each in the buffer solution.

$$0.100 \text{ mol} - x = 0.1(0.100 \text{ mol} + x) = 0.0100 \text{ mol} + 0.1x$$

$$0.100 \text{ mol} - 0.0100 \text{ mol} = 0.090 \text{ mol} = x + 0.1x = 1.1x$$

$$x = (0.090 \text{ mol}/1.1) = 0.082 \text{ mol} = \text{amount of HCl required}$$

The solution of HCl is 1.00 M, and the volume added is $x/(1.00 \text{ mol·L}^{-1})$ = 0.082 L = 82 mL. pH = 4.75 – 1.00 = 3.75 as in Example 11.4b

Note: The buffer solution in Example 11.4c has a *greater* buffer capacity than that in Example 11.4b, because the concentration of the acid-base pair is twice as large.

Titrations (Sections 11.5-11.7)

Key Concepts

acid-base titration, redox titration, titrant, analyte, stoichiometric point, strong acid-strong base titration, pH curve, strong acid-weak base titration, weak acid-strong base titration, acid-base indicator, acid dissociation constant for an indicator K_{In}

Overview

- **Titration**
  - → Important technique in chemistry that is commonly used to determine the concentration of a solute (see Section L in the text)
  - → A solution (*titrant*) in a buret is added to a sample (*analyte*) in a flask.
  - → The stage in a titration when exactly the right volume of solution needed to complete the reaction has been added is called the *stoichiometric point.*
  - → Reactants in the same proportion as their coefficients in the chemical equation are said to have *stoichiometric proportions.*
- **Acid-base titration** → Used to determine the amounts of acids or bases in solution and for environmental monitoring (acid rain)
- **Redox titration** → Used to determine the amounts of oxidizing and reducing agents in solution, and for metal analysis of an ore

11.5 Strong Acid-Strong Base Titrations

- **pH curve**
  - → Plot of the pH of the *analyte* solution as a function of the volume of the *titrant* added during a titration is called a pH curve. The stoichiometric point has a pH value of 7.

→ Titration of a strong base with a strong acid is shown in Fig. 11.4 in the text. The pH *drops slowly* with added titrant until the stoichiometric point is approached. The pH then *drops rapidly* in the stoichiometric point region. At the stoichiometric point, the base is neutralized. The pH approaches the value of the titrant in the region well beyond the stoichiometric point.

→ Titration of a strong acid with a strong base is shown in Fig. 11.5 in the text. The pH *rises slowly* with added titrant until the stoichiometric point is approached. The pH then *rises rapidly* in the stoichiometric point region. At the stoichiometric point, the acid is neutralized. The pH approaches the value of the titrant in the region well beyond the stoichiometric point.

- **pH curve calculations** → Before the stoichiometric point is reached, follow the procedure outlined in Example 11.5 in the text. For a strong acid-strong-base titration, the pH at the stoichiometric point is 7. After the stoichiometric point is passed, follow the procedure outlined in Example 11.6 in the text. If the sample is a solution, always assume that the volumes of the titrant solution added and the analyte solution are additive ($V_{\text{total}} = V_{\text{analyte}} + V_{\text{titrant added}}$). If the sample is a solid, its volume is usually neglected.

Example 11.5a A titration is performed in which 0.260 M HCl(aq) is the titrant and 25.0 mL of 0.130 M KOH(aq) with pH = 13.11 is the analyte. Calculate the pH of the analyte solution after 5.0 mL of titrant is added.

Solution The initial amount of OH^-(aq) is (0.0250 L)(0.130 mol·L^{-1}) = 3.25×10^{-3} mol.
The amount of H_3O^+(aq) added is (0.0050 L)(0.260 mol·L^{-1}) = 1.30×10^{-3} mol.
The final volume of the analyte solution is 0.0250 + 0.0050 = 0.0300 L.
The amount of OH^-(aq) remaining is (3.25 – 1.30) × 10^{-3} = 1.95×10^{-3} mol.
$[OH^-] = (1.95 \times 10^{-3}\text{ mol})/(0.0300\text{ L}) = 0.0650$ M
pOH = –log(0.0650) = 1.19 and pH = 14.00 – pOH = 12.81

Example 11.5b Calculate the pH of the analyte in Example 11.5a after 25.0 mL of titrant is added.

Answer The initial amount of OH^-(aq) is 3.25×10^{-3} mol as given above.
The amount of H_3O^+(aq) added is (0.0250 L)(0.260 mol·L^{-1}) = 6.50×10^{-3} mol.
All the OH^-(aq) is consumed by the acid and the stoichiometric point is passed.
The amount of H_3O^+(aq) remaining is (6.50 – 3.25) × 10^{-3} = 3.25×10^{-3} mol.
The final volume of the analyte solution is 0.0250 + 0.0250 = 0.0500 L.
$[H_3O^+] = (3.25 \times 10^{-3}\text{ mol})/(0.0500\text{ L}) = 0.0650$ M
pH = –log(0.0650) = 1.19

Notes: The pH of the analyte solution approaches the pH of the titrant solution as tritrant is added beyond the stoichiometric point. For the titrant solution, pH = –log(0.260) = 0.58. In practice, titrations are ended when the stoichiometric point is reached.

11.6 Strong Acid-Weak Base and Weak Acid-Strong Base Titrations

- **pH curve**
  → The stoichiometric point in a strong acid-weak base and weak acid-strong base titration has a pH value different from 7. The titrant is the *strong* acid or base, the analyte is the *weak* base or acid.
  → The strong base *dominates* the weak acid and the pH at the stoichiometric point is greater than 7. The strong base converts the weak acid into its conjugate base form, and the pH at the stoichiometric point reflects the value for the resulting basic salt solution.
  → The strong acid *dominates* the weak base and the pH at the stoichiometric point is less than 7. The strong acid converts the weak base into its conjugate acid form, and the pH at the stoichiometric point reflects the value for the resulting acidic salt solution.
  → Shape of the pH curve is similar to that of a strong acid-strong base titration. The differences are that the pH is fairly stable in a region that corresponds to a buffer solution and it does not change as abruptly at the stoichiometric point, which does not have a value of 7.

- **pH curve calculations** → Before the stoichiometric point is reached, follow the procedure outlined in Example 11.8 in the text. To calculate the pH at the stoichiometric point, follow the procedure outlined in Example 11.7 in the text. After the stoichiometric point is passed, the pH of the solution approaches the value of the titrant. If the sample is a solution, always assume that the volumes of the titrant solution added and the analyte solution are additive ($V_{total} = V_{analyte} + V_{titrant\ added}$). If the sample is a solid, its volume is usually neglected.

Example 11.6a Calculate the pH at the stoichiometric point in a titration of 50.0 mL of 0.150 M morphine, $C_{17}H_{19}O_3N$, with 0.100 M HCl. pK_b of morphine is 5.79.

Solution The initial amount of morphine is $(0.0500\ L)(0.150\ mol{\cdot}L^{-1}) = 7.50 \times 10^{-3}$ mol.
The amount of $H_3O^+(aq)$ required is also 7.50×10^{-3} mol.
The volume of HCl required is $(7.50 \times 10^{-3}\ mol)/(0.100\ mol{\cdot}L^{-1}) = 0.0750$ L
The final volume of the analyte solution is $0.0500 + 0.0750 = 0.1250$ L.
The resulting solution is morphine hydrochloride, with the molarity given by
$[C_{17}H_{19}O_3NHCl] = [C_{17}H_{19}O_3NH^+] = (7.50 \times 10^{-3}\ mol)/(0.125\ L) = 0.0600$ M.
The chemical equilibrium is

$$C_{17}H_{19}O_3NH^+(aq) + H_2O(l) \rightleftharpoons H_3O^+(aq) + C_{17}H_{19}O_3N(aq) \qquad K_a = K_w/K_b$$

Let $x = [H_3O^+] = [C_{17}H_{19}O_3N]$, and assume that $0.0600 - x \approx 0.0600$
$K_a = 10^{-14}/10^{-5.79} = 6.17 \times 10^{-9} = (x^2)/(0.0600)$
$x^2 = 3.70 \times 10^{-10}$ and $x = 1.9 \times 10^{-5} << 0.060$ (0.03% reaction)
$pH = -\log(1.9 \times 10^{-5}) = 4.72 < 7$
As expected, the strong acid *dominates* at the stoichiometric point.

Example 11.6b In the titration of Example 11.6a, calculate the pH after 50.0 mL of 0.100 M HCl is added. Compare the value to the analyte before the addition of any acid. pK_b of morphine is 5.79.

Solution As before, the initial amount of morphine is 7.50×10^{-3} mol.
The amount of $H_3O^+(aq)$ added is $(0.0500\ L)(0.100\ mol \cdot L^{-1}) = 5.00 \times 10^{-3}$ mol.
The final volume of the analyte solution is 0.0500 + 0.0500 = 0.1000 L.
The amount of morphine remaining is $(7.50 - 5.00) \times 10^{-3} = 2.50 \times 10^{-3}$ mol.
$[C_{17}H_{19}O_3N] = (2.50 \times 10^{-3}\ mol)/(0.100\ L) = 0.0250$ M
The amount of morphine hydrochloride produced is 5.00×10^{-3} mol.
$[C_{17}H_{19}O_3NHCl] = [C_{17}H_{19}O_3NH^+] = (5.00 \times 10^{-3}\ mol)/(0.100\ L) = 0.0500$ M.
Because we have the concentrations of both acid and base forms, the Henderson-Hasselbalch equation can be used.

$pK_a = 14.00 - pK_b = 14.00 - 5.79 = 8.21$

$pH = pK_a + \log([base]/[acid]) = 8.21 + \log(0.0250/0.0500)$

$= 8.21 - 0.30 = 7.91$ after 50.0 mL of 0.100 M HCl is added

Because $pH = pK_a \pm 1$, this is safely in the buffer region.

The pH of 0.150 M morphine is calculated by the methods of Chapter 10, and the value is 10.69. The titration of the analyte begins with a pH value of 10.69, drops to a value of 8.21 ($pH = pK_a$) at the half-way point, and attains a final value of 4.72 at the stoichiometric point. Normally, a titration ends at the stoichiometric point.

11.7 Acid-Base Indicators

- **Acid-base indicator**
  - → A water-soluble dye with different *characteristic colors* associated with its acid and base form
  - → Exhibits a color change over a *narrow* range of pH values
  - → Present in small concentrations in the analyte so that a titration is primarily that of the analyte, not the indicator
  - → Selected to monitor the stoichiometric or *end point* in a titration
  - → An indicator is a weak acid. It takes part in the following proton-transfer equilibrium:

$$HIn(aq) + H_2O(l) \rightleftharpoons H_3O^+(aq) + In^-(aq) \quad K_{In} = \frac{[H_3O^+][In^-]}{[HIn]}$$

Values of pK_{In} are given in Table 11.2 in the text.

  - → The color change is usually most apparent when $[HIn] = [In^-]$ and $pH = pK_{In}$. The color change begins typically within one pH unit before pK_{In} and is essentially complete about 1 pH unit after pK_{In}. The pK_{In} value of an indicator is usually chosen to be within 1 pH unit of the stoichiometric point.

$$\boxed{pK_{In} \approx pH(\text{stoichiometric point}) \pm 1}$$

Example 11.7a Which indicator(s) from Table 11.2 in the text would be best for a titration with pH = 8.9 at the stoichiometric point?

Solution Choose an indicator for which the midpoint of the pH range is close to that of the stoichiometric point in a titration. The two choices are

thymol blue	pH range = 8.0–9.6	midpoint of range = 8.8
phenolphthalein	pH range = 8.2–10.0	midpoint of range = 9.1

Notes: The titration of a weak acid with a strong base has a stoichiometric point with pH > 7. These indicators may even be used for a strong acid-strong base titration, because the sharp pH change extends over several pH units near the stoichiometric point of pH = 7.

Example 11.7b Which indicator(s) from Table 11.2 in the text would be best for a titration with pH = 5.4 at the stoichiometric point?

Answer Methyl red is the only choice from the table.

Note: The titration of a weak base with a strong acid has a stoichiometric point with pH < 7.

Polyprotic Acid Titrations (Sections 11.8-11.9)

Key Concepts

polyprotic acid titrations, pH changes during titration, additional stoichiometric points and buffer regions

Overview

- **Polyprotic acid** → A polyprotic acid is a species that can donate more than one proton.
- **Titration of a polyprotic acid**
  - → The number of stoichiometric points is equal to the number of *acidic* protons.
  - → The value of K_a for the removal of each successive proton is less than 10^{-3} times the previous value.
  - → The pH curve can be estimated at any point by considering the primary species in solution and the *main* proton transfer equilibrium that determines the pH.

11.8 Stoichiometry of Polyprotic Acid Titration

- **Polyprotic acids**
  - → Review the list of the common polyprotic acids in Table 10.9 of the text.
  - → All the acids in Table 10.9 are *diprotic*, except for phosphoric acid, which is *triprotic*.

→ Phosphorus acid, H_3PO_3, is *diprotic.* The major Lewis structures are

$$\text{H}-\ddot{\text{O}}-\underset{\text{H}}{\overset{:\ddot{\text{O}}:}{\text{P}}}-\ddot{\text{O}}-\text{H} \longleftrightarrow \text{H}-\ddot{\text{O}}-\underset{\text{H}}{\overset{:\text{O}}{\overset{\|}{\text{P}}}}-\ddot{\text{O}}-\text{H}$$

- **Titration of a polyprotic acid**
  - → The titration of a polyprotic acid has a stoichiometric point corresponding to the removal of each acidic hydrogen atom. The product species at each stoichiometric point is the salt of a conjugate base of the acid or one of its hydrogen-bearing anions.
  - → See the examples of pH curves for H_3PO_4 in Fig. 11.15 and H_2CO_3 in Fig. 11.16 of the text.

Example 11.8 What volume of 0.010 M KOH(aq) is required to reach (1) the first stoichiometric point and (2) the second stoichiometric point in a titration of 50 mL of 0.020 M H_3PO_3(aq)?

Solution The amount of H_3PO_3 is $(0.050\ \text{L})(0.020\ \text{mol·L}^{-1}) = 1.0 \times 10^{-3}$ mol. There are two acidic hydrogen atoms. The amount of OH^- required to remove the first acidic hydrogen atom is 1.0×10^{-3} mol. The amount required to remove two hydrogen atoms is 2.0×10^{-3} mol.

(1) To reach the first stoichiometric point, 1.0×10^{-3} mol of OH^- is used. The volume of KOH required is

$$(1.0 \times 10^{-3}\ \text{mol})/(0.010\ \text{mol·L}^{-1}) = 1.0 \times 10^{-1}\ \text{L} = 100\ \text{mL}.$$

(2) To reach the second stoichiometric point, 2.0×10^{-3} mol of OH^- is used. The volume of KOH required is

$$(2.0 \times 10^{-3}\ \text{mol})/(0.010\ \text{mol·L}^{-1}) = 2.0 \times 10^{-1}\ \text{L} = 200\ \text{mL}.$$

11.9 pH Changes During Titration

- **pH calculations**
  - → Follow the procedure outlined in Toolbox 11.1 in the text.
  - → The pH curve can be estimated at any point by considering the primary species in solution and the *main* proton-transfer equilibrium that determines the pH. See Table 11.4 in the text for the *main* equilibra in each portion of the pH curve.

Example 11.9a Calculate the pH of 0.020 M H_3PO_3(aq). $pK_{a1} = 2.00$ and $pK_{a2} = 6.59$.

Solution The main equilibrium is K_{a1}. Let $x = [H_3O^+] = [H_2PO_3^-]$.

$K_{a1} = 10^{-2.00} = 0.010$, which is too large for the 5% approximation method.

$$\frac{x^2}{0.020 - x} = 0.010 \Rightarrow x^2 + 0.010x - 0.000\,20 = 0 \text{ (quadratic equation)}$$

$$x = \frac{-0.010 \pm \sqrt{(-0.010)^2 - 4(1)(-0.000\,20)}}{2} = \frac{-0.010 \pm \sqrt{0.000\,90}}{2} = \frac{-0.010 + 0.030}{2}$$

$[H_3O^+] = 0.010$ M and $\text{pH} = -\log(1.0 \times 10^{-2}) = 2.00$

Example 11.9b For the titration in Example 11.8, calculate the pH of the solution at the two stoichiometric points.

Solution (1) At the first stoichiometric point, the acid is converted into its conjugate base, $H_2PO_3^-$. There are two possible equilibrium reactions:

$$H_2PO_3^-(aq) + H_2O(l) \rightleftharpoons H_3O^+(aq) + HPO_3^{2-}(aq) \quad K_{a2} = 10^{-6.59} = 2.6 \times 10^{-7}$$

$$H_2PO_3^-(aq) + H_2O(l) \rightleftharpoons H_3PO_3(aq) + OH^-(aq) \quad K = K_w/K_{a1} = 1.0 \times 10^{-12}$$

The reaction of $H_2PO_3^-$ as an acid (proton donor) has a larger equilibrium constant and therefore, the first reaction is the main equilibrium.

From Example 11.8, the concentration of $H_2PO_3^-$ is

$$(1.0 \times 10^{-3} \text{ mol})/(0.050 + 0.100 \text{ L}) = 6.7 \times 10^{-3} \text{ M} = [H_2PO_3^-]$$

Let $x = [H_3O^+] = [HPO_3^{2-}]$. The 5% approximation method should work in this case.

$$x = \sqrt{[H_2PO_3^-]K_{a2}} = [(6.7 \times 10^{-3})(2.6 \times 10^{-7})]^{0.5} = 4.2 \times 10^{-5} = [H_3O^+]$$

$$pH = -\log(4.2 \times 10^{-5}) = 4.38$$

Approximation check: $(4.1 \times 10^{-5})/(6.7 \times 10^{-3}) \times 100\% = 0.6\% << 5\%$

Note: Using the formula from Section 10.16, we obtain

$$pH = \tfrac{1}{2}(pK_{a1} + pK_{a2}) = (0.5)(2.00 + 6.59) = 4.30$$

(2) At the second stoichiometric point, the acid, $H_2PO_3^-$ is converted into its conjugate base, HPO_3^{2-}. There is only one possible equilibrium reaction:

$$HPO_3^{2-}(aq) + H_2O(l) \rightleftharpoons H_2PO_3^-(aq) + OH^-(aq) \quad K = K_w/K_{a2} = 3.9 \times 10^{-8}$$

From Example 11.8, the concentration of HPO_3^{2-} is

$$(1.0 \times 10^{-3} \text{ mol})/(0.050 + 0.200 \text{ L}) = 4.0 \times 10^{-3} \text{ M} = [H_2PO_3^-]$$

Let $x = [OH^-] = [H_2PO_3^-]$. The 5% approximation method should work in this case.

$$x = \sqrt{[HPO_3^{2-}](K_w/K_{a2})} = [(4.0 \times 10^{-3})(3.9 \times 10^{-8})]^{0.5} = 1.2 \times 10^{-5} = [OH^-]$$

$$pOH = -\log(1.2 \times 10^{-5}) = 4.92 \text{ and } pH = 14.00 - 4.92 = 9.08$$

Approximation check: $(1.2 \times 10^{-5})/(4.0 \times 10^{-3}) \times 100\% = 0.3\% << 5\%$

Note: The two equilibria for the second stoichiometric point of a H_3PO_4 titration are

$$HPO_4^{2-}(aq) + H_2O(l) \rightleftharpoons H_3O^+(aq) + PO_4^{3-}(aq) \quad K_{a3} = 2.1 \times 10^{-13}$$

$$HPO_4^{2-}(aq) + H_2O(l) \rightleftharpoons H_2PO_4^-(aq) + OH^-(aq) \quad K = K_w/K_{a2} = 1.6 \times 10^{-7}$$

The reaction of HPO_4^{2-} as a base (proton acceptor) has a larger equilibrium constant and therefore, the second reaction is the main equilibrium.

Solubility Equilibria (Sections 11.10-11.16)

Key Concepts

solubility equilibria, solubility product K_{sp}, molar solubility, common-ion effect, predicting precipitation, selective precipitation, dissolving precipitates, complex ion formation, formation constant K_f, qualitative analysis

Overview

- **Solubility equilibria**
  - → Equilibrium between a solid salt and its dissolved ions in a saturated solution
  - → Using equilibrium arguments and calculations to predict the solubility of a salt and to control precipitate formation
  - → Formation of complex ions by Lewis acid-base reactions
  - → Laboratory method to separate and analyze mixtures of salts
  - → Application to municipal wastewater treatment and extraction of minerals from seawater

11.10 The Solubility Product

- **Solubility product, K_{sp}**
  - → Equilibrium constant for the solubility equilibrium between a solid and its dissolved form is the *solubility product* of the solute.
  - → For example, $Sb_2S_3(s) \rightleftharpoons 2\,Sb^{3+}(aq) + 3\,S^{2-}(aq)$ $\quad K_{sp} = \dfrac{a_{Sb^{3+}}^{\;2}\, a_{S^{2-}}^{\;3}}{a_{Sb_2S_3(s)}}$
  - → Activity of a pure solid is 1, and for dilute solutions, activity of a solute species is replaced by its molarity.
  - → To a fair approximation, $K_{sp} = [Sb^{3+}]^2[S^{2-}]^3$ (saturated solution)
- **Molar solubility, *s*** → Maximum amount of solute that dissolves in enough water to make a liter of solution
- **Relationship between *s* and K_{sp}**
  - → The stoichiometric relationships are
    $1 \text{ mol } Sb_2S_3 \simeq 2 \text{ mol } Sb^{3+}$ and $1 \text{ mol } Sb_2S_3 \simeq 3 \text{ mol } S^{2-}$
  - → The molar solubility, *s*, of solute $Sb_2S_3(s)$ is related to the ion concentrations at equilibrium as follows:
    $[Sb^{3+}] = 2s$ and $[S^{2-}] = 3s$.
  - → The relationship between *s* and K_{sp} is then
    $K_{sp} = [Sb^{3+}]^2[S^{2-}]^3 = (2s)^2(3s)^3 = (4s^2)(27s^3) = 108s^5$ and $s = (K_{sp}/108)^{1/5}$

Example 11.10a The solubility of lead chloride at 20°C is 0.44 g per 100 mL of water. Estimate K_{sp} of $PbCl_2$ and compare the value to the one in Table 11.5 in the text.

Solution The molar mass of lead chloride is 207.19 + 2(35.45) = 278.09 g·mol^{-1}.
The amount of lead chloride is (0.44 g/278.09 g·mol^{-1}) = 1.6×10^{-3} mol.
The molar solubility of lead chloride is (1.6×10^{-3} mol/0.100 L) = 1.6×10^{-2}.
Thus, $s = 1.6 \times 10^{-2}$ mol·L^{-1}. The solubility equilibrium is

$$PbCl_2(s) \rightleftharpoons Pb^{2+}(aq) + 2\,Cl^-(aq) \qquad K_{sp} = [Pb^{2+}][Cl^-]^2$$

$$1 \text{ mol } PbCl_2 \simeq 1 \text{ mol } Pb^{2+} \text{ and } 1 \text{ mol } PbCl_2 \simeq 2 \text{ mol } Cl^-$$

$$[Pb^{2+}] = s \text{ and } [Cl^-] = 2s$$

The relationship between K_{sp} and s is $K_{sp} = (s)(2s)^2 = 4s^3$.
$K_{sp} = 4(1.6 \times 10^{-2})^3 = 1.6 \times 10^{-5}$, which is in agreement with the table entry.

Note: The solubility product expression is approximate because these aqueous solutions are often too concentrated in ionic solute to fully justify replacing activities with molarities. Also, note that the values of K_{sp} in Table 11.5 apply to 25°C.

Example 11.10b Estimate the molar solubility of silver carbonate, Ag_2CO_3, at 25°C. K_{sp} of silver carbonate is 6.2×10^{-12}.

Solution The solubility equilibrium is

$$Ag_2CO_3(s) \rightleftharpoons 2\,Ag^+(aq) + CO_3^{2-}(aq) \qquad K_{sp} = [Ag^+]^2[CO_3^{2-}]$$

$$1 \text{ mol } Ag_2CO_3 \simeq 2 \text{ mol } Ag^+ \text{ and } 1 \text{ mol } Ag_2CO_3 \simeq 1 \text{ mol } CO_3^{2-}$$

$$[Ag^+] = 2s \text{ and } [CO_3^{2-}] = s$$

The relationship between K_{sp} and s is $K_{sp} = [Ag^+]^2[CO_3^{2-}] = (2s)^2(s) = 4s^3$
This expression is easily solved for s:

$$s = (K_{sp}/4)^{1/3} = (6.2 \times 10^{-12}/4)^{1/3} = 1.2 \times 10^{-4}$$

The molar solubility of silver carbonate is therefore 1.2×10^{-4} mol·L^{-1}.

Note: The solubility of 1.2×10^{-4} mol·L^{-1} corresponds to 0.032 g per 100 mL of solution.

11.11 The Common-Ion Effect

- **Common-ion effect:**

  Qualitative features

  → According to Le Chatelier's principle, addition of any product (that is not a pure solid or a pure liquid) to a dynamic equilibrium forces the equilibrium to shift towards reactants.

  → For the equilibrium $Sb_2S_3(s) \rightleftharpoons 2\,Sb^{3+}(aq) + 3\,S^{2-}(aq)$, addition of Sb^{3+} ion (by the addition of $Sb(NO_3)_3$, for example) will cause some precipitation of $Sb_2S_3(s)$. Thus, less $Sb_2S_3(s)$ is dissolved in the presence of an external source of Sb^{3+} ion or S^{2-} ion than in pure water.

  → The *reduction* in the solubility of a sparingly soluble salt by the addition of a soluble salt that has an ion in common with it is called the *common-ion effect.*

Quantitative features

→ The solubility product expression is used to estimate the reduction in molar solubility when soluble salts with a common ion are added.

→ For the equilibrium $Sb_2S_3(s) \rightleftharpoons 2\,Sb^{3+}(aq) + 3\,S^{2-}(aq)$, $K_{sp} = [Sb^{3+}]^2[S^{2-}]^3$. If $Sb_2S_3(s)$ is added to a solution containing Sb^{3+} ions with a concentration $[Sb^{3+}]$, then the dissolution of solid is assumed to add a negligible amount to the cation concentration. The solubility, *s*, and its relationship to K_{sp} is then

$$K_{sp} = [Sb^{3+}]^2\,(3s)^3 \text{ and } s = (K_{sp}/27[Sb^{3+}]^2)^{1/3}$$

Example 11.11a Estimate the molar solubility of BaF_2 in pure water and in 0.20 M NaF at 25°C. K_{sp} of BaF_2 is 1.7×10^{-6}.

Solution The solubility equilibrium is

$$BaF(s) \rightleftharpoons Ba^{2+}(aq) + 2\,F^{-}(aq) \qquad K_{sp} = [Ba^{2+}][F^{-}]^2$$

$$1 \text{ mol } BaF_2 \simeq 1 \text{ mol } Ba^{2+} \text{ and } 1 \text{ mol } BaF_2 \simeq 2 \text{ mol } F^{-}$$

In pure water, $[Ba^{2+}] = s$ and $[F^{-}] = 2s$.

The relationship between K_{sp} and s is $K_{sp} = (s)(2s)^2 = 4s^3$.

$$s = (K_{sp}/4)^{1/3} = (1.7 \times 10^{-6}/4)^{1/3} = 7.5 \times 10^{-3} \text{ mol·L}^{-1}$$

In 0.20 M NaF, $[Ba^{2+}] = s$ and $[F^{-}] = 0.20 + 2s \approx 0.20$.

The relationship between K_{sp} and s is $K_{sp} = (s)(0.20)^2 = (0.04)s$.

$$s = K_{sp}/(0.040) = 1.7 \times 10^{-6}/0.040 = 4.3 \times 10^{-5} \text{ mol·L}^{-1}$$

Example 11.11b Estimate the molar solubility of CuI in pure water and in 0.30 M NaI at 25°C. K_{sp} of CuI is 1.0×10^{-6}.

Answer In pure water, $s = 1.0 \times 10^{-3}$ mol·L^{-1}.
In 0.30 M NaI, $s = 3.3 \times 10^{-6}$ mol·L^{-1}.

11.12 Predicting Precipitation

- **Procedure used to predict whether precipitation occurs**

→ Calculate Q_{sp} using the values of actual or predicted concentrations of the ions in solution.

→ Compare the magnitude of Q_{sp} to K_{sp}.

→ If $Q_{sp} > K_{sp}$, the solid precipitates from solution.

→ If $Q_{sp} = K_{sp}$, the solution is saturated in the concentrations of the ions. Any additional amount of ions will cause precipitation to occur.

→ If $Q_{sp} < K_{sp}$, the solution is unsaturated in the concentration of the ions. No precipitate will form.

Example 11.12a Does a precipitate of calcium carbonate form when 200 mL of 1.0×10^{-3} M $Ca(NO_3)_2$ is added to 100 mL of 1.0×10^{-4} M Na_2CO_3? K_{sp} of $CaCO_3$ is 8.7×10^{-9}.

Solution The solubility equilibrium is

$$CaCO_3(s) \rightleftharpoons Ca^{2+}(aq) + CO_3^{2-}(aq) \qquad K_{sp} = [Ca^{2+}][CO_3^{2-}]$$

The amount of Ca^{2+} is $(0.200\ L)(1.0 \times 10^{-3}\ mol{\cdot}L^{-1}) = 2.0 \times 10^{-4}$ mol.
The amount of CO_3^{2-} is $(0.100\ L)(1.0 \times 10^{-4}\ mol{\cdot}L^{-1}) = 1.0 \times 10^{-5}$ mol.
After the solutions are mixed, the new volume is assumed to be 0.300 L.
$[Ca^{2+}] = (2.0 \times 10^{-4}\ mol)/(0.300\ L) = 6.7 \times 10^{-4}\ mol{\cdot}L^{-1}$
$[CO_3^{2-}] = (1.0 \times 10^{-5}\ mol)/((0.300\ L) = 3.3 \times 10^{-5}\ mol{\cdot}L^{-1}$
$Q_{sp} = [Ca^{2+}][CO_3^{2-}] = (6.7 \times 10^{-4})(3.3 \times 10^{-5}) = 2.2 \times 10^{-8} > K_{sp}$
A precipitate of $CaCO_3$ will form.

Example 11.12b Does a precipitate of calcium sulfate form when 200 mL of 1.0×10^{-3} M $Ca(NO_3)_2$ is added to 100 mL of 1.0×10^{-4} M Na_2SO_4? K_{sp} of $CaSO_4$ is 2.4×10^{-5}.

Answer $Q_{sp} = [Ca^{2+}][SO_4^{2-}] = (6.7 \times 10^{-4})(3.3 \times 10^{-5}) = 2.2 \times 10^{-8} < K_{sp}$
A precipitate of $CaSO_4$ will *not* form.

11.13 Selective Precipitation

- **Separation of different cations**
  - → Adding a soluble salt containing an anion with which the cations form insoluble salts
  - → The insoluble salts have different solubility values and will precipitate at different anion concentrations.

Example 11.13a A solution contains the following concentrations of soluble cations: 0.050 M $Cu^{+}(aq)$ and 0.010 M $Ag^{+}(aq)$. Determine the order in which each ion precipitates as solid NaBr is added and give the concentration of Br^{-} when each solid begins to precipitate. Assume no volume change on addition of NaBr. K_{sp} of CuBr is 4.2×10^{-8} and K_{sp} of AgBr is 7.7×10^{-13}.

Solution The concentration of Br^{-} required to precipitate the salts is calculated for each insoluble salt from the two K_{sp} expressions:

$$[Br^{-}] = K_{sp}/[Cu^{+}] = (4.2 \times 10^{-8})/(0.050) = 8.4 \times 10^{-7}\ mol{\cdot}L^{-1}$$
$$[Br^{-}] = K_{sp}/[Ag^{+}] = (7.7 \times 10^{-13})/(0.010) = 7.7 \times 10^{-11}\ mol{\cdot}L^{-1}$$

AgBr precipitates first when the Br^{-} concentration reaches 7.7×10^{-11} M.
CuBr precipitates next when the Br^{-} concentration reaches 8.4×10^{-7} M.

Example 11.13b A solution contains the following concentrations of soluble cations: 0.050 M $Cu^{+}(aq)$ and 0.010 M $Ag^{+}(aq)$. Give the concentration of S^{2-} when each solid begins to precipitate as solid Na_2S is added. K_{sp} of Cu_2S is 2.0×10^{-47} and K_{sp} of Ag_2S is 6.3×10^{-51}.

Answer $[S^{2-}] = 8.0 \times 10^{-45}$ M for Cu_2S and $[S^{2-}] = 6.3 \times 10^{-47}$ M for Ag_2S.

11.14 Dissolving Precipitates

- **Method of anion removal**
  → Removing an anion increases the solubility of the solid.
  → Precipitate completely dissolves if a sufficient number of anions are removed.

- **Removal of basic anions**
  → Addition of a strong acid, such as HCl, neutralizes the basic anion.
  For example, hydroxides react to form water:
  $$H_3O^+(aq) + OH^-(aq) \rightarrow 2\,H_2O(l)$$
  → In some cases, gases are evolved.
  For example, carbonates react to form carbon dioxide gas:
  $$2\,H_3O^+(aq) + CO_3^{2-}(aq) \rightarrow CO_2(g) + 3\,H_2O(l)$$

- **Removal by oxidation of anions**
  → Oxidizing the sulfide ion to elemental sulfur
  → Addition of nitric acid to a metal sulfide solution produces the following reaction:
  $$3\,S^{2-}(aq) + 8\,H_3O^+(aq) + 2\,NO_3^-(aq) \rightarrow 3\,S(s) + 2\,NO(g) + 12\,H_2O(l)$$

Example 11.14a Devise a method to dissolve the insoluble salt cobalt(II) cyanide, $Co(CN)_2$. Write the reactions involved.

Solution The cyanide ion is the conjugate base of a weak acid, HCN. If an acid, such as HCl, is added to the solution, the cyanide ion will disappear from the solution by its reaction to form HCN. Thus, $Co(CN)_2$ will dissolve. The two reactions are

$H_3O^+(aq) + CN^-(aq) \rightarrow HCN(aq) + H_2O(l)$ and CN^- is removed.

$Co(CN)_2(s) \rightleftharpoons Co^{2+}(aq) + 2\,CN^-(aq)$ and the equilibrium shifts to the right.

Example 11.14b Devise a method to dissolve copper(I) sulfite, Cu_2SO_3. Write the equations involved.

Answer The two reactions are

$2\,H_3O^+(aq) + SO_3^{2-}(aq) \rightarrow SO_2(g) + 3\,H_2O(l)$ and SO_3^{2-} is removed.

$Cu_2SO_3(s) \rightleftharpoons 2\,Cu^+(aq) + SO_3^{2-}(aq)$ and the equilibrium shifts to the right.

- **Summary** → Solubility of a solid may be increased by removing an anion from solution. An acid can be used to dissolve a hydroxide, sulfide, sulfite, or carbonate precipitate. Nitric acid can be used to oxidize metal sulfides to sulfur and a soluble salt.

11.15 Complex Ion Formation

- **Complex-ion formation:**
  Qualitative features
  → Complexes are species formed by a Lewis acid-base reaction. They were first introduced in Section 2.13.

→ In this section, we consider Lewis acids that are metal cations.

→ The *formation reactions* of complex ions are given in Table 11.6 in the text.

→ When it is possible to form a complex ion of one of the ions in a salt, the ion will be removed from solution and the salt will dissolve.

Example 11.15a What reagent might be used to dissolve solid nickel(II) sulfide through complex formation? Write the reactions involved.

Solution In Table 11.6, the reaction of Ni^{2+} with ammonia to form a complex ion is listed. The addition of NH_3 will result in the formation of $Ni(NH_3)_6^{2+}$ and removal of Ni^{2+} from solution. Thus, the solubility equilibrium will shift to the right and NiS dissolves.
The two reactions are

$Ni^{2+}(aq) + 6\,NH_3(aq) \rightleftharpoons Ni(NH_3)_6^{2+}$ and Ni^{2+} is removed.

$NiS(s) \rightleftharpoons Ni^{2+}(aq) + S^{2-}(aq)$ and the equilibrium shifts to the right.

Example 11.15b What reagent might be used to dissolve solid gold(III) bromide through complex formation? Write the reactions involved.

Answer The reagent NaCN could be used. The two reactions are

$Au^{3+}(aq) + 2\,CN^-(aq) \rightleftharpoons Au(CN)_2^-$ and Au^{3+} is removed.

$AuBr_3(s) \rightleftharpoons Au^{3+}(aq) + 3\,Br^-(aq)$ and the equilibrium shifts to the right.

Quantitative features

→ Equilibrium constant for the formation reaction of a complex ion is the *formation constant*, K_f.

For example, $Fe^{2+}(aq) + 6\,CN^-(aq) \rightleftharpoons Fe(CN)_6^{4-}(aq)$ $\quad K_f = \dfrac{[Fe(CN)_6^{4-}]}{[Fe^{2+}][CN^-]^6}$

In the preceding expression for K, the molarity of a solute species has replaced its activity.

→ Values of K_f at 25°C are given in Table 11.6 in the text.

→ Formation and solubility reactions may be combined to predict the solubility of a salt in the presence of a species that forms a complex with the metal ion.
For example, $FeS(s) \rightleftharpoons Fe^{2+}(aq) + S^{2-}(aq)$ $\quad K_{sp} = [Fe^{2+}][S^{2-}]$
Combining the solubility reaction with the preceding formation reaction yields

$$FeS(s) + 6\,CN^-(aq) \rightleftharpoons Fe(CN)_6^{4-}(aq) + S^{2-}(aq) \quad K = K_{sp} \times K_f = \frac{[Fe(CN)_6^{4-}][S^{2-}]}{[CN^-]^6}$$

For this reaction, $K = (6.3 \times 10^{-18}) \times (7.7 \times 10^{36}) = 4.9 \times 10^{19} >> 1$. In this case, the reaction goes to completion if there is sufficient solid present. All the CN^- in solution will be consumed. Because 1 mol FeS ≙ 1 mol S^{2-}, the molar solubility, *s*, is equal to the sulfide ion concentration, which is in turn equal to 1/6 the initial CN^- concentration.

→ The following examples illustrate how to handle equilibrium problems when $K < 1$. Also, see Example 11.15 in the text.

Example 11.15c Estimate the molar solubility of AgBr in pure water and in 0.20 M NaCN at 25°C. K_{sp} of AgBr is 7.7×10^{-13} and K_f of $Ag(CN)_2^-$ is 5.6×10^8.

Solution In pure water, $K_{sp} = [Ag^+][Br^-] = s^2$ and $s = (K_{sp})^{1/2} = 8.8 \times 10^{-7}\ mol{\cdot}L^{-1}$.
In 0.20 M NaCN, we must combine the solubility and formation reactions.

$$AgBr(s) + 2\,CN^-(aq) \rightleftharpoons Ag(CN)_2^-(aq) + Br^-(aq)$$

$$K = K_{sp} \times K_f = (7.7 \times 10^{-13}) \times (5.6 \times 10^8) = 4.3 \times 10^{-4}$$

The table of concentrations is

	Species		
	CN^-	$Ag(CN)_2^-$	Br^-
1. Initial concentration	0.20 M	0	0
2. Change	$-2x$	$+x$	$+x$
3. Equilibrium	0.20 M $- 2x$	x	x

$$K = \frac{(x)(x)}{(0.20-2x)^2} = 4.3 \times 10^{-4}$$

This equation is solved by taking the square root of both sides.

$$x = (0.20 - 2x)(2.1 \times 10^{-2}) = 4.2 \times 10^{-3} - (4.2 \times 10^{-2})x$$

$$(1.042)x = 4.2 \times 10^{-3} \text{ and } x = 4.0 \times 10^{-3}\ mol{\cdot}L^{-1}$$

Because 1 mol AgBr $\simeq$ 1 mol Br^-, the molar solubility of AgBr is given by the equation $s = [Br^-]$. Thus, $s = x = [Br^-] = 4.0 \times 10^{-3}\ mol{\cdot}L^{-1}$.
As expected, the solubility increases in the presence of a species that forms a complex with Ag^+.

Example 11.15d Estimate the molar solubility of AgBr in pure water and in 0.20 M NH_3 at 25°C. K_{sp} of AgBr is 7.7×10^{-13} and K_f of $Ag(NH_3)_2^+$ is 1.6×10^7.

Answer In pure water, $s = 8.8 \times 10^{-7}\ mol{\cdot}L^{-1}$
In 0.20 M NH_3, $s = 7.0 \times 10^{-4}\ mol{\cdot}L^{-1}$

11.16 Qualitative Analysis

- **Separation and identification of ions**
  - → Method employing selective precipitation, complex formation, and the control of pH
  - → Using solubility equilibria to remove and identify ions selectively
  - → See the qualitative analysis scheme in Table 11.7 in the text.

Example 11.16 For a mixture of halide ions with $[Cl^-] = [Br^-] = [I^-] = 0.010$ M, calculate the value of $[Ag^+]$ required to begin the precipitation of each silver halide salt. What are the values of $[Br^-]$ and $[I^-]$ when Cl^- begins to precipitate?

Answer For AgI(s), $[Ag^+] = K_{sp}/[I^-] = (1.5 \times 10^{-16}/0.010) = 1.5 \times 10^{-14}$ M
For AgBr(s), $[Ag^+] = K_{sp}/[Br^-] = (7.7 \times 10^{-13}/0.010) = 7.7 \times 10^{-11}$ M
For AgCl(s), $[Ag^+] = K_{sp}/[Cl^-] = (1.6 \times 10^{-10}/0.010) = 1.6 \times 10^{-8}$ M

When Cl^- begins to precipitate, $[Ag^+] = 1.6 \times 10^{-8}$ M. At this point,
$[Br^-] = K_{sp}/[Ag^+] = (7.7 \times 10^{-13}/1.6 \times 10^{-8}) = 4.8 \times 10^{-5}$ M
$[I^-] = K_{sp}/[Ag^+] = (1.5 \times 10^{-16}/1.6 \times 10^{-8}) = 9.4 \times 10^{-9}$ M

Chapter 12 Electrochemistry

Redox Reactions (Sections 12.1-12.2)

Key Concepts

electrochemistry, electrochemical techniques, redox equation, half reaction, redox couple, balancing redox equations

Overview

- **Electrochemistry**
  - → Use of chemical reactions to produce electricity (*galvanic cell*)
  - → Use of electricity to drive nonspontaneous chemical reactions (*electrolysis*)
  - → Important to help meet future energy needs
  - → Links chemistry of oxidation-reduction (*redox*) reactions to the physics of charge flow
  - → Provides an important way of using free energy available in spontaneous chemical reactions to perform useful work
- **Electrochemical techniques**
  - → Procedures based on electrochemistry
  - → Permit us to monitor the composition of solutions, to measure pH, to determine the concentration of ions, to measure K_a of acids, and to measure K_{sp} of salts
- **Goal** → To develop insight into the use of chemical change to generate electricity and the use of electricity to effect chemical change

12.1 Half-Reactions

- **Half-reaction**
  - → *Conceptual* way of reporting an oxidation or reduction process separately
  - → Important for writing and balancing oxidation-reduction (*redox*) reactions
- **Oxidation**
  - → *Removal* of electrons from a species in a redox reaction
  - → Represented by a separate half-reaction
- **Reduction**
  - → *Gain* of electrons by a species in a redox reaction
  - → Represented by a separate half-reaction

Example 12.1a Describe the oxidation half-reaction $Zn(s) \rightarrow Zn^{2+}(aq) + 2\,e^-$.

Solution Electrons are removed from zinc metal to form aqueous zinc(II) ions. The electrons removed are not free but directly transferred to another species in a separate reduction process. Zinc metal is oxidized. The oxidized and reduced species in the half-reaction jointly form a *redox couple,* Zn^{2+}/Zn, in which the oxidized form is listed first by convention.

Example 12.1b Describe the reduction half-reaction $Cu^{2+}(aq) + 2e^- \rightarrow Cu(s)$.

Solution Copper(II) ions gain electrons to form copper metal. The electrons are obtained from another species in a separate oxidation process. Copper(II) ions are reduced. The oxidized and reduced species in the half-reaction jointly form a *redox couple,* Cu^{2+}/Cu.

- **Half-reactions**
  - → Express separately the oxidation and reduction contributions to a redox reaction
  - → May be added to obtain an overall redox reaction, if the electrons cancel

Example 12.1c Combine the oxidation and reduction half-reactions in Examples 12.1a and 12.1b to obtain an overall redox reaction.

Solution The overall reaction is obtained by adding the half-reactions, making sure that the electrons cancel.

Oxidation half-reaction: $Zn(s) \rightarrow Zn^{2+}(aq) + 2e^-$

Reduction half-reaction: $Cu^{2+}(aq) + 2e^- \rightarrow Cu(s)$

Redox reaction: $Zn(s) + Cu^{2+}(aq) \rightarrow Zn^{2+}(aq) + Cu(s)$

12.2 Balancing Redox Equations

- **Balancing methods**
  - → Two methods are commonly used for balancing redox equations.
  - → The half-reaction method is introduced in Toolbox 12.1 of the text.
  - → An alternative method employs oxidation numbers. The basic methodology is the same: essentially, both mass and charge are balanced. Mass is first balanced by insuring that there are equal numbers of each type of atom undergoing oxidation or reduction on each side of the equation. Balancing both charge and mass of other atoms is obtained through use of oxidation numbers.
  - → A convention used in either method is the representation of hydronium ions, $H_3O^+(aq)$, by the shorthand symbol, $H^+(aq)$. The "conversion" equation is

$$H^+(aq) + H_2O(l) \rightarrow H_3O^+(aq) \quad \Delta G_r{}^\circ = 0 \text{ at } 25°C$$

- **Half-reaction method**
  - → A 6-step procedure outlined in Toolbox 12.1 and reproduced here
    1. Identify all species being oxidized and reduced from the changes in their oxidation numbers.
    2. Write unbalanced skeletal equations for each half-reaction.
    3. Balance all elements in each half-reaction except O and H.
    4. For an *acidic* solution, balance O atoms by using H_2O, then balance H atoms by adding H^+. For a *basic* solution, balance O atoms by using H_2O, then balance H atoms by adding H_2O to the side of each half-reaction that needs H and adding OH^- to the other side.

Note: The last part (adding H_2O to one side and OH^- to the other side) has the *net* effect of adding H atoms to balance hydrogen, if charge is ignored. Charge is balanced in the next step.

5. Balance electron charges by adding electrons to the appropriate side of each half-reaction. Add electrons to the left side for reduction and to the right side for oxidation.
6. The number of electrons in each half-reaction should be the same. Multiply the half-reactions by factors that give an equal number of electrons in each reaction. Finally, add the two half-reactions together to obtain the balanced redox equation.

→ An alternative procedure for *basic* solutions is to balance the redox equation for *acid* solutions first. Then, add the appropriately balanced equation $n\,H^+(aq) + n\,OH^-(aq) \rightarrow n\,H_2O(l)$ *or* $n\,H_2O(l) \rightarrow n\,H^+(aq) + n\,OH^-(aq)$ to the redox equation to eliminate $n\,H^+$ from the product *or* reactant side, respectively.

Example 12.2a In acid solution, tin metal is oxidized by the nitrate ion to produce Sn(IV) ions and dinitrogen monoxide gas. Write a balanced equation for this redox reaction.

Solution Following the procedure in step 1, we obtain

$$\underset{0}{Sn(s)} + \underset{+5}{NO_3^-(aq)} \rightarrow \underset{+4}{Sn^{4+}(aq)} + \underset{+1}{N_2O(g)} \qquad \textit{oxidation numbers}$$

The oxidation numbers beneath the Sn and N species show that Sn is oxidized (4 electrons transferred per Sn atom) and N is reduced (4 electrons transferred per N atom).

We then determine in step 2 that the unbalanced skeletal half-reactions are

Reduction: $NO_3^-(aq) \rightarrow N_2O(g)$
Oxidation: $Sn(s) \rightarrow Sn^{4+}(aq)$

According to step 3, we must balance all elements except O and H. The oxidation reaction is already balanced, but the reduction reaction is not. A coefficient of 2 is required on the nitrate ion to balance the N atoms. The result is

Reduction: $2\,NO_3^-(aq) \rightarrow N_2O(g)$
Oxidation: $Sn(s) \rightarrow Sn^{4+}(aq)$

Following the procedure for acidic solutions in step 4, we obtain

Reduction: $2\,NO_3^-(aq) \rightarrow N_2O(g) + 5\,H_2O(l)$ (*balance* O with H_2O)
$10\,H^+(aq) + 2\,NO_3^-(aq) \rightarrow N_2O(g) + 5\,H_2O(l)$ (*balance* H with H^+)
Oxidation: $Sn(s) \rightarrow Sn^{4+}(aq)$ (O and H are not present)

Balancing charge with electrons (step 5) is sometimes tricky.

Reduction: On the left side, $10(+1) + 2(-1) = +8$ and on the right side, $1(0) + 5(0) = 0$. Thus, 8 electrons are required on the left side to reduce the charge to 0.
$8\,e^- + 10\,H^+(aq) + 2\,NO_3^-(aq) \rightarrow N_2O(g) + 5\,H_2O(l)$

Oxidation: On the left side, $1(0) = 0$ and on the right side, $1(+4) = +4$. Thus, 4 electrons are required on the right side to reduce the charge to 0.
$Sn(s) \rightarrow Sn^{4+}(aq) + 4\,e^-$

The number of electrons in each half-reaction should be the same before they are combined (step 6). If we multiply the oxidation step by 2, then 8 electrons are transferred in each half-reaction. The two half-reactions are

Reduction: $8\,e^- + 10\,H^+(aq) + 2\,NO_3^-(aq) \rightarrow N_2O(g) + 5\,H_2O(l)$

Oxidation: $2\,Sn(s) \rightarrow 2\,Sn^{4+}(aq) + 8\,e^-$

The addition of the two half-reactions yields the final balanced redox equation:

$$10\,H^+(aq) + 2\,Sn(s) + 2\,NO_3^-(aq) \rightarrow 2\,Sn^{4+}(aq) + N_2O(g) + 5\,H_2O(l)$$

Example 12.2b Balance the following skeletal equation for a reaction that occurs in basic solution: $HXeO_4^-(aq) \rightarrow XeO_6^{4-}(aq) + Xe(g)$

+6 +8 0 *oxidation numbers*

Solution The oxidation numbers beneath the Xe species show that Xe is both oxidized and reduced, a type of reaction that is called *disproportionation*. The skeletal equations are

Reduction: $HXeO_4^-(aq) \rightarrow Xe(g)$

Oxidation: $HXeO_4^-(aq) \rightarrow XeO_6^{4-}(aq)$

Because Xe is already balanced in both half-reactions, we jump to step 4.

Add $4\,H_2O$ to the right side to balance the 4 O atoms on the left side of the reduction reaction. For a basic solution, add 7 H_2O to the left side and $7\,OH^-$ to the right side to balance the 7 excess H atoms on the right side. The net result is $3\,H_2O$ on the left side of the reduction reaction.

Add $2\,H_2O$ to the left side to balance the 2 excess O atoms on the right side of the oxidation reaction. For a basic solution, add 5 H_2O to the right side and 5 OH^- to the left side to balance the 5 H atoms on the left side. The net result is $3\,H_2O$ on the right side of the oxidation reaction. The result for each half-reaction is

Reduction: $3\,H_2O(l) + HXeO_4^-(aq) \rightarrow Xe(g) + 7\,OH^-(aq)$

Oxidation: $5\,OH^-(aq) + HXeO_4^-(aq) \rightarrow XeO_6^{4-}(aq) + 3\,H_2O(l)$

Next, balance charge with electrons. The reduction reaction requires 6 electrons on the left side to produce a total charge of –7 on both sides. The oxidation reaction requires 2 electrons on the right side to produce a total charge of –6 on both sides. If the oxidation equation is multiplied by a factor of 3, the number of electrons transferred in each half-reaction will be identical. Then combine the half-reactions to produce a balanced redox equation in a basic solution.

Reduction: $6\,e^- + 3\,H_2O(l) + HXeO_4^-(aq) \rightarrow Xe(g) + 7\,OH^-(aq)$

Oxidation: $15\,OH^-(aq) + 3\,HXeO_4^-(aq) \rightarrow 3\,XeO_6^{4-}(aq) + 9\,H_2O(l) + 6\,e^-$

Redox: $8\,OH^-(aq) + 4\,HXeO_4^-(aq) \rightarrow Xe(g) + 3\,XeO_6^{4-}(aq) + 6\,H_2O(l)$

Note: It is a good practice to check the balanced equation at the finish. In the above equation, there are 4 Xe atoms, 12 H atoms, and 24 O atoms on each side. Also, the charge on each side sums to –12. Therefore, the final equation is balanced correctly and is expressed in the smallest whole number ratio of stoichiometric integers as well.

Galvanic Cells (Sections 12.3-12.13)

Key Concepts

electrochemical cell, galvanic cell, anode, cathode, Daniell cell, electrode, hydrogen electrode, salt bridge, cell diagram, cell notation, cell potential, cell potential and reaction free energy, standard electrode potential (or standard potential), protective oxide coating, passivation, electrochemical series, disproportionation, standard potentials and equilibrium constants, concentration dependence of cell potential, Nernst equation, ion-selective electrode, corrosion, galvanize, cathodic protection, sacrificial anode

Overview

- **Electrochemical cell**
  - → Device that utilizes a *spontaneous* chemical reaction to produce an electric current *or* one that utilizes an electric current to bring about a *nonspontaneous* chemical reaction
  - → Chemical reaction ⇔ electric current
- **Galvanic cell**
  - → Electrochemical cell in which a spontaneous chemical reaction is used to produce an electric current
  - → Spontaneous chemical reaction ⇒ electric current
  - → Galvanic cells are also called *voltaic cells*.
- **Battery**
  - → The technical definition is a collection of electrochemical cells joined in a series to produce a higher voltage than a single cell.
  - → The common definition is a single galvanic cell or several galvanic cells connected in series.

12.3 Examples of Galvanic Cells

- **Galvanic cell components**
  - → Two *electrodes* (metallic conductors and/or conducting solids) make electrical contact with the cell contents, the conducting medium.
  - → An *electrolyte* (an ionic solution, paste or crystal) is the conducting medium.
  - → The electrode at which oxidation occurs is called the *anode*. In a commercial cell, the anode is labeled (–).
  - → The electrode at which reduction occurs is called the *cathode*. In a commercial cell, the cathode is labeled (+).
  - → Normally, the oxidation reaction and reduction reaction compartments are separated to prevent the mixing of ions in the electrolyte solutions. Electrical contact is maintained by a *porous pot* in the Daniell cell (see Fig. 12.4 in the text) or by a more effective *salt bridge* in modern cells. The salt bridge is often a gel containing KCl.

- **Example of a galvanic cell**
  - → Zinc metal reacts spontaneously with Cu(II) ions in aqueous solution. Copper metal precipitates and Zn(II) ions enter the solution. The reaction is exothermic, and if Zn metal is added directly to a solution containing Cu(II), heat is evolved as suggested by the standard enthalpy of reaction

$$Zn(s) + Cu^{2+}(aq) \rightarrow Zn^{2+}(aq) + Cu(s) \qquad \Delta H_r° = -225.56 \text{ kJ·mol}^{-1}$$

  - → A galvanic cell that uses this chemical reaction to produce electrical work is shown schematically (also see Fig. 12.5 in the text) and described below.

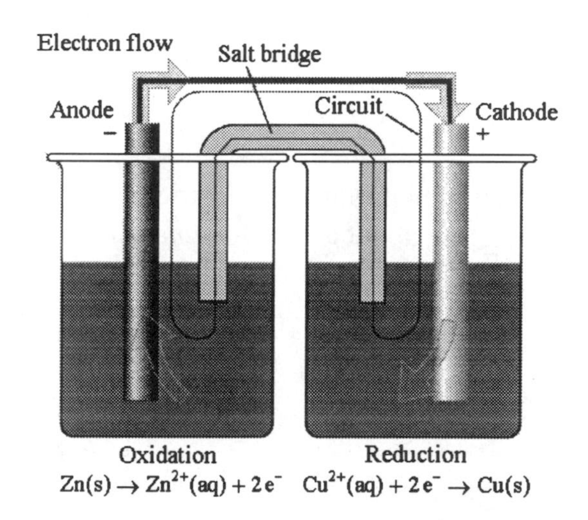

The cell is composed of an *anode compartment* and a *cathode compartment* linked by a salt bridge to prevent the mixing of ions. The anode compartment contains a zinc electrode in a solution of an electrolyte such as KCl. The cathode compartment contains an electrode (not necessarily Cu) in a solution of an electrolyte with $Cu^{2+}(aq)$ ions. The salt bridge contains an electrolyte such as KCl in an aqueous gelatinous medium to control the flow of $K^+(aq)$ and $Cl^-(aq)$ ions. When current is drawn from the cell, the electrodes are linked by an external electrical wire to a load (examples include a light bulb or motor). Current (electrons) flows through the wire as indicated and the spontaneous chemical *cell reaction* $Zn(s) + Cu^{2+}(aq) \rightarrow Cu(s) + Zn^{2+}(aq)$ occurs. As the reaction proceeds, note that Zn(s) is oxidized to $Zn^{2+}(aq)$ in the anode compartment (oxidation) and $Cu^{2+}(aq)$ is reduced to Cu(s) in the cathode compartment (reduction). The Cu(s) deposits on the cathode. The closed line labeled *circuit* is a schematic representation of the current that flows in the completed cell. The current is carried by *electrons* in the electrodes and external connections, and by *ions* in the conducting medium. Both conducting solutions separately maintain electrical neutrality.

Example 12.3a Describe the motion of charge in the preceding galvanic cell. Write the anode and cathode reactions. Combine the half-reactions to yield the cell reaction.

Solution Begin with the zinc electrode (anode) and follow a clockwise motion around the cell. At the zinc electrode, Zn(s) is oxidized to $Zn^{2+}(aq)$. The electrons released to the metal flow through the external circuit (load) and produce excess negative charge on the electrode in the cathode compartment. Excess electrons on the cathode reduce $Cu^{2+}(aq)$ to Cu(s). Physically, copper plates out on the electrode. To maintain charge balance, $K^+(aq)$ ions in the salt bridge flow to the cathode compartment to compensate for the loss of $Cu^{2+}(aq)$. Also, $Cl^-(aq)$ ions in the salt bridge flow to the anode compartment to counteract the gain of positive charge by the release of $Zn^{2+}(aq)$ ions. This flow completes the circuit.

The half-reactions and overall cell reaction are

Oxidation (anode):	$Zn(s) \rightarrow Zn^{2+}(aq) + 2\,e^-$
Reduction (cathode):	$Cu^{2+}(aq) + 2\,e^- \rightarrow Cu(s)$
Cell reaction:	$Zn(s) + Cu^{2+}(aq) \rightarrow Zn^{2+}(aq) + Cu(s)$

Example 12.3b The reaction $H_2(g) + NiCl_2(aq) \rightarrow 2\,HCl(aq) + Ni(s)$ is *not* spontaneous. Could this reaction be used in a galvanic cell?

Solution A galvanic cell cannot be constructed using $H_2(g)$ and $NiCl_2(aq)$ as reactants. Only spontaneous reactions can be used in galvanic cells.

Example 12.3c The cell reaction $2\,HCl(aq) + Ni(s) \rightarrow H_2(g) + NiCl_2(aq)$ is spontaneous. Note that this is the reverse of the preceding reaction. Write anode and cathode reactions for a galvanic cell utilizing this chemical reaction.

Solution From the cell reaction, Ni(s) is oxidized (anode) to $Ni^{2+}(aq)$ and $H^+(aq)$ is reduced (cathode) to $H_2(g)$. The half-reactions are

Oxidation (anode):	$Ni(s) \rightarrow Ni^{2+}(aq) + 2\,e^-$
Reduction (cathode):	$2\,H^+(aq) + 2\,e^- \rightarrow H_2(g)$

Note: The anode electrode is Ni(s). The cathode may be any inert electrode, such as platinum wire or graphite rod. Any electrolyte that does not react with Ni(s) may be used in the anode compartment. The cell reaction suggests that hydrochloric acid is the likely electrolyte in the cathode compartment. Because $H_2(g)$ is relatively insoluble in water (see Henry's law), it will form bubbles at the cathode and leave the cell.

12.4 The Notation for Cells

- **Cell diagram**
  - → A symbolic representation of the components of a cell
  - → A single vertical line | represents a phase boundary.
  - → A double vertical line || represents a salt bridge.
  - → Reactants and products are represented by chemical symbols.
  - → Components in the same phase are separated by commas.
  - → The anode components are written first, followed by the salt bridge, if present, and the cathode components. Phase symbols are normally used.

Example 12.4a Write a cell diagram for a galvanic cell for which the cell reaction is $Mg(s) + Zn^{2+}(aq) \rightarrow Zn(s) + Mg^{2+}(aq)$.

Solution The half-reactions are

Oxidation (anode):	$Mg(s) \rightarrow Mg^{2+}(aq) + 2\,e^-$
Reduction (cathode):	$Zn^{2+}(aq) + 2\,e^- \rightarrow Zn(s)$

The diagram for the anode is $Mg(s)\,|\,Mg^{2+}(aq)$. The electrode is Mg(s).
The diagram for the cathode is $Zn^{2+}(aq)\,|\,Zn(s)$. The electrode may be Zn(s).
The cell diagram with a salt bridge is $Mg(s)\,|\,Mg^{2+}(aq)\,||\,Zn^{2+}(aq)\,|\,Zn(s)$.

Note: In each compartment, the reactant in the cell equation is written first.

Example 12.4b Write a cell diagram for a galvanic cell for which the cell reaction is
$3\,OCl^-(aq) + I^-(aq) \rightarrow IO_3^-(aq) + 3\,Cl^-(aq)$ (basic solution).

Solution The unbalanced half-reactions are

Oxidation (anode): $I^-(aq) \rightarrow IO_3^-(aq) + 6\,e^-$

Reduction (cathode): $OCl^-(aq) + 2\,e^- \rightarrow Cl^-(aq)$

The electrode material is not specified. The conducting solid, graphite, is a reasonable choice. A commonly used symbol for graphite is C(gr).

The diagram for the anode is $C(gr) \mid I^-(aq), IO_3^-(aq)$.

The diagram for the cathode is $OCl^-(aq), Cl^-(aq) \mid C(gr)$.

The cell diagram is $C(gr) \mid I^-(aq), IO_3^-(aq) \parallel OCl^-(aq), Cl^-(aq) \mid C(gr)$.

- **Cell diagrams with concentrations of reactants and products**
  - → Often the concentrations of reactants and products are included in the cell diagram to give a more complete description of the cell contents.
  - → An example is the following cell reaction and cell diagram:

$$Mg(s) + Zn^{2+}(aq, 1\ \text{M}) \rightarrow Zn(s) + Mg^{2+}(aq, 0.5\ \text{M})$$
$$Mg(s) \mid Mg^{2+}(aq, 0.5\ \text{M}) \parallel Zn^{2+}(aq, 1\ \text{M}) \mid Zn(s)$$

12.5 Cell Potential

- **Cell potential**
  - → The *cell potential, E,* is a measure of the ability of a cell reaction to move electrons through the external circuit of a cell.
  - → Other names for the cell potential are *electromotive force* (emf) and *voltage*. Other symbols commonly used are $\mathcal{E}$ and ΔV.
  - → The useful work a cell can perform depends on the cell potential and the amount of the limiting reactant present. A comparison of the work a cell can perform to the gravitational potential energy, *mgh*, is informative. The mass, *m*, is similar to the mass of the limiting reactant in a galvanic cell and the gravitational potential energy per unit mass, *gh*, is similar to the cell potential, *E*.
- **Units of the cell potential**
  - → SI base unit of electric current is the ampere, A. The SI derived unit of charge is the coulomb, $C \equiv A{\cdot}s$. The SI derived unit of energy is the joule, $J \equiv kg{\cdot}m^2{\cdot}s^{-2}$.
  - → SI derived unit of the cell potential is the volt, $V \equiv J{\cdot}C^{-1}$. A charge of one coulomb falling through a potential difference of one volt produces one joule of energy.
  - → Cell potential is measured with an electronic voltmeter connected to the two electrodes of a cell as shown in Fig. 12.6 in the text.
  - → By convention, galvanic cells are written with the reduction compartment (cathode) on the right side. The cell potential will then have a positive value.

- **Examples of galvanic cell reactions with measured cell potentials**
  - → $Mg(s) + Zn^{2+}(aq, 1\ \text{M}) \rightarrow Zn(s) + Mg^{2+}(aq, 1\ \text{M})$ $E = 1.60$ V
  - → $2H^{+}(aq, 1\ \text{M}) + Zn(s) \rightarrow Zn^{2+}(aq, 1\ \text{M}) + H_2(g, 1\ \text{atm})$ $E = 0.76$ V

12.6 Cell Potential and Reaction Free Energy

- **Derivation of the relationship between ΔG_r and E**
  - → From Section 7.13, the reaction free energy, ΔG_r, corresponds to the *maximum nonexpansion* work, w_e, that a reaction can do at constant temperature, T, and pressure, P. Assuming maximum nonexpansion work, $\Delta G_r = w_e$.
  - → The work done when n electrons in the cell reaction move through a potential difference, E, is the total electron charge times the potential difference. The charge per mole of electrons is N_A times the charge of a single electron, $-e = -1.602177 \times 10^{-19}$ C. The total electron charge is then $-neN_A$.
  - → At this point, it is customary to introduce a new constant for the *magnitude* of the charge of *one mole of electrons*, $F = eN_A = 9.648531 \times 10^4\ \text{C·mol}^{-1}$. The new constant, F, is called the *Faraday constant* and is often given a special symbol, $\mathcal{F}$.
  - → The work done and reaction free energy are now given by the product of the total charge, $-nF$, and the cell potential, E. The result is $\Delta G_r = w_e = -nFE$, which is Eq. 2 in Section 12.6 of the text. The value of n in the expression is the stoichiometric coefficient of the electrons in the oxidation and reduction half-reactions that are combined to determine the balanced cell reaction.
  - → The underlying assumption in Eq. 2 that *maximum* nonexpansion work is done requires the cell to operate in a *reversible* way. The potential E is only realized when the applied potential balances the potential generated by the cell. A working cell that generates current is *irreversible* in nature and produces a potential that is smaller than Eq. 2 predicts. Some of the energy produced is wasted as heat. Galvanic cells are, however, more efficient than combustion reactions.
  - → In summation, the relationship between the free energy of the cell reaction and the cell potential for *reversible* electrochemical cells is one of the most important concepts in this chapter and is given by

$$\boxed{\Delta G_r = -nFE}$$

Note: The cell potential E is an *intensive property* (independent of the balanced cell reaction), whereas ΔG_r is an *extensive property* (dependent on the balanced cell reaction). Doubling the stoichiometric coefficients in the cell reaction doubles the values of n and ΔG_r but does *not* change the value of E.

Note: The conditions for spontaneity and equilibrium that apply to ΔG_r also apply to E. If a cell reaction is spontaneous as written, $\Delta G_r < 0$ and $E > 0$. If a cell reaction is at equilibrium, $\Delta G_r = 0$ and $E = 0$. If a cell reaction is nonspontaneous as written, $\Delta G_r > 0$ and $E < 0$. A reaction that is nonspontaneous in one direction is always spontaneous in the other. By convention, a galvanic cell is set up in the spontaneous direction, and $E > 0$.

- **Standard cell potential, $E°$**

  → Recall the definition of the standard state of a species. The standard state of a solid, liquid, or gaseous substance is the pure substance at a pressure of 1 bar. For solute in solution, we take a molar concentration of 1 $mol{\cdot}L^{-1}$ as the standard-state activity.

  → The standard cell potential, $E°$, is the cell potential measured when all reactants and products are in their standard states. The resulting equation is Eq. 3 in Section 12.6.

$$\boxed{\Delta G_r° = -nFE°}$$

Example 12.6a A standard galvanic cell is constructed and operated at 25°C. The cell diagram is

$$Zn(s) \mid Zn^{2+}(aq,\ 1\ \text{M}) \parallel Cu^{2+}(aq,\ 1\ \text{M}) \mid Cu(s)$$

Determine the maximum electrical work done by the cell. Calculate the standard cell potential. Assume that all concentrations remain at the standard state value throughout the process. Use Appendix 2A for all required thermodynamic data.

Solution In the cell diagram, reduction occurs at the cathode on the right side and oxidation occurs at the anode on the left. The half-reactions and cell reaction are

$$\begin{array}{rll} Cu^{2+}(aq) + 2\,e^- \rightarrow Cu(s) & & \textit{cathode} \\ Zn(s) \rightarrow Zn^{2+}(aq) + 2\,e^- & & \textit{anode} \\ \hline Zn(s) + Cu^{2+}(aq) \rightarrow Zn^{2+}(aq) + Cu(s) & & \textit{cell reaction} \end{array}$$

The maximum electrical work is equal to the standard free energy of the cell reaction. Both Zn(s) and Cu(s) are the most stable forms of the pure elements at 25°C and 1 atm, thus the standard free energy of formation is 0 in each case. The free energy change will be

$$\begin{aligned} \Delta G_r° &= \sum n\Delta G_f°(\text{products}) - \sum n\Delta G_f°(\text{reactants}) \\ &= \Delta G_f°[Zn^{2+}(aq,\ 1\ \text{M})] - \Delta G_f°[Cu^{2+}(aq,\ 1\ \text{M})] = -147.06 - (+65.49)\ \text{kJ}{\cdot}\text{mol}^{-1} \\ &= -212.55\ \text{kJ}{\cdot}\text{mol}^{-1} = w_e \end{aligned}$$

Work with a negative sign is that done by the system on the surroundings. The work is done per mole of Cu^{2+} ions and Zn(s) that react and per mole of Zn^{2+} ions and Cu(s) that are formed. Two moles of electrons are transferred in the process. When calculating the cell potential, be careful to use consistent units. The stoichiometric coefficient of electrons in the cell reaction is n, which is unitless.

$$E° = \frac{-\Delta G_r°}{nF} = \frac{-(-212\,550\ \text{J}{\cdot}\text{mol}^{-1})}{(2)(96\,485\ \text{C}{\cdot}\text{mol}^{-1})} = 1.101\ \text{J}{\cdot}\text{C}^{-1} = 1.101\ \text{V}$$

Example 12.6b For the galvanic cell in Example 12.6a, calculate the maximum electrical work done when 2.000 g of copper are plated out. What is the standard cell potential? Again, assume that all concentrations remain at the standard state value.

Solution Work done in Example 12.6a is calculated for one mole of Cu(s) plated out. For 2.000 g of copper,

$n = m/M = (2.000\ \text{g})/(63.54\ \text{g}{\cdot}\text{mol}^{-1}) = 3.1476 \times 10^{-2}$ mol Cu(s).

$w_e = (-212.55\ \text{kJ}{\cdot}\text{mol}^{-1})(3.1476 \times 10^{-2}\ \text{mol}) = -6.6902$ kJ

The cell potential is an intensive property and $E° = 1.101$ V as in the preceding example.

12.7 Standard Electrode Potentials

- **Cell potential**
  → *Difference* between the cathode and anode potentials of an electrochemical cell
  → For cells with standard state components, the standard cell potential, $E°$, is the difference between the standard potentials of the cathode and the anode. The standard potential of an electrode is sometimes called the *standard reduction potential* or the *standard electrode potential.*

$$E° = E°(\text{cathode}) - E°(\text{anode})$$

Example 12.7a Consider the standard cell diagram

$$Zn(s) \mid Zn^{2+}(aq, 1\ \text{M}) \parallel Cu^{2+}(aq, 1\ \text{M}) \mid Cu(s)$$

Write the cell potential in terms of the standard reduction potentials of the half-reactions.

Solution The cathode reaction (*reduction*) is on the right side of the diagram and the anode reaction (*oxidation*) is on the left. The half-reaction and standard reduction potentials are

cathode:	$Cu^{2+}(aq, 1\ \text{M}) + 2\ e^- \rightarrow Cu(s)$	$E°(Cu^{2+}, Cu)$
anode:	$Zn(s) \rightarrow Zn^{2+}(aq, 1\ \text{M}) + 2\ e^-$	$-E°(Zn^{2+}, Zn)$

Note the sign change in the potential for the anode, because the electrode potentials correspond to reduction. The standard cell potential is then

$$E° = E°(Cu^{2+}, Cu) - E°(Zn^{2+}, Zn)$$

One way to remember this result is to note that the cell potential is given by

$$E° = E°(\text{right electrode}) - E°(\text{left electrode})$$

and right ≡ reduction.

- **Standard electrode potentials, $E°$**
  → The potential of a single electrode cannot be measured.
  → A relative scale is required, and by convention, the standard electrode potential of the hydrogen electrode is assigned a value of 0 at all temperatures.
  → For the standard hydrogen electrode, $E°(H^+, H_2) \equiv 0$. The half-reaction for reduction is $2\,H^+(aq, 1\ \text{M}) + 2\,e^- \rightarrow H_2(g, 1\ \text{bar})$, and the half-cell diagram is $H^+(aq, 1\ \text{M}) \mid H_2(g, 1\ \text{bar}) \mid Pt(s)$.
  → The electrode of interest is called the *test electrode.* If the test electrode is found to be the anode in combination with the standard hydrogen electrode, then it is assigned a negative potential; if it is the cathode, then its standard potential is positive. Values of standard electrode potentials measured at 25°C are given in Table 12.1 and Appendix 2B in the text.

Example 12.7b Describe how the standard electrode potential of the (Cu^{2+}, Cu) couple, $E°(Cu^{2+}, Cu)$, can be determined.

Solution Construct a standard galvanic cell consisting of the (Cu^{2+}, Cu) half-cell and the standard hydrogen electrode. If the test electrode is the cathode, then its standard potential is positive. The value of the standard electrode potential in Table 12.1 is +0.34 V; therefore we write the cell as

$$Pt(s) \mid H_2(g, 1\ bar) \mid H^+(aq, 1\ \mathrm{M}) \parallel Cu^{2+}(aq, 1\ \mathrm{M}) \mid Cu(s)$$

cathode: $Cu^{2+}(aq, 1\ \mathrm{M}) + 2\,e^- \rightarrow Cu(s)$ $\qquad E°(Cu^{2+}, Cu)$

anode: $H_2(g, 1\ bar) \rightarrow 2\,H^+(aq, 1\ \mathrm{M}) + 2\,e^-$ $\qquad -E°(H^+, H_2)$

The standard cell potential is

$$E° = E°(Cu^{2+}, Cu) - E°(H^+, H_2) = +0.34\ V$$

Because $E°(H^+, H_2) \equiv 0$, the value of the cell potential is assigned to (Cu^{2+}, Cu).

Example 12.7c Describe how the standard electrode potential of the (Al^{3+}, Al) couple, $E°(Al^{3+}, Al)$, can be determined.

Solution Construct a standard galvanic cell consisting of the (Al^{3+}, Al) half-cell and the standard hydrogen electrode. If the test electrode is the anode, then its standard potential is negative. The value of the standard electrode potential in Table 12.1 is –1.66 V, therefore we write the cell as

$$Al(s) \mid Al^{3+}(aq, 1\ \mathrm{M}) \parallel H^+(aq, 1\ \mathrm{M}) \mid H_2(g, 1\ bar) \mid Pt(s)$$

cathode: $2\,H^+(aq, 1\ \mathrm{M}) + 2\,e^- \rightarrow H_2(g, 1\ bar)$ $\qquad E°(H^+, H_2)$

anode: $Al(s) \rightarrow Al^{3+}(aq, 1\ \mathrm{M}) + 3\,e^-$ $\qquad -E°(Al^{3+}, Al)$

The standard cell potential is

$$E° = E°(H^+, H_2) - E°(Al^{3+}, Al) = +1.66\ V$$

Because $E°(H^+, H_2) \equiv 0$, the negative value of the cell potential is assigned to (Al^{3+}, Al).

- **Standard electrode potentials and free energy**
  → Eq. 3 in the text also applies to standard electrode potentials.
  → $\Delta G_f°$ values in Appendix 2A in combination with Eq. 3 can be used to determine an unknown standard potential.

Example 12.7d Under standard conditions, zinc metal is oxidized to Zn^{2+} in acid solution and a gas is evolved. Set up a galvanic cell for this reaction. Calculate the cell potential from the free energy of formation values in Appendix 2A. Determine the value of the standard electrode potential for (Zn^{2+}, Zn) and compare the value to the one listed in Table 12.1.

Solution Because the reaction is spontaneous, the galvanic cell diagram is

$$Zn(s) \mid Zn^{2+}(aq, 1\ \mathrm{M}) \parallel H^+(aq, 1\ \mathrm{M}) \mid H_2(g, 1\ bar) \mid Pt(s)$$

The cell reaction is $Zn(s) + 2\,H^+(aq, 1\ \mathrm{M}) \rightarrow Zn^{2+}(aq, 1\ \mathrm{M}) + H_2(g, 1\ bar)$.

$$\Delta G_r° = \sum n\Delta G_f°(\text{products}) - \sum n\Delta G_f°(\text{reactants})$$
$$= \Delta G_f°[Zn^{2+}(aq)] + \Delta G_f°[H_2(g)] - \{\Delta G_f°[Zn(s)] + 2\Delta G_f°[H^+(aq)]\}$$

$= -147.06\ \text{kJ}\cdot\text{mol}^{-1} + 0 - \{0 + 0\} = -147.06\ \text{kJ}\cdot\text{mol}^{-1}$ at 25°C

The stoichiometric coefficient of the electrons in each half-reaction is 2.

$$E° = \frac{-\Delta G_r°}{nF} = \frac{-(-147\,060\ \text{J}\cdot\text{mol}^{-1})}{(2)(96\,485\ \text{C}\cdot\text{mol}^{-1})} = 0.762\ \text{V}$$

$E° = E°(H^+, H_2) - E°(Zn^{2+}, Zn) = +0.762$ V and $E°(Zn^{2+}, Zn) = -0.762$ V, which is in excellent agreement with the value of –0.76 V in Table 12.1.

12.8 The Significance of Standard Potentials

- **Standard electrode potentials of half-reactions**
  - → Construct a cell made up of a standard *test* half-reaction at one electrode and the standard hydrogen electrode at the other.
  - → For a *test* half-reaction having a standard electrode potential less than zero, the half-reaction will be the oxidation (*anode*) step (see Example 12.7c). The H^+ (1 M) ions will be reduced to H_2 (1 bar) gas at the cathode.
  - → For a *test* half-reaction having a standard electrode potential greater than zero, the half-reaction will be the reduction (*cathode*) step (see Example 12.7b). The H_2 (1 bar) gas will be oxidized to H^+ (1 M) ions at the anode.
  - → The more negative the standard electrode potential, the greater the tendency to reduce H^+. The more positive the standard electrode potential, the greater the tendency to oxidize H_2.

- **Passivation**
  - → When a layer forms on a metal surface that prevents further reaction, the metal has been *passivated.*
  - → Aluminum has a large, negative standard electrode potential and therefore a great tendency to be oxidized by reaction with a strong acid. Reaction with oxidizing nitric acid produces an oxide layer (alumina, Al_2O_3) on aluminum metal that resists further reaction, whereas aluminum readily reacts with hydrochloric acid to form Al^{3+} that replaces H^+ ions in solution. The tendency of aluminum metal for passivation is an important property that allows the otherwise reactive metal to be used commercially.
  - → Metals that form *protective oxide* layers may be plated onto metals that are not readily passivated. Chromium, nickel, or zinc coatings are often used to protect iron from corrosion.

12.9 The Electrochemical Series

- **Electrochemical series**
  - → Standard half-reactions arranged in order of *decreasing* standard electrode potential (see Table 12.1 and Appendix 2B in the text)
  - → Table of relative strengths of oxidizing and reducing agents
  - → Note that *standard* electrode potentials have values that range from about +3 V to –3 V. Thus, the range of potentials in the electrochemical series is about 6 V. No

single *standard* galvanic cell may have a potential larger than this value, because there are no half-reactions that give a larger potential.

- **Understanding the electrochemical series**
  → The importance of the electrochemical series can be understood with the help of the following table of *standard* electrode potentials.

	Reduction Half-Reaction		**$E°$ (V)**
↑	$F_2(g) + 2e^- \rightarrow 2F^-(aq)$	reducing strength	+2.87
	$Mn^{3+}(aq) + e^- \rightarrow Mn^{2+}(aq)$		+1.51
	$I_2(s) + 2e^- \rightarrow 2I^-(aq)$	↓	+0.54
	$2H^+(aq) + 2e^- \rightarrow H_2(g)$		0
	$Fe^{2+}(aq) + 2e^- \rightarrow Fe(s)$		−0.44
	$Al^{3+}(aq) + 3e^- \rightarrow Al(s)$		−1.66
oxidizing strength	$Na^+(aq) + e^- \rightarrow Na(s)$		−2.71
	$Li^+(aq) + e^- \rightarrow Li(s)$		−3.05

  → On the *left side* of the table, oxidizing strength *increases* from bottom to top. $F_2(g)$ is an exceptionally strong oxidizing agent, because it has a large tendency to gain electrons, and be reduced. $F_2(g)$ will thereby oxidize all the product species on the right side of the table in the reactions listed below it.
  → On the *right side* of the table, reducing strength *increases* from top to bottom. Li(s) is an exceptionally strong reducing agent, because it has a large tendency to lose electrons, and be oxidized. Li(s) will thereby reduce all the reactant species on the left side of the table in the reactions listed above it.

- **Viewing half-reactions as conjugate pairs**
  → For a Brønsted-Lowry acid-base conjugate pair, the stronger the conjugate acid, the weaker the conjugate base.
  → For a given electrochemical half-reaction, the stronger the *conjugate* oxidizing agent, the weaker the *conjugate* reducing agent. $F_2(g)$ is an extremely powerful oxidizing agent, whereas $F^-(aq)$ is an extremely weak reducing agent.

- **Uses of the electrochemical series**
  → Predicting *relative* reducing and oxidizing strengths
  → Predicting which reactants may react spontaneously in a redox reaction
  → Calculating standard cell potentials

Example 12.9a Which one is the stronger oxidizing agent, Co^{3+} or Hg^{2+}?

Solution The following half-reactions and standard electrode potentials are taken from Appendix 2B.

$$Co^{3+}(aq) + e^- \rightarrow Co^{2+}(aq) \qquad E°(Co^{3+}, Co^{2+}) = +1.81\ V$$
$$Hg^{2+}(aq) + 2\,e^- \rightarrow Hg(s) \qquad E°(Hg^{2+}, Hg) = +1.62\ V$$

Because $E°(Co^{3+}, Co^{2+}) > E°(Hg^{2+}, Hg)$, Co^{3+} has a greater tendency to be reduced than does Hg^{2+}, so it is the stronger oxidizing agent.

Example 12.9b Is silver metal able to reduce chlorine gas? Calculate the standard cell potential.

Solution The half-reactions are

cathode: $Cl_2(g) + 2\,e^- \rightarrow 2\,Cl^-(aq)$ $\qquad E° = +1.36\ V$

anode: $Ag(s) \rightarrow Ag^+(aq) + e^-$ $\qquad -E° = -(+0.80\ V)$

For the standard cell reaction, $E° = E°(\text{cathode}) - E°(\text{anode}) = +0.56\ V$.
A *positive* cell potential corresponds to a *negative* free energy change. Therefore, the reaction is spontaneous, and $Cl_2(g)$ is expected to be able reduce $Ag(s)$.

- **Calculating the standard potential of a half-reaction from related half-reactions**
  - → Combining two half-reactions (A and B) to obtain a third half-reaction (C) is accomplished by addition of the associated free energy changes. If a half-reaction is an oxidation step, the potential for that step changes sign.
  - → The equation, $\Delta G_r° = -nFE°$, applies to each half-reaction, but $E°$ may now be associated with *either* a reduction *or* an oxidation potential for a half-reaction. Recall that the sign of $\Delta G_r°$ changes if the reactants and products are interchanged.
  - → In general, $\Delta G_r°(C) = \Delta G_r°(A) + \Delta G_r°(B)$ and $-n_C FE°(C) = -n_A FE°(A) - n_B FE°(B)$.

$$E°(C) = \frac{n_A E°(A) + n_B E°(B)}{n_C}$$

Example 12.9c Calculate $E°$ for the half-reaction $Cu^{2+}(aq) + e^- \rightarrow Cu^+(aq)$. Find appropriate half-reactions listed in Appendix 2B. Compare your calculated value to the one listed in the appendix.

Solution The half-reactions in Appendix 2B that contain Cu^{2+} and Cu^+ and one other species in common are

$$Cu^{2+}(aq) + 2\,e^- \rightarrow Cu(s) \qquad E° = +0.34\ V$$
$$Cu^+(aq) + e^- \rightarrow Cu(s) \qquad E° = +0.52\ V$$

Because Cu^+ is a product in the desired reaction, rewrite the second half-reaction as an oxidation step.

$$Cu(s) \rightarrow Cu^+(aq) + e^- \qquad E° = -0.52\ V$$

The reactions may be combined as follows.

$Cu^{2+}(aq) + 2\,e^- \rightarrow Cu(s)$	$E°(A) = +0.34\ V$	$n_A = 2$
$Cu(s) \rightarrow Cu^+(aq) + e^-$	$E°(B) = -0.52\ V$	$n_B = 1$
$Cu^{2+}(aq) + e^- \rightarrow Cu^+(aq)$	$E°(C) = \quad ?$	$n_C = 1$

$$E°(C) = \frac{n_A E°(A) + n_B E°(B)}{n_C} = \frac{(2)(+0.34\ V) + (1)(-0.52\ V)}{(1)} = +0.16\ V$$

The value of +0.15 V in the appendix is within the round-off errors. Note that adding the potential $E°(A)$ directly to $E°(B)$ gives an erroneous answer of −0.18 V. Also, note that $n_C \neq n_A + n_B$. If both reactions A and B are reduction *or* oxidation steps, then $n_C = n_A + n_B$.

12.10 Standard Potentials and Equilibrium Constants

- **Equilibrium constants from standard cell potentials**
  - → Redox reactions
  - → Acid-base reactions
  - → Dissolution/precipitation reactions
  - → Dilution reactions (reactions involving a change in concentration)
- **Quantitative aspects**
  - → Combining the equations $\Delta G_r° = -RT\ln K$ from Chapter 9 and $\Delta G_r° = -nFE°$ from this chapter leads to

$$\ln K = \frac{nFE°}{RT} = \frac{nE°}{0.025\,693\text{ V}} \quad \text{at } T = 298.15\text{ K}$$

  - → Standard electrode potentials are used to calculate the value of $E°$.
  - → Because $\Delta G_r°$ also applies to reactions that are not redox in character, the equation shown above may be used for acid-base reactions, precipitation reactions and dilution reactions. The overall "cell" reaction will not show explicitly the electron transfer that occurs in the half-reactions.

Example 12.10 Calculate the equilibrium constant for the reaction

$$Ni^{2+}(aq) + Zn(s) \rightarrow Ni(s) + Zn^{2+}(aq)$$

at 298.15 K.

Solution Write the reaction as the sum of two half-reactions and obtain the half-reaction potentials from Appendix 2B. Calculate the cell potential and use the equation given above to determine K.

cathode:	$Ni^{2+}(aq) + 2\,e^- \rightarrow Ni(s)$	$E° = -0.23$ V
anode:	$Zn(s) \rightarrow Zn^{2+}(aq) + 2\,e^-$	$-E° = -(-0.76$ V)
cell:	$Ni^{2+}(aq) + Zn(s) \rightleftharpoons Ni(s) + Zn^{2+}(aq)$	$E° = +0.53$ V

For the overall cell reaction, $n = 2$.

$$\ln K = \frac{(2)(+0.53\text{ V})}{(0.025\,693\text{ V})} = 41.256 \quad \text{and} \quad K = e^{41.256} = 8.3 \times 10^{17}$$

Notes: As expected, the equilibrium position is close to the product side of the reaction. Example 12.7 in the text demonstrates the application to precipitation reactions. The overall "cell" reaction is not then an *explicit* redox reaction. Also, a cell reaction may be set up such that $\Delta G_r° > 0$ and $E° < 0$. In this case, $K < 1$, and the cathode reaction (reduction) will actually occur at the left electrode.

12.11 The Nernst Equation

- **Properties of a galvanic cell**
  - → The cell potential, E, decreases as the cell discharges.
  - → The reactants form products as the cell discharges. The cell potential is zero when the cell is completely discharged.
  - → The cell reaction is at equilibrium when the cell is completely discharged.
- **Quantitative aspects**
  - → The relationship between the cell potential, E, and the chemical composition of the cell (concentration of reactants and products) is obtained from the properties of ΔG_r presented in Chapter 9.
  - → From Chapter 9, $\Delta G_r = \Delta G_r° + RT\ln Q$, where Q is the reaction quotient. Note that in Section 12.16, the *identical* symbol Q is used for the quantity of electricity.
  - → From this chapter, $\Delta G_r = -nFE$ and $\Delta G_r° = -nFE°$.
  - → Combining the above equations leads to the *Nernst equation*.

$$E = E° - \frac{RT}{nF}\ln Q = E° - \frac{(0.025\,693\ \text{V})}{n}\ln Q \quad \text{at } 298.15\ \text{K}$$

- **The Nernst equation**
  - → Describes the quantitative relationship between the cell potential and the chemical composition of the cell
  - → Is used to estimate the potential of a cell if its chemical composition is known

Example 12.11 Calculate the potential of the cell

$Pt(s) \mid In^{2+}(aq, 2.0\ M), In^{3+}(aq, 0.1\ M) \parallel Sn^{4+}(aq, 2.0\ M), Sn^{2+}(aq, 0.1\ M) \mid Pt(s)$

If the concentration of $Sn^{2+}(aq)$ increases, describe the effect on the cell potential.

Solution Determine the half-reactions and their standard potentials. Write the cell reaction and calculate $E°$. Then calculate Q, the reaction quotient, from the given concentrations of the reactants and products. Substitute $E°$ and Q into the Nernst equation and calculate E.

The half-reactions are

cathode:	$Sn^{4+}(aq) + 2e^- \rightarrow Sn^{2+}(aq)$	$E° = +0.15$ V
anode:	$In^{3+}(aq) + e^- \rightarrow In^{2+}(aq)$	$E° = -0.49$ V

Since the cathode reaction requires 2 electrons, the half-reactions become

cathode:	$Sn^{4+}(aq) + 2e^- \rightarrow Sn^{2+}(aq)$	$E° = +0.15$ V
anode:	$2\,In^{2+}(aq) \rightarrow 2\,In^{3+}(aq) + 2e^-$	$-E° = -(-0.49$ V)
cell:	$Sn^{4+}(aq) + 2\,In^{2+}(aq) \rightarrow Sn^{2+}(aq) + 2\,In^{3+}(aq)$	$E° = +0.64$ V

$$Q = \frac{[Sn^{2+}][In^{3+}]^2}{[Sn^{4+}][In^{2+}]^2} = \frac{(0.10)(0.10)^2}{(2.0)(2.0)^2} = 1.25 \times 10^{-4}$$

Finally, substitute the values of $E°$ and Q into the Nernst equation, with $n = 2$.

$$E = E° - \frac{0.025\,693\text{ V}}{n} \ln Q = +0.64\text{ V} - \frac{0.025\,693\text{ V}}{2} \ln(1.25 \times 10^{-4})$$

$$E = +0.64\text{ V} - (-0.115\text{ V}) = +0.76\text{ V}$$

Note: The cell potential is greater than the standard cell potential. This situation occurs when $Q < 1$ in the Nernst equation. As the cell discharges, reactants will be converted to products and Q will increase until equilibrium is attained. At equilibrium, $E = 0$ and $Q = K$.

12.12 Ion-Selective Electrodes

- **Ion-selective electrode**
  - → An electrode sensitive to the concentration of a particular ion in solution
  - → A simple example is a metal wire inserted into a solution containing the corresponding metal ion. The electrode potential for $E(Ag^+, Ag)$ or $E(Cu^{2+}, Cu)$ is sensitive to concentration of the metal ion Ag^+ or Cu^{2+} in solution.
- **Measurement of pH**
  - → An important application of the Nernst equation
  - → A galvanic cell that includes the solution of interest employs an ion-selective electrode sensitive to the concentration of H^+.
  - → An example given in the text is that of a hydrogen electrode connected by a salt bride to the calomel electrode, $Hg_2Cl_2(s) + 2e^- \rightarrow 2Hg(l) + 2Cl^-(aq)$, $E° = +0.27$ V. The cell reaction is $Hg_2Cl_2(s) + H_2(g) \rightarrow 2H^+(aq) + 2Hg(l) + 2Cl^-(aq)$, $E° = +0.27$ V. The Nernst equation can be written for this cell reaction in terms of the unknown concentration of H^+. The Cl^- concentration in the calomel electrode is fixed at a value determined for a saturated solution of KCl and the pressure of H_2 gas is usually 1 bar.
- **Glass electrode**
  - → A thin-walled glass bulb containing an electrolyte with a potential that is proportional to pH is called a *glass electrode*. Usually, a calomel electrode is built into the probe that contacts the test solution through a small salt bridge. The probe contains a complete electrochemical cell and is much easier to use than a hydrogen electrode.
  - → An electrode of this type may be used in *pX meters* that are sensitive to other ions in industrial applications and pollution control.

12.13 Corrosion

- **Corrosion**
  - → Unwanted oxidation of a metal
  - → An electrochemical process that is extremely destructive
  - → Corrosion may be slowed by use of metal coatings or a block of a more strongly reducing metal.

- **Mechanism of corrosion**
  - → Exposed metal surfaces can act as anodes or cathodes, points where oxidation or reduction can occur.
  - → If the metal is wet (from dew if above ground or groundwater if below), a film of water containing dissolved ions may link the anode and cathode, acting as a salt bridge.
  - → The circuit is completed by electrons flowing through the metal, which acts as a wire in the external circuit of a galvanic cell.
- **Mechanism of rust formation on iron metal**
  - → At the *anode*, iron is oxidized to iron(II) ions.

    (1) $Fe(s) \rightarrow Fe^{2+}(aq) + 2\,e^-$ $\quad -E° = -(-0.44\text{ V}) = +0.44\text{ V}$
  - → At the *cathode*, oxygen is reduced to water.

    (2) $O_2(g) + 4\,H^+(aq) + 4\,e^- \rightarrow 2\,H_2O(l)$ $\quad E° = +1.23\text{ V}$
  - → Iron(II) migrates to the cathode where it may be oxidized by oxygen to iron(III).

    (3) $Fe^{2+}(aq) \rightarrow Fe^{3+}(aq) + e^-$ $\quad -E° = -(+0.77\text{ V}) = -0.77\text{ V}$
  - → The overall redox equation is obtained by adding equation (1) four times, equation (3) four times, and equation (2) three times. The resulting cell reaction has $n = 12$.

    (4) $4\,Fe(s) + 3\,O_2(g) + 12\,H^+(aq) \rightarrow 4\,Fe^{3+}(aq) + 6\,H_2O(l)$ $\quad E° = +1.27\text{ V}$
  - → Next, the iron(III) ions in equation (4) precipitate as a hydrated iron(III) oxide or rust. The coefficient, x, of the waters of hydration in the oxide is not well defined.

    (5) $4\,Fe^{3+}(aq) + 2(3 + x)\,H_2O(l) \rightarrow 2\,Fe_2O_3 \cdot xH_2O(s) + 12\,H^+(aq)$
  - → Finally, the overall reaction for the formation of rust is obtained by adding equation (4) and (5).

    (6) $4\,Fe(s) + 3\,O_2(g) + 2x\,H_2O(l) \rightarrow 2\,Fe_2O_3 \cdot xH_2O(s)$
  - → From this mechanism of rust formation, we expect corrosion to be accelerated by moisture, oxygen, and the presence of a salt, which dissolves in water and enhances the rate of ion transport in the reaction.
- **Combating corrosion**
  - → The surface of the metal may be protected by *painting*. This process is not uniform and the bonding process is subject to imperfections that lead to deterioration of the coating.
  - → A superior method that provides greater protection is to *galvanize* the metal, a process that deposits an unbroken film of zinc metal on the surface. Zinc is preferentially oxidized, since its reduction potential is more negative than iron. The zinc oxide that forms provides a protective coating because of the passivation process.
  - → A method that is used for large structures is *cathodic protection*. Instead of coating the metal of interest with zinc or another suitable metal, a block of metal with a more negative reduction potential is attached to the metal to be protected. The block of metal is more readily oxidized and serves as a *sacrificial anode*.

Example 12.13 Consult Appendix 2B to determine a list of possible sacrificial anodes for iron.

Solution Any metal in Appendix 2B with an electrode potential value more negative than that of iron is a candidate. Very reactive alkali metals are not practical, however. The relevant electrode potentials for iron are

$$Fe^{2+}(aq) + 2\,e^- \rightarrow Fe(s) \quad E° = -0.44\ V$$
$$Fe^{3+}(aq) + 3\,e^- \rightarrow Fe(s) \quad E° = -0.04\ V$$

Thus, a candidate for a successful sacrificial anode must have an electrode potential value more negative than –0.44 V.

Except for the very reactive alkali metals, the metals in Appendix 2B that qualify are Ga, Cr, Zn, Mn, V, Ti, Al, <u>U</u>, Be, Mg, Ce, La, Ca, Sr, Ba, and <u>Ra</u>. Radioactive metals (underlined) and metals of high cost are not feasible choices.

Electrolysis (Sections 12.14-12.17)

Key Concepts

electrolytic cell, electrolysis, potential needed for electrolysis, the products of electrolysis, Faraday's law of electrolysis, electroplating

Overview

- **Electrolysis**
  - → Process of driving a reaction in a nonspontaneous direction using electric current supplied from an external source
  - → Conducted in an electrolytic cell
  - → Important for the synthesis of chemical species, and for plating metals or other conducting surfaces

12.14 Electrolytic Cells

- **Electrolytic cell**
  - → An electrochemical cell in which electrolysis occurs
  - → Different from a galvanic cell in design
  - → Both electrodes normally share the same compartment.
  - → As in a galvanic cell, oxidation occurs at the anode and reduction occurs at the cathode.
  - → Unlike a galvanic cell, the anode has a *positive* charge and the cathode is *negative*.
- **Example of an electrolytic cell**
  - → Sodium metal is produced mainly by the electrolysis of molten sodium chloride to which some calcium chloride has been added. Calcium chloride lowers the melting point from 801°C (pure NaCl) to about 600°C (mixture of NaCl and $CaCl_2$, recall freezing point depression), increasing the energy efficiency of the process.
  - → An electrolytic cell that uses electrical energy to produce Na(l) and Cl_2(g) is shown schematically (also see Fig. 12.18 in the text) and described below.

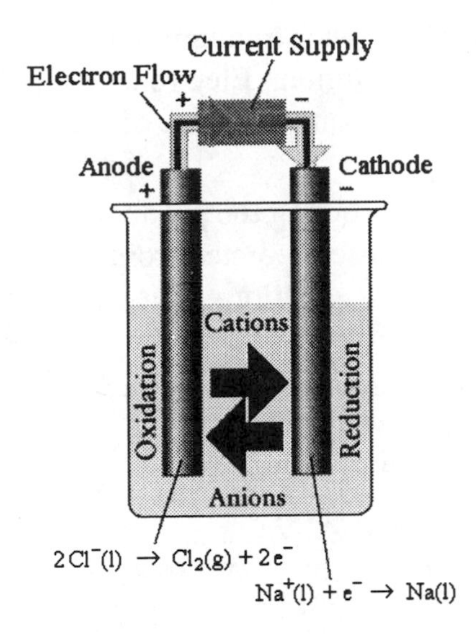

The cell is composed of one compartment that contains both the anode and cathode. The electrode material is usually an inert metal such as platinum. To drive the nonspontaneous reaction, electrical current is supplied by an external power source, such as a battery or power supply. The current forces reduction to occur at the cathode and oxidation at the anode. The cathode is placed on the right side as in a galvanic cell. The reactions that occur are

cathode: $Na^+(l) + e^- \rightarrow Na(l)$

anode: $2\,Cl^-(l) \rightarrow Cl_2(g) + 2\,e^-$

Note that liquid sodium metal is formed in the electrolysis because the cell temperature of 600°C is above the melting point of sodium (98°C).

Example 12.14 Consider an electrolytic cell in which a solution of NaCl(aq) at 25°C is used in place of the molten salt. Sodium metal is very reactive with water and will not form at the cathode. Chlorine gas, however, may form at the anode. Consider the half-reactions that form Na(s) and $Cl_2(g)$. Determine the standard electrode potential for each half-reaction and the standard cell potential with Na(s) forming at the cathode.

Solution The half-reactions and standard electrode potentials are

cathode: $Na^+(aq) + e^- \rightarrow Na(s)$ $\quad E° = -2.71\text{ V}$

anode: $2\,Cl^-(aq) \rightarrow Cl_2(g) + 2\,e^-$ $\quad -E° = -(+1.36\text{ V}) = -1.36\text{ V}$

cell: $2\,Na^+(aq) + 2\,Cl^-(aq) \rightarrow 2\,Na(s) + Cl_2(g)$ $\quad E° = -4.07\text{ V}$

The cell potential is unfavorable as written and an external power source with a voltage greater than 4.07 V is required to drive the reaction. Actually, the reduction of water will occur rather than of Na^+. The cathode reaction is

cathode: $2\,H_2O(l) + 2\,e^- \rightarrow H_2(g) + 2\,OH^-(aq)$ $\quad E° = -0.83\text{ V}$

The *standard* cell potential is now –2.19 V. The *actual* cell potential is even less negative at a pH of 7. Note that as the reaction proceeds, the OH^- ion replaces the Cl^- ion in solution. The oxidation of water is more electrochemically favorable than that of the chloride ion. Other factors to be described in the next section account for the fact that the chloride ion is often oxidized as readily as water.

In aqueous solution, it is more difficult to reduce calcium ions than sodium ions. In the molten salt mixture, Na^+ is reduced preferentially.

12.15 The Potential Needed for Electrolysis

- **Factors that govern the potential needed for electrolysis**
  - → The potential supplied to an electrolytic cell must be *at least as great as* the potential of the spontaneous cell reaction that is to be reversed.
  - → The *actual* potential required is often *greater* than the minimum value. The value of the potential in excess of the minimum is called the *overpotential*.

→ The underlying cause of the overpotential effect is complex. The structure of a solution near an electrode is different from that in the bulk solution. Electrons are transferred across an electrode-solution interface that depends on the nature of the solution and the condition of the electrode.

→ Because of the overpotential effect, care must be taken in predicting the products of electrolysis. The products observed will often differ from those predicted from an analysis based on standard electrode potentials and the Nernst equation.

Example 12.15a Determine the minimum potential required in the electrolysis of pure water (pH = 7) to form $H_2(g)$ and $O_2(g)$, each at 1 bar pressure.

Solution The appropriate half-reactions at a pH of 7 are

cathode: $2\,H_2O(l) + 2\,e^- \rightarrow H_2(g) + 2\,OH^-(aq)$ $\quad E = -0.42\text{ V}$

anode: $2\,H_2O(l) \rightarrow O_2(g) + 4\,H^+(aq) + 4\,e^-$ $\quad -E = -(+0.82\text{ V}) = -0.82\text{ V}$

cell: $6\,H_2O(l) \rightarrow 2\,H_2(g) + O_2(g) + 4\,H^+(aq) + 4\,OH^-(aq)$ $\quad E = -1.24\text{ V}$

$4\,H^+(aq) + 4\,OH^-(aq) \rightarrow 4\,H_2O(l)$ $\quad E = 0\text{ V}$

cell: $2\,H_2O(l) \rightarrow 2\,H_2(g) + O_2(g)$ $\quad E = -1.24\text{ V}$

The solution remains neutral as the reaction proceeds. A minimum potential of +1.24 V is required to drive the reaction. Because of the overvoltage effect, the actual potential required is around +1.8 V.

Example 12.15b A solution of 1 M NaBr(aq) is electrolyzed. Predict the initial products of the electrolysis and determine the overall cell reaction. Ignore the overpotential effect. Assume any gas formed is at a pressure of 1 bar and the pH of the solution is 7.

Solution Because standard conditions apply for NaBr(aq), standard electrode potentials are used for Na^+ and Br^- reactions. For the half-reactions of water, electrode potentials at a pH of 7 are used.

The two candidates for the cathode reaction are the reduction of $Na^+(aq)$ and the reduction of $H_2O(l)$. The reaction with the most positive (least negative) standard electrode potential is favored. The half-reactions are

$Na^+(aq) + e^- \rightarrow Na(s)$ $\quad E° = -2.71\text{ V}$

$2\,H_2O(l) + 2\,e^- \rightarrow H_2(g) + 2\,OH^-(aq)$ $\quad E = -0.42\text{ V}$

The least negative potential is that of the reduction of water.

The two candidates for the anode reaction are the oxidation of $Br^-(aq)$ and the oxidation of $H_2O(l)$. The reaction with the least positive (most negative) standard electrode potential is favored. The half-reactions are

$2\,Br^-(aq) \rightarrow Br_2(l) + 2\,e^-$ $\quad -E° = -(+1.09\text{ V}) = -1.09\text{ V}$

$2\,H_2O(l) \rightarrow O_2(g) + 4\,H^+(aq) + 4\,e^-$ $\quad -E = -(+0.82\text{ V}) = -0.82\text{ V}$

The least positive (reduction) potential is that of water. The overall cell reaction is predicted to be the same as in the electrolysis of pure water in Example 12.15a.

12.16 The Products of Electrolysis

- **Faraday's law of electrolysis**
  - → The number of moles of product formed by an electric current is stoichiometrically equivalent to the number of moles of electrons supplied.
  - → Once the amount of product is known, the mass of product may be calculated.
  - → The amount of product formed in an electrolysis depends on the current used, the time the electrolysis runs, and the number of moles of electrons in the half-reaction.

- **Quantitative aspects of Faraday's law**
  - → Determine the stoichiometry of the half-reaction. For a typical reduction process of a metal ion to a metal solid, the half-reaction has the form

    $$M^{n+} + n\,e^- \rightarrow M(s)$$

  - → The quantity of electricity, Q, passing through the electrolysis cell is determined by the electric current, I, and the time, t, of current flow. *Note that the symbol Q is also used for the reaction quotient.* Recall that 1 ampere (A) = 1 coulomb (C) s^{-1}.

Charge supplied (C) = current (A) × time (s) or $Q = It$

  - → Because $Q = nF$, where n is the number of moles of electrons and F is the Faraday constant, it follows that

    $$\text{Moles of electrons} = \frac{\text{charge supplied (C)}}{F} = \frac{\text{current (I)} \times \text{time (t)}}{F} \quad \text{or} \quad n(e^-) = \frac{It}{F}$$

Example 12.16a In an electrolysis cell, cobalt metal is formed at the cathode in the reduction of $Co^{2+}(aq)$ ions. Determine the mass of cobalt deposited by a current of 10.00 A passing through the cell for 5.000 hours.

Solution The half-reaction is $Co^{2+}(aq) + 2\,e^- \rightarrow Co(s)$.
The number of moles of electrons passing through the cell is

$$n(e^-) = \frac{It}{F} = \frac{(10.00\text{ A})(5.000\text{ hr})(3600\text{ s}\cdot\text{hr}^{-1})}{96\,485\text{ C}\cdot\text{mol}} = 1.866\text{ mol e}^-$$

The stoichiometric relationship is 1 mol Co ≙ 2 mol e^- and the molar mass of Co is 58.93 g·mol^{-1}. The mass of Co is then

$$m(\text{Co}) = (1.866\text{ mol e}^-)\left(\frac{1\text{ mol Co}}{2\text{ mol e}^-}\right)(58.93\text{ g}\cdot\text{mol}^{-1}) = 54.98\text{ g Co}$$

Example 12.16b Electrolysis of an aqueous solution of potassium iodide produces $I_2(s)$ at the anode. If a current of 50.0 A is used, how much time is required to produce 10.0 mol of iodine solid?

Solution The half-reaction is $2\,I^-(aq) \rightarrow I_2(s) + 2\,e^-$.
The stoichiometric relationship is 1 mol I_2 ≙ 2 mol e^-.
The number of moles of electrons required is

$$n(\text{e}^-) = (10.0 \text{ mol I}_2)\left(\frac{2 \text{ mol e}^-}{1 \text{ mol I}_2}\right) = 20.0 \text{ mol e}^-$$

$$t = \frac{n(\text{e}^-)F}{I} = \frac{(20.0 \text{ mol})(96\,485 \text{ C}\cdot\text{mol}^{-1})}{(50.0 \text{ C}\cdot\text{s}^{-1})} = 3.86 \times 10^4 \text{ s} = 10.7 \text{ hr}$$

12.17 Electrolysis in Action

- **Uses of electrolysis**
  - → Extracting metals from their salts is a major application. Examples include aluminum, sodium (see Fig. 12.22 in the text), and magnesium.
  - → Preparation of fluorine, chlorine, and sodium hydroxide
  - → Refining copper
  - → Electroplating metals
- **Electroplating**
  - → The electrolytic deposition of a thin film of metal on an object
  - → Object to be plated is made the cathode of an electrolysis cell.
  - → Cell electrolyte is a salt of the metal to be plated.
  - → The cations are supplied either by the added salt or from oxidation of the anode, which is made of the plating metal.
  - → Examples include silver, gold, and chromium plating (see Fig. 12.23 in the text).

Chapter 13 Chemical Kinetics

Reaction Rates (Sections 13.1-13.3)

Key Concepts

chemical kinetics, reaction rate, average reaction rate, unique average rate, quenching, stopped-flow technique, instantaneous reaction rate, initial reaction rate, rate constant, rate law, reaction order, first-order reaction, second-order reaction, zero-order reaction, overall order of reaction, pseudo-first-order reaction

Overview

- **Thermodynamics**
  → Predicts the direction of spontaneous chemical change
  → Yields the extent of chemical change in terms of the equilibrium constant, K
  → Deals only with the initial and final states of a system
  → Cannot determine how long a reaction will take to occur
- **Chemical kinetics**
  → Time is the domain of chemical kinetics. From an experimentally determined rate law, the time required to produce a given amount of product can be predicted.
  → Intermediate steps in a chemical reaction (possible pathways) are of interest as well.
  → Addresses the following questions:
  Exactly how do reactants rearrange to form products?
  How are bonds broken and formed?
  Which species collide?
  What is the mechanism of a chemical change?

13.1 Concentration and Reaction Rate

- **Definition of rate**

  General → Rate is the change in a property divided by the time required for the change to occur.

  Chemical kinetics → Reaction rate is the change in molar concentration of a reactant or product divided by the time required for the change to occur.
- **Average *vs.* instantaneous rates: an analogy to travel**
  → The *average* speed of an automobile trip is the length of the journey divided by the total time for the journey.
  → The *instantaneous* speed is obtained if the car is timed over a very short distance at some point in the journey.

Example 13.1a A person drives by car from city A to city B, stopping an hour for lunch halfway through the trip. The distance between the cities is 300 km and the total time required for the trip is 4 h. Calculate the average rate of travel (speed) for the trip including and excluding lunch. Also, calculate the instantaneous rate when the driver is halfway through lunch.

Solution The average rate of travel is the total distance divided by the total time of the trip.

average rate = 300 km / 4 hr = 75 $km \cdot h^{-1}$

The average rate of travel excluding lunch is the same total distance divided by the total time excluding lunch (3 hr).

average rate = 300 km / 3 hr = 100 $km \cdot h^{-1}$

The instantaneous rate halfway through lunch is zero because the car is not moving. Because the average rate excluding lunch is 100 $km \cdot h^{-1}$, this will be the instantaneous rate at some point during the trip when the car is in motion. Speedometer readings approximate instantaneous values.

- **Average reaction rate**
  - → For a general chemical reaction, $aA + bB \rightarrow cC + dD$, the average reaction rate is the *change* in molar concentration of a reactant or product divided by the time interval required for the change to occur.
  - → The average reaction rate may be determined for *any* reactant or product.
  - → Rates for different reactants and products may have different numerical values.

Example 13.1b Write expressions for the average reaction rate in terms of each reactant and product in the general reaction given above.

Solution

$$\text{Average rate of disappearance of reactant A} = -\frac{\Delta[A]}{\Delta t}$$

$$\text{Average rate of disappearance of reactant B} = -\frac{\Delta[B]}{\Delta t}$$

Reaction rates are always written as *positive* values. The minus sign in the rate equations arise because the concentration of both reactants decreases with time. Note the units of reaction rate are typically $mol \cdot L^{-1} \cdot s^{-1}$. Other units of time commonly used are minutes, m, and hours, h.

$$\text{Average rate of formation of product C} = \frac{\Delta[C]}{\Delta t}$$

$$\text{Average rate of formation of product D} = \frac{\Delta[D]}{\Delta t}$$

From the reaction stoichiometry, when *a* mol of A disappears, *b* mol of B disappears, *c* mol of C appears, and *d* mol of D appears. Thus, for a given time interval, $\Delta[A] \neq \Delta[B]$ unless $a = b$.

- **Unique average reaction rate**
  - → For a general chemical reaction, $aA + bB \rightarrow cC + dD$, the unique average reaction rate is the average reaction rate of reactant or product divided by its stoichiometric coefficient used as a *pure number*.
  - → The unique average reaction rate may be determined from *any* reactant or product.
  - → The unique average reaction rate is the same for any reactant or product.

Example 13.1c Write expressions for the unique average reaction rate in terms of each reactant and product in the general reaction given above.

Solution Unique average reaction rate $= -\frac{1}{a}\frac{\Delta[A]}{\Delta t} = -\frac{1}{b}\frac{\Delta[B]}{\Delta t} = \frac{1}{c}\frac{\Delta[C]}{\Delta t} = \frac{1}{d}\frac{\Delta[D]}{\Delta t}$

Example 13.1d Dinitrogen pentoxide decomposes in carbon tetrachloride at 45°C as follows:

$$2\,N_2O_5(\text{solution}) \rightarrow 4\,NO_2(\text{solution}) + O_2(g)$$

At $t = 0$, the starting concentration of N_2O_5 is 2.332 mol·L^{-1} and no products are present. At $t = 184$ min, the concentration of N_2O_5 is 2.082 mol·L^{-1}. Calculate the average reaction rate of disappearance of N_2O_5, the average reaction rate of formation of NO_2 and of O_2, and the unique average reaction rate for the interval $\Delta t = 184$ min.

Solution $\Delta[N_2O_5] = 2.082 - 2.332 = -\,0.250$ mol·L^{-1}

Average rate of disappearance of $N_2O_5 = -\frac{\Delta[N_2O_5]}{\Delta t} = -\frac{-0.250 \text{ mol·L}^{-1}}{184 \text{ min}}$

$= 1.36 \times 10^{-3}$ mol·L^{-1}·min^{-1}

From the stoichiometry of the reaction, for every 2 moles of N_2O_5 that disappear, 4 moles of NO_2 and 1 mole of O_2 appear, respectively. The rate of appearance of NO_2 is twice (4 mol NO_2/2 mol N_2O_5) the rate of disappearance of N_2O_5, and the rate of appearance of O_2 is one-half (1 mol O_2/2 mol N_2O_5) the rate of disappearance of N_2O_5.

Average rate of appearance of $NO_2 = 2(1.36 \times 10^{-3}) = 2.72 \times 10^{-3}$ mol·L^{-1}·min^{-1}
Average rate of appearance of $O_2 = (0.5)(1.36 \times 10^{-3}) = 6.80 \times 10^{-4}$ mol·L^{-1}·min^{-1}
The unique average reaction rate is obtained from any reactant or product.

Unique average rate $= -\frac{1}{2}\frac{\Delta[N_2O_5]}{\Delta t} = \frac{1}{4}\frac{\Delta[NO_2]}{\Delta t} = \frac{\Delta[O_2]}{\Delta t}$

$= 6.80 \times 10^{-4}$ mol·L^{-1}·min^{-1}

Convert the units of rate to mol·L^{-1}·s^{-1} by multiplying by (1 min/60 s).

Unique average rate $= 1.13 \times 10^{-5}$ mol·L^{-1}·s^{-1}

- **Reaction rates**
  - → The time required to measure the concentration changes of reactants and products varies considerably according to the reaction.
  - → Spectroscopic techniques are often used to monitor concentration, particularly for fast reactions. An important example is the *stopped-flow technique* shown in Fig. 13.3 in the text. The fastest reactions occur on a time scale of femtoseconds (10^{-15} s).

13.2 The Instantaneous Rate of Reaction

- **Reaction rates**
  - → Most reactions slow down as they proceed and reactants are depleted.
  - → If equilibrium is reached, the forward and reverse reaction rates are equal.
  - → The reaction rate at a specific time is called the *instantaneous reaction rate.*
- **Instantaneous reaction rate**
  - → For a product, the *slope* of a tangent line of a graph of concentration *vs.* time
  - → For a reactant, the *negative of the slope* of a tangent line on a graph of concentration *vs.* time

Example 13.2 Kinetic data for the gas-phase ozonolysis of ethene are as follows:

$$O_3(g) + C_2H_4(g) \rightarrow C_2H_4O(g) + O_2(g)$$

$[O_3]$ ($mol{\cdot}L^{-1}$)	t (s)
3.20×10^{-5}	0
2.42×10^{-5}	10
1.23×10^{-5}	50
1.10×10^{-5}	60

Determine the average reaction rate for the first and second time intervals. How does the average rate change with time? Plot a graph of $[O_3]$ *vs.* time and estimate the instantaneous rate of reaction at $t = 5$ s and $t = 30$ s.

Solution The first time interval is centered at 5 s and is 10 s in duration. The values of t_2 and t_1 are 10 s and 0 s, respectively. The values of $[O_3]_2$ and $[O_3]_1$ are 2.42×10^{-5} $mol{\cdot}L^{-1}$ and 3.20×10^{-5} $mol{\cdot}L^{-1}$, respectively. The average rate of disappearance of ozone in this time interval is then

$$-\frac{\Delta[O_3]}{\Delta t} = -\frac{[O_3]_2 - [O_2]_1}{t_2 - t_1} = -\frac{(2.42 - 3.20) \times 10^{-5}\ mol \cdot L^{-1}}{(10 - 0)\ s}$$

$$= 7.8 \times 10^{-7}\ mol{\cdot}L^{-1}{\cdot}s^{-1}$$

The second time interval is centered at 30 s and is 40 s in duration. The values of t_2 and t_1 are 50 s and 10 s, respectively. The values of $[O_3]_2$ and $[O_3]_1$ are 1.23×10^{-5} $mol{\cdot}L^{-1}$ and 2.42×10^{-5} $mol{\cdot}L^{-1}$, respectively. The average rate of disappearance of ozone in this time interval is then

$$-\frac{\Delta[O_3]}{\Delta t} = -\frac{[O_3]_2 - [O_2]_1}{t_2 - t_1} = -\frac{(1.23 - 2.42) \times 10^{-5}\ mol \cdot L^{-1}}{(50 - 10)\ s}$$

$$= 3.0 \times 10^{-7}\ mol{\cdot}L^{-1}{\cdot}s^{-1}$$

The average rate of reaction decreases as time increases.

The data given above are displayed in the following plot. A polynomial curve-fitting program is used to construct a continuous curve connecting the four points. The first tangent line at 5 s is drawn between 0 and 20 s. The y-axis intercepts are roughly 3.15×10^{-5} and 1.61×10^{-5} $mol{\cdot}L^{-1}$, respectively. The second tangent line at 30 s is drawn between 20 and 40 s. The y-axis intercepts are roughly 1.82×10^{-5} and 1.30×10^{-5} $mol{\cdot}L^{-1}$. The two contact points are indicated by the symbol ⊠.

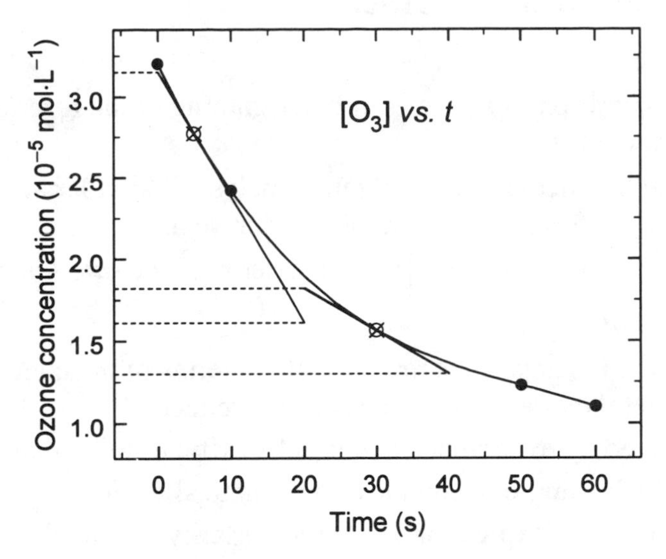

The instantaneous rates are the two slopes of the tangent lines shown above.

Negative slope of the first time interval

$$= -\frac{(1.61 - 3.15) \times 10^{-5}\ \text{mol} \cdot \text{L}^{-1}}{(20 - 0)\ \text{s}} = 7.7 \times 10^{-7}\ \text{mol} \cdot \text{L}^{-1} \cdot \text{s}^{-1}$$

Negative slope of the second time interval

$$= -\frac{(1.30 - 1.82) \times 10^{-5}\ \text{mol} \cdot \text{L}^{-1}}{(40 - 20)\ \text{s}} = 2.6 \times 10^{-7}\ \text{mol} \cdot \text{L}^{-1} \cdot \text{s}^{-1}$$

Estimated instantaneous rates closely approximate the average rate values.

Note: This method of determining instantaneous rates is relatively inaccurate because it is difficult to determine the slope precisely. A better method that avoids the use of tangents is given in Section 13.4.

- **Mathematical form of the instantaneous rate**
  - → The slope of the concentration *vs.* time curve is the derivative of the concentration with respect to time.
  - → For the instantaneous disappearance of a reactant, R,

$$\text{Reaction rate} = -\frac{\text{d}[\text{R}]}{\text{d}t}$$

  - → For the instantaneous appearance of a product, P,

$$\text{Reaction rate} = \frac{\text{d}[\text{P}]}{\text{d}t}$$

  - → For the general reaction, $a\text{A} + b\text{B} \rightarrow c\text{C} + d\text{D}$,

$$\boxed{\text{Unique instantaneous reaction rate} = -\frac{1}{a}\frac{\text{d}[\text{A}]}{\text{d}t} = -\frac{1}{b}\frac{\text{d}[\text{B}]}{\text{d}t} = \frac{1}{c}\frac{\text{d}[\text{C}]}{\text{d}t} = \frac{1}{d}\frac{\text{d}[\text{D}]}{\text{d}t}}$$

  - → In the remainder of the chapter, the term reaction rate is used specifically to mean the *unique* instantaneous reaction rate written here.

13.3 Rate Laws and Reaction Order

- **Initial reaction rate**
  - → Change in concentration of a species at the beginning of the reaction is the instantaneous rate measured at $t = 0$.
  - → In what is called the method of initial rates, analysis is simpler at $t = 0$ because products that may affect reaction rates are not present.
  - → The method of initial rates is often used to determine the *rate law* of a reaction.
- **Rate law of a reaction**
  - → Expression for the instantaneous reaction rate in terms of the concentrations of species that affect the rate such as reactants or products
  - → Always determined by experiment, and usually difficult to predict
  - → May have relatively simple or complex mathematical form
  - → The exponents in a rate expression are *not* necessarily the stoichiometric coefficients of a reaction; therefore, we use m and n in place of a and b in rate equations.
- **Simple rate laws**
  - → Form of a simple rate law: $\boxed{\text{Rate} = k[\text{A}]^m}$
  - → k = rate constant, which is the reaction rate when all the concentrations appearing in the rate law are 1 M.
  - → k is *independent* of concentration(s).
  - → k is *dependent* on temperature, $k(T)$.
  - → A refers to the reactant, and m, the power of [A], is called the order in reactant A and is determined experimentally. Order is *not related to the reaction stoichiometry*. Exceptions include a special class of reactions, which are called *elementary* (see Section 13.10).
  - → Typical orders for reactions with one reactant are 0, 1, and 2. Fractional values are also common.

    Zero-order reaction ($m = 0$): Rate = k
    First-order reaction ($m = 1$): Rate = $k[\text{A}]$
    Second-order reaction ($m = 2$): Rate = $k[\text{A}]^2$
- **More complex rate laws**
  - → Many reactions have experimental rate laws containing concentrations of more than one species.
  - → Form of a complex rate law: $\boxed{\text{Rate} = k[\text{A}]^m[\text{B}]^n\ldots}$

    The overall order is the sum of the powers $m + n + \ldots$.
  - → The powers may also be negative values, and forms of even greater complexity with no overall order are found.
  - → More complex rate laws are treated in Section 13.11.
  - → The following table lists experimental rate laws for a variety of reactions:

Reaction	Rate law	Order	Units of k
$2NH_3(g) \rightarrow N_2(g) + 3H_2(g)$	rate = k	zero	$mol \cdot L^{-1} \cdot s^{-1}$
$2N_2O_5(g) \rightarrow 4NO_2(g) + O_2(g)$	rate = $k[N_2O_5]$	first	s^{-1}
$2NO_2(g) \rightarrow NO(g) + NO_3(g)$	rate = $k[NO_2]^2$	second	$L \cdot mol^{-1} \cdot s^{-1}$
$2NO(g) + O_2(g) \rightarrow 2NO_2(g)$	rate = $k[NO]^2[O_2]$	first in O_2 second in NO third overall	$L^2 \cdot mol^{-2} \cdot s^{-1}$
$I_3^-(aq) + 2N_3^-(aq) \rightarrow 3I^-(aq) + 3N_2(g)$ in the presence of $CS_2(aq)$	rate = $k[CS_2][N_3^-]$	first in CS_2 first in N_3^- second overall	$L \cdot mol^{-1} \cdot s^{-1}$
$2O_3(g) \rightarrow 3O_2(g)$	rate = $k\dfrac{[O_3]^2}{[O_2]}$ $= k\,[O_3]^2[O_2]^{-1}$	second in O_3 minus one in O_2 first overall	s^{-1}

→ In general, the reaction order does not follow from the stoichiometry of the chemical equation. In the case of the *catalytic* decomposition of ammonia on a hot platinum wire, the reaction order initially is zero because the reaction occurs on the surface of the wire, and the surface coverage is independent of concentration. The rate of a zero-order reaction is independent of concentration until the reactant is nearly exhausted or until equilibrium is reached.

→ Reaction rates may depend on species that do not appear in the overall chemical equation. The reaction of triodide ion with nitride ion, for example, is accelerated by the presence of carbon disulfide, neither a reactant nor product. The concentration of carbon disulfide, a *catalyst*, appears in the rate law. The topic of catalysis is presented in Section 13.14.

→ In the last example listed above, the order with respect to the product O_2 is -1. Negative orders often arise from reverse reactions in which products reform reactants, thereby slowing the overall reaction rate. Oxygen is present at the beginning of this reaction as commonly studied, because ozone is made by an electrical discharge in oxygen gas. It is possible to study the kinetics of the decomposition of pure ozone. In the absence of oxygen, the initial rate law is quite different from the one listed in the above table.

- **Procedure for determining the reaction orders and rate law**

  → The reaction order for a given species is determined by varying its initial concentration and measuring the initial reaction rate, while keeping the concentrations of all other species constant. The procedure is repeated for the other species until all the reaction orders are obtained.

→ Once the form of the rate law is known, a value of the rate constant is calculated for each set of initial concentrations. The reported rate constant is typically the average value for all the measurements.

Example 13.3a Use the data given in the following table to obtain the form of the rate law and a value of the rate constant for the reaction at 50°C

$$2\,NO(g) + Cl_2(g) \rightarrow 2\,NOCl(g)$$

Predict a value of the initial rate of disappearance of Cl_2 if [NO] = 0.200 M and $[Cl_2]$ = 0.750 M.

Experiment	[NO]	$[Cl_2]$	Formation rate of NOCl
1	0.500 M	0.500 M	1.14×10^{-5} mol·L^{-1}·s^{-1}
2	1.00 M	0.500 M	4.56×10^{-5} mol·L^{-1}·s^{-1}
3	1.00 M	1.00 M	9.12×10^{-5} mol·L^{-1}·s^{-1}

Solution The concentration of Cl_2 is constant in experiments 1 and 2. Thus, experiments 1 and 2 can be used to determine the reaction order in NO. In this example, the ratios of concentrations and rates are simple integer values, and the math is straightforward. When the concentration of NO is doubled, the measured initial formation rate of NOCl quadruples. The order in NO is then +2.

A similar analysis for Cl_2 with NO held constant yields the following result. As seen for experiments 2 and 3, when the concentration of Cl_2 is doubled, the measured initial rate of formation of NOCl also doubles. The order in Cl_2 is +1.

The rate law is then

Rate of formation of NOCl = $k\,[NO]^2[Cl]$

After rearranging the rate law expression in terms of k and substituting the values from experiment 1, we obtain

$$k = \frac{\text{initial rate (NOCl)}}{[NO]^2[Cl]} = \frac{(1.14 \times 10^{-5}\ \text{mol}\cdot\text{L}^{-1}\cdot\text{s}^{-1})}{(0.500\ \text{mol}\cdot\text{L}^{-1})^2(0.500\ \text{mol}\cdot\text{L}^{-1})}$$

$$= 9.12 \times 10^{-5}\ \text{L}^2\cdot\text{mol}^{-2}\cdot\text{s}^{-1}$$

Experiments 2 and 3 yield the same value, so the average value of k is 9.12×10^{-5} L^2·mol^{-2}·s^{-1}.

If [NO] = 0.200 M and $[Cl_2]$ = 0.750 M, then the rate of formation of NOCl is (9.12×10^{-5} L^2·mol^{-2}·s^{-1})(0.200 mol·L^{-1})^2(0.750 mol·L^{-1}) = 2.74×10^{-6} mol·L^{-1}·s^{-1}. According to the reaction stoichiometry, the *unique* reaction rate is equal to the rate of disappearance of Cl_2.

$$-\frac{d[Cl_2]}{dt} = \frac{1}{2}\frac{d[NOCl]}{dt} = \frac{2.74 \times 10^{-6}\,\text{mol}\cdot\text{L}^{-1}\cdot\text{s}^{-1}}{2} = 1.37 \times 10^{-6}\,\text{mol}\cdot\text{L}^{-1}\cdot\text{s}^{-1}$$

The *unique* reaction rate constant is then one-half the value of k.

Note: In this example, the ratios of concentrations and rates are simple. In the next example, a more general procedure is required because the ratios are not as straightforward to analyze.

Example 13.3b Use the data given in the following table to obtain the form of the rate law and a value of the rate constant for the reaction at constant temperature:

$$2NO(g) + H_2(g) \rightarrow N_2O(g) + H_2O(g)$$

Experiment	[NO]	$[H_2]$	Formation rate of N_2O
1	0.140 M	0.140 M	4.39 $mol \cdot L^{-1} \cdot s^{-1}$
2	0.370 M	0.140 M	30.66 $mol \cdot L^{-1} \cdot s^{-1}$
3	0.370 M	0.480 M	105.14 $mol \cdot L^{-1} \cdot s^{-1}$

Solution The general form of the rate law is

$$\text{Rate} = k\,[NO]^m[H_2]^n$$

To solve for the unknown order, *m*, take the ratio of the rate-law expressions for experiments 2 and 1.

$$\frac{\text{Rate}_2}{\text{Rate}_1} = \frac{k[NO]_2^m[H_2]_2^n}{k[NO]_1^m[H_2]_1^n} = \frac{(0.370)^m(0.140)^n}{(0.140)^m(0.140)^n} = \left(\frac{0.370}{0.140}\right)^m = \frac{30.66}{4.39}$$

$$(2.643)^m = 6.984$$

$$\ln(2.643)^m = m\ln(2.643) = \ln(6.984)$$

$$m\,(0.9719) = 1.944 \text{ and } m = 1.9998 = 2.00$$

The same procedure applied to experiments 3 and 2 yields $n = 1.00$.

The form of the rate law is then

$$\text{Rate} = k\,[NO]^2[H_2]$$

The rate constant values for each of the three experiments are the same, all equal to $1.60 \times 10^3\ L^2 \cdot mol^{-2} \cdot s^{-1}$. The reaction rate and rate constant are *unique* values.

Note: These examples are slightly unrealistic. Actual kinetics experiments normally produce rate constant values that vary somewhat among different data sets.

- **Pseudo-order reactions**
  - → If a species in the rate-law equation is present in great excess, its concentration is effectively constant as the reaction proceeds. The constant concentration may be incorporated into the rate constant and the apparent order of the reaction changes.
  - → Suppose the concentration of NO(g) is 1000 times that of $H_2(g)$ in Example 13.3b. The rate law may then be written as: $\text{Rate} = k\,[NO]^2[H_2] = k'\,[H_2]$
    Under these conditions, the overall third-order reaction is now *pseudo*-first-order.
  - → Reactions in which the solvent appears in the rate law often show pseudo-order behavior, because the solvent is usually present in large excess.
  - → A technique called the *isolation method* is often used to determine the order in a particular reactant by keeping the concentration of other reactants in excess. The method is repeated for all reactants and the complete rate law is obtained. A hidden danger is that occasionally the rate law is *not* the same in the extreme ranges of the concentrations of reactants. Different reaction mechanisms that yield different rate laws may occur under the extreme conditions.

Concentration and Time (Sections 13.4-13.6)

Key Concepts

integrated rate law, first-order integrated rate law, exponential decay, half-life $t_{1/2}$, second-order integrated rate law

Overview

- **Concentration and time**
  → Actual kinetic measurements do not directly yield the rate of reaction.
  → The concentrations of reactants and/or products are measured at selected times after the reaction has begun. The measurement of concentration is "instantaneous" or very fast with respect to the reaction time.
  → To follow the time evolution of concentrations, we require an *integrated* form of the rate law that gives the time dependence of the concentrations.

- **Differential rate law**
  → Instantaneous reaction rate law is a *differential equation.*
  → Example of a simple rate law: $-\frac{1}{a}\frac{d[A]}{dt} = k[A]^m$
  → Rearranged differential equation: $-\frac{d[A]}{[A]^m} = akdt$

- **Integrated rate law**
  → Differential equations may be integrated to obtain the *explicit* time dependence of the concentrations of reactants.
  → Integration of differential equation: $-\int \frac{d[A]}{[A]^m} = \int akdt$
  → Integration yields [A] as an explicit function of time. The time dependence of [A] is independent of the stoichiometric coefficient (a), so assume $k \equiv$ "ak" for simplicity.

13.4 First-order Integrated Rate Laws

- **Integration of a first-order rate law**
  → First-order rate law: $-\frac{d[A]}{dt} = k[A]$
  → Integration with limits: $-\int_{[A]_0}^{[A]_t} \frac{d[A]}{[A]} = \int_0^t kdt$
  → Result in logarithmic form: $\ln\left(\frac{[A]_t}{[A]_0}\right) = -kt$
  → Result in exponential form: $[A]_t = [A]_0 e^{-kt}$

→ If a plot of $\ln[A]_t$ *vs.* t gives a straight line, the rate law is first-order in [A].
→ The rearranged logarithmic expression, $\ln[A]_t = \ln[A]_0 - kt$, shows the straight-line behavior, y = intercept + (slope × x), of a plot of $\ln[A]_t$ *vs.* t.
→ Plot of $\ln[A]_t$ *vs.* t: slope = $-k$ *and* intercept = $\ln[A]_0$

- **Integrated first-order rate law**
  → [A] decays exponentially with time.
  → Used to confirm that a reaction is first order and to measure its rate constant
  → Known k and $[A]_0$ values can be used to predict the value of [A] at any time t.

Example 13.4 Dinitrogen pentoxide, N_2O_5, decomposes in CCl_4 at 45°C as follows:

$$2N_2O_5(soln) \rightarrow 4NO_2(soln) + O_2(g)$$

Using the experimental data given in the table, confirm that the reaction is first order, determine the value of the rate constant, and predict the concentration of N_2O_5 at a time of 2000 min.

t	0	184	319	526	867	1198	1877 min
$[N_2O_5]$	2.33	2.08	1.91	1.67	1.35	1.11	0.720 mol·L^{-1}
$\ln[N_2O_5]$	0.846	0.732	0.647	0.513	0.300	0.104	−0.328

Solution The simplest method to confirm that the reaction is first order is to plot $\ln[N_2O_5]$ *vs.* t. The values of $\ln[N_2O_5]$ are given in the last row of the table. The plot shown below is obtained using the *Tool - Curve Fitter* program on the CD-ROM accompanying the text.

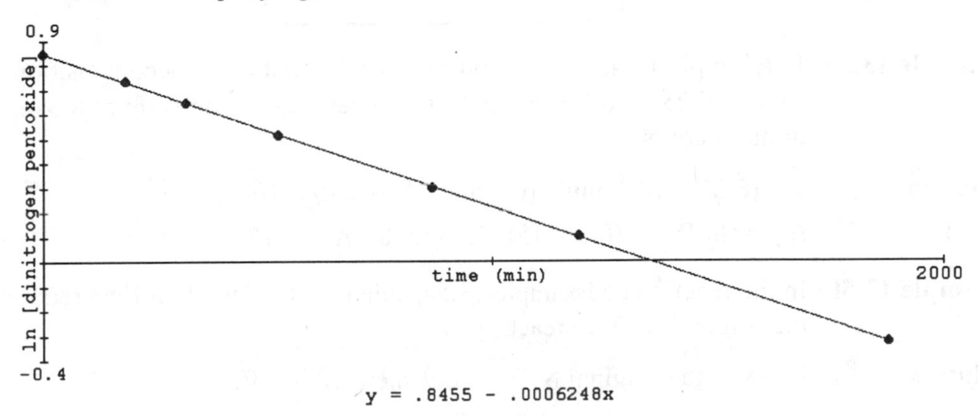

This linear plot has a slope and intercept, obtained from the fitting program, of -6.248×10^{-4} min^{-1} and 0.8455, respectively. The rate constant is equal to the negative value of the slope. The intercept may be used to check how well the best fit reproduces the initial concentration of 2.33 mol·L^{-1}. The predicted concentration at t = 2000 min is obtained by the substitution of this value into the equation $\ln[A]_t = 0.8455 - (6.248 \times 10^{-4}\ \text{min}^{-1})t$. The final values are reported to three significant figures. (It is customary to keep at least one extra significant figure in the best-fit equation.)

$k = 6.25 \times 10^{-4}$ min^{-1} (a = 2 and the *unique* $k = 3.12 \times 10^{-4}$ min^{-1})

$\ln[A]_0 = 0.8455$ and $[A]_0 = e^{+0.8455} = 2.33$ mol·L^{-1} (*check*)

$\ln[A]_{2000} = 0.8455 - (6.248 \times 10^{-4})(2000) = -0.4041$ and

$[A]_{2000} = e^{-0.4041} = 0.668$ mol·L^{-1}

13.5 Half-Lives for First-Order Reactions

- **Half-life expression**
  → The half-life, $t_{1/2}$, of a substance is the time required for its concentration to fall to one-half its initial value.
  → At $t = t_{1/2}$, $[A]_{t_{1/2}} = \frac{1}{2}[A]_0$. The ratio of concentrations is $\dfrac{[A]_0}{[A]_{t_{1/2}}} = \dfrac{[A]_0}{\frac{1}{2}[A]_0} = 2$.
  → Solve for $t_{1/2}$ by rearranging the logarithmic form of the integrated rate law to give $t_{1/2} = \dfrac{1}{k}\ln\left(\dfrac{[A]_0}{[A]_{t_{1/2}}}\right)$. Then substitute $\dfrac{[A]_0}{[A]_{t_{1/2}}} = 2$.
  → Half-life expression: $\boxed{t_{1/2} = \dfrac{\ln 2}{k}}$

- **Half-life of a first-order reaction**
  → Independent of the initial concentration
  → Characteristic of the reaction
  → The rate constant and $t_{1/2}$ are inversely proportional.
  → The logarithmic form of the rate law may be modified as follows:

$$\boxed{\ln\left(\frac{[A]_0}{[A]_t}\right) = kt = (\ln 2)\left(\frac{t}{t_{1/2}}\right)}$$

Example 13.5a In Example 13.4, the rate constant for the first-order decomposition of N_2O_5 in CCl_4 is $6.25 \times 10^{-4}\ \text{min}^{-1}$ at 45°C. Determine the half-life of the reaction in the units of hours.

Solution $k = (6.25 \times 10^{-4}\ \text{min}^{-1})(60\ \text{min·h}^{-1}) = 3.75 \times 10^{-2}\ \text{h}^{-1}$

$t_{1/2} = (\ln 2)/k = (0.693\,15)/(3.75 \times 10^{-2}\ \text{h}^{-1}) = 18.5\ \text{h}$

Example 13.5b In the reaction of Example 13.5a, calculate the length of time required for 7/8 of the original N_2O_5 to react.

Solution If 7/8 of the original N_2O_5 reacts, then 1/8 is left.

$$t = \frac{1}{k}\ln\left(\frac{[A]_0}{[A]_t}\right) = \frac{1}{k}\ln\left(\frac{[A]_0}{\frac{1}{8}[A]_0}\right) = \frac{1}{k}\ln 8 = (2.079\,4)/(3.75 \times 10^{-2}\ \text{h}^{-1}) = 55.5\ \text{h}$$

An alternative procedure is to note that at the end of each half-life period, one-half of the original concentration is left. At the point when one-eighth is left, the time elapsed is exactly three half-lives, $(\frac{1}{2})^3 = \frac{1}{8}$. The time for three half-lives is $3 \times 18.5\ \text{h} = 55.5\ \text{h}$.

Example 13.5c In the reaction of Example 13.5a, calculate the fraction of the original N_2O_5 that remains after 1000 h.

Solution The exponential form of the integrated rate law is appropriate in this case:

$[A]_t/[A]_0 = e^{-kt} = e^{-(3.75\times10^{-2}\ \text{h}^{-1})(1000\ \text{h})} = e^{-37.5} = 5.18\times10^{-17}$ *(not much remains)*

13.6 Second-Order Integrated Rate Laws

- **Integration of a second-order rate law**

→ Second-order rate law: $$-\frac{d[A]}{dt} = k[A]^2$$

→ Integration with limits: $$-\int_{[A]_0}^{[A]_t} \frac{d[A]}{[A]^2} = \int_0^t k\,dt$$

→ Integrated form: $$\frac{1}{[A]_t} - \frac{1}{[A]_0} = kt$$

→ Alternative form: $$[A]_t = \frac{[A]_0}{1 + [A]_0 kt}$$

→ If a plot of $1/[A]_t$ *vs.* t gives a straight line, the rate law is second-order in [A].

→ The rearranged expression, $1/[A]_t = 1/[A]_0 + kt$, reveals the linear behavior of a plot of $1/[A]_t$ *vs.* t.

→ Plot of $1/[A]_t$ *vs. t:* slope = $+k$ *and* intercept = $1/[A]_0$

- **Integrated second-order rate law**

→ For the same initial rates, [A] decays more slowly with time if it is a second-order process than if it is first-order (see Fig. 13.13 in the text).

→ Used to confirm that a reaction is second order and to measure its rate constant

→ If k and$[A]_0$ are known, the value of [A] at any time t can be predicted.

Example 13.6a The reaction $2NO_2(g) \rightarrow 2NO(g) + O_2(g)$ is second order. The rate equals $k[NO_2]^2$. From a table of data (not given) listing $[NO_2]$ at various times during the reaction, explain how one can verify whether the reaction is indeed second order. Also, plot the two data points given below to estimate a value for the rate constant.

t	0	1000 s
$[NO_2]$	1.00×10^{-2}	5.00×10^{-3} mol·L^{-1}
$1/[NO_2]$	1.00×10^{2}	2.00×10^{2} L·mol^{-1}

Solution To verify that the reaction is second order, select values from a table of data and plot $1/[NO_2]$ *vs. t*. A second-order reaction has the following form:

$$\frac{1}{[NO_2]_t} = \frac{1}{[NO_2]_0} + kt$$

If the plot is linear, the reaction is second order, and the slope is equal to the rate constant. Because the problem states that the reaction is indeed second order, we may plot the two points given and connect them with a straight line.
The plot shown on the next page is obtained using the *Tool - Curve Fitter* program.

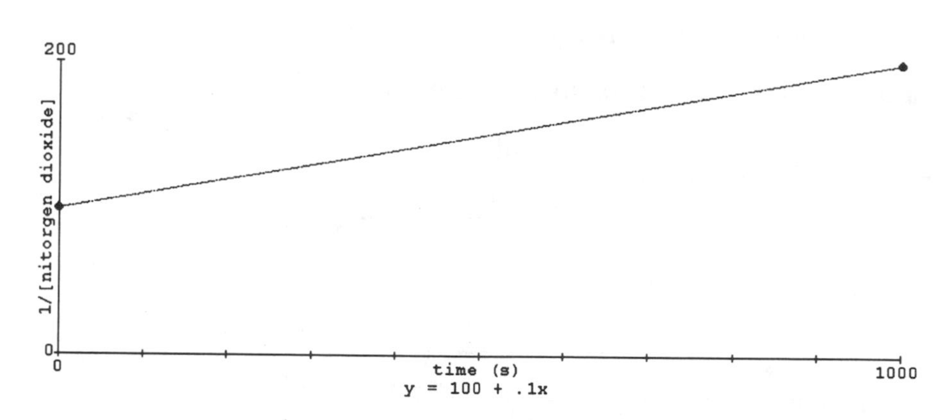

$$k = \frac{(200-100)\ \text{L·mol}^{-1}}{(1000-0)\ \text{s}} = 0.100\ \text{L·mol}^{-1}\text{·s}^{-1}$$ is the slope of the straight line.

Note that $a = 2$ and the *unique* $k = (0.100/2) = 0.0500\ \text{L·mol}^{-1}\text{s}^{-1}$.

Example 13.6b The decomposition of HI(g) to form H_2(g) and I_2(g) is second order with a rate constant $k = 0.060\ \text{L·mol}^{-1}\text{·s}^{-1}$ at 630 K. If the initial concentration of HI is 0.20 mol·L^{-1}, what is the half-life of the reaction and the concentration of HI after 250 s? Assume $a = 1$ for HI.

Solution The second-order expression for this reaction is

$$\frac{1}{[\text{HI}]_t} = \frac{1}{[\text{HI}]_0} + kt \quad \text{and} \quad t = \frac{1}{k}\left(\frac{1}{[\text{HI}]_t} - \frac{1}{[\text{HI}]_0}\right)$$

The initial concentration is 0.20 mol·L^{-1}, and the value after the half-life period is 0.10 mol·L^{-1}.

$$t_{1/2} = \frac{1}{k}\left(\frac{1}{\frac{1}{2}[\text{HI}]_0} - \frac{1}{[\text{HI}]_0}\right) = \frac{1}{k}(2-1)\left(\frac{1}{[\text{HI}]_0}\right) = \frac{1}{k[\text{HI}]_0}$$

$t_{1/2} = 1/(0.060\ \text{L·mol}^{-1}\text{·s}^{-1})(0.20\ \text{mol·L}^{-1}) = 83\ \text{s}$

Note that the half-life for a second-order reaction depends on the initial concentration.

The concentration of HI after 250 s may be obtained from the alternative form of the integrated rate expression.

$$[\text{HI}]_t = \frac{[\text{HI}]_0}{1 + [\text{HI}]_0 kt}$$

$$= \frac{0.20\ \text{mol·L}^{-1}}{1 + (0.20\ \text{mol·L}^{-1})(0.060\ \text{L·mol}^{-1}\text{·s}^{-1})(250\ \text{s})} = 0.050\ \text{mol·L}^{-1}$$

The concentration is then one-fourth its initial value.

Note: Compare the answers to the two parts. It takes 83 s for the concentration to decrease to one-half its initial value, and 250 s for the concentration to decrease to one-fourth its initial value. Because the half-life of a second-order reaction depends on the initial concentration, it is a much less useful concept than it is for a first-order reaction.

Models of Reactions (Sections 13.7-13.9)

Key Concepts

Arrhenius equation, Arrhenius parameters, pre-exponential factor, activation energy, Arrhenius behavior, collision theory of reactions, collision cross-section σ, Boltzmann distribution, steric requirement (factor) of reaction, activated complex theory of reactions, activated complex, reaction profile, potential energy surface

Overview

- **Fundamental questions**
  → Why do rate laws have different mathematical forms?
  → Why do rate constants have different magnitudes?
  → Why do rate constants depend on temperature?
  → What is the nature of the molecular processes (bond breaking and bond formation) that accompany chemical reactions?

13.7 The Effect of Temperature

- **Qualitative aspects**
  → A change in temperature results in a change of the rate constant of a reaction, and, therefore, its rate.
  → For *most* reactions, *increasing* the temperature results in a *larger value* of the rate constant.
  → For *many* reactions in organic chemistry, reaction rates *approximately* double for a 10°C increase in temperature.

- **Quantitative aspects**
  → The temperature dependence of the rate constant of many reactions is given by the Arrhenius equation shown here in two forms:

Exponential form: $$k = Ae^{-\frac{E_a}{RT}}$$

Logarithmic form: $$\ln k = \ln A - \frac{E_a}{RT}$$

R is the universal gas constant (8.314 51 $J{\cdot}K^{-1}{\cdot}mol^{-1}$)
T is the absolute temperature (K)

  → The two constants, A and E_a, called the *Arrhenius parameters*, are *nearly* independent of temperature.
  → The parameter A is the *pre-exponential factor*. The parameter E_a is the *activation energy*. The two parameters depend on the reaction being studied.
  → The Arrhenius equation applies to all types of reactions, gas phase or in solution.

- **Obtaining Arrhenius parameters**
  - → Plot $\ln k$ *vs.* $1/T$. If the plot displays straight-line behavior, the *slope* of the line is $-E_a/R$ and the *intercept* is $\ln A$. Note that the *intercept* at $(1/T = 0)$ implies $T = \infty$.
  - → The parameter A *always* has a positive value. The *larger* the value of A, the *larger* the value of k.
  - → The parameter E_a *usually* has a positive value. The *larger* the value of E_a, the *smaller* the value of k. If $E_a > 0$, the value of k *increases* as temperature *increases*. The rate constant increases more quickly with increasing temperature if E_a is large.
  - → If the value of the activation energy and also a value for k at a given temperature, T, are known, the logarithmic form of the Arrhenius equation can be used to calculate the rate constant, k', at another temperature, T'. Or alternatively, a value of E_a can be estimated from values of k' at T' and k at T.

$$\ln\frac{k'}{k} = \frac{E_a}{R}\left(\frac{1}{T} - \frac{1}{T'}\right)$$

- **Units of the Arrhenius parameters**
  - → A has the same units as the rate constant of the reaction.
  - → E_a has the units of energy, and values are usually reported in $kJ{\cdot}mol^{-1}$.

Note: An equivalent SI unit for the liter is dm^3. Rate constants and Arrhenius pre-exponential factors incorporating volume units are sometimes reported using dm^3 or cm^3 instead of liters. Those incorporating amount units are sometimes reported in units of molecules instead of moles. For example, a second-order rate constant of 1.2×10^{11} $L{\cdot}mol^{-1}{\cdot}s^{-1}$ is equivalent to 1.2×10^{11} $dm^3{\cdot}mol^{-1}{\cdot}s^{-1}$ or 2.0×10^{-10} $cm^3{\cdot}molecule^{-1}{\cdot}s^{-1}$.

Example 13.7a How to determine Arrhenius parameters from rate constant data.

Bromide ions react with iodomethane in methanol solution as follows:

$$CH_3I(soln) + Br^-(soln) \rightarrow CH_3Br(soln) + I^-(soln)$$

Use the kinetic data given below to determine values for the Arrhenius parameters of this reaction:

T	3	13	23	33°C
k	5.0×10^{-6}	1.7×10^{-5}	6.1×10^{-5}	2.0×10^{-4} $L{\cdot}mol^{-1}{\cdot}s^{-1}$

Solution The rate constant has the units for a second-order reaction. The value of A has the same units. For a plot of $\ln k$ *vs.* $1/T$, the data values are

T	276	286	296	306 K
$1/T$	3.62×10^{-3}	3.50×10^{-3}	3.38×10^{-3}	3.27×10^{-3} K^{-1}
$\ln k$	−12.21	−10.98	−9.70	−8.52

The plot shown on the next page, obtained using the *Tool – Curve Fitter* program, displays linear behavior.

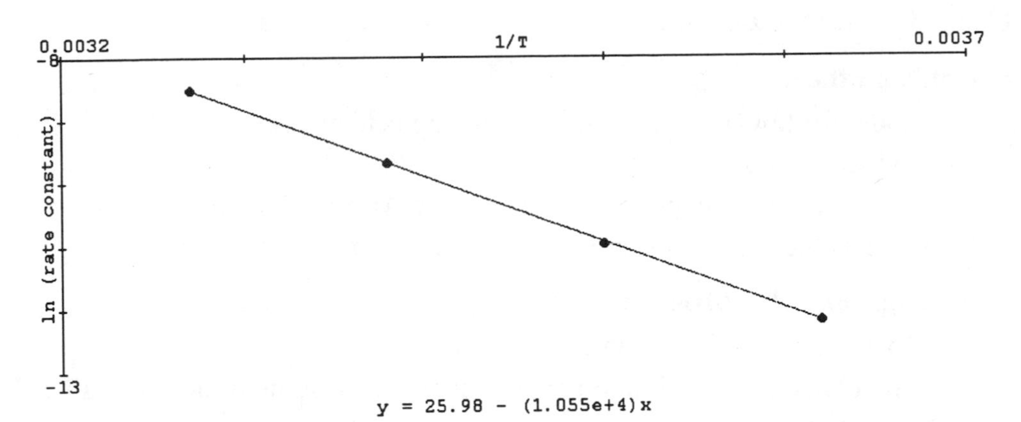

Slope = $-E_a/R = -1.055 \times 10^4$ K
(Note that in some computer programs, $e+n \equiv \times 10^n$ and $e-n \equiv \times 10^{-n}$.)
$E_a = -(-1.055 \times 10^4\ \text{K})(8.31451\ \text{J·K}^{-1}\text{·mol}^{-1})(10^{-3}\ \text{kJ·J}^{-1}) = 87.7\ \text{kJ·mol}^{-1}$
Intercept = $\ln A = 25.98$
$A = e^{25.98} = 1.9 \times 10^{11}\ \text{L·mol}^{-1}\text{·s}^{-1}$

Example 13.7b How to determine the rate constant at any temperature from E_a and k at T.
The gas-phase formation reaction of hydrogen iodide is

$$H_2(g) + I_2(g) \rightarrow 2\,HI(g)$$

For this reaction, $E_a = 163\ \text{kJ·mol}^{-1}$ and $k = 1.32 \times 10^{-4}\ \text{L·mol}^{-1}\text{·s}^{-1}$ at $T = 575$ K. Calculate the value of the rate constant, k', at $T' = 525$ K.

Solution Use the expression: $\ln\frac{k'}{k} = \frac{E_a}{R}\left(\frac{1}{T} - \frac{1}{T'}\right)$

$$\ln\frac{k'}{k} = \frac{(163\,000\ \text{J·mol}^{-1})}{(8.31451\ \text{J·K}^{-1}\text{·mol}^{-1})}\left(\frac{1}{575\ \text{K}} - \frac{1}{525\ \text{K}}\right) = -3.247$$

$$\frac{k'}{k} = e^{-3.247} = 3.889 \times 10^{-2}$$

$k' = (3.889 \times 10^{-2})(1.32 \times 10^{-4}\ \text{L·mol}^{-1}\text{·s}^{-1}) = 5.13 \times 10^{-6}\ \text{L·mol}^{-1}\text{·s}^{-1}$

Note: As expected, the rate constant *decreased* because the temperature *decreased.*

Example 13.7c How to determine the activation energy from k at T and k' at T'.
Calculate the activation energy for a reaction whose rate constant doubles when the temperature is raised from 290 K to 300 K.

Solution $\frac{k'}{k} = 2$ $\quad T = 290$ K $\quad T' = 300$ K

$$\ln 2 = \frac{E_a}{(8.31451\ \text{J·K}^{-1}\text{·mol}^{-1})}\left(\frac{1}{290\ \text{K}} - \frac{1}{300\ \text{K}}\right) = (1.382 \times 10^{-5}\ \text{J}^{-1}\text{·mol})E_a$$

$$E_a = \frac{\ln 2}{(1.382 \times 10^{-5}\ \text{J}^{-1}\text{·mol})}(10^{-3}\ \text{kJ·J}^{-1}) = 50.2\ \text{kJ·mol}^{-1}$$

13.8 Collision Theory

- **Collision theory**
  - → Model for how reactions occur at the molecular level
  - → Applies to gases
  - → Accounts for the exponential form of the Arrhenius equation
  - → Reveals the significance of the Arrhenius parameters A and E_a
- **Assumptions of collision theory**
  - → Molecules must collide in order to react.
  - → Only collisions with kinetic energy in excess of some minimum value, E_{min}, lead to reaction.
  - → Only molecules with the correct orientation with respect to each other may react.
- **Total rate of collisions in a gas mixture**
  - → Determined quantitatively given from the kinetic model of a gas
  - → Rate of collision = total number of collisions per unit volume per second

$$\boxed{\text{Rate of collision} = \sigma c N_A^2[\text{A}][\text{B}]}$$

σ = collision cross-section (area a molecule presents as a target during collision)
Larger molecules are more likely to collide with other molecules than are smaller ones.

c = mean speed at which molecules approach each other in a gas
For a gas mixture of A and B with molar masses M_A and M_B, respectively,

$$\boxed{c = \sqrt{\frac{8RT}{\pi\mu}} \qquad \mu = \frac{M_A M_B}{M_A + M_B}}$$

Molecules at high temperature collide more often than at low temperature.

N_A = Avogadro constant

[A] and [B] are the molar concentrations of A and B, respectively.

- **Total rate of reaction in a gas mixture**
  - → Rate of reaction = number of collisions that lead to reaction per unit volume per second

$$\boxed{\text{Rate of reaction} = P\sigma c N_A^2[\text{A}][\text{B}] \times e^{-E_{min}/RT}}$$

E_{min} = minimum kinetic energy required for a collision to lead to reaction
Collisions with energy less than the minimum do not lead to reaction.

$e^{-E_{min}/RT}$ = fraction of collisions with at least the energy E_{min}
Derived from the *Boltzmann distribution* (see Fig. 13.17 in the text)

P = steric factor < 1
Reactive collisions often require preferred directions of approach.

- **Reaction rate constant**
  → Rate constant = $k = \dfrac{\text{rate of reaction}}{[\text{A}][\text{B}]}$

$$\boxed{k = P\sigma c N_A^{\,2} e^{-E_{\min}/RT}}$$

  → Comparison with the Arrhenius equation in exponential form, $k = Ae^{-\frac{E_a}{RT}}$:

  $E_a \approx E_{\min}$ and $A \approx P\sigma c N_A^{\,2}$

  Because the speed is proportional to $\sqrt{T}$, the model predicts that the pre-exponential factor *and* activation energy both have a *slight* temperature dependence.

- **Summary**
  → According to collision theory, reaction occurs only if reactant molecules collide with a kinetic energy equal to or greater than the Arrhenius activation energy.
  → The reaction rate constant is greatest if the size of the molecules and their mean speed are large.
  → The steric factor accounts for collisions that are unreactive because the orientation of the molecules during such a collision is unfavorable.

13.9 Activated Complex Theory

- **Activated complex theory**
  → Applies to gas phase and solution reactions
  → Reacting molecules collide and distort.
  → During the encounter, the kinetic energy, E_K, of reactants decreases, the potential energy increases, the original chemical bonds lengthen, and new bonds begin to form.
  → Products may form if E_K for the collision is equal to or greater than E_a, and if the reactants have the proper orientation.
  → Reactants reform if E_K is less than E_a regardless of orientation.
  → If E_K is equal to or greater than E_a, the molecules form either reactants or products. The lowest kinetic energy that may produce products is equal to E_a. An *activated complex* is a combination of the two reacting molecules that are at the transition point between reactants and products.

- **Reaction profile**
  → Shows how energy changes as colliding reactants form an *activated complex* and then products along the *minimum energy pathway.* (An analogous process is hiking through a mountain range and choosing the pathway of lowest elevation through it.)
  → Progress of reaction refers to the relevant spatial coordinates of the molecules that show breaking of a bond as the energy maximum is approached and the formation of a new bond as the products form.

→ Reaction profiles for exothermic and endothermic reactions are shown in the following schematic drawings.

Exothermic reaction

Activated complex
Activation energy, E_a
Activation barrier
Reactants
Products
Potential energy
Progress of reaction

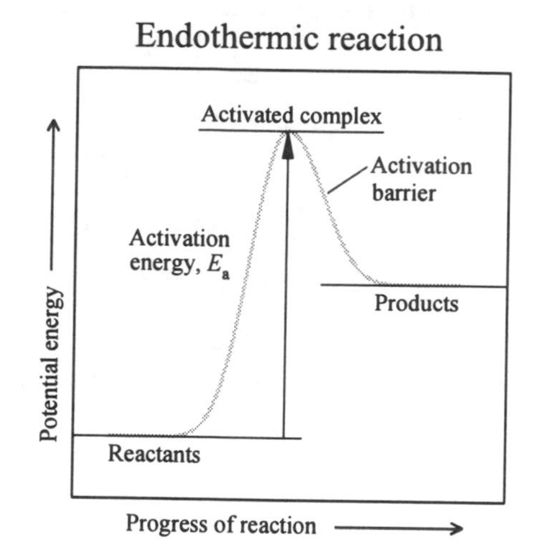

- **Potential energy surface**
  - → Three-dimensional plot in which the potential energy is plotted on the *z*-axis, and the other two axes correspond to interatomic distances
  - → Similar to a mountain pass, the *saddle point* corresponds to an activated complex with the minimum activation energy, E_a, required for reactants to form products.

Example 13.9 Describe a journey on the potential energy surface shown below (Fig. 13.21 in the text) in which H atoms collide with HBr molecules to form Br atoms and H_2 molecules. Follow the minimum energy pathway from reactants to products:

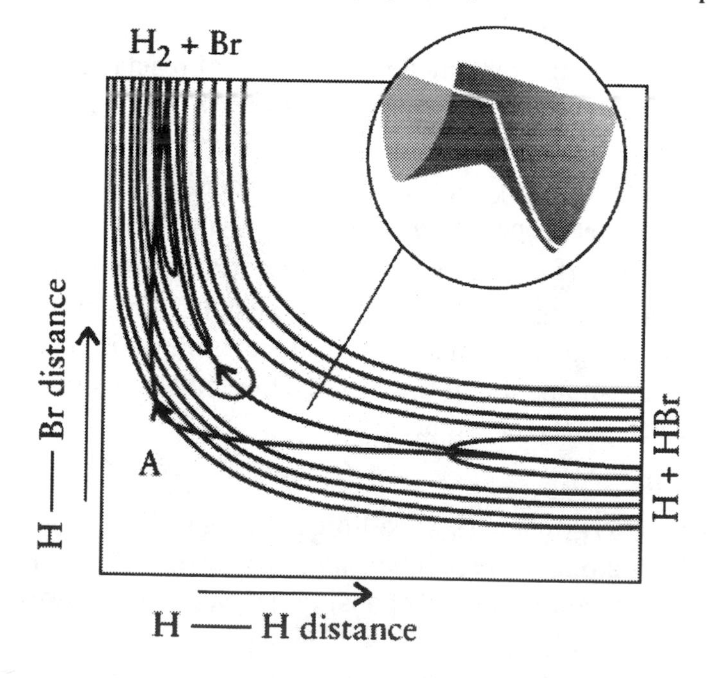

Solution From a great distance, an H atom approaches the H atom end of an HBr molecule from the right side of the diagram. At this distance, the minimum in the potential energy surface represents the bond distance of 141 pm in HBr. As the H atom approaches the molecule more closely, the HBr bond stretches, the potential energy increases, and eventually a maximum point (saddle point) is reached along the minimum energy pathway at the geometry of the activated complex. As the activated complex decays to products, the potential energy decreases as the new bond in H_2 is formed, and the Br atom departs. At large separation, the H_2 molecule attains its normal bond distance of 74 pm. The potential energy surface shown is one in which all three atoms are restricted to a straight line and is said to be a *collinear* surface. The second path (A) shown on the figure is one in which the initial kinetic energy is much higher and the minimum potential energy pathway is not followed. The *saddle point* region is identified by a line that leads to an inset figure showing a representation of the three-dimensional surface.

Reaction Mechanisms (Sections 13.10-13.13)

Key Concepts

elementary reaction, reaction mechanism, reaction intermediate, unimolecular reaction, bimolecular reaction, molecularity, termolecular reaction, steady-state approximation, pre-equilibrium condition, rate-determining step, chain reaction, chain carrier, radical chain reaction, initiation step, propagation step, termination step, chain branching, equilibrium constant for an elementary reaction

Overview

- **Fundamental questions**
  - → How do reactions occur on the molecular level?
  - → Can reactions be described in terms of separable elementary reactions?
  - → Can a sequence of elementary steps, a *reaction mechanism*, be found that reproduces both the overall stoichiometry and experimental rate law of a reaction?
  - → Can a reaction mechanism be proved?
  - → What role do intermediates play in a chemical reaction?
  - → How do chain reactions differ from ordinary ones?
  - → How does the equilibrium constant of the overall reaction relate to the rate constants for the forward and reverse elementary reactions in a reaction mechanism?
- **Focus** → In the previous sections, models for understanding *rate constants* are described. Here, models for understanding the *concentration dependence* of reaction rates are introduced.

13.10 Elementary Reactions

- **Reactions**
  - → A net chemical reaction is assumed to occur at the molecular level in a series of separable steps, called *elementary reactions*.
  - → A *reaction mechanism* is a proposed group of elementary reactions that accounts for the reaction's overall stoichiometry and experimental rate law.
- **Elementary reaction**
  - → A single-step reaction, written *without* state symbols, that describes the behavior of *individual* atoms and molecules taking part in a chemical reaction
  - → The *molecularity* of an elementary reaction specifies the number of *reactant* molecules that take part in it.
  - → *Unimolecular:* Only one reactant species is written (molecularity = 1).
    C_3H_6(cyclopropane) → $CH_2{=}CH{-}CH_3$(propene)
    A cyclopropane molecule decomposes spontaneously to form a propene molecule.
  - → *Bimolecular:* Two reactant species combine to form products (molecularity = 2).
    $H_2 + Br \rightarrow HBr + H$
    A hydrogen molecule and a bromine atom collide to form a hydrogen bromide molecule and a hydrogen atom.
  - → *Termolecular:* Three reactant species combine to form products (molecularity = 3).
    $I + I + H_2 \rightarrow HI + HI$
    Two iodine atoms and a hydrogen molecule collide simultaneously to form two hydrogen iodide molecules.
- **Reaction mechanism**
  - → *Sequence* of elementary reactions which, when added together, gives the net chemical reaction and reproduces its rate law
  - → *Reaction intermediate, not* a reactant or a product, is a species that appears in one or more of the elementary reactions in a proposed mechanism. It usually has a small concentration, and does not appear in the overall rate law.
  - → The plausibility of a reaction mechanism may be tested, but it cannot be proved. The presence and behavior of a reaction intermediate are sometimes testable features of a proposed reaction mechanism.

Example 13.10a Write a one-step and a two-step reaction mechanism for the decomposition of ozone to form oxygen: $2\,O_3(g) \rightarrow 3\,O_2(g)$.

Solution A possible one-step mechanism is a simple bimolecular reaction between two molecules of ozone.

Mechanism: $O_3 + O_3 \rightarrow O_2 + O_2 + O_2$

Net reaction: $2\,O_3(g) \rightarrow 3\,O_2(g)$

A possible two-step mechanism is the unimolecular decomposition of ozone in which one bond is broken to form an oxygen atom and diatomic molecule. The second step is the bimolecular reaction of an oxygen atom with ozone to form two molecules of oxygen. The oxygen atom is a *reaction intermediate.*

Mechanism: $O_3 \rightarrow O_2 + O$

$O + O_3 \rightarrow O_2 + O_2$

Net reaction: $2\,O_3(g) \rightarrow 3\,O_2(g)$

Note: See the note following Example 13.10b. The molecularity of dissociation and formation reactions of small molecules is more complicated than indicated in this example.

Example 13.10b Devise a reaction mechanism for the formation of ozone from oxygen molecules assuming that the first step is the dissociation of oxygen molecules into oxygen atoms. Give the molecularity of each step and name any reaction intermediate.

Solution The net reaction is $3\,O_2(g) \rightarrow 2\,O_3(g)$.

Step 1 is the unimolecular dissociation of an oxygen molecule into two oxygen atoms. Because an oxygen atom does not appear as a reactant or product, it must be a reaction intermediate. A possible second step is the reaction of an oxygen atom with an oxygen molecule to produce a molecule of ozone. Because two oxygen atoms are generated in step 1, two molecules of O_3 must react for each occurrence of step 1.

Mechanism: $O_2 \rightarrow O + O$ Step 1 (unimolecular)

$O + O_2 \rightarrow O_3$ Step 2 (bimolecular)

$O + O_2 \rightarrow O_3$ Step 3 (bimolecular)

Net reaction: $3\,O_2(g) \rightarrow 2\,O_3(g)$

Note: Step 3 is a repetition of step 2. In writing the mechanism, step 3 is often omitted and its occurrence is implicitly assumed to account for material balance. For small molecules, the molecularity of steps 1 and 2 is more complicated than indicated here. In step 1, the molecule actually decomposes after a bimolecular collision that imparts energy in excess of its dissociation limit. In step 2, the incipient ozone molecule quickly dissociates back into an oxygen atom and oxygen molecule unless energy is rapidly removed. For these reasons, the reactions are often written as follows.

$O_2 + M \rightarrow O + O + M$ Step 1 (bimolecular)

$O + O_2 + M \rightarrow O_3 + M$ Step 2 (termolecular)

The species M adds energy to O_2 in step 1, and removes energy in a nearly simultaneous collision of three species in step 2. If only molecular oxygen is present initially, the unidentified species M is molecular oxygen at the beginning of the reaction. It may also be an added inert gas, such as nitrogen or a noble gas. Larger polyatomic molecules that undergo dissociation have more ways to store energy internally. For these molecules, species M is not needed and the molecularity is reduced. For simplicity, the authors of the text assume that elementary reactions in which a single molecule undergoes dissociation are all *unimolecular*, and the elementary reactions between two species that stick together to form a new molecule are *bimolecular*. Step 1 may also occur by photodissociation, the absorption of light of sufficient energy to break a bond in the molecule, vacuum ultraviolet light for O_2. This process is bimolecular.

13.11 The Rate Laws of Elementary Reactions

- **Forms of rate laws for elementary reactions**
  - → Order follows from the stoichiometry of the *reactants* in an *elementary* reaction. Products do *not* appear in the rate law of an *elementary* reaction.
  - → *Unimolecular* elementary reactions are first-order.
    Reaction: $A \rightarrow$ products   Rate $= k[A]$
  - → *Bimolecular* elementary reactions are second-order overall. If there are two different species present, the reaction is first-order in each one.
    Reaction: $A + A \rightarrow$ products   Rate $= k[A]^2$
    Reaction: $A + B \rightarrow$ products   Rate $= k[A][B]$
  - → *Termolecular* elementary reactions are third-order overall. If there are two different species, the reaction is second-order in one of them, and first-order in the other. If there are three different species, the reaction is first-order in each one.
    Reaction: $A + A + A \rightarrow$ products   Rate $= k[A]^3$
    Reaction: $A + A + B \rightarrow$ products   Rate $= k[A]^2[B]$
    Reaction: $A + B + C \rightarrow$ products   Rate $= k[A][B][C]$
  - → The molecularity of an elementary reaction that is written in the reverse direction also follows from its stoichiometry. The *products* have become *reactants*.
    Forward reaction: $A + B \rightarrow P + Q + R$ Forward rate $= k[A][B]$
    Reverse reaction: $P + Q + R \rightarrow A + B$ Reverse rate $= k'[P][Q][R]$

Note: The prime symbol, ′, is used here and in the following sections to denote a rate constant written for a *reverse* elementary reaction. The same symbol is used in Section 13.7 to denote a rate constant for the *same* reaction (elementary or otherwise) measured at a *different* temperature. *Do not confuse these two very different meanings.*

Example 13.11a For each of the following gas-phase elementary reactions, determine the molecularity of the reaction, and write the rate expression for the formation of the product. Use numerical subscripts on the rate constants to identify them with a particular reaction.

1. $N_2O_4 \rightarrow 2\,NO_2$
2. $HO + NO_2 + Ar \rightarrow HNO_3 + Ar$
3. $2\,NO_2 \rightarrow N_2O_4$

Solution

1. An energized molecule of N_2O_4 decomposes spontaneously by the rupture of an N–N bond to form two molecules of NO_2. The elementary reaction is unimolecular and the rate law is first order in N_2O_4. The reaction is shown schematically in terms of relevant Lewis structures.

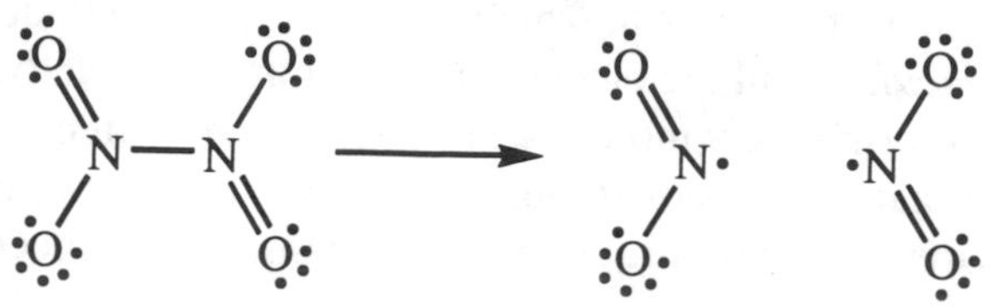

$$\text{Rate} = \frac{1}{2}\frac{d[NO_2]}{dt} = k_1[N_2O_4]$$

Note: Because resonance occurs in these molecules, the Lewis structure selected for N_2O_4 is one of four equivalent structures. Two equivalent Lewis structures may be drawn for each NO_2 product if the odd electron is located on the nitrogen atom.

2. The three species, HO, NO_2, and Ar, are combined in a termolecular reaction. The rate law is first-order in each species. The role of argon is to carry away energy created by the formation of an N–O bond. The reaction is shown schematically in terms of relevant Lewis structures.

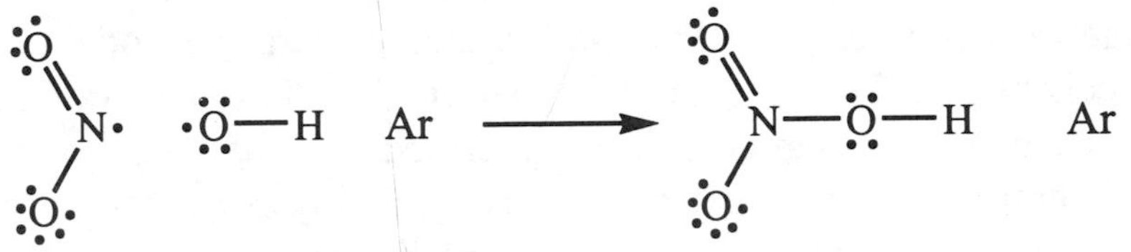

$$O{=}\dot{N}{-}O \quad \cdot\ddot{O}{-}H \quad Ar \longrightarrow O{=}N(O){-}\ddot{O}{-}H \quad Ar$$

$$\text{Rate} = \frac{d[HNO_3]}{dt} = k_2[HO][NO_2][Ar]$$

3. This reaction is the reverse of the elementary reaction in part 1. Here, two NO_2 radicals react to form the N–N bond in N_2O_4. The elementary reaction is bimolecular and the rate law is second-order in NO_2.

$$\text{Rate} = \frac{d[N_2O_4]}{dt} = k_3[NO_2]^2 = k_1'[NO_2]^2$$

Note: Because this reaction is the reverse of the elementary reaction in part 1, $k_3 = k_1'$.

- **Rate laws from reaction mechanisms**
  - → The time evolution of each reactant, intermediate, and product may be determined by integrating the rate laws for the elementary reactions in a proposed reaction mechanism. The results are compared to experiment. Recall that a successful mechanism must also account for the stoichiometry of the overall reaction.
  - → The mathematical solution of kinetic equations often requires numerical methods for solving simultaneous differential equations. In a few cases, exact analytical solutions exist.
  - → For multi-step mechanisms, approximate methods are often used to determine a rate law consistent with a proposed reaction mechanism. If the derived rate law is in agreement with the experimental one, the plausibility of the mechanism is tested by further experimentation, if possible.

- **Approximate methods for determining rate laws**
  - → Characteristics of reactions commonly encountered allow for the determination of rate laws with the use of three types of approximations.
  - → The *rate-determining step* approximation is made to determine a rate law for a mechanism in which one step occurs at a rate *substantially* slower than any others.

The slow step is a bottleneck, and the overall reaction rate cannot be larger than the slow step. The rate law for the rate-determining step is written first. If a reaction intermediate appears as a reactant in this step, its concentration term must be eliminated from the rate law. The final rate law only has concentration terms for reactants and products.

→ The *pre-equilibrium condition* describes situations in which reaction intermediates are formed and removed in steps prior to the rate-determining one. If the formation and removal of the intermediate in prior steps is rapid, an equilibrium concentration is established. If the intermediate appears in the rate-determining step, its relatively slow reaction in that step does not change its equilibrium concentration. The pre-equilibrium condition is sometimes called a *fast equilibrium*.

→ The *steady-state approximation* is a more general method for solving reaction mechanisms. The net rate of formation of any intermediate in the reaction mechanism is set equal to 0. An intermediate is assumed to attain its steady-state concentration instantaneously, decaying slowly as reactants are consumed. An expression is obtained for the steady-state concentration of each intermediate in terms of the rate constants of elementary reactions and the concentrations of reactants and products. The rate law for an elementary step that leads directly to product formation is usually chosen. The concentrations of all intermediates are removed from the chosen rate law, and a final rate law for the formation of product that reflects the concentrations of reactants and products is obtained.

→ Often a predicted rate law does not quite agree with the experimental one. Reaction conditions that modify the form of the rate law predicted by the mechanism may lead to final agreement (see Section 13.15 for an example). If complete exploration of conditions fails to reproduce the experimental result, the proposed reaction mechanism is rejected.

→ Applications of the approximate methods are explored in the following examples.

Example 13.11b How to use the rate-determining step approximation.
Predict the rate law for the following overall reaction from the proposed mechanism.

Overall reaction: $2\,NO(g) + Cl_2(g) \rightarrow 2\,NOCl(g)$

Proposed mechanism:

Step 1	$NO + Cl_2 \rightarrow NOCl_2$	(slow step)
Step 2	$NOCl_2 + NO \rightarrow 2\,NOCl$	(fast step)

Solution Adding the reaction for step 1 to step 2 reproduces the reaction stoichiometry. The first step is the slow one, and that limits the reaction rate. The rate of this step is written as

$$\text{Rate of step 1} = \text{rate of reaction} = -\frac{d[Cl_2]}{dt} = k_1[NO][Cl_2]$$

The predicted rate law is first-order in NO and first-order in Cl_2. The predicted rate constant is equal to that of the elementary reaction corresponding to step 1. Because step 2 is fast, its rate constant does not appear in the overall rate law.

Example 13.11c How to use the pre-equilibrium condition.
Predict the rate law for the following overall reaction from the proposed mechanism.

Overall reaction: $H_2(g) + I_2(g) \rightarrow 2\,HI(g)$

Proposed mechanism:

Step 1	$I_2 \rightleftharpoons 2\,I$	(pre-equilibrium)
Step 2	$2\,I + H_2 \rightarrow 2\,HI$	(slow step)

Solution Adding the reaction for step 1 to step 2 reproduces the reaction stoichiometry. In this case, add only the *forward* direction for step 1. The second step is the slow one limiting the reaction rate. The rate of this step is written as

$$\text{Rate of step 2} = \text{rate of reaction} = -\frac{d[H_2]}{dt} = k_2[I]^2[H_2]$$

The iodine atom is a reaction intermediate, and its concentration must be eliminated. The concentration of iodine atoms is obtained from the pre-equilibrium step. When step 1 is at equilibrium, the rate of its forward reaction is equal to the rate of its reverse reaction.

$$\text{Forward rate of step 1} = k_1\,[I_2] = \text{reverse rate of step 1} = k_1'\,[I]^2$$

Solve for $[I]^2$ and substitute into the equation for the rate of step 2.

$$[I]^2 = \frac{k_1}{k_1'}[I_2] \quad \text{and} \quad \text{rate of reaction} = -\frac{d[H_2]}{dt} = k_2\left(\frac{k_1}{k_1'}\right)[H_2][I_2]$$

The predicted rate law is first-order in H_2 and first-order in I_2. The predicted rate constant is equal to that of the elementary reaction of step 2 times the ratio of the forward and reverse reaction rate constants of step 1.

Note: The same rate law is obtained from a one-step mechanism in which the reaction is bimolecular, $H_2 + I_2 \rightarrow 2\,HI$. There is experimental evidence that both mechanisms contribute to the observed rate law for this reaction. Because step 1 is at equilibrium, we may write the following equilibrium constant expression (see Section 13.13):

$$K_1 = \frac{k_1}{k_1'}$$

Example 13.11d How to use the steady-state approximation.
Predict the rate law for the following overall reaction from the proposed mechanism. Compare the result to the observed rate law, which is first-order in NO_2Cl.

Overall reaction: $2\,NO_2Cl(g) \rightarrow 2\,NO_2(g) + Cl_2(g)$

Proposed mechanism:

Step 1	$NO_2Cl \rightarrow NO_2 + Cl$
Step 2	$Cl + NO_2Cl \rightarrow NO_2 + Cl_2$

Solution Adding the reaction for step 1 to step 2 reproduces the reaction stoichiometry. The product Cl_2 is formed only in step 2. Therefore, the rate of reaction is equal to the reaction rate of step 2.

$$\text{Rate of step 2} = \text{rate of reaction} = \frac{d[Cl_2]}{dt} = k_2[Cl][NO_2Cl]$$

The chlorine atom is an intermediate, and its concentration is obtained by the *steady-state approximation*. Assume that the rate of formation of Cl in step 1 is equal to the rate of its removal in step 2.

$$\frac{d[Cl]}{dt} = k_1[NO_2Cl] \quad \text{and} \quad -\frac{d[Cl]}{dt} = k_2[Cl][NO_2Cl]$$

$$\frac{d[Cl]}{dt} = k_1[NO_2Cl] - k_2[Cl][NO_2Cl] = 0 \quad \text{for a } \textit{steady state}$$

$k_1[NO_2Cl] = k_2[Cl][NO_2Cl]$ and $[Cl] = k_1/k_2$

The predicted rate law is then

$$\frac{d[Cl_2]}{dt} = k_2\left(\frac{k_1}{k_2}\right)[NO_2Cl] = k_1[NO_2Cl]$$

The predicted result of first-order behavior in NO_2Cl is in agreement with experiment.

Example 13.11e How to use the steady-state approximation.
Predict the rate law for the same overall reaction as in Example 13.11d, using a different proposed mechanism.

Overall reaction: $2\,NO_2Cl(g) \rightarrow 2\,NO_2(g) + Cl_2(g)$

Proposed mechanism:

Step 1 $2\,NO_2Cl \rightarrow N_2O_4 + Cl_2$

Step 1′ $N_2O_4 + Cl_2 \rightarrow 2\,NO_2Cl$

Step 2 $N_2O_4 \rightarrow 2\,NO_2$

Solution Adding the reactions for step 1 and step 2 reproduces the reaction stoichiometry. Step 1′ is listed separately to indicate that a pre-equilibrium condition may not exist in step 1. The product NO_2 is formed only in step 2. Therefore, the rate of reaction is equal to the reaction rate of step 2.

$$\text{Rate of step 2} = \text{rate of reaction} = \frac{1}{2}\frac{d[NO_2]}{dt} = k_2[N_2O_4]$$

The N_2O_4 molecule is an intermediate, and its concentration is obtained by the *steady-state approximation*. Assume that the rate of formation of N_2O_4 in step 1 is equal to the rate of its removal in step 1′ and step 2.

$$\frac{d[N_2O_4]}{dt} = k_1[NO_2Cl]^2 \quad \text{and} \quad -\frac{d[N_2O_4]}{dt} = k_1'[N_2O_4][Cl_2] + k_2[N_2O_4]$$

$$\frac{d[N_2O_4]}{dt} = k_1[NO_2Cl]^2 - \left(k_1'[N_2O_4][Cl_2] + k_2[N_2O_4]\right) = 0 \quad \textit{steady state}$$

$k_1[NO_2Cl]^2 = k_1'[N_2O_4][Cl_2] + k_2[N_2O_4]$

Solving for the steady-state concentration of N_2O_4 leads to

$$[N_2O_4] = \frac{k_1[NO_2Cl]^2}{k_1'[Cl_2] + k_2}$$

The predicted rate law is then

$$\frac{1}{2}\frac{d[NO_2]}{dt} = k_2[N_2O_4] = \frac{k_2k_1[NO_2Cl]^2}{k_1'[Cl_2] + k_2}$$

There are two limiting forms of this rate law. If step 1′ has a greater rate than step 2, $k_1'[Cl_2] >> k_2$ and the rate law is second-order in NO_2Cl and negative first-order in Cl_2. As Cl_2 forms, the rate of reaction slows. This type of behavior is called *product inhibition*. This result is equivalent to a pre-equilibrium condition in step 1 and a rate determining condition in step 2. If step 2 has a greater rate than step 1′, $k_2 >> k_1'[Cl_2]$ and the rate law is second-order in NO_2Cl. Neither form of the limiting behavior is in agreement with the observed rate law that is first-order in NO_2Cl, and the mechanism is rejected.

13.12 Chain Reactions

- **Chain reaction**
  - → Series of linked elementary reactions that *propagate* in chain cycles
  - → A reaction intermediate is produced in an *initiation* step.
  - → Reaction intermediates are the *chain carriers*, which are often radicals.
  - → In one cycle, a reaction intermediate typically reacts to form a different intermediate, which in turn reacts to regenerate the original one.
  - → In each cycle, some product is usually formed.
  - → The cycle is eventually broken by a *termination* step that consumes the intermediate.
  - → In *chain branching*, a chain carrier reacts to form two or more carriers in a single step. Chain branching often leads to chemical explosions.

Example 13.12a Identify the steps in the gas-phase mechanism for the chain reaction of hydrogen and bromine, the product of which is hydrogen bromide.

Overall reaction: $H_2(g) + Br_2(g) \rightarrow 2\,HBr(g)$

Chain mechanism:

Step 1	$Br_2 \rightarrow 2\,Br\cdot$
Step 1′	$2\,Br\cdot \rightarrow Br_2$
Step 2	$Br\cdot + H_2 \rightarrow HBr + H\cdot$
Step 3	$H\cdot + Br_2 \rightarrow HBr + Br\cdot$

Solution Step 1 is the *initiation* step. The initial radical is a bromine atom. The dissociation of bromine molecules is achieved by thermal (heat) or photochemical (light) techniques. Steps 2 and 3 are the chain *propagation* steps. The bromine atom reacts with hydrogen molecules to form product and a new intermediate, the hydrogen atom. The hydrogen atom reacts with bromine molecules to form product and the initial radical, the bromine atom. The chain is broken by a *termination* step (step 1′), in which the initial intermediate is removed.

Note: As the concentration of HBr product increases, another reaction (step 2′) occurs. This step slows the chain cycle process but does not terminate it. It is often called an *inhibition* step.

Step 2′ $H\cdot + HBr \rightarrow H_2 + Br\cdot$

Radicals may react on the walls of the container in other possible *termination* steps. The dissociation of H_2 to form H atoms is not as likely an *initiation* process, because the bond enthalpy for H_2 (436 kJ·mol^{-1}) is much larger than the value for Br_2 (193 kJ·mol^{-1}).

Example 13.12b In the explosive reaction of hydrogen molecules with oxygen molecules to form water, *chain branching* is a significant process. Determine the most probable *initiation* step and two likely *chain branching* steps.

Solution The dissociation of hydrogen molecules or oxygen molecules to produce atoms (radicals) is the likely initiation process. The bond enthalpy of H_2 (436 kJ·mol^{-1}) is somewhat less than that of O_2 (496 kJ·mol^{-1}), and the likely *initiation* step is

Step 1 $\quad H_2 \rightarrow 2H\cdot$

A possible *chain branching* process (step 2) is the reaction of hydrogen atoms with oxygen molecules to form hydroxyl radicals and oxygen atoms. Another *chain branching* process (step 3) is the reaction of oxygen atoms with hydrogen molecules to form hydroxyl radicals and hydrogen atoms. In each case, a single *chain carrier* produces two new *chain carriers*. The oxygen atom is a biradical, and its symbol is $\cdot O\cdot$.

Step 2 $\quad H\cdot + O_2 \rightarrow HO\cdot + \cdot O\cdot$

Step 3 $\quad \cdot O\cdot + H_2 \rightarrow HO\cdot + H\cdot$

13.13 Rates and Equilibrium

- **Elementary reaction**
  - → At *equilibrium*, the forward and reverse reaction rates for an *elementary* reaction are equal.

    For a given elementary reaction at equilibrium, $A + B \rightleftharpoons P + Q$

    forward rate = k[A][B] = reverse rate = k'[P][Q]
  - → The *equilibrium constant* for an *elementary* reaction is equal to the ratio of the rate constants for the forward and reverse reactions.

    In the above example, $K = \dfrac{[P][Q]}{[A][B]} = \dfrac{(\text{reverse rate})/k'}{(\text{forward rate})/k} = \dfrac{k}{k'}$

$$\boxed{K = \frac{k}{k'}}$$ for any elementary reaction

Example 13.13 For the following reaction occurring in aqueous solution, write an expression for the forward reaction rate and the reverse reaction rate, assuming that the two reactions are *elementary*.

$$H_2S(aq) + H_2O(l) \rightleftharpoons H_3O^+(aq) + HS^-(aq)$$

Determine an expression for the equilibrium constant K in terms of the forward and reverse rate constants. Assume that the concentration of water is fixed and it may be incorporated into the forward reaction rate constant.

Solution The forward reaction and its rate law are

$H_2S + H_2O \rightarrow H_3O^+ + HS^-$ $\qquad$ forward rate = $k[H_2S][H_2O] = k[H_2S]$

The reverse reaction and its rate law are

$H_3O^+ + HS^- \rightarrow H_2S + H_2O$ reverse rate = $k'[H_3O^+][HS^-]$

At equilibrium, the forward and reverse rates are equal.

$$k[H_2S] = k'[H_3O^+][HS^-]$$

Solve for the ratio of forward to reverse rate constants.

$$\frac{k}{k'} = \frac{[H_3O^+][HS^-]}{[H_2S]} = K_a$$

The ratio of the forward reaction rate constant to the reverse value is equal to the acidity constant for the weak acid, H_2S. The equilibrium constant for the dissociation of a weak acid is consistent with proton transfer taking place in a single elementary step.

- **Equilibrium constants from reaction mechanisms**
  - → Consider an overall reaction with a multi-step mechanism in which the rate constants for the elementary reactions are k_1, k_2, k_3, … in the forward direction and k_1', k_2', k_3', … in the reverse direction.
  - → The equilibrium constant for an overall reaction is related to the individual rate constants by the following equation:

$$K = \frac{k_1}{k_1'} \times \frac{k_2}{k_2'} \times \frac{k_3}{k_3'} \times \ \ldots$$

- **Temperature dependence of the equilibrium constant**
  - → The temperature dependence of the equilibrium constant of an elementary reaction follows from the temperature dependence of the ratio of the rate constant for the forward reaction to that of the reverse reaction, with both expressed in Arrhenius form.
  - → If the reaction is *exothermic*, the activation energy in the forward direction is smaller than in the reverse one (see the two upper figures in the Study Guide on p. 274). The rate constant with the larger activation energy increases more rapidly as temperature increases than the one with smaller energy. As temperature is *increased,* k' is predicted to increase faster than k and the equilibrium constant *decreases*. *Reactants* are favored.
  - → If the reaction is *endothermic*, the activation energy in the forward direction is larger than in the reverse one. As temperature is *increased,* k is predicted to increase faster than k' and the equilibrium constant *increases*. *Products* are favored.

Accelerating Reactions (Sections 13.14-13.15)

Key Concepts

catalysis, catalyst, homogeneous catalyst, heterogeneous catalyst, poisoning a catalyst, enzymes, substrate, induced-fit mechanism, Michaelis-Menten mechanism, Michaelis constant K_M

Overview

- **Factors that increase reaction rates**
  - → An increase in temperature *usually* results in an increase in the rate constant because more molecules have energy in *excess* of the activation energy and may react.
  - → An increase in the concentration of a species that appears with positive order in the rate law leads to an increase in reaction rate (greater rate of collisions).
  - → Use of an appropriate catalyst may increase the rate of reaction by changing the mechanism to a more favorable one.

13.14 Catalysis

- **Nature of a catalyst**
  - → Accelerates a reaction rate without being consumed
  - → Does not appear in the overall reaction
  - → Lowers the activation energy of a reaction by changing the pathway (mechanism)
  - → Accelerates *both* the forward and reverse reaction rates
  - → Does *not* change the final equilibrium composition of the reaction
  - → May appear in the rate equation, but usually does not
- **Homogeneous catalyst** → Present in the same phase as the reactants
- **Heterogeneous catalyst** → Present in a different phase from the reactants
- **Poisoning a catalyst**
  - → Catalysts can be poisoned or inactivated.
  - → For a heterogeneous catalyst, irreversible adsorption of a substance on the catalyst's surface may prevent reactants from reaching the surface and reacting on it.
  - → For a homogenous catalyst, a reactive site on a molecule may be similarly "poisoned" by the binding of a substance directly to the site or nearby either in a permanent or reversible way. The properties of the site are significantly altered to prevent catalytic action.

Example 13.14a How to recognize a catalyst.
Two steps in the mechanism for an electron-transfer reaction that occurs in aqueous solution are

$$V^{3+}(aq) + Cu^{2+}(aq) \rightarrow V^{4+}(aq) + Cu^{+}(aq)$$
$$Cu^{+}(aq) + Fe^{3+}(aq) \rightarrow Cu^{2+}(aq) + Fe^{2+}(aq)$$

What is the overall reaction? What is the catalyst in the reaction? Is the catalyst homogeneous or heterogeneous?

Solution The overall reaction is the sum of the two individual steps.

$$V^{3+}(aq) + Fe^{3+}(aq) \rightarrow V^{4+}(aq) + Fe^{2+}(aq)$$

The copper(II) ion, Cu^{2+}, is the catalyst. It is consumed in the first step, but regenerated in the second one, so a small amount can convert a much larger amount of reactants to products. Because the Cu^{2+} catalyst is dissolved in water along with the reactants and products, it is a homogeneous catalyst.

Example 13.14b How to recognize the poisoning of a catalyst.
The decomposition of ammonia is catalyzed by finely divided tungsten metal.

$$2\,NH_3(g) \rightarrow N_2(g) + 3\,H_2(g)$$

Is the catalyst homogeneous or heterogeneous? Addition of H_2 to the reaction vessel decreases the reaction rate, whereas addition of N_2 does not. Explain this observation.

Solution The tungsten catalyst is in a different phase (solid) from the reactants and products (gas phase), so it is a heterogeneous catalyst. It is likely that the catalyst functions by adsorbing the only reactant NH_3 on its surface. Because the addition of N_2 has no effect on the reaction rate, we assume that it is not simply the presence of product that reduces the effectiveness of the catalyst, but rather something to do with H_2. Because the addition of H_2 leads to a decrease in the reaction rate, it is likely that H_2 is adsorbed more strongly on a tungsten surface than is NH_3, and therefore, H_2 poisons the catalyst by preferential binding.

Example 13.14c How to treat a catalyst appearing in the rate law.
For a homogeneous reaction, the concentration of a catalyst may appear in the rate law. In a reaction catalyzed by carbon disulfide, the triiodide ion reacts with the azide ion to form iodide ions and nitrogen gas. The steps in the proposed mechanism are

Step 1 $CS_2 + N_3^- \rightarrow S_2CN_3^-$

Step 2 $2\,S_2CN_3^- + I_3^- \rightarrow 2\,CS_2 + 3\,N_2 + 3\,I^-$

Write the overall reaction, and generate an expression for the rate of disappearance of I_3^- assuming the concentration of the intermediate is determined by the steady-state approximation.

Solution The overall reaction is the sum of the individual steps. Step 1 must occur twice each time step 2 occurs in order to generate the net reaction:

$$\begin{array}{l} CS_2 + N_3^- \rightarrow S_2CN_3^- \\ CS_2 + N_3^- \rightarrow S_2CN_3^- \\ 2\,S_2CN_3^- + I_3^- \rightarrow 2\,CS_2 + 3\,N_2 + 3\,I^- \\ \hline I_3^-(aq) + 2\,N_3^-(aq) \rightarrow 3\,N_2(g) + 3\,I^-(aq) \end{array}$$

The rate of disappearance of I_3^- is determined from step 2.

$$-\frac{d[I_3^-]}{dt} = k_2[S_2CN_3^-]^2[I_3^-]$$

The concentration of the intermediate is determined from the steady-state approximation.

$$\frac{d[S_2CN_3^-]}{dt} = k_1[CS_2][N_3^-] - 2k_2[S_2CN_3^-]^2[I_3^-] = 0$$

The factor of 2 in the second term arises from the fact that step 1 occurs twice as often as step 2. Rearranging, the expression leads to

$$k_1[CS_2][N_3^-] = 2k_2[S_2CN_3^-]^2[I_3^-]$$

The overall rate law is then

$$-\frac{d[I_3^-]}{dt} = k_2[S_2CN_3^-]^2[I_3^-] = \frac{k_1}{2}[CS_2][N_3^-]$$

The reaction is first order in the nitride ion reactant and first order in the catalyst.

13.15 Living Catalysts: Enzymes

- **Enzyme, E**
  - → A biological catalyst
  - → Typically, a large molecule (protein) with crevices, pockets, and ridges on its surface (see Fig. 13.33 in the text)
  - → Catalyzes the reaction of a reactant known as the *substrate*, S, that produces a *product*, P: E + S → E + P
  - → Capable of increasing the rate of specific reactions to an enormous extent

Example 13.15a The enzyme catalase increases the rate of decomposition of H_2O_2 to form H_2O by a factor of 10^{15}. Assume this rate increase arises from a lowering of the activation energy with no accompanying change in the pre-exponential factor. If the activation energy for the uncatalyzed reaction is 94 kJ·mol^{-1} at 25°C, calculate the activation energy for the catalyzed reaction.

Solution Start with the exponential form of the Arrhenius equation and assume the A factor is the same for both mechanisms. The ratio of catalyzed to uncatalyzed rate constants is then:

$$\frac{k_{cat}}{k_{uncat}} = 10^{15} = \frac{e^{-E_a(\text{cat})/RT}}{e^{-E_a(\text{uncat})/RT}} = e^{-[E_a(\text{cat})-E_a(\text{uncat})]/RT}$$

$$\ln\left(\frac{k_{cat}}{k_{uncat}}\right) = 36.841 = -\frac{E_a(\text{cat}) - E_a(\text{uncat})}{RT}$$

$$E_a(\text{cat}) = E_a(\text{uncat}) - (36.841)RT$$

$$E_a(\text{cat}) = 94\ \text{kJ·mol}^{-1} - (36.841)(8.314\,51 \times 10^{-3}\ \text{kJ·K}^{-1}\text{·mol}^{-1})(298.15\ \text{K})$$
$$= 3\ \text{kJ·mol}^{-1}$$

The catalyzed mechanism is characterized by an activation energy much smaller than the uncatalyzed one.

- **Induced-fit mechanism**
  - → Enzymes are highly specific catalysts that function by an *induced-fit mechanism* in which the substrate approaches a correctly-shaped pocket of the enzyme known as the *active site*.

→ On binding, the enzyme distorts to accommodate the substrate allowing the reaction to occur.

→ The shape of the pocket determines which substrate can react.

Example 13.14b How to understand the nature of enzyme activity.
Succinic dehydrogenase catalyzes the reduction of succinic acid to fumaric acid.

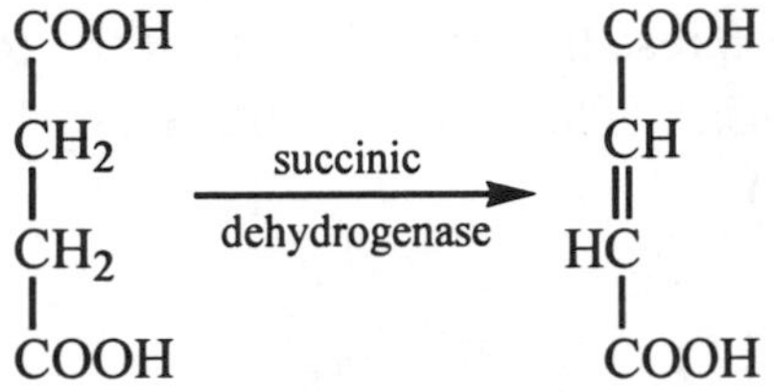

$$\begin{array}{c} COOH \\ | \\ CH_2 \\ | \\ CH_2 \\ | \\ COOH \end{array} \xrightarrow[\text{dehydrogenase}]{\text{succinic}} \begin{array}{c} COOH \\ | \\ CH \\ \| \\ HC \\ | \\ COOH \end{array}$$

Only one of the following molecules acts as an inhibitor (modest poison) for this enzyme. An inhibitor binds to or near the active site in an enzyme. Which one is it?

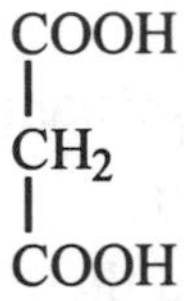

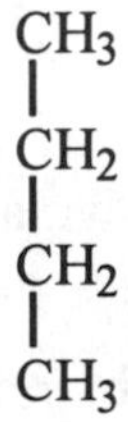

$$\underset{\text{I}}{\begin{array}{c} COOH \\ | \\ CH_2 \\ | \\ COOH \end{array}} \qquad \underset{\text{II}}{\text{(benzene ring)}} \qquad \underset{\text{III}}{\begin{array}{c} CH_3 \\ | \\ CH_2 \\ | \\ CH_2 \\ | \\ CH_3 \end{array}}$$

Solution One way to poison an enzyme is for a molecule other than the substrate to occupy the active site, preventing the substrate from reacting. It is reasonable to expect that a molecule that is chemically similar to the substrate would act as an inhibitor. Molecule I (malonic acid) has two carboxylic acid (–COOH) groups and a methylene ($-CH_2-$) group, and it most closely resembles the substrate chemically. It is expected to be an inhibitor.

- **Michaelis-Menten mechanism of enzyme reaction**

→ A two-step mechanism accounts for the observed dependence of the reaction rate of many enzyme-substrate reactions.

→ For an enzyme, E, and a substrate, S, the overall reaction and the Michaelis-Menten mechanism are

Overall reaction: $E + S \rightarrow E + P$

Mechanism:

Step 1 $E + S \rightleftharpoons ES$

Step 2 $ES \rightarrow E + P$

→ In step 1, the enzyme and substrate form an enzyme-substrate complex, ES, that can dissociate to reform substrate (reverse of step 1) or decay to form product P (step 2). The rate of formation of product is given by step 2.

Rate of formation of P = $k_2[ES]$

→ The steady-state approximation may be applied to the intermediate, ES, and its concentration is

$$[\mathrm{ES}] = \frac{k_1[\mathrm{E}][\mathrm{S}]}{k_1' + k_2}$$

→ Because the free enzyme concentration may be substantially diminished during the reaction, it is customary to express the rate in terms of the total enzyme concentration, $[\mathrm{E}]_0 = [\mathrm{E}] + [\mathrm{ES}]$. Then, $[\mathrm{E}]$ is replaced by $[\mathrm{E}]_0 - [\mathrm{ES}]$ in the steady-state result, leading to

$$[\mathrm{ES}] = \frac{k_1[\mathrm{E}]_0[\mathrm{S}]}{k_1' + k_2 + k_1[\mathrm{S}]}$$

$$\text{Rate of formation of P} = \frac{k_1 k_2[\mathrm{E}]_0[\mathrm{S}]}{k_1' + k_2 + k_1[\mathrm{S}]} = \frac{k_2[\mathrm{E}]_0[\mathrm{S}]}{K_\mathrm{M} + [\mathrm{S}]}$$

$K_\mathrm{M} = \dfrac{k_1' + k_2}{k_1}$ is known as the Michaelis constant

→ At low concentration of substrate, $[\mathrm{S}] \ll K_\mathrm{M}$, and the rate law is given by

$$\text{Rate of formation of P} = \frac{k_2[\mathrm{E}]_0[\mathrm{S}]}{K_\mathrm{M}}$$

The rate varies *linearly* with the concentration of substrate in this region. *Physically, in this region, most enzyme sites lack substrate molecules, so doubling the substrate concentration doubles the number of occupied sites and the rate as well.*

→ At high concentration of substrate, $[\mathrm{S}] \gg K_\mathrm{M}$ and the rate law becomes

$$\text{Rate of formation of P} = \frac{k_2[\mathrm{E}]_0[\mathrm{S}]}{[\mathrm{S}]} = k_2[\mathrm{E}]_0$$

In this region, the rate is *independent* of the substrate concentration (see Fig. 13.35 in the text).

Physically, all the enzyme sites are occupied with substrate molecules, so adding substrate has no effect on the rate.

- **Review of the Michaelis-Menten mechanism of enzyme catalysis**

→ $\boxed{E + S \longrightarrow E + P}$ *overall reaction*

→ | Step 1 | $E + S \rightleftharpoons ES$ |
| --- | --- |
| Step 2 | $ES \rightarrow E + P$ |

mechanism

→ $\boxed{[E]_0 = [E] + [ES]}$ *total enzyme concentration*

→ $$\boxed{\frac{d[P]}{dt} = \frac{k_2[E]_0[S]}{K_M + [S]}}$$ *rate of formation of product*

→ $$\boxed{K_M = \frac{k_1' + k_2}{k_1}}$$ *Michaelis constant*

- **Summary**
  → *Enzymes* are large proteins that function as *biological catalysts*.
  → *Substrate* molecules (reactants) bind to *active sites* on enzyme molecules.
  → The model of enzyme *binding* and *action* is called the *induced-fit mechanism*.
  → Substrate molecules undergo reaction at the active site to form *product* molecules.
  → Product molecules are released and the active site is restored.

Chapter 14 The Elements: The First Four Main Groups

Periodic Trends (Sections 14.1-14.4)

Key Concepts

descriptive chemistry, periodic trends, congeners, atomic properties, atomic radii, first ionization energy, electron affinity, electronegativity, polarizability, bonding trends, chemical properties, hydrides, saline hydride, metallic hydride, molecular hydride, oxides, ionic oxides, acid anhydride, formal anhydride

Overview

- **Descriptive chemistry**
  - → Compilation of the behavior of elements and their compounds
  - → Details of their preparation, properties, and applications
- **Trends in the properties of elements in the periodic table**
  - → A trend across a period correlates with adding electrons to the same valence shell.
  - → A group or family trend correlates with the value of the *n* quantum number of the valence electron configuration. A *congener* is a member of a group. For example, sodium and potassium are *congeners.*
  - → *Diagonal trends* are particularly important between elements in Period 2 and 3, and in separating metals from nonmetals.

14.1 Atomic Properties

- **Atomic radii**
  - → Atomic radii typically *decrease* from left to right across a period and *increase* down a group. Ionic radii follow similar trends (see Sections 1.15 and 1.16).
  - → The radius of an atom helps determine the maximum number of atoms that may bond to it. Small atoms have low valences, because only a few atoms can pack around them. With few exceptions, only Period 2 elements have the ability to form multiple bonds with themselves or other atoms. Their small size allows for the side-by-side overlap of *p*-orbitals in π-bonding.
- **First ionization energies**
  - → First ionization energies typically *increase* from left to right across a period and *decrease* down a group (see Section 1.17).
  - → The increase across a period is caused by increasing attraction of the nuclear charge to the electrons in the same valence shell. The trend down a group is ascribed to the concentric shells of electrons that are progressively more distant from the nucleus.

- **Electron affinities**
  - → The *highest* electron affinities are found at the top right of the periodic table (see Section 1.20). Recall that a chlorine atom has a higher *electron affinity* than a fluorine atom, but fluorine is the element with the highest *electronegativity*.
  - → Electron affinity is a measure of the energy *released* when an anion is formed. The greater the electron affinity, the more stable the anion formed. Elements with high electron affinities are present as *anions* in compounds with *metallic* elements, and commonly have *negative* oxidation states in the covalent compounds they form with other *nonmetallic* elements. For example, fluorine has an oxidation state of −1 in the compound sulfur hexafluoride, SF_6.
- **Electronegativities**
  - → *Electronegativity* is a measure of the tendency of an atom to attract bonding electrons when part of a compound. The electronegativity of an element is a useful guide to the type of bond it may form. It is also a predictor of the *polarity* of a chemical bond formed between nonmetallic elements.
  - → Electronegativities typically *increase* from left to right across a period and *decrease* down a group (see Section 2.14).
- **Polarizabilities**
  - → *Polarizability* typically *decreases* from left to right across a period and *increases* down a group (see Section 2.15).
  - → *Polarizability* is a measure of the ease with which an electron cloud of an atom may be distorted. It is the greatest for the more massive atoms in a group. Anions are more polarizable than the parent atoms, and polarizability decreases from left to right across a period. High polarizing power is associated with ions of small size and high charge. Cations with high polarizing power tend to have covalent character in the bonds formed with highly polarizable anions. Diagonal neighbors in the periodic table have similar polarizability or polarizing power and exhibit a similar amount of covalent character in the bonds they form.

Example 14.1a Which atom is *smaller*, thallium or lead?

Solution Thallium, Tl, and lead, Pb, are in the same period, and Pb is to the right of Tl. The atomic radius decreases across a period from left to right, so we expect Pb to be smaller than Tl.

Example 14.1b Which atom is more *electropositive*, oxygen or selenium?

Solution *Electropositive* is the opposite of *electronegative*. The *more* electronegative atom is the *less* electropositive atom. Oxygen, O, and selenium, Se, are congeners, members of the same group. Electronegativity typically decreases in going down a group, so we expect Se to be *less* electronegative and *more* electropositive than O.

Example 14.1c Which atom is the most *polarizable*: bromine, fluorine, or oxygen?

Solution The most polarizable atom has the greatest polarizability. Oxygen, O, and fluorine, F, are located in the upper right corner of the periodic table in Period 2. They are small atoms with tightly bound electrons and are expected to have low polarizability. Bromine, Br, with its larger size and its electrons further removed from the nucleus, is expected to be more *polarizable* than either O or F.

14.2 Bonding Trends

- **Elements bonded to other elements**
  - → The valence of an element in Period 2 is determined by the number of electrons in the valence shell and the octet rule.
  - → The valence of an element in Period 3 or higher ($Z \geq 14$) may be increased through octet expansion allowed by access to empty *d*-orbitals.
  - → Elements at the bottom of the *p* block may display an oxidation number 2 less than the group number suggests (*inert-pair effect*, see Section 1.18). For example, lead, Pb, has an oxidation number of +2 in PbO and +4 in PbO_2.

Example 14.2 Give a chemical formula and write a Lewis structure of a phosphorus-containing compound and of a sulfur-containing compound that exhibit expanded valence through octet expansion.

Solution Both PCl_5 and SF_6 are compounds that require octet expansion. In PCl_5, phosphorus is in the +5 oxidation state and in SF_6, sulfur is in the +6 oxidation state. The Lewis structures are given below and clearly show expanded octets about the central P and S atoms (10 and 12 electrons, respectively).

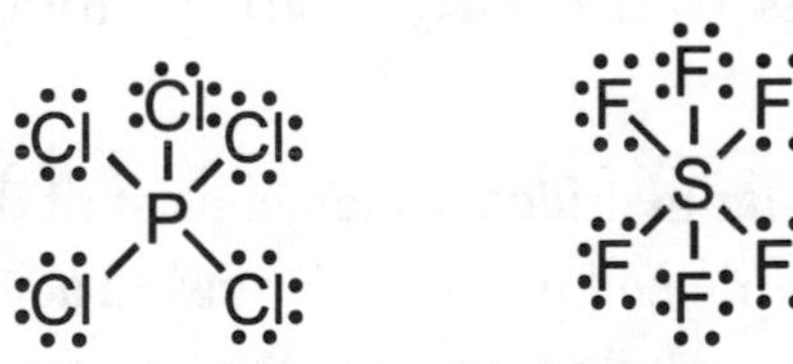

- **Elements bonded to one another in the pure elemental state**
  - → Elements with *low* ionization energies tend to have *metallic* bonds. Some examples are lithium, Li, and magnesium, Mg, solids.
  - → Elements with high ionization energies are typically *molecular* and form *covalent* bonds. Some examples are fluorine, F_2, and oxygen, O_2, gases.
  - → Elements in or near the diagonal band of metalloids (intermediate ionization energies and three, four, or five valence electrons) tend to form network structures in which each atom is bonded to three or more other atoms. Some examples are the *network* structures of *black phosphorus*, in which each phosphorus atom is bonded to three others in a network structure, and *diamond*, in which each carbon atom is bonded to four others.

14.3 Chemical Properties: Hydrides

- **Binary hydrides**
  - → Binary compounds with hydrogen are formed by almost all main-group elements (exceptions include the noble gases, and possibly indium and thallium) and several *d*-block elements.
  - → The binary compounds of elements with hydrogen are called *hydrides*, and their properties show periodic behavior.
  - → The formulas are directly related to the group number. For example, SiH_4 in Group 14, PH_3 in Group 15, H_2S in Group 16, and HCl in Group 17.
- **Saline hydrides**
  - → Strongly electropositive metallic elements form ionic compounds with hydrogen, present as the hydride ion, H^-. They are formed by all members of the *s* block except beryllium.
  - → Saline (salt-like) hydrides are white, high melting point solids with crystal structures resembling the corresponding halides. They are made by heating the metal in hydrogen. The crystallographic radius of H^- is between those of F^- and Cl^-.
  - → Saline hydrides readily react with water to form a basic solution releasing hydrogen gas. The hydride ion is a strong base: $H^-(aq) + H_2O(l) \rightarrow H_2(g) + OH^-(aq)$.
- **Metallic hydrides**
  - → Certain elements in the *d* block form hydrides that are black, powdery, electrically conducting solids.
  - → Metallic hydrides release hydrogen when heated or when treated with acids. A typical reaction in acid is $CuH(s) + H_3O^+(aq) \rightarrow Cu^+(aq) + H_2(g) + H_2O(l)$.
- **Molecular hydrides**
  - → Nonmetals form molecular hydrides, which consist of discrete molecules.
  - → The compounds are volatile and many are Brønsted acids or bases.
  - → The gaseous compounds include ammonia, the hydrogen halides, and the lighter hydrocarbons such as methane, CH_4. The liquid compounds include water and heavier hydrocarbons, such as benzene, C_6H_6, and octane, C_8H_{18}.

Example 14.3a Name a molecular hydride that behaves as an acid in water and one that behaves as a base in water. Write balanced chemical equations for the reaction of each with water.

Solution Ammonia, NH_3, is a molecular hydride that is a weak base in water:

$$NH_3(aq) + H_2O(l) \rightleftharpoons NH_4^+(aq) + OH^-(aq)$$

Hydrogen sulfide, H_2S, is a molecular hydride that is a weak acid in water:

$$H_2S(aq) + H_2O(l) \rightleftharpoons HS^-(aq) + H_3O^+(aq)$$

Example 14.3b Write balanced equations for the formation of barium hydride from the elements and for its reactions with water, with hydrobromic acid and with aqueous zinc sulfate.

Solution Barium is a member of Group 2, so barium hydride is a saline hydride of formula, BaH_2 with hydrogen in the −1 oxidation state. Recall that saline hydrides are high-melting solids:

$$Ba(s) + H_2(g) \rightarrow BaH_2(s)$$

The hydride ion is a strong base and should react with water to form H_2 and OH^-:

$$BaH_2(s) + 2\,H_2O(l) \rightarrow Ba^{2+}(aq) + 2\,H_2(g) + 2\,OH^-(aq)$$

With a strong acid such as HBr, H^+ from the acid reacts with H^- to form H_2:

$$BaH_2(s) + 2\,HBr(aq) \rightarrow Ba^{2+}(aq) + 2\,H_2(g) + 2\,Br^-(aq)$$

With zinc sulfate, we expect the hydride ion to react with water as above to give H_2 and OH^-. Because zinc hydroxide and barium sulfate are insoluble salts, we also expect precipitates to form:

$$BaH_2(s) + 2\,H_2O(l) \rightarrow Ba^{2+}(aq) + 2\,H_2(g) + 2\,OH^-(aq)$$
$$Ba^{2+}(aq) + SO_4^{2-}(aq) \rightarrow BaSO_4(s)$$
$$Zn^{2+}(aq) + 2\,OH^-(aq) \rightarrow Zn(OH)_2(s)$$

$$BaH_2(s) + Zn^{2+}(aq) + SO_4^{2-}(aq) + 2\,H_2O(l) \rightarrow 2\,H_2(g) + BaSO_4(s) + Zn(OH)_2(s)$$

(*overall reaction*)

14.4 Chemical Properties: Oxides

- **Binary oxides**
  - → All the main-group elements except the noble gases react with oxygen.
  - → The oxides of main-group *metals* are basic and the oxides of *nonmetals* are acidic. An exception is Al_2O_3, which is *amphoteric*. Other *amphoteric oxides* include BeO, Ga_2O_3, SnO_2, and PbO_2.
  - → Soluble ionic oxides are formed from elements on the left side of the periodic table. Insoluble oxides with high melting points are formed from elements in the left of the *p* block; oxides with low melting points that are often gaseous are formed from elements on the right of the block.
- **Metallic elements and metalloids**
  - → Ionic oxides are formed from metallic elements with low ionization energies. Typically, these oxides form basic solutions in water.
  - → Amphoteric oxides are formed from metallic elements with intermediate ionization energies (Be, B, and Al) and from the metalloids. These oxides do not react with water but do dissolve in both acidic and basic solutions.
- **Nonmetallic elements**
  - → Many oxides of nonmetals are gaseous molecular compounds. Most can act as Lewis acids.
  - → Many oxides of nonmetals form acidic solutions in water and are called *acid anhydrides*. The acids HNO_3 and H_2SO_4 are derived from the oxides N_2O_5 and SO_3, respectively.
  - → Oxides that do *not* react with water are called *formal anhydrides* of acids. The formal anhydride is obtained by removing the elements of water (H, H, and O) from

the molecular formula of the acid. For example, the formal anhydride of acetic acid, CH_3COOH, is CH_2CO (ketene), whose Lewis structure is

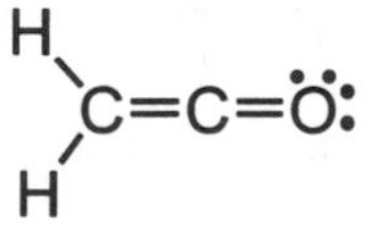

$$\mathrm{H_2C{=}C{=}\ddot{O}\!:}$$

Example 14.4 Identify each of the following as an *acid anhydride* or a *base anhydride* and write a balance equation for its reaction with water: $SrO(s)$, $Cl_2O_7(l)$, $N_2O_5(s)$.

Solution The oxides of main-group metals tend to be basic whereas those of the nonmetals tend to be acidic. On this basis, we expect that SrO is a *base* anhydride and both Cl_2O_7 and N_2O_5 are *acid* anhydrides. The corresponding reactions are:

$$SrO(s) + H_2O(l) \rightarrow Sr^{2+}(aq) + 2\,OH^-(aq)$$

The *base anhydride* of strontium hydroxide, $Sr(OH)_2$, is SrO.

$$Cl_2O_7(l) + 3\,H_2O(l) \rightarrow 2\,H_3O^+(aq) + 2\,ClO_4^-(aq)$$

The *acid anhydride* of perchloric acid, $HClO_4$, is Cl_2O_7.

$$N_2O_5(s) + 3\,H_2O(l) \rightarrow 2\,H_3O^+(aq) + 2\,NO_3^-(aq)$$

The *acid anhydride* of nitric acid, HNO_3, is N_2O_5.

- **Trends in acidity for the oxides of main-group elements**

Increasing acidity →

1	2	13	14	15	16	17
Li_2O	BeO	B_2O_3	CO_2	N_2O_5	(O_2)	OF_2
Na_2O	MgO	Al_2O_3	SiO_2	P_4O_{10}	SO_3	Cl_2O_7
K_2O	CaO	Ga_2O_3	GeO_2	As_2O_5	SeO_3	Br_2O_7
Rb_2O	SrO	In_2O_3	SnO_2	Sb_2O_5	TeO_3	I_2O_7
Cs_2O	BaO	Tl_2O_3	PbO_2	Bi_2O_5	PoO_3	At_2O_7

Increasing acidity ↑

Note: Acidity in water tends to increase from left to right and from bottom to top for the oxides of the elements in the periodic table. Amphoteric oxides are in the *shaded* region. Basic oxides are on the left and acid oxides on the right of the *shaded* region. Exceptions are O_2, which is neutral, and OF_2, which is only weakly acidic.

Hydrogen (Sections 14.5-14.6)

Key Concepts

uses and manufacture of hydrogen, re-forming reaction, synthesis gas, shift reaction, active metal with strong acid, hydride ion properties, hydrogen bonds

Overview

- **Structure of the H atom**
  → Hydrogen atoms are the simplest of all the elements. Most consist of only two particles, a proton and an electron.
  → Although it has the same outer electronic configuration as Group 1 metals, hydrogen is a nonmetal and exists as a diatomic molecule, H_2.
  → Many chemists assign hydrogen to the Group 1 family based on its electron configuration or to Group 17, or even both groups. The authors of this text make a cogent argument for its uniqueness and do not assign it to any group.
  → Some of the physical properties of hydrogen are given in Table 14.1 in the text; some chemical properties are found in Table 14.2.

14.5 The Element

- **Commercial production of hydrogen**
  → Formed in a sequence of two catalyzed reactions as a by-product of the *refining* of petroleum and by the *electrolysis* of water
  → The first step in the refining process, the *re-forming reaction*, employs a nickel catalyst. The products are called *synthesis gas.*
  $$CH_4(g) + H_2O(g) \rightarrow CO(g) + 3\,H_2(g)$$
  The second step in the process, the *shift reaction*, employs an iron/copper catalyst.
  $$CO(g) + H_2O(g) \rightarrow CO_2(g) + H_2(g)$$
  → The second process, the *electrolysis of water*, employs a salt solution to improve conductivity: $2\,H_2O(l) \rightarrow 2\,H_2(g) + O_2(g)$

- **Laboratory production and uses of hydrogen**
  → Produced by reaction of an active metal with a strong acid
  $$Zn(s) + 2\,H_3O^+(aq) \rightarrow Zn^{2+}(aq) + H_2(g) + H_2O(l)$$
  → Produced by reaction of a hydride with water
  $$CaH_2(s) + 2\,H_2O(l) \rightarrow Ca^{2+}(aq) + 2\,OH^-(aq) + 2\,H_2(g)$$

Uses of hydrogen	Chemical equations
rocket fuel	$2\,H_2(l) + O_2(l) \rightarrow 2\,H_2O(g)$ *(liquid fuel)*
conversion to ammonia	$3\,H_2(g) + N_2(g) \rightarrow 2\,NH_3(g)$
conversion to methanol	$2\,H_2(g) + CO(g) \rightarrow CH_3OH(l)$
hydrometallurgical extraction	$H_2(g) + Cu^{2+}(aq) \rightarrow Cu(s) + 2\,H^+(aq)$
hydrogenation	$H_2(g) + \cdots C{=}C \cdots \rightarrow \cdots CH{-}CH \cdots$

14.6 Compounds of Hydrogen

- **Unusual properties of hydrogen**
  - → Hydrogen can form a cation, H^+, or an anion, H^-.
  - → Because of its intermediate electronegativity value (2.2), hydrogen tends to form covalent bonds with many other elements.
  - → The hydrides are introduced in Section 14.3.
- **Saline hydrides**
  - → The hydride ion is large, with a radius (154 pm) falling between the fluoride (133 pm) and the chloride (181 pm) ions. It is highly polarizable, which adds covalent character in its bonds to cations.
  - → The two electrons in the H^- ion are only weakly bound to the single proton, and an electron is easily lost. Saline hydrides are very strong reducing agents. The reduction half-reaction and standard reduction potential are $H_2(g) + 2e^- \rightarrow 2H^-(aq)$ and $E° = -2.25$ V, respectively. The reduction potentials of the alkali metals are similar to the hydride value. The alkali metals and saline hydrides are both strong reducing agents.
  - → Hydride ions in saline hydrides reduce water upon contact.
    $H^-(aq) + H_2O(l) \rightarrow H_2(g) + OH^-(aq)$
- **Hydrogen bonds**
  - → The hydrides of nitrogen, oxygen, and fluorine can form relatively strong intermolecular bonds known as *hydrogen bonds* (see Section 5.5).
  - → A simple explanation for the strength of the hydrogen bond is in terms of the coulombic attraction between the partial positive charge on a hydrogen atom and the partial negative charge of another atom. The hydrogen atom is covalently bonded to a very electronegative nitrogen, oxygen, or fluorine atom. The partial negative charge on an atom usually in a neighboring molecule is, in fact, the lone-pair electrons on a different nitrogen, oxygen, or fluorine atom.
  - → A typical hydrogen bond is represented as three dots between atoms that are *not* covalently bonded to each other [O–H···:O].
  - → Molecular orbital theory provides a more detailed framework for hydrogen bonding. If the two atoms in the hydrogen bond and the one atom covalently bonded to hydrogen each provide one atomic orbital, three molecular orbitals then arise from the three atomic orbitals (see Fig. 14.13 in the text). The three molecular orbitals are bonding, nonbonding, and antibonding in character. The four electrons, two in the σ-bond plus the two lone-pair electrons, are placed in the molecular orbitals. There are then two electrons in the bonding molecular orbital and two electrons in the nonbonding molecular orbital. The antibonding orbital is vacant, because it is the highest in energy. There is a net bonding character to this combination. A hydrogen bond is typically about

5% as strong as a covalent bond between the same types of atoms. An example is the hydrogen bonding in hydrogen fluoride:

$$H{-}\ddot{\underset{\cdot\cdot}{F}}: \cdots H{-}\ddot{\underset{\cdot\cdot}{F}}: \cdots H{-}\ddot{\underset{\cdot\cdot}{F}}:$$

Example 14.6 Propose a structure for the gas-phase tetramer of methanol, $(CH_3OH)_4$, which is held together by hydrogen bonds.

Solution The Lewis structure of methanol reveals a methyl group, CH_3, bonded to a hydroxyl group, OH. Only the OH group in methanol can form hydrogen bonds and these are expected to be of the –O–H···:O– type. Four such bonds could bind four CH_3OH molecules together. An eight-member ring (probably not planar) is a reasonable suggestion for the structure:

$CH_3{-}O{-}H$

Methanol

CH_3 O–H ··· O(CH_3)–H ··· O(CH_3)–H ··· O(CH_3)–H ··· (ring)

Methanol tetramer

Note: The two pairs of nonbonding electrons on each oxygen atom are not shown. The four hydrogen bonds are indicated by dashed lines.

Group 1: The Alkali Metals (Sections 14.7-14.9)

Key Concepts

valence electron configuration, ease of electron removal, reactivity of metals, physical properties, preparation, chemical properties, compounds of lithium, sodium, and potassium

Overview

- **Similarity in properties of the alkali metals**
  - → The valence electron configuration of all alkali metals is ns^1, where n is the period number. Physical and chemical properties follow from the ease of removal of the single valence electron. These metals are the most reactive of all the metals.
  - → Some physical properties of the alkali metals are given in Table 14.3 in the text, some chemical properties are found in Table 14.4, and some properties of Group 1 compounds are listed in Table 14.5.

14.7 The Group 1 Elements

- **Preparation**
  - → Pure metals are obtained by electrolysis of the molten salts (see Section 12.17).
  - → Special method for the production of potassium is the exposure of molten potassium chloride to sodium vapor at 750°C.

$$KCl(l) + Na(g) \rightarrow NaCl(s) + K(g)$$

  Because the equilibrium constant is unfavorable, the reaction is driven to form products by the selective condensation (removal) of potassium vapor.

- **Properties**
  - → Soft, silver-gray metals with low melting points, boiling points, and densities.
  - → Melting and boiling points decrease down the group. Cesium, which has a melting point of 28°C and is very reactive, is transported in sealed ampoules. It is easily melted by holding an ampoule in your hand.

- **Applications**
  - → Lithium metal is used in the lithium-ion battery, which holds a charge over a long period of time and is rechargeable.
  - → Lithium is used in thermonuclear weapons.

14.8 Chemical Properties of the Alkali Metals

- **Behavior as reducing agents**
  - → Low ionization energies of the alkali metals lead to their ease of oxidation and their use as *strong reducing agents*.
  - → Alkali metals reduce water to form basic solutions and the release of hydrogen gas. Alkali metals dissolve in liquid ammonia to produce solvated electrons, which are used to reduce organic compounds.
  - → Only one alkali metal, lithium, reacts with nitrogen to form the nitride.

- **Formation of compounds containing oxygen**
  - → Lithium forms mainly the oxide, Li_2O.
  - → Sodium forms a pale yellow peroxide, Na_2O_2.
  - → Potassium, rubidium, and cesium form mainly the superoxide, for example, KO_2.

Example 14.8 Potassium is called a sliver gray metal in textbooks; but, in the laboratory, it often has a dull white appearance. Devise a possible explanation for this observation.

Solution The Group 1 metals are highly reactive and easily oxidized by atmospheric oxygen even at room temperature. The dull white color is caused by surface oxidation, resulting for potassium in the formation of potassium superoxide, KO_2. When the superoxide coating is removed, the silver gray metal under the surface is revealed, at least until a new superoxide coating is formed.

14.9 Compounds of Lithium, Sodium, and Potassium

- **Lithium compounds**
  - → Lithium differs significantly from the other Group 1 elements. The difference is typical for an element at the head of its group and originates in part from the small size of Li^+.
  - → The cation of lithium has strong polarizing power and a tendency toward covalence in its bonding.
  - → Lithium has a diagonal relationship with magnesium; it is found in the minerals of magnesium.
  - → Lithium is found in compounds with oxidation number +1.
  - → Lithium compounds are used in ceramics, lubricants, and medicine. Lithium carbonate is an effective treatment for manic-depressive (bipolar) behavior. Lithium soaps, which have higher melting points than sodium and potassium soaps, are used as thickeners in lubricating greases for high-temperature applications.

- **Sodium compounds**
  - → Sodium compounds are important in large part because they are plentiful, inexpensive, and soluble in water. Sodium chloride, NaCl, is mined as rock salt and is used in large quantities in the electrolytic production of chlorine, Cl_2, and sodium hydroxide. Sodium hydroxide, NaOH, is an inexpensive starting material for the production of other sodium salts.
  - → Sodium hydrogen carbonate, $NaHCO_3$ (sodium bicarbonate), is also known as *bicarbonate of soda* or *baking soda.* Weak acids (lactic acid from sour milk or buttermilk, citric acid from lemons, or acetic acid in vinegar) react with hydrogen carbonate to release carbon dioxide gas. Double-acting *baking powder* contains a solid weak acid as well as the hydrogen carbonate ion.
  - → Sodium carbonate decahydrate, $Na_2CO_3 \cdot 10H_2O$, was once used as *washing soda.*

- **Potassium compounds**
  - → Mineral sources of potassium are *sylvite*, KCl, and *carnallite*, $KCl \cdot MgCl_2 \cdot 6\,H_2O$. Potassium chloride is used directly in some fertilizers as a source of essential potassium, but potassium nitrate is required for some crops that cannot tolerate high chloride ion concentrations.
  - → Potassium compounds are usually more expensive than the corresponding sodium compounds, but often the expense is outweighed by an important advantage. Potassium compounds are generally less hygroscopic (water absorbing) than sodium compounds. The K^+ cation is larger and is less strongly hydrated by H_2O molecules. Potassium nitrate, KNO_3, releases oxygen when heated and is used to facilitate the ignition of matches. Potassium nitrate is also the oxidizing agent in black gunpowder (75% potassium nitrate, 15% charcoal, and 10% sulfur).

Group 2: The Alkaline Earth Metals (Sections 14.10-14.12)

Key Concepts

valence electron configuration, ease of electron removal, reactivity of metals, physical properties, preparation, chemical properties, compounds of beryllium, magnesium, and calcium

Overview

- **Similarity in properties of the alkaline earth metals**
  - → The valence electron configuration of all alkaline earth metals is ns^2, where n is the period number. Physical and chemical properties follow from the ease of removal of the two valence electrons. All elements are found in compounds with the oxidation number of +2.
  - → Beryllium shows some nonmetallic character, but the other Group 2 elements are all typical metals.
  - → The oxides (earths) of magnesium, calcium, strontium, and barium are alkaline or basic. Beryllium oxide is amphoteric and dissolves in acidic or basic solutions. The name alkaline earth metals is often extended to include all members of Group 2.
  - → Magnesium and calcium are the two most important members. Magnesium occurs in every chlorophyll molecule, and enables photosynthesis to take place. The calcium cation occurs in the bones of skeletons, the shells of shellfish, and the concrete, mortar and limestone of buildings.
  - → Some physical properties of the alkaline earth metals are given in Table 14.6 in the text, some chemical properties are found in Table 14.7, and some properties of the compounds of Group 2 elements are listed in Table 14.8.

14.10 The Group 2 Elements

- **Preparation**
  - → Pure metals are obtained by electrolysis of the molten salts (see Section 12.17) or by chemical reduction with the exception of beryllium.
  - → Three alkaline earth metals (calcium, strontium, and barium) may be obtained by reduction with aluminum in a variation of the thermite reaction. For example,

$$3\,CaO(s) + 2\,Al(s) \rightarrow Al_2O_3(s) + 3\,Ca(s)$$

  - → The element beryllium occurs in nature as *beryl*, $3\,BeO{\cdot}Al_2O_3{\cdot}6\,SiO_2$. Magnesium is found in seawater as $Mg^{2+}(aq)$ and in the mineral dolomite, $CaCO_3{\cdot}MgCO_3$.
- **Properties**
  - → Silver-gray metals with values of melting points, boiling points, and densities that are much higher than those of the preceding alkali metals in the same period.
  - → Melting points decrease down the group with the exception of magnesium, which has the lowest value. The trend in boiling point is irregular.

→ All the Group 2 elements with the exception of beryllium reduce water. For example,

$$Sr(s) + 2\,H_2O(l) \rightarrow Sr^{2+}(aq) + 2\,OH^-(aq) + H_2(g)$$

Beryllium does not react with water, even when the water is extremely hot. Magnesium reacts with hot water, and calcium reacts with cold water. Beryllium and magnesium do not dissolve in nitric acid because they develop a protective oxide film. Magnesium burns vigorously in air because it reacts with oxygen, nitrogen, and carbon dioxide.

→ The reactivity of the Group 2 metals with water and oxygen increases down the group.

- **Applications**

→ Beryllium has a low density that makes it useful for the fabrication of missiles and satellites. Beryllium is used as a "window" for x-ray tubes and is added in small amounts to copper to increase its rigidity without appreciably diminishing its conductivity. Beryllium-copper alloys are used in nonsparking tools (oil refineries and grain elevators), and in tiny nonmagnetic parts and contacts that resist deformation and corrosion (electronics industry).

→ Magnesium is a silver-white metal that is protected from extensive oxidation in air by a film of white oxide and appears dull gray for that reason. It is light and very soft, but its alloys have great strength. The alloys are used in applications where lightness and toughness are desired (airplanes). The limitation in widespread usage is related to its expense, low melting temperature, and difficulty in machining.

→ The Group 2 metals can be detected in burning compounds by the colors they give to flames. Calcium burns orange-red, strontium crimson, and barium yellow-green. Fireworks are made from their salts (usually nitrates and chlorates, which supply an additional source of oxygen).

Example 14.10 Beryllium metal is more strongly passivated than magnesium metal. Suggest a reason for this observation based on the lattice enthalpy of their respective oxides. The lattice enthalpy of BeO is 4290 $kJ\cdot mol^{-1}$; that of MgO is 3850 $kJ\cdot mol^{-1}$.

Solution The difference in lattice enthalpy between BeO and MgO is largely due to the smaller size of the Be^{2+} ion relative to the Mg^{2+} ion. The higher the lattice enthalpy, the stronger the bond between the metal and oxygen in the crystal. Stronger bonds make it more difficult for a reactant such as water to penetrate the oxide coating.

14.11 Compounds of Beryllium and Magnesium

- **Beryllium compounds**

→ Beryllium differs significantly from the other Group 2 elements. The difference is typical of an element at the head of its group. The difference originates in part from the small size of the Be^{2+} cation. The cation has strong polarizing power and a tendency toward covalence in its bonding. Beryllium has a diagonal relationship with aluminum. It is the only member of the group that, like aluminum, reacts in aqueous sodium hydroxide. In a strongly basic solution, it forms the *beryllate ion*, $[Be(OH)_4]^{2-}$.

Example 14.11a Like aluminum, beryllium metal is amphoteric and dissolves in both acid and base. Write net ionic equations for the dissolution of Be(s) in NaOH(aq) and HCl(aq).

Answer

$$Be(s) + 2\,OH^-(aq) + 2\,H_2O(l) \rightarrow [Be(OH)_4]^{2-} + H_2(g)$$
$$Be(s) + 2\,H_3O^+(aq) \rightarrow Be^{2+}(aq) + H_2(g) + 2\,H_2O(l)$$

→ Beryllium compounds are very toxic and dangerous to handle. The high polarizing power of Be^{2+} produces moderately covalent compounds; the small size of Be^{2+} limits to four the number of groups that can be attached to the ion. These two features account for the prominence of the tetrahedral BeX_4 unit. Gas phase molecules of BeH_2 and $BeCl_2$ condense to form solids with chains of tetrahedral BeH_4 and $BeCl_4$ units, respectively. In beryllium chloride, the Be atoms act as Lewis acids and accept electron pairs from the Cl atoms, forming a chain of tetrahedral $BeCl_4$ units in the solid.

Example 14.11b The empirical formula of beryllium hydride is BeH_2, yet, in the solid state, the structure consists of BeH_4 tetrahedral units. Draw a structure for BeH_2 solid that resolves this apparent inconsistency.

Solution A tetrahedral BeH_4 unit has hydrogen in excess of the amount given by the stoichiometry of its empirical formula. This "excess" can be decreased by sharing H atoms between Be atoms. If each H atom is shared between two Be atoms, the stoichiometry is accounted for. A structure consistent with this analysis is a linear arrangement of chains of tetrahedral BeH_4 units with bridging H atoms:

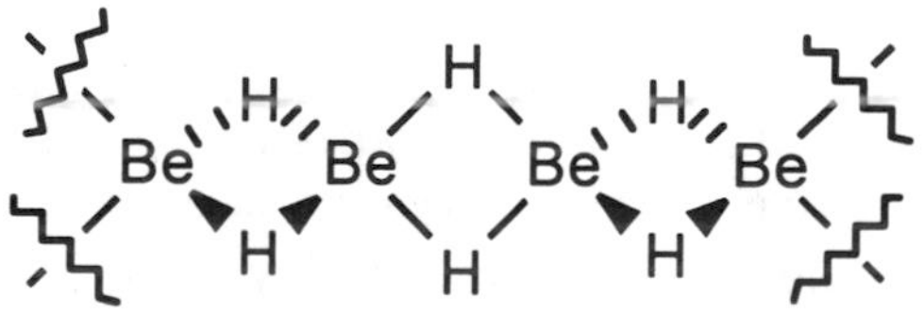

H H H
Be Be Be Be
H H H

- **Magnesium compounds**

→ Magnesium has more metallic properties than does beryllium; its compounds are primarily ionic with some covalent character. Magnesium oxide, MgO, and magnesium nitride, Mg_3N_2, are formed when magnesium burns in air. Magnesium oxide dissolves very slowly and only slightly in water. It is able to withstand high temperatures (*refractory*) because it melts at 2800°C. The oxide conducts heat well but electricity only poorly, and is therefore used as an insulator in electric heaters.

→ Magnesium hydroxide, $Mg(OH)_2$, is a base. It is not very soluble in water but forms instead a white colloidal suspension, which is a stomach antacid called *milk of magnesia*. An unfortunate side effect of stomach neutralization is the formation of magnesium chloride, which acts as a purgative. Stomach neutralization with hydroxides does not lead to the formation of carbon dioxide and the result of belching as in hydrogen carbonate antacids. Magnesium sulfate, $MgSO_4$, is also a common purgative and is called *Epsom salts*.

→ Chlorophyll is the most important compound of magnesium. Magnesium also plays an important role in energy generation in living cells. It is involved in the contraction of muscles.

14.12 Compounds of Calcium

- **Carbonate, oxide, and hydroxide**

→ Calcium is more metallic in character than magnesium, but its compounds share some similar properties. Calcium carbonate, $CaCO_3$, occurs naturally as chalk and limestone. Marble is a very dense form with colored impurities, most commonly iron cations. The two most common forms of pure calcium carbonate are *calcite* and *aragonite*. All carbonates are the fossilized remains of marine life.

→ Calcium oxide, CaO, is formed by heating calcium carbonate.

$$CaCO_3(s) \rightarrow CaO(s) + CO_2(g)$$

Calcium oxide is called *quicklime* because it reacts rapidly and exothermically with water:

$$CaO(s) + H_2O(l) \rightarrow Ca^{2+}(aq) + 2\,OH^-(aq)$$

→ The product of the reaction of calcium oxide and water, calcium hydroxide, is known as *slaked lime* because in this form the thirst for water has been quenched or slaked. An aqueous solution of calcium hydroxide, which is only slightly soluble in water, is called *lime water*, which is used as a test for carbon dioxide.

$$Ca(OH)_2(aq) + CO_2(g) \rightarrow CaCO_3(s) + H_2O(l)$$

Example 14.12a The decomposition of calcium carbonate to form calcium oxide and carbon dioxide is endothermic. Use a thermodynamic argument to suggest a reason why $CaCO_3$ must be heated to make it decompose.

Solution The reaction is

$$CaCO_3(s) \rightarrow CaO(s) + CO_2(g)$$

The reaction will occur spontaneously if $\Delta G° = \Delta H° - T\Delta S° < 0$. The net formation of gas-phase molecules in a reaction produces a large, positive $\Delta S°$. Because $\Delta H°$ is positive, spontaneity will be controlled by temperature in the entropy term. As T increases $\Delta G°$ decreases, and product formation is favored.

- **Carbide, sulfate, and phosphate**

→ Calcium oxide may be converted into calcium carbide:

$$CaO(s) + 3\,C(s) \rightarrow CaC_2(s) + CO(g)$$

The carbide ion, $[:C{\equiv}C:]^{2-}$, is a Brønsted base and is readily protonated by water, resulting in the formation of the organic compound ethyne (acetylene), H–C≡C–H *or* C_2H_2:

$$CaC_2(s) + 2\,H_2O(l) \rightarrow C_2H_2(g) + Ca(OH)_2(aq)$$

→ Another calcium-containing compound of use in construction materials is calcium sulfate dihydrate (*gypsum*), $CaSO_4{\cdot}2\,H_2O$.

→ Calcium is found in the rigid structural components of living organisms, either as calcium carbonate of the shells of shellfish or the calcium phosphate, $Ca_3(PO_4)_2$, of bone. Tooth enamel is a *hydroxyapatite*, $Ca_5(PO_4)_3OH$, which is subject to attack by acids produced when bacteria act on food residues. A more resistant coating is formed when the OH^- ions are replaced by F^- ions. The resulting mineral is called *fluoroapatite.*

$$Ca_5(PO_4)_3OH(s) + F^-(aq) \rightarrow Ca_5(PO_4)_3F(s) + OH^-(aq)$$

Tooth enamel is strengthened by the addition of fluorides (fluoridation) to drinking water, and by the use of fluoridated toothpaste containing tin(II) fluoride, SnF_2, or sodium monofluorophosphate (MFP), Na_2FPO_3.

Example 14.12b Suggest two reasons why fluoroapatite is more resistant to attack by acids than is hydroxyapatite.

Answer One reason is that OH^- is a stronger base than F^-, so OH^- reacts more readily with acid. A second reason is that F^- is smaller than OH^-, so F^- interacts more strongly with its cation neighbors and holds the enamel together in a tighter, stronger structure.

Group 13: The Boron Family (Sections 14.13-14.16)

Key Concepts

valence electron configuration, intermediate ease of electron removal, physical properties, preparation, applications, chemical properties, compounds of boron and aluminum, oxides, carbides, nitrides, halides, boranes, borohydrides, borides

Overview

- **Properties of the boron family**
  - → The boron family is the first group that is a member of the *p* block. Physical and chemical properties follow from the *intermediate* difficulty of removing the valence electrons. The elements are neither so electropositive that they easily lose electrons nor so electronegative that they easily gain electrons.
  - → The valence electron configuration of all Group 13 elements is ns^2np^1, where n is the period number. The elements are usually found in compounds with an oxidation number of +3. The heavier elements are more likely to keep their *s*-electrons (inert-pair effect); therefore, they also form compounds with an oxidation number of +1, as for example, thallium(I) bromide, TlBr.
  - → Boron is grouped with the metalloids because it exhibits properties characteristic of both the metals and the nonmetals. It is best regarded as a nonmetal in most of its chemical properties. The other Group 13 elements are all typical metals, but aluminum shows a trace of nonmetallic character. Boron and aluminum halides have incomplete octets and act as Lewis acids.

→ Some physical properties of the Group 13 elements are given in Table 14.9 in the text, and some chemical properties are listed in Table 14.10. The text focuses on the two most important members of the group, boron and aluminum.

14.13 The Group 13 Elements

- **Preparation**

→ Boron is mined as the hydrates *borax* and *kernite*, $Na_2B_4O_7 \cdot xH_2O$, with $x = 10$ and 4, respectively. It is converted to the oxide, B_2O_3, and extracted in its amorphous elemental form by reaction of the oxide with magnesium. A purer form is obtained by reduction of a volatile boron chloride or bromide with hydrogen on a heated tantalum filament.

→ Aluminum is mined as *bauxite*, $Al_2O_3 \cdot xH_2O$, where x ranges to a maximum of 3. The ore contains considerable iron oxide impurity. Bauxite ore is processed to obtain alumina, Al_2O_3. Aluminum is then extracted by an electrolytic process (*Hall process*). Alumina is mixed with the mineral *cryolite*, Na_3AlF_6, to obtain a lower melting mixture, 950°C instead of 2050°C in pure alumina. The electrolysis half-reactions and cell reaction are

Cathode:	$Al^{3+}(melt) + 3\,e^- \rightarrow Al(l)$
Anode:	$2\,O^{2-}(melt) + C(s,gr) \rightarrow CO_2(g) + 4\,e^-$
Cell:	$4\,Al^{3+}(melt) + 6\,O^{2-}(melt) + 3\,C(s,gr) \rightarrow 4\,Al(l) + 3\,CO_2(g)$

- **Properties**

→ Boron exists in a variety of allotropic forms in which the atom attempts to share eight electrons, despite having a small size and the ability to contribute only three electrons. In this sense, it is regarded as electron deficient. One common form of boron is a gray-black nonmetallic, high-melting solid. Another common form is a dark brown powder with a structure (icosahedral with 20 faces) that has clusters of 12 atoms (see Margin Figure (**6**) on page 672 of the text). The bonds create a very hard, three-dimensional structure. The element is attacked by only the strongest of oxidizing agents. It is a nonmetal in most of its chemical properties. It has acidic oxides and forms a fascinating array of binary molecular hydrides.

→ Although aluminum has a low density, it is a strong metal with excellent electrical conductivity. It is easily oxidized, but its surface is passivated by a protective oxide film that prevents further oxidation. If the thickness of the oxide layer is increased by an electrolytic process, *anodized aluminum* results. Aluminum is amphoteric, reacting with both *nonoxidizing* acids and hot aqueous bases. The reactions are

$$2\,Al(s) + 6\,H^+(aq) \rightarrow 2\,Al^{3+}(aq) + 3\,H_2(g)$$

$$2\,Al(s) + 2\,OH^-(aq) + 6\,H_2O(l) \rightarrow 2\,Al(OH)_4^-(aq) + 3\,H_2(g)$$

- **Applications**

→ Boron fibers are incorporated into plastics to form a very resilient material that is stiffer than steel and lighter than aluminum. Boron carbide is similar in hardness to diamond, and boron nitride is similar in structure and mechanical properties to graphite. However, unlike graphite, boron nitride does not conduct electricity.

→ Aluminum has widespread use in construction and aerospace industries. Because it is a soft metal, its strength is improved by formation of an alloy with copper and silicon. Because of its high electrical conductivity, aluminum is used in overhead power lines. Its high negative electrode potential has led to its use in fuel cells.

14.14 Group 13 Oxides

- **Boron oxide**

→ Boron behaves as a nonmetal in its chemical properties and, like most nonmetals, has *acidic* oxides. When B_2O_3 is hydrated, it forms boric acid, H_3BO_3 or $B(OH)_3$. Because boron in boric acid has an incomplete octet, it can act as a Lewis acid.

$$(OH)_3B + {:}OH_2 \rightarrow (OH)_3B{-}OH_2$$

The complex formed is a weak *monoprotic* acid with $pK_a = 9.14$.

→ Boric acid is a mild antiseptic and pesticide. It is also used as a fire retardant in home insulation and clothing. The main use of boric acid is as a source of boron oxide. Heating boric acid removes water and the anhydride, B_2O_3, is formed:

$$2\,H_3BO_3(s) \xrightarrow{\text{heat}} B_2O_3(s) + 3\,H_2O(l)$$

Because it liquefies at 450°C and dissolves many metal oxides, boron oxide is used as a flux for soldering or welding. It is also used in the manufacture of fiberglass and borosilicate glass, one example of which is the glass with low thermal expansion, Pyrex.

- **Aluminum oxide**

→ Aluminum oxide, Al_2O_3, is known as *alumina.* Alumina exists in a variety of different crystal structures, and each structure finds uses in science and in commerce. In the form of α-alumina, it is a very hard substance, *corundum.* In an impure microcrystalline state, corundum is used as an abrasive known as *emery.*

→ One of the less dense forms, γ-alumina, is very absorbent and is used as the stationary phase in chromatography. γ-alumina is produced by heating aluminum hydroxide. It is moderately reactive and is amphoteric, reacting with both acids and bases:

$$Al_2O_3(s) + 6\,H_3O^+(aq) + 3\,H_2O(l) \rightarrow 2\,[Al(H_2O)_6{}^{3+}](aq)$$
$$Al_2O_3(s) + 2\,OH^-(aq) + 3\,H_2O(l) \rightarrow 2\,Al(OH)_4{}^-(aq)$$

The strong polarizing effect of the small, highly charged Al^{3+} ion on the water molecules surrounding it gives the $[Al(H_2O)_6{}^{3+}]$ ion acidic properties.

→ An important aluminum salt is aluminum sulfate, which is prepared by reaction of aluminum oxide with sulfuric acid.

$$Al_2O_3(s) + 3\,H_2SO_4(aq) \rightarrow Al_2(SO_4)_3(aq) + 3\,H_2O(l)$$

Aluminum sulfate (*papermaker's alum*) is used in the preparation of paper. True alums are hydrated sulfates containing two metallic cations, a +1 cation and Al^{3+}. Two examples are $KAl(SO_4)_2$ and $NH_4Al(SO_4)_2 \cdot 12H_2O$. Sodium aluminate, $NaAl(OH)_4$, is used along with aluminum sulfate in water purification.

Example 14.14 A coworker shows you a jar containing colorless crystals and claims that they are a new alum-like compound of formula $CaAl(SO_4)_2 \cdot 12\,H_2O$. You are skeptical regarding the existence of this material and tell your colleague so. What makes you skeptical?

Solution From your knowledge of chemistry, you realize that the oxidation states of calcium and aluminum are +2 and +3, respectively, and that sulfate, SO_4^{2-}, is a doubly negatively charged ion. The calcium ion, Ca^{2+}, and the aluminum ion, Al^{3+}, contribute a total charge of +5 to the formula of the compound; the two sulfates, SO_4^{2-}, contribute −4. Thus, the net charge on the proposed formula is +1, which is not allowed for a compound whose net charge must be 0.

14.15 Carbides, Nitrides, and Halides

- **Carbides**
  - → When boron is heated to high temperatures with graphite, it forms boron carbide:

    $$2\,B(s) + C(s,gr) \rightarrow B_{12}C_3(s)$$

    The solid consists of B_{12} icosahedra held together by C atoms.
  - → Boron carbide is a solid with a high melting point; it is almost as hard as diamond.
- **Nitrides**
  - → When boron is heated to white heat in ammonia, it forms boron nitride:

    $$2\,B(s) + 2\,NH_3(g) \rightarrow 2\,BN(s) + 3\,H_2(g)$$

    The structure of the solid is similar to graphite, but the planes of hexagons of carbon atoms are replaced by hexagons of alternating B and N atoms (see Fig.14.31 in the text).
  - → Boron nitride is a fluffy, slippery powder. It can be pressed into a solid rod that is easy to machine, and boron nitride, unlike graphite, is a good insulator. Under high pressure, boron nitride is converted to a diamondlike crystalline material called Borazon. Boron nitride nanotubes similar to those formed by carbon have been found to have semiconducting properties.
- **Halides**
  - → Boron halides are made either by direct reaction of the elements or from the oxide. Two reactions that use the oxide to form halides are

    $$B_2O_3(s) + 3\,CaF_2(s) + 3\,H_2SO_4(l) \xrightarrow{\text{heat}} 2\,BF_3(g) + 3\,CaSO_4(s) + 3\,H_2O(l)$$

    $$B_2O_3(s) + 3\,C(s, gr) + 3\,Cl_2(g) \xrightarrow{500°C} 2\,BCl_3(g) + 3\,CO(g)$$
  - → All the boron halides are covalently bonded and electron deficient, with an incomplete octet on boron. The molecules are trigonal planar and, because of the incomplete octet, are strong Lewis acids.

→ Aluminum chloride, $AlCl_3$, is formed as an ionic compound by the reaction of chlorine with aluminum or alumina with chlorine in the presence of carbon:

$$2\,Al(s) + 3\,Cl_2(g) \rightarrow 2\,AlCl_3(s)$$

$$Al_2O_3(s) + 3\,Cl_2(g) + 3\,C(s, gr) \rightarrow 2\,AlCl_3(s) + 3\,CO(g)$$

Heating ionic $AlCl_3$ to its melting point of 192°C results in a rearrangement to form a molecular liquid, Al_2Cl_6. Molecular Al_2Cl_6 consists of units in which an unshared pair of electrons on chlorine (a Lewis base) is donated to a vacant orbital on an adjacent aluminum atom (a Lewis acid). This dimer of $AlCl_3$ contains $AlCl_4$ tetrahedral units and two bridging Cl atoms (see Margin Figure (**7**) on page 677 of the text).

→ Aluminum halides react with water in a highly exothermic reaction.

→ The white solid, lithium aluminum hydride, is an important reducing agent in organic chemistry. It is prepared from aluminum chloride in diethyl ether solvent:

$$4\,LiH + AlCl_3 \rightarrow LiAlH_4 + 3\,LiCl$$

14.16 Boranes, Borohydrides, and Borides

- **Boranes**

→ Boranes are compounds of boron and hydrogen, such as B_2H_6 and $B_{10}H_{14}$. Diborane, B_2H_6, is an extremely reactive compound. When heated to high temperature, it decomposes to hydrogen and pure boron.

$$B_2H_6(g) \rightarrow 2\,B(s) + 3\,H_2(g)$$

Diborane immediately reacts with water, reducing hydrogen and forming boric acid.

$$B_2H_6(g) + 6\,H_2O(l) \rightarrow 2\,B(OH)_3(aq) + 6\,H_2(g)$$

→ Boranes are electron-deficient compounds with three-center, two-electron B–H–B bonds. Valid Lewis structures cannot be written. The molecular orbital theory provides a framework for understanding the delocalized behavior of an electron pair associated with three atoms (see Margin Figure (**9**) on page 679 of the text).

Example 14.16 In the structure of tetraborane (10), B_4H_{10}, there are six B–H single bonds, a single B–B bond, and four B–H–B bridging hydrogen atoms. The sketch given below shows the atom *connectivity*. Describe the distribution of valence electrons in this compound.

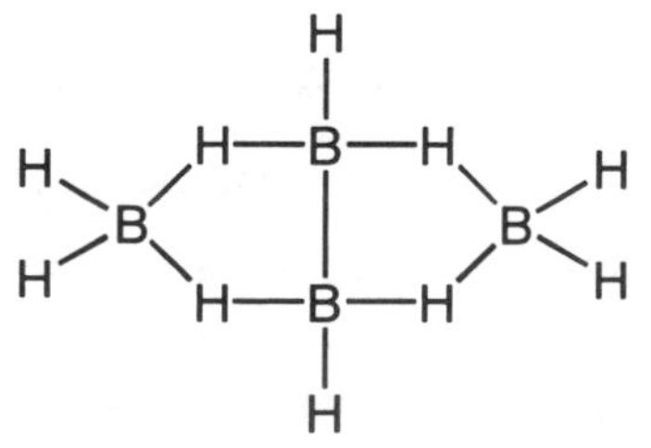

```
                H
                |
H        H——B——H        H
  \    /    |    \    /
    B       |       B
  /    \    |    /    \
H        H——B——H        H
                |
                H
```

Solution In B_4H_{10}, there is a total of 22 valence electrons, 12 from the four boron atoms and 10 from the hydrogen atoms. Twelve of the valence electrons are used to form six B–H covalent single bonds and two are used to form the B–B covalent single bond. This leaves 22 – 14 = 8 valence electrons, which are used to form *four* three-center, two electron bonds. Each delocalized three-center bond links a bridging hydrogen atom and its two adjacent boron atoms.

- **Borohydrides**
  - → Borohydrides are anionic versions of boranes. The most important borohydride is BH_4^- in sodium borohydride, $NaBH_4$. It is produced by the reaction of sodium hydride with boron trichloride in a nonaqueous solvent.

    $$4\,NaH + BCl_3 \rightarrow NaBH_4 + 3\,NaCl$$

    The borohydride ion is an effective and important reducing agent.
  - → Diborane is produced by the reaction of sodium borohydride with boron trifluoride in an organic solvent.

    $$3\,BH_4^- + 4\,BF_3 \rightarrow 3\,BF_4^- + 4\,B_2H_6$$

- **Borides**
  - → Many borides of the metals and nonmetals are known. Formulas of borides are usually unrelated to the position of the two elements in the periodic table. Examples given in the text are AlB_2, CaB_6, $B_{13}C_2$, $B_{12}S_2$, Ti_3B_4, TiB, and TiB_2.
  - → Boron atoms in borides commonly form extended structures such as zigzag chains, branched chains, or networks of hexagonal rings of boron atoms.

Group 14: The Carbon Family (Sections 14.17-14.22)

Key Concepts

valence electron configuration, physical properties, preparation, applications, chemical properties, oxides of carbon, oxides of silicon, silicates, carbides, saline carbides, covalent carbides, interstitial carbides, tetrachlorides, cyanides, silanes

Overview

- **The carbon family**
  - → The carbon family is the second group that is a member of the *p* block. Carbon is at the center of life and *natural intelligence*. Silicon and germanium are important to electronic technology and help make *artificial intelligence* possible.
  - → The valence electron configuration of all Group 14 elements is ns^2np^2, where n is the period number. The half-filled valence shell leads to special properties that span the range from metallic to nonmetallic. The elements show increasing metallic character down the group. Carbon is a nonmetal, silicon and germanium are metalloids, and tin and lead are metals.

→ Carbon forms numerous compounds and has its own branch of chemistry, *organic chemistry* (see Chapters 18 and 19).

→ Some physical properties of the Group 14 elements are given in Table 14.11 in the text, and some chemical properties are listed in Table 14.12.

14.17 The Group 14 Elements

- **General properties**

→ Carbon forms covalent compounds with nonmetals and ionic compounds with metals. The oxides of carbon and silicon are acidic. Germanium, a metalloid, exhibits metallic or nonmetallic properties, depending upon the compound. Tin and lead are classified as metals, but tin has some intermediate properties. Tin is amphoteric and reacts with both hot concentrated hydrochloric acid and hot alkali.

$$Sn(s) + 2\,H_3O^+(aq) \rightarrow Sn^{2+}(aq) + H_2(g) + 2\,H_2O(l)$$
$$Sn(s) + 2\,OH^-(aq) + 2\,H_2O(l) \rightarrow Sn(OH)_4^{2-}(aq) + H_2(g)$$

→ Carbon at the top of the group differs greatly in its properties from the other members. For example, carbon tends to form multiple bonds, whereas silicon does not. Carbon dioxide is a molecular gas, whereas silicon dioxide (silica) consists of networks of –O–Si–O– groups and is a mineral. Silicon compounds can act as Lewis acids, whereas carbon compounds usually cannot. Silicon may expand its octet, whereas carbon does not.

→ Tin and lead are commonly found with oxidation number of +2 because of the inert-pair effect.

14.18 The Different Forms of Carbon

- **Graphite**

→ Graphite is the thermodynamically most stable allotrope of carbon. Soot and carbon black contain small crystals of graphite.

→ Pure graphite is produced commercially by heating carbon rods to high temperature in an electric furnace for several days.

→ Carbon in graphite is sp^2 hybridzed; structurally, graphite consists of large sheets of interconnected benzene-like hexagonal units (see Fig. 14.34 in the text). The π-bonding network of delocalized electrons accounts for the high electrical conductivity of graphite. When certain impurities are present, the large sheets in graphite slip past one another, and under these conditions, graphite is an excellent dry lubricant.

→ Soot and carbon black have commercial applications in rubber and inks. Activated charcoal is an important and versatile purifier. Unwanted compounds are adsorbed onto its microcrystalline surface.

→ Graphite is only soluble in a few liquid metals.

- **Diamond**
  - → Diamond is the hardest substance known and an excellent conductor of heat. Its hardness makes it an excellent abrasive, and synthetic diamonds are widely used as abrasives. The heat generated by friction is rapidly conducted away.
  - → Natural sources of diamonds are uncommon (see Section 5.15). Synthetic diamonds are produced from graphite at pressures of over 80 kbar and temperatures above 1500°C, and by thermal decomposition of methane.
  - → Carbon in diamond is sp^3 hybridized, and each carbon atom in a diamond crystal is directly bonded to four other carbon atoms and indirectly interconnected to all of the other carbon atoms in the crystal through C–C single σ-bonds (see Fig. 14.35 in the text). The σ-bonding network of localized electrons accounts for the electrical insulating properties of diamond.
  - → Diamond is soluble in several liquid metals, such as chromium and iron, but less soluble than graphite. The differing solubility of graphite and diamond in liquid metals is utilized in the synthesis of diamond at high pressure and high temperature (see Section 5.15).
- **Fullerenes**
  - → Buckminsterfullerene, C_{60}, is a soccer-ball-like molecule that was identified in 1985. The discovery of many other fullerenes, containing 44 to 84 carbon atoms, quickly followed. Fullerenes contain an even number of carbon atoms, differ by a C_2 unit, and have pentagonal and hexagonal ring structures. Solid samples of fullerenes are called *fullerites* (see Fig. 14.36 in the text).
  - → C_{60} has a geometrical structure, called a *truncated icosahedron*, with 32 faces, 12 pentagons and 20 hexagons (see Margin Figure (**10**) in the text). The interior of the molecule is quite large (internal radius of 352 pm), and other atoms may be inserted to produce compounds with unusual properties. The crystal structure of C_{60} is face-centered cubic and is shown on p. 156 of this study guide.
  - → Fullerenes are molecular and are electrical insulators. They have relatively low ionization energies *and* large electron affinities, so they readily lose or gain electrons to form ions. The anion C_{60}^{3-} is very stable, and K_3C_{60} is a superconductor below 18 K.
  - → Fullerenes are soluble in organic solvents.

Example 14.18 From the sketch of buckminsterfullerene, C_{60}, on page 682 of the text, determine how many other carbon atoms each carbon atom is bonded to, and the total number of C–C single bonds in the molecule. Given that each carbon atom has four valence electrons, how many valence electrons are available to form multiple bonds. If all the remaining electrons are used to form localized π-bonds, how many double bonds would buckminsterfullerene have?

Solution The sketch reveals that each C atom is bonded to *three* others in C_{60}. There are 60 C atoms in buckminsterfullerene. Because two atoms are required for each bond, there are $3(60)/2 = 90$ C–C *single σ-bonds*. Because there are two electrons per bond, 180 valence electrons are required for the single-bond framework. The total number of valence electrons is (4 valence electrons per atom)(60 atoms) = 240 electrons. The number of electrons available to form π-bonds is then 240 – 180 = 60 electrons. If they all form double bonds, there would be 30 *double bonds*, because two electrons are required in addition to those in a single bond to form a double bond. For C_{60}, a large number of resonance Lewis structures can be written; therefore, the π-electrons are *delocalized*.

14.19 Silicon, Germanium, Tin, and Lead

- **Silicon**
  - → Silicon occurs naturally as silicon dioxide and as silicates, which are compounds containing SiO_4^{4-}. Quartz, quartzite, and sand are forms of SiO_2.
  - → Pure silicon, which is widely used in semiconductors, is obtained from quartzite in a three-step process. Silicon produced in the first step is crude and requires further purification.

$$SiO_2(s) + 2\,C(s) \rightarrow Si(s) + 2\,CO_2(g)$$
$$Si(s) + 2\,Cl_2(g) \rightarrow SiCl_4(l)$$
$$SiCl_4(l) + 2\,H_2(g) \rightarrow Si(s) + 4\,HCl(g)$$

- **Germanium**
  - → Mendeleev predicted the existence of germanium, "eka-silicon", long before its discovery (see Box 1.2 in the text). It occurs as a natural impurity in zinc ores.
  - → Germanium is used mainly in the semiconductor industry.
- **Tin**
  - → Tin is reduced from its ore *cassiterite*, SnO_2, through reaction with carbon at 1200°C in a process that requires great care in order to avoid contamination with iron.

$$SnO_2(s) + C(s) \rightarrow Sn(l) + CO_2(g)$$

  - → Tin is used mainly in plating other materials and for the production of alloys.

Example 14.19 Explain why the iron metal in a tin-plated iron can will corrode with moisture present in preference to the tin coating.

Solution Corrosion is an oxidation reaction. The oxidation of iron is thermodynamically favored over the oxidation of tin:

$$Fe(s) \rightarrow Fe^{2+}(aq) + 2\,e^- \qquad E° = +0.44\text{ V}$$
$$Sn(s) \rightarrow Sn^{2+}(aq) + 2\,e^- \qquad E° = +0.14\text{ V}$$

Thus, if moisture permeates through the coating to the iron and there is electron conduction to a cathode, the iron will be oxidized first. Note that oxidation reactions are listed, so the sign of the reduction potential is changed.

- **Lead**
  - → The principal ore of lead is *galena*, PbS, which is converted to the oxide in air. The oxide in turn is reduced with coke.

$$2\,PbS(s) + 3\,O_2(g) \rightarrow 2\,PbO(s) + 2\,SO_2(g)$$

$$PbO(s) + C(s) \rightarrow Pb(s) + CO(g)$$

  - → Lead is very dense, malleable, and chemically inert. It is used mainly as the electrode material in car batteries.

14.20 Oxides of Carbon

- **Carbon dioxide, CO_2**
  - → Carbon dioxide, a nonmetal oxide, is the acid anhydride of carbonic acid, H_2CO_3. The equilibrium between CO_2 and H_2CO_3 favors CO_2. A solution of CO_2 in water is an equilibrium mixture of CO_2, H_2CO_3, HCO_3^-, and a very small amount of CO_3^{2-}.
  - → Solid CO_2, commonly known as *dry ice*, sublimes directly to the gas phase, a property that makes it useful as a refrigerant and cold pack.
- **Carbon monoxide, CO**
  - → Carbon monoxide is a gas that is colorless, odorless, flammable, nearly insoluble in water, and extremely toxic. It has an extremely large bond enthalpy of 1074 $kJ \cdot mol^{-1}$.
  - → Carbon monoxide is produced when an organic compound or carbon itself is burned in a limited amount of oxygen. It is commercially produced as *synthesis gas*.

$$CH_4(g) + H_2O(g) \rightarrow CO(g) + 3\,H_2(g)$$

  - → Carbon monoxide is the *formal* anhydride of formic acid and is produced in the laboratory by the dehydration of formic acid with hot, concentrated sulfuric acid.

$$HCOOH(l) \rightarrow CO(g) + H_2O(l)$$

  - → Carbon monoxide is very toxic and is a moderately good Lewis base. For example, in a typical reaction, CO reacts with a *d*-block metal or ion.

$$Ni(s) + 4\,CO(g) \rightarrow Ni(CO)_4(l)$$

  - → Complex formation is responsible for the toxicity of carbon monoxide. It attaches more strongly than oxygen to iron in hemoglobin, thereby blocking the acceptance of oxygen by red blood cells.

Example 14.20 On the basis of its Lewis structure, explain why carbon monoxide is a Lewis base.

Solution The Lewis structure of carbon monoxide, :C≡O:, reveals two lone pairs of electrons, one on the carbon atom and one on oxygen. The unshared electron pairs make it a Lewis base. In most of its reactions as a Lewis base, the unshared pair of electrons on carbon is used in bonding. In nickel tetracarbonyl, $Ni(CO)_4$, the molecule has a tetrahedral shape and contains four Ni–CO bonds.

14.21 Oxides of Silicon: The Silicates

- **Silica or silicon dioxide, SiO_2**
  → Silica gets its strength from its covalently bonded network structure.
  → Silica occurs naturally in pure form only as quartz and sand.
- **Silicates, compounds containing SiO_4^{4-} units**
  → *Orthosilicates* are the simplest silicates and contain isolated SiO_4^{4-} ions. Examples are Na_4SiO_4 and $ZrSiO_4$, *zircon*.
  → *Pyroxenes* consist of chains of tetrahedral SiO_4^{4-} units in which one oxygen atom bridges two silicon atoms , Si–O–Si, from adjacent tetrahedra, so the "average" unit is a SiO_3^{2-} unit (see Fig.14.42 in the text). Cations placed along the chain in more or less random order provide electrical neutrality. Examples are $NaAl(SiO_3)_2$ and $CaMg(SiO_3)_2$.
  → *Amphiboles* consist of chains of silicate ions linked to form a cross-linked ladderlike structure containing $Si_4O_{11}^{6-}$ units. An example of a fibrous mineral is *temolite*, $Ca_2Mg_5(Si_4O_{11})_2(OH)_2$. Temolite is one form of *asbestos*, a material that withstands extreme heat but is a known carcinogen. Almost all amphiboles contain hydroxide ions attached to the metal.
  → Other types of silicates are made up of sheets containing $Si_2O_5^{2-}$ units, with cations lying between the sheets and linking them together. An example is *talc*, $Mg_3(Si_2O_5)_2(OH)_2$. Additional structural types of silicon oxide exist.
- **Aluminosilicates and cements**
  → *Aluminosilicates* form when a Si^{4+} ion is replaced by an Al^{3+} ion plus additional cations to make up for the charge difference. An example is *mica*, $KMg_3(Si_3AlO_{10})_2(OH)_2$. The extra cations hold the sheets of tetrahedra together, which accounts for hardness of the material. *Feldspar* is a silicate material in which more than half the silicon is replaced by aluminum.
  → *Cements* are produced by melting aluminosilicates with limestone and other materials and then allowing them to solidify. One of the materials in concrete is cement.
- **Silicones**
  → *Silicones* are made of long –O–Si–O– chains with the remaining two silicon bonding positions occupied by organic groups, such as the methyl group, $-CH_3$.
  → Silicones possess both hydrophobic and hydrophilic properties, which makes them useful in waterproofing fabrics and in biological applications, such as surgical and cosmetic implants.

14.22 Other Important Group 14 Compounds

- **Carbides**
  → *Saline carbides* are formed most commonly from Group 1 and 2 metals, aluminum, and a few other metals. The *s*-block metals form saline carbides when their oxides are heated with carbon. Anions present in carbides are normally *methanide*, C^{4-}, or

acetylide, C_2^{2-}, which are both very strong Brønsted bases. Methanides reacting with water produce a basic solution and release methane gas. Acetylides release acetylene, HC≡CH, upon reaction with water.

$$Al_4C_3(s) + 12\,H_2O(l) \rightarrow 4\,Al(OH)_3(s) + 3\,CH_4(g)$$
$$CaC_2(s) + 2\,H_2O(l) \rightarrow Ca(OH)_2(s) + C_2H_2(g)$$

Example 14.22a What are the oxidation numbers of carbon in methane, CH_4, in the methanide ion, C^{4-}, and in the acetylide ion, C_2^{2-}?

Solution In hydrocarbons, an H atom has an oxidation number of +1. Thus, the C atom has an oxidation number of −4 in CH_4.

The oxidation number of an atomic ion is equal to the charge on the ion, so the C atom has an oxidation number of −4 in C^{4-}.

The oxidation number of a C atom in C_2^{2-} can be determined by appealing to the Lewis structure, $[:C{\equiv}C:]^{2-}$. Each C atom has five valence electrons, one more than a neutral atom, so the oxidation number of each carbon atom is −1.

→ *Covalent carbides* are carbides formed by metals with electronegativity values similar to that of carbon, such as silicon and boron. Silicon carbide, SiC, known as *carborundum*, is an extremely hard, covalent carbide and is an excellent abrasive. It is produced by the reaction of silicon dioxide with carbon at 2000°C.

$$SiO_2(s) + 3\,C(s) \rightarrow SiC(s) + 2\,CO(g)$$

→ *Interstitial carbides* are compounds formed by the direct reaction of a *d*-block metal with carbon at temperatures above 2000°C. Carbon atoms lie in holes in close-packed arrays of metal atoms. Bonds between the carbon atoms and metal atoms stabilize the metal lattice and help to hold the metal atoms together. Examples are WC, Cr_3C and Fe_3C, an important component of steel.

- **Tetrachlorides**

→ All Group 14 elements form *liquid* molecular tetrachlorides, of which $PbCl_4$ is the least stable. Carbon tetrachloride is formed by the reaction of chlorine and methane.

$$CH_4(g) + 4\,Cl_2(g) \rightarrow CCl_4(l) + 4\,HCl(g)$$

→ Silicon reacts directly with chlorine to form silicon tetrachloride. Unlike CCl_4, $SiCl_4$ reacts with water as a Lewis acid, accepting a lone pair of electrons from H_2O.

$$SiCl_4(l) + 2\,H_2O(l) \rightarrow SiO_2(s) + 4\,HCl(aq)$$

- **Cyanides**

→ The cyanide ion, CN^-, is a conjugate base of hydrogen cyanide, HCN. The acid is made by heating methane, ammonia, and air in the presence of a platinum catalyst at 1100°C.

$$2\,CH_4(g) + 2\,NH_3(g) + 3\,O_2(g) \rightarrow 2\,HCN(g) + 6\,H_2O(g)$$

→ *Acrylonitrile*, $H_2C{=}CHCN$, is produced by the reaction of HCN with HC≡CH, using a copper(I) chloride catalyst. Acrylonitrile is the starting point for the production of the synthetic fibers Acrilan and Orlon.

→ Cyanides are strong Lewis bases that form a range of complexes with *d*-block metal ions. They are extremely poisonous, combining with cytochromes involved with the transfer of electrons and the supply of energy in cells.

- **Methane and silane**

→ Carbon bonds readily with itself and there are numerous hydrocarbons. Silicon forms a much smaller number of compounds with hydrogen; the simplest is the analogue of methane, *silane*, SiH_4. Silane is formed by the reaction of lithium aluminum hydride with silicon halides in ether.

$$SiCl_4 + LiAlH_4 \rightarrow SiH_4 + LiCl + AlCl_3$$

Example 14.22b The reaction directly above is a redox reaction. Which elements change oxidation number? Give their oxidation numbers both in the reactant and in the product. What is the reducing agent? What is the oxidizing agent?

Solution In $SiCl_4$, the oxidation number of the Si atom is +4 because that of the Cl atom is −1; in SiH_4, it is −4 just like the C atom in methane. So, $SiCl_4$ is reduced; it is the *oxidizing agent*.

In $LiAlH_4$, the oxidation number of H is −1 because Li is +1 and Al +3; in SiH_4, it is +1. So, $LiAlH_4$ is oxidized; it is the *reducing agent*.

→ Silane is much more reactive than methane and ignites spontaneously in air. More complicated silanes are unstable and decompose on standing.

→ Methane reacts with sulfur in the presence of aluminum oxide at 600°C to form the sulfur analogue of carbon dioxide, carbon disulfide, CS_2. Carbon disulfide is a volatile, poisonous liquid, which is used as a nonpolar solvent. The spontaneous ignition of carbon disulfide by dissolved phosphorus on filter paper placed above a graduated cylinder produces a barking sound. The "Barking Dogs" demonstration may be viewed with the CD-ROM that accompanies the text.

The Impact on Materials (Sections 14.23-14.24)

Key Concepts

glasses, soda-lime glass, borosilicate glass, chemicals that attack glass, ceramics, clay

Overview

- **Modern materials**

→ Rapid changes are occurring in the development of new materials.

→ Advanced materials of major importance are glasses and ceramics.

→ Optical fibers made of glasses will control computers, and automobile parts made of ceramics will be much lighter and more economical.

→ Silicon compounds are used to make both glasses and ceramics.

14.23 Glasses

- **Preparation and properties**
  - → A *glass* is an ionic solid with an amorphous structure similar to a liquid. It is characterized by a network structure based on a nonmetal oxide, commonly silica, SiO_2, melted together with metal oxides that function as *network modifiers*. Heating silica causes the rupture of many Si–O bonds, which enables metal ions to form ionic bonds with some of the oxygen atoms of the silica tetrahedra. The nature of the glass depends upon the metal added.
  - → *Soda-lime glass*, the common glass used for windows and bottles, is formed by addition of Na^+ (12% Na_2O) and Ca^{2+} (12% CaO). Heating sodium carbonate (soda) produces the sodium oxide, whereas heating calcium carbonate (lime) produces the calcium oxide.
  - → *Borosilicate glass*, such as Pyrex, is produced by reducing the amounts of soda and lime and adding 16% B_2O_3. This type of glass expands only slightly on heating, and it is used for ovenware and laboratory glassware.

- **Reaction with acids and bases**
  - → Glass is resistant to attack by most chemicals.
  - → Although it is very stable, the silica in glass is subject to attack by HF.

$$SiO_2(s) + 6\,HF(aq) \rightarrow SiF_6^{2-}(aq) + 2\,H_3O^+(aq)$$

  - → Silica is also attacked by thc Lcwis basc OH^- in molten NaOH and by the carbonate anion in molten Na_2CO_3 at 1400°C.

$$SiO_2(s) + 2\,Na_2CO_3(l) \rightarrow Na_4SiO_4(l) + 2\,CO_2(g)$$

  - → The process by which silica is removed from glass by the ions F^- from HF, OH^-, and CO_3^{2-} is called *etching*.

14.24 Ceramics

- **Preparation and properties**
  - → *Ceramics* are inorganic materials that have been hardened by heating to a high temperature. A typical ceramic is very hard, insoluble in water, and stable to corrosion and high temperatures. Ceramics can be used at high temperatures without failing, and they resist deformation. Many tend to be brittle, however.
  - → *Ceramics* are often oxides of elements on the border between metals and nonmetals. Some *d*-metal oxides and compounds of boron and silicon with carbon and nitrogen are also ceramic materials. Most ceramics are electrical insulators, but some are semiconductors and superconductors (see Section 3.13).
  - → Many *aluminosilicate ceramics* are formed by heating clays to drive out the water trapped between sheets of tetrahedral aluminosilicate units (see Section 15.13). The result is a rigid heterogeneous mass of small interlocking crystals bound together by glassy silica. One form is *china clay* that is used to make porcelain and china. It is free of iron impurities that tend to make clays reddish brown.

- → Methods developed to make ceramics less brittle are the *sol-gel process* and the *composite material technique*.
- → Ceramics are used in many automobile parts including spark plugs, pressure and vibration sensors, brake linings, and catalytic converters.

Chapter 15 The Elements: The Last Four Main Groups

Group 15: The Nitrogen Family (Sections 15.1-15.4)

Key Concepts

valence electron configuration, physical properties, preparation, applications, chemical properties, elements, nitrogen fixation, Haber synthesis, white and red phosphorus, hydrogen-containing compounds, nitrides and phosphides, azides, halides, nitrogen oxides and oxoacids, phosphorus oxides and oxoacids, ATP

Overview

- **Nitrogen family**
  - → The families on the *right* side of the periodic table have electron configurations only a few electrons short of a closed-shell, and the effective nuclear charge is high. The chemistry of these elements reflects their high ionization energies and high electronegativity values (with the exception of the noble gases).
  - → The valence electron configuration of all Group 15 elements is ns^2np^3, where n is the period number. The elements show increasing metallic character down the group. Nitrogen and phosphorus are nonmetals, arsenic and antimony are metalloids, and bismuth is a metal.
  - → Elemental nitrogen is unreactive, whereas white phosphorus spontaneously ignites in air. Arsenic and antimony are semiconducting materials. Like ice, solid bismuth is less dense than the liquid. It is used in alloys that serve as fire detectors.
  - → Some physical properties of the Group 15 elements are given in Table 15.1 in the text, some chemical properties are listed in Table 15.2, and the oxides and oxoacids of nitrogen are described in Table 15.3.

15.1 The Group 15 Elements

- **Nitrogen**
  - → *Nitrogen* differs greatly from other group members owing to its high electronegativity ($\chi = 3.0$), small size (radius of 74 pm), ability to form multiple bonds, and lack of available *d*-orbitals. It is found with oxidation numbers from −3 (in NH_3) to +5 (in HNO_3 and nitrates).
  - → *Elemental nitrogen*, N_2, is found in air (78.1% by volume) and is prepared by fractional distillation of liquid air. The strong triple bond in N_2 ($\Delta H_B = 944\ kJ{\cdot}mol^{-1}$) makes it very stable.
  - → *Nitrogen fixation* is the process in which N_2 is converted into compounds usable by plants. The *Haber synthesis* of ammonia is the major industrial method for fixing nitrogen, but it requires high temperatures and pressures and is, therefore, expensive (see Section 9.11). Lightning produces some oxides of nitrogen, which are washed into soil by rain. Bacteria found in the root nodules of legumes also fix N_2. An

important area of current research is the search for catalysts that can mimic bacteria and fix nitrogen under ordinary temperatures.

Example 15.1 If approximately 100 lightning bolts strike the earth every second and each bolt produces approximately 1500 moles of NO from N_2 in the air, how many kilograms of nitrogen would be fixed worldwide per year (365 days)?

Solution The number of lightning bolts striking the earth per year is

$$\left(100\frac{\text{bolts}}{\text{s}}\right)\left(3600\frac{\text{s}}{\text{h}}\right)\left(24\frac{\text{h}}{\text{d}}\right)\left(365\frac{\text{d}}{\text{y}}\right) = 3.15\times10^{9}\ \text{bolts}\cdot\text{y}^{-1}$$

The mass of nitrogen in kilograms produced per bolt is

$$\left(1500\frac{\text{mol NO}}{\text{bolt}}\right)\left(1\frac{\text{mol N}}{\text{mol NO}}\right)\left(14.01\frac{\text{g}}{\text{mol N}}\right)\left(10^{-3}\frac{\text{kg}}{\text{g}}\right) = 21.0\ \text{kg}\cdot\text{bolt}^{-1}$$

The mass of nitrogen fixed per year (2 significant figures) is

$$\left(3.15\times10^{9}\frac{\text{bolts}}{\text{y}}\right)\left(21.0\frac{\text{kg}}{\text{bolt}}\right) = 6.6\times10^{10}\ \text{kg}\cdot\text{y}^{-1}\ \text{or about 66 billion kg}$$

- **Phosphorus**
  - → *Phosphorus* differs from nitrogen in its lower electronegativity ($\chi = 2.2$), larger size (radius of 110 pm), and the presence of available *d*-orbitals. Phosphorus may form as many as six bonds, whereas nitrogen can form a maximum of four.
  - → *Phosphorus* is prepared from the *apatites*, mineral forms of *calcium phosphate*, $Ca_3(PO_4)_2$. The rocks are heated in an electric furnace with carbon and sand.

    $$2\,Ca_3(PO_4)_2(s) + 6\,SiO_2(s) + 10\,C(s) \rightarrow P_4(g) + 6\,CaSiO_3(l) + 10\,CO(g)$$

    The phosphorus vapor condenses as *white phosphorus*, a soft, white, poisonous molecular solid consisting of tetrahedral P_4 molecules (see Margin Figure (**3**) on page 703 in the text). It bursts into flame upon exposure to air and is normally stored under water.
  - → *Red phosphorus* is less reactive than white phosphorus and is used in match heads because it can be ignited by friction. It is a network solid consisting, most likely, of linked P_4 tetrahedral units.
  - → *Red phosphorus* is the most thermodynamically stable form of phosphorus, yet *white phosphorus* crystallizes from the vapor or from the liquid phase. It is the form chosen to have an enthalpy and free energy of formation equal to 0. For *red phosphorus*, $\Delta H_f° = -17.6$ kJ·mol^{-1} and $\Delta G_f° = -12.1$ kJ·mol^{-1} at 25°C.
  - → *Black phosphorus* is a third form of the element; it has a network structure in which each P atom is bonded to three others at approximately right angles.

- **Arsenic and antimony**
  - → *Arsenic* and *antimony* are metalloids. They are elements known since ancient times because they are easily reduced from their ores.
  - → As elements, *arsenic* and *antimony* are used in lead alloys in the electrodes of storage batteries and in the semiconductor industry. *Gallium arsenide* is used as a light detecting material with near infrared response and in lasers, including ones used for CD players.

- **Bismuth**
  - → *Bismuth* is a metal with large, weakly bonded atoms. Bismuth has a low melting point, and it is used in alloys that serve as fire detectors in sprinkler systems.
  - → *Bismuth* is also used to make low temperature castings. Like ice, solid bismuth is less dense than the liquid. Bismuth ($Z = 83$) is the last element in the periodic table with a stable isotope, ^{209}Bi.

15.2 Compounds with Hydrogen and the Halogens

- **Ammonia, NH_3, and phosphine, PH_3**
  - → *Ammonia* is prepared in large amounts by the Haber process. Ammonia is a pungent, toxic gas that condenses to a colorless liquid at −33°C. It acts as a solvent for a wide range of substances. Autoprotolysis occurs to a much smaller extent in liquid ammonia (am) than in pure water.

    $$2NH_3(am) \rightleftharpoons NH_4^+(am) + NH_2^-(am) \qquad K_{am} = 1 \times 10^{-33} \text{ at } -35°C$$

    Very strong bases that are protonated by water survive in liquid ammonia.
  - → *Ammonia* is a weak Brønsted base in water; it is also a fairly strong Lewis base, particularly toward *d*-block elements.

    $$Cu^{2+}(aq) + 4NH_3(aq) \rightarrow Cu(NH_3)_4^{2+}(aq)$$

    In this reaction, unshared electron pairs on nitrogen interact with the Lewis acid Cu^{2+} to form four Cu–NH_3 bonds.
  - → *Ammonium salts* decompose when heated; if the salt contains a nonoxidizing anion, NH_3 is produced.

    $$(NH_4)_2SO_4(s) \rightarrow 2NH_3(g) + H_2SO_4(l)$$

    Decomposing ammonium carbonate has a pungent odor and is used as a "smelling salt" to revive people who have fainted.
  - → *Phosphine* is a poisonous gas that smells faintly of garlic and bursts into flame in air if it is impure. It is much less stable than ammonia. Because it does not form hydrogen bonds, it is not very soluble in water and is a very weak base ($pK_b = 27.4$). Its aqueous solution is nearly neutral.

- **Hydrazine, N_2H_4**
  - → *Hydrazine* is an oily, colorless liquid. It is prepared by gentle oxidation of ammonia with alkaline hypochlorite solution.

    $$2NH_3(aq) + ClO^-(aq) \rightarrow N_2H_4(aq) + Cl^-(aq) + H_2O(l)$$

    Its physical properties are similar to water; its melting point is 1.5°C and its boiling point is 113°C. It is dangerously explosive and is stored in aqueous solution.
  - → A mixture of *methylhydrazine* and liquid N_2O_4 is used as a rocket fuel because these two liquids ignite on contact and produce a large volume of gas.

    $$4CH_3NHNH_2(l) + 5N_2O_4(l) \rightarrow 9N_2(g) + 12H_2O(g) + 4CO_2(g)$$

Example 15.2 Draw the Lewis structure for hydrazine. Is hydrazine expected to exhibit acid or base properties in aqueous solution? What is the oxidation number of nitrogen in hydrazine?

Solution We expect hydrazine to follow the octet rule and its Lewis structure is

$$H-\ddot{N}(H)-\ddot{N}(H)-H$$

Note that hydrazine molecules contain an N–N single bond and that each N atom has a lone pair of electrons. Because of the lone pairs on N, in analogy with ammonia, we expect hydrazine to behave as a weak base in water. The H atoms in H_2NNH_2 are assigned an oxidation number of +1 (the compound is not a hydride and H is less electronegative than N), so each N atom must have an oxidation number of −2. Compare this result with ammonia, NH_3, in which the oxidation number of N is −3.

- **Nitrides, N^{3-}, and phosphides, P^{3-}**
  - → *Nitrides* are solids that contain the nitride ion. Nitrides are stable only for small cations such as lithium, magnesium, or aluminum. Nitrides dissolve in water to form ammonia and the corresponding hydroxide.

    $$Zn_3N_2(s) + 6\,H_2O(l) \rightarrow 2\,NH_3(g) + 3\,Zn(OH)_2(aq)$$
  - → *Phosphides* are solids that contain the phosphide ion. Because PH_3 is a very weak parent acid of the strong Brønsted base P^{3-}, water is a sufficiently strong proton donor to react with phosphides to form phosphine.

    $$2\,P^{3-}(s) + 6\,H_2O(l) \rightarrow 2\,PH_3(g) + 6\,OH^-(aq)$$

- **Azides, N_3^-**
  - → The *azide ion* is a highly reactive polyatomic anion of nitrogen. Sodium azide is prepared by the reaction of dinitrogen oxide with molten sodium amide, $NaNH_2$.

    $$N_2O(g) + 2\,NaNH_2(l) \rightarrow NaN_3(l) + NH_3(g) + NaOH(l)$$
  - → Some *azides*, such as AgN_3, $Cu(N_3)_2$, and $Pb(N_3)_2$, are shock sensitive. The azide ion is a weak base, and its conjugate acid, hydrazoic acid, HN_3, is a weak acid similar in strength to acetic acid.

- **Halides**
  - → *Nitrogen* has an oxidation number of +3 in the *nitrogen halides*. Nitrogen trifluoride, NF_3, is the most stable and does not react with water. Nitrogen trichloride, however, reacts with water to form ammonia and hypochlorous acid, HClO. Nitrogen triodide is so unstable that it decomposes explosively upon contact with light.
  - → *Phosphorus* forms chlorides with an oxidation number of +3 in PCl_3 and +5 in PCl_5. *Phosphorus trichloride*, a liquid, is formed by direct chlorination of phosphorus. It is a major intermediate in the production of pesticides, oil additives, and flame retardants. *Phosphorus pentachloride*, a solid, is made by allowing the trichloride to react with additional chlorine. It exists in the solid as tetrahedral PCl_4^+ cations and octahedral PCl_6^- anions, but it vaporizes to a gas of trigonal bipyramidal PCl_5 molecules. *Phosphorus pentabromide* is also molecular in the vapor and ionic as a solid; but in the solid, the anions are Br^- anions, presumably because of the difficulty in fitting six large Br atoms around a central P atom.
  - → Both *phosphorus trichloride* and *phosphorus pentachloride* react with water in a *hydrolysis reaction*, a reaction with water in which new element-oxygen bonds are

formed and there is no change in oxidation state. An *oxoacid* and *hydrogen chloride* gas are formed in the reaction.

$$PCl_3(l) + 3\,H_2O(l) \rightarrow H_3PO_3(s) + 3\,HCl(g)$$
$$PCl_5(s) + 4\,H_2O(l) \rightarrow H_3PO_4(l) + 5\,HCl(g)$$

15.3 Nitrogen Oxides and Oxoacids

- **Dinitrogen oxide, N_2O**

→ *Dinitrogen oxide* is commonly called *nitrous oxide* (laughing gas), and it is the oxide of nitrogen with the lowest oxidation number for nitrogen (+1). It is formed by gentle heating of ammonium nitrate.

$$NH_4NO_3(s) \rightarrow N_2O(g) + 2\,H_2O(g)$$

→ N_2O is fairly unreactive, but it is toxic if inhaled in large amounts.

- **Nitrogen oxide, NO**

→ *Nitrogen oxide* is commonly called *nitric oxide*; the nitrogen atom has an oxidation number of +2. Nitrogen oxide, a colorless gas, is produced industrially by the catalytic oxidation of ammonia at 1000°C.

$$4\,NH_3(g) + 5\,O_2(g) \rightarrow 4\,NO(g) + 6\,H_2O(g)$$

→ The endothermic formation of NO from the oxidation of N_2 occurs readily at the high temperatures that exist in automobile engines and turbine engine exhausts.

$$N_2(g) + O_2(g) \rightarrow 2\,NO(g)$$

→ NO is further oxidized to NO_2 upon exposure to air. In this form, it contributes to smog, acid rain, and the destruction of the stratospheric ozone layer.

- **Nitrogen dioxide, NO_2**

→ *Nitrogen dioxide* is a choking, poisonous, brown gas that contributes to the color and odor of smog. The oxidation number of nitrogen in NO_2 is +4. NO_2, like NO, is paramagnetic. It disproportionates in water to form nitric acid and nitrogen oxide.

$$3\,NO_2(g) + H_2O(l) \rightarrow 2\,HNO_3(aq) + NO(g)$$

→ *Nitrogen dioxide* is prepared in the laboratory by heating lead(II) nitrate:

$$2\,Pb(NO_3)_2(s) \rightarrow 4\,NO_2(g) + 2\,PbO(s) + O_2(g)$$

and exists in equilibrium with its colorless dimer, dinitrogen tetroxide:

$$2\,NO_2(g) \rightleftharpoons N_2O_4(g)$$

Only the dimer exists in the solid, so the brown gas condenses to a colorless solid.

- **Nitrous acid, HNO_2**

→ *Nitrous acid* can be produced in aqueous solution by mixing its anhydride, dinitrogen trioxide, with water. The oxidation number of nitrogen in HNO_2 is +3.

$$N_2O_3(g) + H_2O(l) \rightarrow 2\,HNO_2(aq)$$

→ *Nitrous a*cid has not been isolated as a pure compound, but it has many uses in aqueous solution. It is a weak acid with $pK_a = 3.4$. The conjugate base, nitrite ion, NO_2^-, forms ionic solids that are soluble in water and mildly toxic.

- **Nitric acid, HNO_3**

  → *Nitric acid*, a widely used industrial and laboratory acid, is produced by the three-step *Ostwald process*. The oxidation number of nitrogen in HNO_3 is +5.

  $$4NH_3(g) + 2O_2(g) \rightarrow 4NO(g) + 6H_2(g)$$
  $$2NO(g) + O_2(g) \rightarrow 2NO_2(g)$$
  $$3NO_2(g) + H_2O(l) \rightarrow 3HNO_3(aq) + NO(g)$$

  Nitric acid has been isolated as a pure compound, and it is a colorless liquid with a boiling point of 83°C. It is usually used in aqueous solution. It is an excellent oxidizing agent as well as a strong acid.

Example 15.3 What is the acid anhydride of nitric acid?

Solution An anhydride is formed from an acid by removal of water, H_2O. Nitric acid, HNO_3, contains only one H atom, so two molecules of the acid are required to remove one H_2O molecule. Removal of H_2O from two molecules of HNO_3 leaves dinitrogen pentoxide, N_2O_5, which is the acid anhydride:

$$2HNO_3(l) \rightarrow N_2O_5(s) + H_2O(l)$$

Dinitrogen pentoxide contains NO_2^+ cations and NO_3^- anions in the solid, but vaporizes to a molecular gas.

15.4 Phosphorus Oxides and Oxoacids

- **Phosphorus(III) oxide, P_4O_6, and phosphorous acid, H_3PO_3**

  → *Phosphorus(III) oxide* is made by heating white phosphorus in a limited supply of air. P_4O_6 molecules consist of P_4 tetrahedral units, with each oxygen atom lying between two phosphorus corner atoms. Six oxygen atoms lie along the six edges of the tetrahedron, one per edge. The edge bonds are represented as –P–O–P–.

  → P_4O_6 is the anhydride of *phosphorous acid:*

  $$4H_3PO_3(aq) \rightarrow P_4O_6(s) + 6H_2O(l)$$

  Phosphorous acid is a *diprotic* acid.

- **Phosphorus(V) oxide, P_4O_{10}, and phosphoric acid, H_3PO_4**

  → *Phosphorus(V) oxide* is made by heating white phosphorus in an excess supply of air. P_4O_{10} molecules are similar to P_4O_6, but have an additional oxygen attached to each phosphorus at the apices of the tetrahedron.

  → P_4O_{10} is the *anhydride* of *phosphoric acid*, H_3PO_4. It traps and reacts with water very efficiently and is widely used as a drying agent. Phosphoric acid is a *triprotic acid.* It is only mildly oxidizing, despite the fact that phosphorus is in a high oxidation state. Phosphoric acid is the parent acid of phosphate salts that contain the tetrahedral phosphate ion, PO_4^{3-}. Phosphates are generally not very soluble, which makes them suitable structural material for bones and teeth.

Example 15.4 Explain why phosphorous acid, H_3PO_3, is *diprotic* whereas phosphoric acid, H_3PO_4, is *triprotic*.

Solution Phosphorous acid contains three H atoms and, if it is diprotic, only two H atoms may dissociate in water or react with a base. The observation that

phosphorous acid is diprotic suggests that one of the three H atoms in the molecule is *not* bonded to an oxygen atom.
Phosphoric acid also contains three H atoms, but it is triprotic and all three atoms may dissociate in water or react with a base. The observation that phosphoric acid is triprotic suggests that each one of the three H atoms in the molecule is bonded to an oxygen atom.
Two possible Lewis structures of the acids are

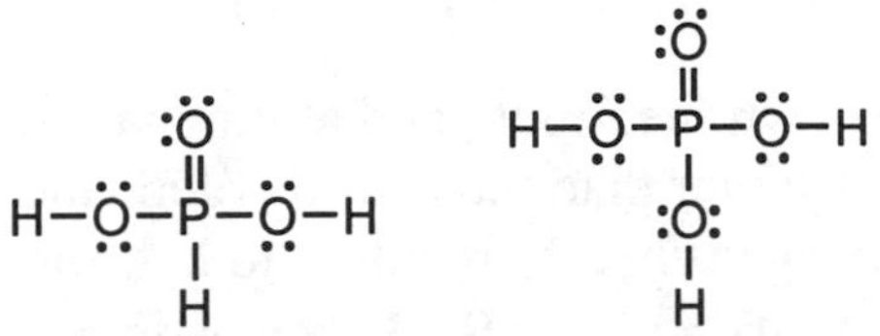

```
      :Ö                 :Ö
       ‖                  ‖
  H—Ö—P—Ö—H          H—Ö—P—Ö—H
       |                  |
       H                 :O:
                          |
                          H
```

Phosphorous acid   Phosphoric acid

Note that each structure shown contains an expanded octet on the central phosphorus atom.

- **Polyphosphates**
  → *Polyphosphates* are compounds made up of linked PO_4^{3-} tetrahedral units. The simplest structure is the *pyrophosphate ion*, $P_2O_7^{4-}$, in which two phosphate ions are linked by an oxygen atom through –O–P–O– bonds.

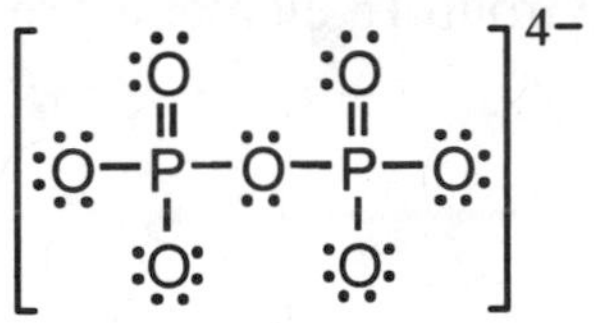

```
 [    :Ö      :Ö      ]4−
 [     ‖       ‖      ]
 [ :Ö—P—Ö—P—Ö: ]
 [     |       |      ]
 [    :O:     :O:     ]
```

  → More complicated structures that form with longer chains or rings also exist. The biochemically most important polyphosphate is *adenosine triphosphate*, ATP, which contains three phosphorus tetrahedral units linked by –O–P–O– bonds. The hydrolysis of ATP to *adenosine diphosphate*, ADP, by the rupture of an O–P bond releases energy that is used by cells to drive biochemical reactions within the cell.

$$ATP + H_2O \rightarrow ADP + HPO_4^{2-} \qquad \Delta H = -41 \text{ kJ}$$

ATP is often called the "fuel of life," because it provides the energy for many of the biochemical reactions occurring in cells.

Group 16: The Oxygen Family (Sections 15.5-15.8)

Key Concepts

valence electron configuration, chalcogens, elements, physical properties, allotropes, preparation, applications, chemical properties, *compounds with hydrogen:* water, hydrogen sulfide, hydrogen peroxide, and polysulfanes; *sulfur oxides and oxoacids:* sulfur dioxide, sulfurous acid, sulfites, sulfur trioxide, sulfuric acid, and sulfates; *sulfur halides:* sulfur hexafluoride, disulfur chloride, sulfur dichloride, mustard gas

Overview

- **Oxygen family**
  - → The valence electron configuration of all Group 16 elements is ns^2np^4, where n is the period number. These elements need only two more electrons to complete their valence shells. Metallic character increases down the group, but none in the group is a metal. Oxygen, sulfur, and selenium are nonmetals, whereas tellurium and polonium are metalloids.
  - → Electronegativities decrease down the group, and atomic and ionic radii increase. Sulfur has atoms that are 58% larger than those of oxygen, and it has a lower first ionization energy and electronegativity. In bonding to hydrogen, sulfur forms less polar bonds than does oxygen. As a result, S–H groups do not form hydrogen bonds and H_2S is a gas, whereas O–H groups form strong hydrogen bonds and H_2O is a liquid. Sulfur has less of a tendency to form multiple bonds; instead, it tends to form additional single bonds to as many as six other atoms, making use of its *d*-orbitals to do so.
  - → The members of Group 16 are called collectively the *chalcogens*, a Greek word meaning "brass giver." The elements are found in the ores of copper, which is, in addition to zinc, a major component of brass.
  - → Some physical properties of the Group 16 elements are given in Table 15.4 in the text.

15.5 The Group 16 Elements

- **Oxygen**
  - → *Elemental oxygen*, O_2, is found in air (21.0% by volume) and is prepared by fractional distillation of liquid air. It is a colorless, odorless, paramagnetic gas. A satisfactory Lewis structure cannot be drawn. The Molecular Orbital theory accounts for the paramagnetic properties (see Section 3.10).
  - → *Elemental oxygen* exists in two allotropic forms, O_2 and O_3 (*ozone*). Inorganic chemists often use the term *dioxygen* to refer to molecular oxygen. Ozone is produced by an electric discharge in oxygen gas; its pungent odor is apparent near sparking electrical equipment and lightning strikes. Some properties of oxygen and ozone are given in the following table.

Property	O_2	O_3
gas color	colorless	blue
boiling point	−183°C	−112°C
liquid color	pale blue	deep blue
melting point	−218°C	−192°C
solid color	pale blue	black-violet
atmosphere	23% by mass	small amounts
magnetism	paramagnetic	diamagnetic

Example 15.5a Use the VSEPR theory to predict the shape of the ozone molecule and the hybridization of the central oxygen atom.

Solution The ozone molecule contains 18 valence electrons and is represented by two resonance Lewis structures:

$$:\ddot{O}=\ddot{O}-\ddot{\underset{\cdot\cdot}{O}}: \longleftrightarrow :\ddot{\underset{\cdot\cdot}{O}}-\ddot{O}=\ddot{O}:$$

The central atom has three regions of high concentration of electrons around it, resulting in a trigonal orientation for the electron pairs and sp^2 hybridization. Because one of the regions of high concentration of electrons is a lone pair, the molecule has a bent or angular shape.

- **Sulfur**

→ *Sulfur* is found in ores and as elemental sulfur. It is widely distributed as sulfide ores, which include *galena*, PbS; *cinnabar*, HgS; *iron pyrite*, FeS_2 (fool's gold), and *spahlerite*, ZnS. Sulfur is found in elemental form as deposits, called *brimstone*, which are created by bacterial action on H_2S.

→ *Elemental sulfur* is a yellow, tasteless, almost odorless, insoluble, nonmetallic molecular solid of crownlike S_8 rings (see Margin Figure **(9)** on page 717 in the text). The more stable allotrope, *rhombic sulfur*, forms beautiful yellow crystals, whereas the less stable allotrope, *monoclinic sulfur*, forms needlelike crystals. The two allotropes differ in the manner in which the S_8 rings are stacked together. For simplicity, the elemental form of sulfur is often represented as S(s) rather than $S_8(s)$.

Example 15.5b Use your knowledge of bonding and your chemical intuition to suggest a reason why the eight-member rings in sulfur are puckered rather than planar.

Solution A planar structure would correspond to a regular octagon of S atoms, in which the S–S–S angle, α in the diagram, is 135° (see if you can prove this). A bond angle of 135° is unusually large and does not correspond to the angles between atomic orbitals or typical hybrid orbitals that are used to form the S–S bonds. It is more reasonable to expect that the rings pucker leading to smaller S–S–S angles. In fact, in sulfur, the S–S–S angles are 108°, near the tetrahedral angle of 109.5°. The nearly tetrahedral angles are consistent with VSEPR; in S_8 each sulfur atom has two lone pairs of electrons and forms two bonds to adjacent S atoms. Four electron pairs around sulfur (two bonding and two nonbonding) lead to a tetrahedral geometry, and the presence of lone pairs would give a bent S–S–S shape and an angle somewhat less than 109.5°.

S–S, S, S, S, S, S–S; α = 135°

→ One method of recovering sulfur is the *Claus process*, in which some of the H_2S that occurs in oil and natural gas is first oxidized to sulfur dioxide.

$$2\,H_2S(g) + 3\,O_2(g) \rightarrow 2\,SO_2(g) + 2\,H_2O(l)$$

Sulfur dioxide oxidizes the remaining hydrogen sulfide and both are converted to elemental sulfur at 300°C in the presence of an alumina catalyst.

$$2\,H_2S(g) + SO_2(g) \rightarrow 3\,S(s) + 2\,H_2O(l)$$

→ *Sulfur* is used primarily to produce *sulfuric acid* and to vulcanize rubber. Vulcanization increases the toughness of rubber by introducing cross-links between the polymer chains of natural rubber.

→ An important property of sulfur is its ability to form chains of atoms, *catenation*. The –S–S– links that connect different parts of the chains of amino acids in proteins are an important example. These *disulfide links* contribute to the shapes of proteins (see Section 19.13).

- **Selenium and tellurium**

→ *Selenium* and *tellurium* are found in sulfide ores; they are also recovered from the anode sludge formed during the electrolytic refining of copper.

→ Both elements have several allotropic forms; the most stable one consists of long *zigzag chains* of atoms. The appearance of the allotropes is that of a silver-white metal, but electrical conductivity is poor. Exposing selenium to light increases its electrical conductivity, so it is used in solar cells.

- **Polonium**

→ *Polonium* is a radioactive, low-melting metalloid. It was identified in uranium ores in 1898 by the Curies. Its most stable isotope, ^{209}Po, has a half-life of 103 y.

→ *Polonium* decays by emission of an alpha particle, $^{4}He^{2+}$. It is used in antistatic devices in textile mills to counteract the buildup of negative charges on fast moving fabric.

15.6 Compounds with Hydrogen

- **Water, H_2O, and hydrogen sulfide, H_2S**

→ *Water* is a remarkable compound possessing a unique set of physical and chemical properties. Water is purified in domestic water supplies in a multi-step procedure that includes *aeration*, addition of *slaked lime*, removal of particles by *coagulation* and *flocculation*, addition of CO_2, *filtration*, reduction of pH, and addition of chlorine *disinfectant* to kill bacteria. High purity water is obtained by *distillation*.

→ *Water* has a higher boiling point than expected on the basis of its molar mass because of the extensive hydrogen bonding between H_2O molecules. Hydrogen bonding also causes a more open structure and, therefore, a lower density for the solid than for the liquid. Because of its high polarity, water is an excellent solvent for ionic compounds.

→ Chemically, water is *amphiprotic*, and it can both donate and accept protons; so it is both a *Brønsted acid* and *Brønsted base*.

$$H_2O(l, \textit{acid}) + NH_3(aq) \rightarrow NH_4^+(aq) + OH^-(aq)$$

$$CH_3COOH(aq) + H_2O(l, \textit{base}) \rightarrow H_3O^+(aq) + CH_3COO^-(aq)$$

→ *Water* can act as a *Lewis base* by donating its unshared pair of electrons on the oxygen atom.

$$Fe^{3+}(aq) + 6\,H_2O(l) \rightarrow Fe(H_2O)_6^{3+}(aq)$$

→ *Water* can act as an *oxidizing agent* and a *reducing agent*.

$$2\,Na(s) + 2\,H_2O(l, \textit{ox. agent}) \rightarrow 2\,NaOH(aq) + H_2(g)$$

$$2\,H_2O(l,\ \textit{red. agent}) + 2\,F_2(g) \rightarrow 4\,HF(aq) + O_2(g)$$

→ A *hydrolysis reaction* is one in which water as a reactant is used to form a bond between oxygen and another element. Hydrolysis can occur *with* or *without* a change in oxidation number.

$$PCl_5(s) + 4\,H_2O(l) \rightarrow H_3PO_4(aq) + 5\,HCl(aq)$$

$$Cl_2(g) + H_2O(l) \rightarrow ClOH(aq) + HCl(aq)$$

The first example given above has *no* change in oxidation number. The second is an example of a *disproportionation reaction*, in which the oxidation number of chlorine changes from 0 in Cl_2 to +1 in HClO and −1 in HCl.

→ *Hydrogen sulfide* is an example of a Group 16 binary compound with hydrogen. It is a toxic gas with an offensive odor (rotten eggs). Egg proteins contain sulfur and release H_2S when they decompose. Iron(II) sulfide sometimes appears as a pale green discoloration at the boundary between the white and the yolk.

→ *Hydrogen sulfide* is prepared by the direct reaction of hydrogen and sulfur at 600°C or by protonation of the sulfide ion, which is a Brønsted base.

$$FeS(s) + 2\,HCl(aq) \rightarrow FeCl_2(aq) + H_2S(g)$$

→ *Hydrogen sulfide* dissolves in water and is slowly oxidized by dissolved air to form a colloidal dispersion of particles of sulfur. Hydrogen sulfide is a *weak diprotic acid*, and the parent acid of the hydrogen sulfides, HS^-, and the sulfides, S^{2-}. Sulfides of *s*-block elements are moderately soluble in water, whereas the sulfides of the heavy *p*- and *d*-block metals are generally very insoluble.

- **Hydrogen peroxide, H_2O_2, and polysulfanes, HS–S_n–SH**

→ *Hydrogen peroxide* is a highly reactive, pale blue liquid that is appreciably more dense (1.44 $g{\cdot}mL^{-1}$) than water. Its melting point is −0.4°C and its boiling point is 152°C.

→ Chemically, *hydrogen peroxide* is more acidic than water ($pK_a = 11.75$). It is a strong oxidizing agent, but it can also function as a reducing agent.

$$2\,Fe^{2+}(aq) + H_2O_2(aq,\ \textit{ox. agent}) + 2\,H^+(aq) \rightarrow 2\,Fe^{3+}(aq) + 2\,H_2O(l)$$

$$Cl_2(g) + H_2O_2(aq,\ \textit{red. agent}) + 2\,OH^-(aq) \rightarrow 2\,Cl^-(aq) + 2\,H_2O(l) + O_2(g)$$

→ The O–O bond in *hydrogen peroxide* is very weak. The oxidation state of oxygen in hydrogen peroxide is −1.

→ A *polysulfane* is a *catenated* molecular compound of composition HS–S_n–SH, where *n* can take values from 0 through 6. The sulfur analogue of hydrogen peroxide occurs with $n = 0$. Two polysulfide ions obtained from polysulfanes are found to occur in the mineral *lapis lazuli* (see Fig.15.16 in the text); its color derives from impurities of S_2^-(blue) and S_3^-(hint of green). The S_2^- ion is an analogue of the *superoxide* ion, O_2^-; and S_3^- is an analogue of the *ozonide* ion, O_3^-.

15.7 Sulfur Oxides and Oxoacids

- **Sulfur dioxide, SO_2, sulfurous acid, H_2SO_3, and sulfite, SO_3^{2-}**

→ *Sulfur dioxide*, a poisonous gas, is made by burning sulfur in air. Volcanic activity, the combustion of fuels contaminated with sulfur, and the oxidation of hydrogen sulfide are the major sources of SO_2 in the atmosphere. The chemical equation for the oxidation of hydrogen sulfide is

$$2H_2S(g) + 2O_2(g) \rightarrow 2SO_2(g) + 2H_2(g)$$

→ *Sulfurous acid* is formed by the reaction of its acid anhydride, SO_2, and water.

$$SO_2(g) + H_2O(l) \rightarrow H_2SO_3(aq)$$

Sulfurous acid is an equilibrium mixture of two molecules, $(HO)SHO_2$ and $(HO)SO(OH)$ (see Margin Figures **(12a)** and **(12b)** on page 721 in the text). These molecules are also in equilibrium with molecules of SO_2, each of which is surrounded by a cage of water molecules. When the solution is cooled, crystals with a composition of roughly $SO_2 \cdot 7H_2O$ form. This substance is an example of a *clathrate*, in which a molecule sits in a cage of other molecules. Methane, carbon dioxide, and the noble gases also form clathrates with water.

→ In *sulfite ions* and *sulfur dioxide*, the oxidation number of the sulfur atom is +4. Because this value is intermediate in sulfur's range of –2 to +6, these compounds can act as either oxidizing agents or reducing agents. The most important reaction of sulfur dioxide is its slow oxidation to sulfur trioxide, in which the oxidation number of the sulfur atom is +6.

$$2SO_2(g) + O_2(g) \rightarrow 2SO_3(g)$$

This reaction is catalyzed by the presence of metal cations in droplets of water or by certain surfaces. Indirect pathways also exist in the atmosphere for the conversion of SO_2 to SO_3.

- **Sulfur trioxide, SO_3, sulfuric acid, H_2SO_4, and sulfate, SO_4^{2-}**

→ At normal temperatures, *sulfur trioxide* is a volatile liquid with a boiling point of 45°C. Its shape is trigonal planar with bond angles of 120°. In the solid, and to some extent in the liquid, the molecules aggregate to form *trimers*, S_3O_9, as well as larger groupings.

→ *Sulfuric acid* is produced by the *contact process*, in which sulfur is first burned in oxygen at 1000°C to form sulfur dioxide, which then reacts with oxygen to form sulfur trioxide over a V_2O_5 catalyst at 500°C. Because sulfur trioxide forms a corrosive acid with water, it is absorbed in 98% concentrated sulfuric acid to give a dense, oily liquid called *oleum*, $H_2S_2O_7$.

$$SO_3(g) + H_2SO_4(l) \rightarrow H_2S_2O_7(l)$$

Example 15.7a Predict the product(s) of the reaction of oleum with water. Suggest a structure for oleum consistent with this reaction.

Solution Formally add H_2O to oleum, $H_2S_2O_7$; this gives $H_4S_2O_8$ which is twice the formula of sulfuric acid, H_2SO_4. Therefore, a reasonable reaction is

$$H_2S_2O_7(l) + H_2O(l) \rightarrow 2H_2SO_4(l)$$

In fact, this reaction occurs and oleum, also known as *disulfuric acid*, may be regarded as the acid anhydride of sulfuric acid.

To determine the structure of disulfuric acid, examine the Lewis structure of H_2SO_4 and think about how H_2O could be eliminated. One way is shown schematically on the following page:

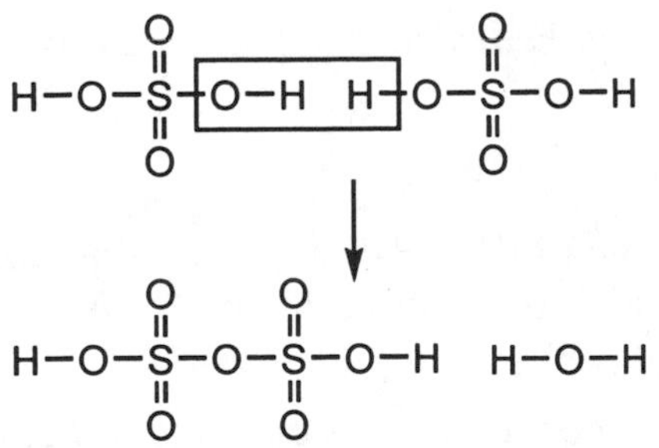

```
       O                O
       ||               ||
H—O—S—[O—H   H]—O—S—O—H
       ||               ||
       O                O

               ↓

       O     O
       ||    ||
H—O—S—O—S—O—H    H—O—H
       ||    ||
       O     O
```

The structure shown above is a reasonable one for disulfuric acid, but it has not been confirmed experimentally, although the structure of the $S_2O_7^{2-}$ anion is known.

→ *Sulfuric acid* is a colorless, corrosive, oily liquid that boils (decomposes) at 300°C. It is a strong *Brønsted acid*, a powerful *dehydrating agent*, and a mild *oxidizing agent*. In water, its first ionization is complete, but HSO_4^- is a *weak acid* with $pK_a = 1.92$. As a *dehydrating agent*, sulfuric acid is able to extract water from a variety of compounds, including sucrose, $C_{12}H_{22}O_{11}$, and formic acid, HCOOH.

$$C_{12}H_{22}O_{11}(s) \rightarrow 12\,C(s) + 11\,H_2O(l)$$

$$HCOOH(l) \rightarrow CO(g) + H_2O(l)$$

Because of the large exothermicity associated with dilution, mixing sulfuric acid with water can cause violent splashing, so for reasons of safety, sulfuric acid (which is more dense than water) is always carefully added to water, not the water to acid. The *sulfate ion* is a mild *oxidizing agent*.

$$4\,H^+(aq) + SO_4^{2-}(aq) + 2\,e^- \rightarrow SO_2(g) + 2\,H_2O(l) \qquad E° = 0.17\ V$$

Example 15.7b Mixing sulfuric acid with solid sodium chloride results in the formation of dry hydrogen chloride. Write an equation for the reaction, state what type of reaction it is, and suggest why the hydrogen chloride is dry.

Solution Because HCl is one of the products formed by the reaction of H_2SO_4 and NaCl, it is reasonable to assume that the Na^+ and SO_4^{2-} that remain form sodium sulfate. The reaction is

$$H_2SO_4(l) + 2\,NaCl(s) \rightarrow 2\,HCl(g) + Na_2SO_4(s)$$

The reaction is a Lewis acid-base reaction in which the Lewis base Cl^- reacts with the Lewis acid H^+ to form HCl, and the salt Na_2SO_4 forms from the remaining ions. The HCl is dry because sulfuric acid is a strong dehydrating agent and removes any stray water present.

15.8 Sulfur Halides

- **Fluorides**

→ Sulfur ignites in fluorine to produce *sulfur hexafluoride*, SF_6, a dense, colorless, odorless, nontoxic, thermally stable ($\Delta G_f° = -1105.3\ kJ\cdot mol^{-1}$), insoluble gas.

→ The oxidation number of sulfur in *sulfur hexafluoride* attains its maximum value of +6, but the large number of strongly electronegative fluorine atoms around the sulfur atom protects the compound from attack. Because it has a high ionization energy, sulfur hexafluoride is a good gas-phase electrical insulator.

→ The predicted shape of SF_6 is *octahedral* and the molecule is *nonpolar*, $\mu = 0$ (see Section 3.3).

- **Chlorides**

→ *Disulfur dichloride,* S_2Cl_2, is one of the products of the reaction of sulfur with chlorine. A yellow, toxic liquid with a nauseating odor, *disulfur dichloride* is used mainly for the vulcanization of rubber.

→ The reaction of *disulfur dichloride* with excess chlorine in the presence of a catalyst, iron(III) chloride, produces a foul smelling, red liquid of *sulfur dichloride*, SCl_2. It reacts with ethene, C_2H_4, to produce *mustard gas*, $S(CH_2CH_2Cl)_2$, which has been used in chemical warfare. *Mustard gas* causes symptoms of severe discomfort, skin blisters, nasal discharges, and vomiting. It also destroys the cornea of the eye, leading to blindness.

Example 15.8 Suggest a structure for the S_2Cl_2 molecule. Draw its Lewis structure.

Solution In the element and in the polysulfanes, sulfur tends to form chains or rings of atoms. Given this, it is reasonable to except that S_2Cl_2 is a derivative of the polysulfane with $n = 0$ (HSSH) and that it contains the disulfide (–S–S–) unit. Chlorine forms single σ-bonds in a variety of compounds (HCl, CH_3Cl) and we might expect the Cl atoms to be bonded terminally in S_2Cl_2. A reasonable Lewis structure is

$$:\ddot{\underset{..}{Cl}}-\ddot{\underset{..}{S}}-\ddot{\underset{..}{S}}-\ddot{\underset{..}{Cl}}:$$

The lone pairs of electrons on sulfur suggest that the molecule is bent at each sulfur atom, rather than linear as shown in the Lewis diagram, and in fact, it is.

Group 17: The Halogens (Sections 15.9-15.10)

Key Concepts

valence electron configuration, halogens, elements, physical properties, preparation, applications, chemical properties, interhalogen compounds, hydrogen halides, metal halides, halogen oxides, halogen oxoacids

Overview

- **Halogen family**

→ The valence electron configuration of all Group 17 elements is ns^2np^5 where n is the period number. These elements need only one additional electron to complete their valence shells, and all form diatomic molecules, such as F_2 and I_2. All members are nonmetals.

→ The members of Group 17 are called collectively the *halogens*, a Greek word with the meaning "that which produces salt."

→ The *halogens* show smooth variations in the properties of the elements and their compounds, with some exceptions for *fluorine*. Many of their physical properties are determined by London forces. *Fluorine* displays only an oxidation number of –1 in its compounds, whereas the other halogens typically show –1, +1, +3, +5, and +7 in their compounds.

→ Some physical properties of the Group 17 elements are given in Table 15.5 in the text, the known *interhalogens* in Table 15.6, some properties of the *hydrogen halides* in Table 15.7, and the *halogen oxoacids* in Table 15.8.

15.9 The Group 17 Elements

- **Fluorine**

→ *Fluorine* is a reactive, almost colorless gas of F_2 molecules. Its properties follow from its high electronegativity, small size, and lack of available *d*-orbitals for bonding. It is the most electronegative element and has an oxidation number of –1 in all its compounds. Elements combined with fluorine are often found with their highest oxidation numbers, such as +7 for I in IF_7. The small size of the fluoride ion results in high lattice enthalpies for fluorides, a property that makes them less soluble than other halides.

→ *Fluorine* occurs widely in many minerals including *fluorspar*, CaF_2; *cryolite*, Na_3AlF_6; and the *fluoroapatites*, $Ca_5F(PO_4)_3$.

→ *Elemental fluorine* is obtained by the electrolysis of an anhydrous, molten KF–HF mixture at 75°C, using a carbon electrode. Fluorine is highly reactive and highly oxidizing; it is used in the production of SF_6 for electrical equipment, Teflon (polytetrafluoroethylene), and UF_6 for isotope enrichment.

→ The strong bonds formed by *fluorine* make many of its compounds relatively inert. Fluorine's ability to form *hydrogen bonds* results in relatively high melting points, boiling points, and enthalpies of vaporization for many of its compounds.

- **Chlorine**

→ *Chlorine* is a reactive, pale yellow-green gas of Cl_2 molecules that condenses at –34°C. It reacts directly with all the elements except carbon, nitrogen, oxygen, and the noble gases. It is used in a large number of industrial processes, including the manufacture of plastics, solvents, and pesticides. It is also used as a disinfectant to treat water supplies.

→ *Elemental chlorine* is obtained by the electrolysis of molten or aqueous NaCl. It is a strong oxidizing agent and oxidizes metals to high oxidation states.

$$2\,Fe(s) + 3\,Cl_2(g) \rightarrow 2\,FeCl_3(s)$$

- **Bromine**

→ *Bromine* is a corrosive, reddish brown liquid of Br_2 molecules. It has a penetrating odor. Organic bromides are used in textiles as fire retardants and in pesticides. Inorganic bromides, particularly silver bromide, are used in photographic emulsions.

→ *Elemental bromine* is produced from brine wells by the oxidation of Br^- by Cl_2.

$$2\,Br^-(aq) + Cl_2(g) \rightarrow Br_2(l) + 2\,Cl^-(aq)$$

Example 15.9 Why is it possible to use Cl_2 to produce Br_2 from Br^-? What is $E°$ for the reaction?

Solution The reaction is an oxidation-reduction process. The reduction potentials for the two half-reactions involved allow us to calculate $E°$ for the reaction:

$$Cl_2(g) + 2e^- \rightarrow 2Cl^-(aq) \qquad E°(\text{reduction}) = +1.36\text{ V}$$
$$2Br^-(aq) \rightarrow Br_2(l) + 2e^- \qquad E°(\text{oxidation}) = -1.07\text{ V}$$
$$Cl_2(g) + 2Br^-(aq) \rightarrow Br_2(l) + 2Cl^-(aq) \qquad E° = +0.29\text{ V}$$

Because the standard reaction potential is positive, the reaction is thermodynamically favored ($\Delta G° = -nFE° < 0$) and can be used to produce Br_2.

- **Iodine**
  - → *Iodine* is a blue-black, lustrous solid of I_2 molecules that readily sublimes to form a purple vapor. It occurs as I^- in seawater and as an impurity in Chile saltpeter, KNO_3. The best source is the brine from oil wells. *Elemental iodine* is produced from brine wells by the oxidation of I^- by Cl_2.

    $$2I^-(aq) + Cl_2(g) \rightarrow I_2(aq) + 2Cl^-(aq)$$
  - → *Elemental iodine* is slightly soluble in water, but it dissolves well in iodide solutions because it reacts with aqueous I^- to form the triodide ion, I_3^-.

    $$I_2(aq) + I^-(aq) \rightarrow I_3^-(aq)$$
  - → *Iodine* is an essential trace element in human nutrition, and *iodides* are often added to table salt. This "iodized salt" prevents iodine deficiency, which leads to the enlargement of the thyroid gland in the neck (a condition called *goiter*).
- **Astatine**
  - → *Astatine* is a radioactive element that occurs in uranium ores, but only to a tiny extent. Its most stable isotope, ^{210}At, has a half-life of 8.3 h. The isotopes formed in uranium ores have much shorter lifetimes. The properties of astatine are surmised from spectroscopic measurements, and found to be similar to those of iodine.
  - → *Astatine* is created by bombarding bismuth with alpha particles in a cyclotron, which accelerates particles to high speed.

15.10 Compounds of the Halogens

- **Interhalogens**
  - → *Interhalogen* compounds consist of a heavier halogen atom at the center of the compound and an odd number of lighter ones bonded to it on the periphery. An exception is I_2Cl_6, which contains two (–I–Cl-I–) bridges between the two central iodine atoms. The two bridging chlorine atoms act as Lewis bases (electron pair donors), in a manner similar to Al_2Cl_6.
  - → *Interhalogens* are typically formed by direct reaction between stoichiometric amounts of the elements.

    $$I_2(g) + 7F_2(g) \rightarrow 2IF_7(g)$$
  - → As the size of the central atom in an *interhalogen* compound increases, the bond enthalpies also increase, *usually* resulting in lower reactivity of the compound.

    (X–F): ClF_5 (154 $kJ\cdot mol^{-1}$) < BrF_5 (187 $kJ\cdot mol^{-1}$) < IF_5 (269 $kJ\cdot mol^{-1}$)
  - → Table 15.6 in the text provides a compilation of the known *interhalogens*.

- **Hydrogen halides and metal halides**
  - → *Hydrogen halides* are prepared by direct reaction of the elements or by the action of nonvolatile acids on *metal halides*:
    $$Cl_2(g) + H_2(g) \rightarrow 2\,HCl(g)$$
    $$CaF_2(s) + H_2SO_4(aq, conc.) \rightarrow Ca(HSO_4)_2(aq) + 2\,HF(g)$$
    Because Br^- and I^- are oxidized by sulfuric acid, phosphoric acid is used in the preparation of HBr and HI.
  - → *Hydrogen halides* are pungent, colorless gases, but *hydrogen fluoride* is a liquid at temperatures below 20°C. Its low volatility is a sign of extensive hydrogen bonding, and short zigzag chains up to about $(HF)_5$ persist in the vapor. The properties of the *hydrogen halides* are given in Table 15.7 in the text.
  - → *Hydrogen halides* dissolve readily in water to produce acidic solutions. *Hydrogen fluoride* has the unusual property of attacking and dissolving silica glasses, and it is used for glass etching (see Section 14.23).
  - → Anhydrous *metal halides* may be formed by direct reaction:
    $$2\,Fe(s) + 3\,Cl_2(g) \rightarrow 2\,FeCl_3(s)$$
    Halides of metals tend to be ionic unless the metal has an oxidation number greater than +2. For example, sodium chloride and copper(II) chloride are ionic compounds with high melting points, whereas titanium(IV) chloride and iron(III) chloride sublime as molecules.

Example 15.10 Arrange the following fluorides in order of increasing melting point:

$$PtF_4,\ CaF_2,\ PtF_6$$

Explain your reasoning.

Solution Metals in low oxidation states form ionic halides, whereas those in higher oxidation states tend to form covalent halides. On this basis, we expect CaF_2 with Ca in the +2 oxidation state to be an ionic solid with a high melting point. We expect PtF_6 with Pt in the +6 oxidation state to be a covalent solid (held together by London forces) with a low melting point. The remaining solid, PtF_4 with Pt in the +4 oxidation state, is expected to have a melting point between the other two. The order of *increasing* melting point is therefore:

$$PtF_6 < PtF_4 < CaF_2$$

The observed melting points are: PtF_6 (61.3°C), PtF_4 (600°C), CaF_2 (1418°C).

- **Halogen oxides and oxoacids**
  - → *Hypohalous acids*, HXO *or* HOX, consist of the halogen atom as a *terminal* atom with an oxidation number of +1. They are prepared by the direct reaction of a halogen with water; the corresponding *hypohalite* (XO^-) salts are prepared by the reaction of a halogen with aqueous alkali solution.
    $$Cl_2(g) + H_2O(l) \rightarrow HOCl(aq) + HCl(aq)$$
    $$Cl_2(g) + NaOH(aq) \rightarrow NaOCl(aq) + HCl(aq)$$
  - → *Hypochlorites* cause the oxidation of organic material in water by producing oxygen in aqueous solution, which readily oxidizes organic materials. The production of O_2 occurs in two steps; the net result is

$$2\,OCl^-(aq) \rightarrow 2\,Cl^-(aq) + O_2(g)$$

→ *Chlorates*, ClO_3^-, contain a *central* chlorine atom with an oxidation number of +5. They are prepared by the reaction of chlorine with hot aqueous alkali:

$$3\,Cl_2(g) + 6\,OH^-(aq) \rightarrow ClO_3^-(aq) + 5\,Cl^-(aq) + 3\,H_2O(l)$$

Chlorates decompose upon heating; the identity of the final product is determined by the presence of a catalyst:

$$4\,KClO_3(l) \rightarrow 3\,KClO_4(s) + KCl(s) \qquad \textit{no catalyst}$$

$$2\,KClO_3(l) \rightarrow 2\,KCl(s) + 3\,O_2(g) \qquad MnO_2\ \textit{catalyst}$$

Chlorates are good oxidizing agents and are also used to produce the important compound, ClO_2. Sulfur dioxide is a convenient reducing agent for this reaction:

$$2\,NaClO_3(l) + SO_2(g) + H_2SO_4(aq,\ dilute) \rightarrow 2\,NaHSO_4(aq) + 2\,ClO_2(g)$$

→ *Chlorine dioxide*, ClO_2, contains a *central* chlorine atom with an oxidation number of +4. It has an odd number of electrons and is a paramagnetic yellow gas that is used to bleach paper pulp. It is highly reactive and may explode violently under the right conditions.

→ *Perchlorates*, ClO_4^-, contain a *central* chlorine atom with an oxidation number of +7. They are prepared by the electrolysis of aqueous chlorates. The half-reaction is

$$ClO_3^-(aq) + H_2O(l) \rightarrow ClO_4^-(aq) + 2\,H^+(aq) + 2\,e^-$$

Perchlorates and *perchloric acid*, $HClO_4$, are powerful oxidizing agents; *perchloric acid* in contact with small amounts of organic materials may explode.

→ The *oxidizing strength* and *acidity* of oxoacids both increase as the oxidation number of the halogen increases. For oxoacids with the same number of oxygen atoms, the oxidizing and acid strength both increase as the electronegativity value of the halogen increases. The general rules for acid strength of oxoacids are summarized in Section 10.9. The known *halogen oxoacids* and their pK_a values are listed in Table 15.8 in the text.

Group 18: The Noble Gases (Sections 15.11-15.12)

Key Concepts

valence electron configuration, noble gases, elements, physical properties, preparation, applications, halogen compounds, oxygen compounds

Overview

- **Noble-gas family**

  → Valence electron configurations of the Group 18 elements are $1s^2$ for helium and ns^2np^6, where n is the period number ($n = 2$–6), for the other members. These elements have completed valence shells, and all form colorless monatomic gases. They are all nonmetals.

→ Members of Group 18 are called collectively the *noble gases*, a name that describes their *low* reactivity. Until the first noble gas compounds were synthesized in 1962, all the noble gases were believed to be *chemically inert.* They are still commonly called the *inert gases* or *rare gases*.

→ Some physical properties of the Group 18 elements are given in Table 15.9 in the text. All the elements occur in the atmosphere, comprising about 1% of its mass. All the *noble gases* except helium and radon are obtained by fractional distillation of liquid air.

15.11 The Group 18 Elements

- **Helium**

→ *Helium* derives its name from the Greek word *helios*, sun. The first evidence for helium was discovered in the solar spectrum taken during a solar eclipse in 1868. *Helium* is the second most abundant element in the universe after hydrogen. Helium atoms are light and travel at velocities sufficient to escape from the earth's atmosphere. Alpha particles, $^4He^{2+}$, are released by nuclear decay of naturally occurring uranium and thorium ores. They are high-energy nuclei of helium-4, which capture two electrons apiece to become helium atoms. The low density and lack of flammability of helium make it an appropriate gas for lighter-than-air airships such as blimps. Helium is found in the atmosphere with a concentration of 5.24 ppm (parts per million by volume).

→ *Helium* has the lowest boiling point, 4.2 K, of any known substance, and it finds extensive use as a *cryogenic* coolant. It cannot be solidified at normal pressure. Below 2 K, it becomes a *superfluid* (helium-II), a fluid that has no viscosity and, therefore, no resistance to flow. Helium is also the only known substance that has two liquid phases, as shown in the phase diagram on the left. Its critical temperature and pressure are 5.2 K and 2.3 atm, respectively. Helium is used to dilute oxygen used in deep-sea diving, to pressurize rocket fuels, to fill party balloons, and in the helium-neon laser as a source for the energy states of neon that show laser action.

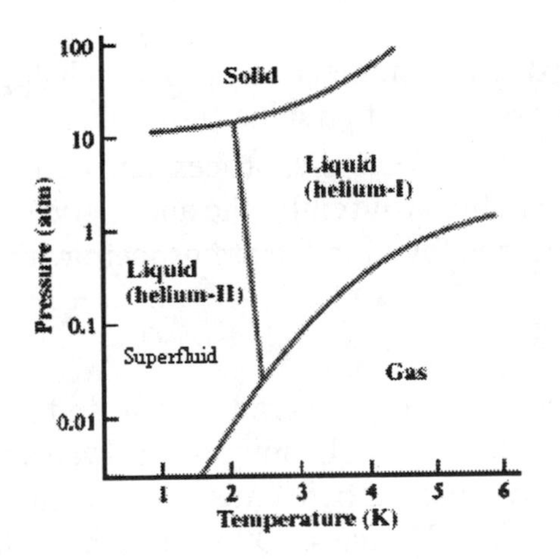

Example 15.11 At what temperature and pressure are the solid, liquid, and vapor phases of helium present at equilibrium? Consult the phase diagram shown above.

Solution On the phase diagram, there is no temperature and pressure at which helium solid, liquid, and vapor are all in equilibrium; in other words, helium does not have a standard triple point.

- **Neon**

→ *Neon* derives its name from the Greek word *neos*, new. It is widely used in advertising display signs because it emits an orange-red glow when an electric current passes through it. Of all the noble gases, the discharge of neon is the most intense at ordinary voltages and currents. The neon atom is the source of laser light in the helium-neon laser.

→ *Neon* has over 40 times more refrigerating capacity per unit volume than liquid helium and more than three times that of liquid hydrogen. It is compact, inert, and less expensive than helium when used as a *cryogenic* coolant. Neon is found in the atmosphere with a concentration of 18.2 ppm.

- **Argon**

→ *Argon* derives its name from the Greek word *argos*, inactive. Its main use is used to fill electric light bulbs at a pressure of about 400 Pa, where it conducts heat away from the filament. It is also used to fill fluorescent tubes.

→ *Argon* is also used as an inert gas shield (to prevent oxidation) for arc welding and cutting, and as a protective atmosphere for growing silicon and germanium crystals. Argon is the most abundant noble gas and is found in the atmosphere with a concentration of about 0.93% (or about 1 part per hundred by volume).

- **Krypton**

→ *Krypton* derives its name from the Greek word *kryptos*, hidden. It gives an intense white light when an electric discharge is passed through it, and it is used in airport runway lighting.

→ *Krypton* is produced in nuclear fission, and its atmospheric abundance is a measure of worldwide nuclear activity. Krypton is found in the atmosphere with a concentration of about 1 ppm.

- **Xenon**

→ *Xenon* derives its name from the Greek word *xenos*, stranger. It is used in halogen lamps for automobile headlights and in high-speed photographic flash tubes.

→ *Xenon* is used in the nuclear energy area in bubble chambers, probes, and other applications where a high atomic mass is desirable. Currently, the anesthetic properties of xenon are being explored. Xenon is found in the atmosphere with a concentration of about 0.087 ppm.

- **Radon**

→ *Radon* derives its name from the element *radium*. The gas is radioactive and is formed by radioactive decay processes deep in the earth. Uranium-238 decays very slowly to radium-226, which further decays by alpha particle emission to radon-222 (see Section 17.4). The half-life of radon-222 is 3.825 days. Other shorter-lived isotopes are formed from the decay of thorium-232 and uranium-235. Every square mile of soil to a depth of 6 inches is estimated to contain about 1 g of radium, which releases radon in tiny amounts into the atmosphere.

→ *Radon* is used in implant seeds for the therapeutic treatment of localized tumors. Radon is found in the atmosphere with a concentration estimated to be 1 in 10^{21} parts of air.

15.12 Compounds of the Noble Gases

- **Helium, neon, and argon**
  → The ionization energies of helium, neon, and argon are very high. They form no stable *neutral* compounds.
  → If electrons are removed from noble gas atoms in ionization processes, the resulting noble gas ions are radicals, and they may form stable molecular ions such as $NeAr^+$, ArH^+, and $HeNe^+$ in the gas phase. If an electron is added to them, these ions rapidly dissociate into neutral atoms.

- **Krypton**
  → *Krypton* forms a thermodynamically unstable neutral compound, KrF_2 (recall that diamond is also thermodynamically unstable). It is a volatile white solid, which decomposes slowly at room temperature.
  → In 1988, a compound with a Kr–N bond was discovered, but it is stable only at temperatures below −50°C.

Example 15.12a What is the oxidation number of krypton in KrF_2? Is KrF_2 expected to be linear?

Solution Fluorine is assigned an oxidation number of −1, so the oxidation number of Kr in KrF_2 is +2. To use the VSEPR theory to predict the shape of the molecule, we first determine the number of valence electrons in the molecule. A Kr atom has eight valence electrons and an F atom has seven. The total number of valence electrons in the molecule is then 8 + 2(7) = 22. Fluorine obeys the octet rule, so in the Lewis diagram for KrF_2 there are five electron pairs about the central Kr atom as shown.

:F̤̈—K̈r̤̈—F̤̈:

The geometry of the five pairs of electrons around Kr is trigonal bipyramidal. To minimize the strong electron repulsion associated with lone pairs, we place the lone pair electrons in the three equatorial positions. The F atoms occupy the two axial positions and the shape of the molecule is *linear*.

- **Xenon**
  → *Xenon* is the noble gas element with the richest chemistry. It forms several compounds with fluorine and oxygen, and even compounds with Xe–N and Xe–C bonds. The compound, XeF_2, is thermodynamically stable. Xenon is found in compounds with oxidation numbers of +2, +4, +6, and +8.
  → Direct reaction of *fluorine* with *xenon* at high temperature results in the formation of compounds with oxidation numbers of +2 (XeF_2), +4 (XeF_4), and +6 (XeF_6). In the case of XeF_6, high pressure is required. All three fluorides are crystalline solids. In the gas phase, all are molecular compounds. Solid *xenon hexafluoride* is an ionic compound with a complex structure of XeF_5^+ cations bridged by F^- anions.
  → *Xenon fluorides* are powerful *fluorinating agents*, reagents that attach fluorine to other substances. *Xenon trioxide* is synthesized by the hydrolysis of *xenon tetrafluoride*:

$$6\,XeF_4(s) + 12\,H_2O(l) \rightarrow 2\,XeO_3(aq) + 4\,Xe(g) + 3\,O_2(g) + 24\,HF(aq)$$

The trioxide is the anhydride of *xenic acid*, H_2XeO_4. In aqueous basic solution, the acid forms the *hydrogen xenate ion*, $HXeO_4^-$, which further disproportionates to the *perxenate ion*, XeO_6^{4-}, in which xenon attains its highest oxidation number of +8.

$$2\,HXeO_4^-(aq) + 2\,OH^-(aq) \rightarrow XeO_6^{4-}(aq) + Xe(g) + O_2(g) + 2\,H_2O(l)$$

Example 15.12b What is the oxidation state of xenon in XeF_4? What shape would you predict for this molecule?

Solution Fluorine is assigned an oxidation number of −1, so the oxidation number of Xe in XeF_4 is +4. To use the VSEPR theory to predict the shape of the molecule, we first determine the number of valence electrons in the molecule. A Xe atom has eight valence electrons and an F atom has seven. The total number of valence electrons in the molecule is then 8 + 4(7) = 36. Fluorine obeys the octet rule, so in the Lewis diagram for XeF_4 there are six electron pairs (two of which are lone pairs) about the central Xe atom as shown.

The geometry of the six pairs of electrons around Xe is octahedral. To minimize the strong electron repulsion associated with lone pairs, we place the lone pair electrons in less confining axial positions (opposite corners). The F atoms occupy the four equatorial positions and the shape of the molecule is *square planar*.

- **Radon**
  - → *Radon* chemistry is difficult to study because all of its isotopes are radioactive.
  - → A *radon fluoride* of unknown composition is found to form readily, but it decomposes rapidly during attempts to vaporize it for further study.

The Impact on Materials (Sections 15.13-15.14)

Key Concepts

colloids, aerosol, sol or gel, emulsion, foam, solid dispersion, solid emulsion, aqueous colloids, hydrophilic and hydrophobic colloids, Brownian motion, clays, incandescence, chemiluminescence, bioluminescence, fluorescence, phosphorescence, triboluminescence, phosphors

Overview

- **Colloids, clays, and gels**
  - → *Colloids* are *dispersions* or *suspensions* of particles with diameters between 1 nm and 1 μm in a gas, liquid, or solid. The particles are large enough to scatter light.
  - → *Clays* used commercially are *colloidal oxides* of silicon, aluminum, and magnesium.
  - → *Gels* are soft, solid colloids.

- **Phosphors and other luminescent materials**
  - → *Incandescence* is light emitted by a hot body; *luminescence* is emission of light by other processes.
  - → *Chemiluminescence* and *bioluminescence* are processes in which light is emitted by the energetically excited products of a chemical reaction.
  - → *Fluorescence* is the *prompt* emission of light from molecules excited by radiation of higher frequency than the light emitted. *Phosphorescence* is the *slow* or *delayed* emission of light from radiated molecules. *Triboluminescence* is the emission of light resulting from crystals subject to mechanical shock.
  - → *Phosphors* are materials that glow when activated by the impact of fast electrons.

15.13 Colloids, Clays, and Gels

- **Colloids**
  - → Particles (*dispersed phase*) with lengths or diameters between 1 nm and 1 μm *dispersed* or *suspended* in a *dispersion medium* (gas, liquid, or solid solvent)
  - → *Classification of colloids* is given in Table 15.10 in the text.
  - → *Aerosols* are solids dispersed in a gas (smoke) *or* liquids dispersed in a gas (hairspray, mist, fog).
  - → *Sols* or *gels* are solids dispersed in a liquid (printing ink, paint).
  - → *Emulsions* are liquids dispersed in a liquid (milk, mayonnaise). *Solid emulsions* are liquids dispersed in a solid (ice cream).
  - → *Foams* are gases dispersed in a liquid (soapsuds). *Solid foams* are gases dispersed in a solid (Styrofoam).
  - → *Solid dispersions* are solids dispersed in a solid (ruby glass, some alloys).

Example 15.13 Suggest a physical method to demonstrate that whole milk is an emulsion (liquid dispersed in a liquid) and not a solution.

Solution One way would be to use a centrifuge. Upon centrifugation, the milk would separate into an aqueous phase and a nonaqueous one. A true solution is not separable by centrifugation.

- **Aqueous colloids**
  - → *Hydrophilic colloids* contain molecules with *polar groups* that are strongly attracted to water. Examples include proteins that form gels and puddings.
  - → *Hydrophobic colloids* contain molecules with *nonpolar groups* that are only weakly attracted to water. Examples include fats that form emulsions (milk and mayonnaise).
  - → Rapid mixing of silver nitrate and sodium bromide may produce a *hydrophobic colloidal suspension* rather than a precipitate of silver bromide. The tiny silver bromide particles are kept from further aggregation by *Brownian motion*, the motion of small particles resulting from constant collisions with solvent molecules. The *sol* is further stabilized by *adsorption* of ions on the surfaces of the particles. The adsorbed ions are hydrated by surrounding water molecules and help prevent further aggregation.

- **Clays**
  - → The structure of a flake of *clay* and its internal structure are like a somewhat disorganized stack of papers. Sheets of tetrahedral silicate units or octahedral units of aluminum or magnesium oxides are separated by water layers that bind the layers of the flake together. Each clay particle is surrounded by a double layer of ions that separate the flakes by repelling the like charges on the other flakes. The layers are able to slide past one another, providing flexibility in response to stress. Clays are easily molded.
  - → When baked in a kiln, *clays* form hard, tough *ceramic materials*. Water is driven out and strong chemical bonds form between the flakes (see Section 14.24).

15.14 Phosphors and Other Luminescent Materials

- **Light emission from materials**
  - → *Incandescence* is light emission from a heated object, such as the filament in a lamp or the particles of hot soot in a candle flame. See the discussion of *black-body radiation* in Section 1.2.
  - → *Luminescence* is light emission from materials caused by other processes, such as light absorption, chemical reaction, impact with electrons, radioactivity, or mechanical shock.
  - → *Chemiluminescence* is light emission from materials caused by chemical reaction. The products of reaction are formed in energetically exited states that decay by light emission. The highly exothermic reaction of hydrogen atoms and fluorine molecules provides an example of *infrared chemiluminescence.* Hydrogen fluoride is produced with considerable *vibrational* energy, and it decays by *emission* in the *infrared* region of the spectrum (see Major Technique 1 for a discussion of *infrared absorption*). *Visible chemiluminescence* is usually produced by *electronic* excitation of product molecules.
  - → *Bioluminescence* is a form of chemiluminescence produced by living organisms, such as fireflies and certain bacteria.

- **Light emission from molecules**
  - → *Fluorescence* is the *prompt* emission of light from molecules excited by radiation of higher frequency. Normally, absorption of ultraviolet radiation by a molecule leads to emission in the visible region; higher frequency *absorption* leads to lower frequency *emission*. [**Note:** With high-power lasers, simultaneous absorption of two or more photons by a molecule may lead to emission of a single photon with a *greater* frequency than that of the individual photons absorbed.]
  - → *Phosphorescence* is the *slow* or *delayed* emission of light from molecules excited by radiation of higher frequency. In this case, the initially excited molecule undergoes a transition to a state that decays more slowly.
  - → *Triboluminescence* (from the Greek word, *tribos*, a rubbing) is luminescence that is produced by a mechanical shock to a crystal. It is readily observed in striking or grinding sugar crystals. Trapped nitrogen gas escapes while in an excited state and produces the radiation.

- **Phosphors**
  - → *Phosphorescent materials* that glow when activated by the impact of fast electrons, as well as high frequency radiation such as ultraviolet or X-ray
  - → Clusters of three *phosphors* are used for each dot in a color television or computer display screen. Commonly used phosphors for this purpose are europium-activated yttrium orthovanadate, YVO_4, for the red color, silver-activated zinc sulfide for blue, and copper-activated zinc sulfide for green.
  - → *Fluorescent lamps* make use of fluorescent materials that are activated by ultraviolet light. Mercury atoms in the lamp are excited by electrons in a discharge. The atoms emit light at 254 and 185 nm, which is absorbed by a phosphor thinly coated on the surface of the lamp. A commonly used phosphor is *calcium halophosphate*, $Ca_5(PO_4)_3F_{1-x}Cl_x$, doped with manganese(II) and antimony(III) ions. An antimony(III)-activated phosphor emits blue light and a manganese(II)-activated phosphor emits yellow light. The net result is light with a spectral range that is approximately white.
  - → *Fluorescent materials* have important applications in medical research. Dyes such as *fluorescein* (see Margin Figure (**17**) in Section 15.14 in the text) are attached to protein molecules to probe biological reactions. *Fluorescent materials*, such as sodium iodide and zinc sulfide, can be activated by radioactivity and are used in scintillation counters to measure radiation (see Box 17.1 in the text).

Chapter 16 The *d* Block: Metals in Transition

The *d*-Block Elements and Their Compounds (Sections 16.1-16.2)

Key Concepts

d-block elements, transition metals, lanthanides, actinides, lanthanide contraction, paramagnetism, ferromagnetism, domains, multiple oxidation states

Overview

- ***d*-Block elements**
  - → Elements from Group 3 through Group 12, corresponding to filling the *nd* atomic orbitals in the periodic table
  - → For example, the electron configuration of the *d*-block elements in Period 5 range from $[Kr]4d^15s^2$ for yttrium, Y, to $[Kr]4d^{10}5s^2$ for cadmium, Cd.
  - → All are metals, many with practical structural uses, such as Fe, Ti, and Cu.
  - → All have incompletely filled *d*-subshells, except for Group 11 and 12 metals.
  - → Many form *paramagnetic* compounds.
  - → Many form brightly-colored compounds.
- **Transition metals**
  - → Metals in Groups 3 to 11
  - → Bridge the gap between the highly reactive Group 1 and 2 metals (*s*-block elements) and the elements beyond Group 11 (Group 12 and the *p*-block elements)
- ***f*-Block elements**
  - → Lie between Groups 2 and 3 of the periodic table in Periods 6 and 7
  - → Consist of 14 elements in which the seven 4*f*-orbitals fill, the *lanthanides* of Period 6, and 14 more in which the 5*f*-orbitals fill, the *actinides* of Period 7, (see figure 16.1 in the text)
  - → The *lanthanides* and *actinides* together are sometimes called *inner transition metals*.
  - → Are responsible for the unusually small size of the *d*-block elements in Period 6 (*lanthanide contraction*)
- **Goals** → To develop insight into the *structure*, *bonding*, *properties*, and *reactivity* of the *d*-block elements and their compounds

16.1 Trends in Physical Properties

- ***d*-Block elements**
  - → All are metals.
  - → Most are good conductors of electricity, particularly silver, copper, and gold.
  - → Most are malleable, ductile (can be drawn into wire), lustrous, and silver-white in color; exceptions include copper (red-brown) and gold (yellow).
  - → Most have higher melting and boiling points than main group elements; a major exception is mercury, which is a liquid at room temperature.
- **Shapes of the *d*-orbitals**
  - → Some of the properties of *d*-block elements follow from the shapes of the *d*-orbitals, which have *lobes* that occupy different regions of space.
  - → The *lobes* are relatively far apart from each other, so electrons in different *d*-orbitals on the same atom repel each other weakly.
  - → Electron density in *d*-orbitals is low near the nucleus. Because *d*-electrons are relatively far from the nucleus, they are not very effective in shielding other electrons from its positive charge.
- **Trends in atomic radii**
  - → *Nuclear charge* and the *number* of *d*-electrons both increase across a row from left to right. The first five electrons added to a *d*-subshell are placed in different orbitals, according to Hund's rule. Because the repulsion between *d*-electrons in different orbitals is relatively small, the increasing *effective nuclear charge* is the dominant factor, so the atoms become smaller. An exception is manganese, shown in Fig. 16.2 in the text.
  - → Further across the block, the radii begin to *increase* slightly as the atomic number is increased. In this case, *electron-electron repulsion* outweighs the effect of increasing *effective nuclear charge*, so the atoms become larger. The attractions and repulsions are finely balanced, and the range of atomic *d*-metal radii is not very large. The *d*-metals form a wide range of alloys, because some of the atoms of one metal can easily replace atoms of another metal in a crystal lattice.
  - → The *d*-metals in Period 5 are typically larger than the ones in Period 4 (see Fig. 16.4 in the text). The effective nuclear charge on the outer electrons in a group of elements is roughly the same, whereas the n quantum number of the outer electrons increases down the group. The usual pattern is an increase in size for group members from the top to the bottom. This pattern is followed in Period 4 and Period 5 for the *d*-metals, except for manganese.
  - → The radii of *d*-metals in Period 6 are roughly the same as in Period 4; thus, they are smaller than expected. This effect is caused by the *lanthanide contraction*, the decrease in radius along the first row of the *f* block. There are 14 *f*-block elements before the first *d*-block element in Period 6 is reached. When the *d* block resumes at lutetium, the atomic radius has fallen from 224 pm for barium to 172 pm for lutetium.

→ A consequence of *lanthanide contraction* is that the *d*-block elements in Period 6 are substantially more dense than ones in Period 5. The radii are about the same, but the masses are almost twice as large. Iridium and osmium are the two elements with the greatest density, approximately 22.6 $g{\cdot}cm^{-3}$ (see Fig. 16.5 in the text). Another effect of the *contraction* is the *low reactivity* of platinum and gold. Their valence electrons are relatively close to the nucleus and are so tightly bound that they are not readily available for chemical reaction.

Example 16.1a The atomic radii of cobalt, Co, and rhodium, Rh, are 125 and 134 pm, respectively. Estimate the atomic radius of the next group member, iridium, Ir.

Solution The elements Co, Rh and Ir are metals from Group 9 in Periods 4, 5, and 6, respectively. Between Co and Rh, there is an increase of 9 pm in the atomic radius. In the absence of the *lanthanide contraction*, we may expect a similar increase between Rh and Ir, making its radius approximately 143 pm. However, the *lanthanide contraction* approximately equalizes the radii between the *d*-block elements in Periods 5 and 6. Therefore, we expect the radius of Ir to be about the same as that of Rh, 134 pm. The measured radius is 136pm, close to our predicted value.

Example 16.1b The densities of Ni, Pd, and Pt metals in Group 10 are 8.91, 12.00, and 21.45 $g{\cdot}cm^{-3}$, respectively. Explain the relatively large increase in density in going from Pd to Pt.

Solution The difference in density between Ni and Pd is 3.09 $g{\cdot}cm^{-3}$. Based on this difference, we may expect the density of Pd to be approximately 12.00 + 3.09 = 15.09 $g{\cdot}cm^{-3}$. It is much larger than this value because of *lanthanide contraction.* The atomic radii of Pd and Pt are each equal to 138 pm; therefore these atoms have approximately equal molar volumes. The molar masses of Pd and Pt are 106.4 and 195.09 $g{\cdot}mol^{-1}$, respectively, approximately the same ratio as their densities. Therefore, platinum packs much more mass into the same volume as palladium, leading to its higher density.

- **Magnetic properties**

  → Many *d*-block elements have unpaired electrons in the *d*-subshell, a property leading to *paramagnetism* or *ferromagnetism*.

  → *Paramagnetism* arises if a substance has at least one unpaired electron. If so, it is attracted into a magnetic field. The unpaired electrons on neighboring atoms or molecules are aligned *randomly* with respect to each other (see Fig. 16.6a in the text), and *paramagnetism* is a very weak property.

  → *Ferromagnetism* arises when unpaired electrons on neighboring atoms or molecules are aligned *in the same direction* with respect to each other in regions of *aligned spins* called *domains* (see Fig. 16.6b in the text). Because of this alignment, *ferromagnetism* is a very strong property.

  → *Ferromagnetism* occurs in metals such as Fe, Co, and Ni, and in some compounds such as the oxides of iron and chromium, which are used in magnetic tape recorders. These materials are *magnetized* by an applied field. Only *ferromagnetic* substances retain their magnetism even when the field is turned off (see Box 3.2 in the text).

16.2 Trends in Chemical Properties

- **Natural occurrence**
  - → Elements on the *left* side of the *d* block are commonly found as oxides, halides, or oxoanion minerals, including carbonates and silicates.
  - → Elements on the *right* side of the *d* block are commonly found as sulfides.
- **Oxidation states (numbers)**
  - → Many chemists commonly use the term *oxidation state* in place of *oxidation number.* We will ignore the subtle distinction between the two terms (Section K in the text).
  - → Most *d*-block elements have more than one common oxidation state and one or more less common oxidation states (see Fig. 16.7 in the text).
  - → Oxidation states range from negative values to +8. In $[Fe(CO)_4]^{2-}$, Fe has an oxidation state of −2, whereas in OsO_4, Os has an oxidation state of +8.
  - → Oxidation states show greatest variability for elements in the middle of the *d* block. As shown in Fig. 16.7, Mn, at the center of its row, has seven oxidation states, whereas Sc, at the beginning of the same row, has only one.
  - → Elements in the second (Period 5) and third row (Period 6) of the *d* block tend to reach higher oxidation states than those in the first row (Period 4). For example, the highest oxidation state of Ni (Group 10, first row) is +4, whereas that of Pt (Group 10, third row) is +6.
- **Chemical properties and oxidation states**
  - → Species containing *d*-block elements with *high oxidation states* tend to be good *oxidizing agents* (MnO_4^- with Mn +7 and CrO_4^{2-} with Cr +6 are examples); to exhibit *covalent bonding* (Mn_2O_7 with Mn +7 is a covalent liquid at room temperature); and to have *oxides* that are *acid anhydrides.*
  - → Species containing *d*-block elements with *low oxidation states* tend to be good *reducing agents* (CrO with Cr +2 and $FeCl_2$ with Fe +2 are examples); to exhibit *ionic bonding* (Mn_3O_4, with Mn +2 and Mn +3 is an ionic solid); and to have *oxides* that are *basic anhydrides.*

Example 16.2a Predict whether the orange solid, V_2O_5, is a good oxidizing agent, reducing agent, or neither. Will it dissolve best in basic or acidic solution?

Solution In V_2O_5, vanadium is in the +5 oxidation state, the *highest common oxidation state* for V in Fig. 16.7. Because vanadium is in a high oxidation state, V_2O_5 is expected to be a good *oxidizing agent* and an *acid anhydride.* As an acid anhydride, it is expected to dissolve in *basic* solution.

Example 16.2b Only one of the following three compounds is basic: CrO_3(s), Mn_2O_7(l), or NiO(s). Which one is it?

Solution The oxidation states of Cr, Mn, and Ni in the three compounds listed above are +6, +7, and +2, respectively. Compounds in low oxidation states tend to form *basic anhydrides* and Ni in NiO(s) has the lowest oxidation number. Therefore, NiO(s) is expected to be the *basic anhydride.*

Selected Elements: A Survey (Sections 16.3-16.4)

Key Concepts

properties of the metals, properties of the compounds, alloy, ferroalloy, bronze, pyrometallurgical process, hydrometallurgical process

Overview

- **Survey of *d*-block elements**
  - → The properties of the *d*-block elements are quite diverse.
  - → A few elements are chosen as representative members of the block.
  - → An important goal is to develop an appreciation of the characteristics and range of behavior of these elements.

16.3 Scandium Through Nickel

- **Properties of the metals (see Table 16.1 in the text)**
  - → Densities increase from scandium (2.99 $g{\cdot}cm^{-3}$) to nickel (8.91 $g{\cdot}cm^{-3}$).
  - → Melting points increase from scandium (1540 K) to vanadium (1920 K), then decrease uniformly to nickel (1455 K), with the exception of the very low value of manganese (1250 K).
  - → Boiling points increase from scandium (2800 K) to vanadium (3400K), then show irregular behavior to nickel (2150 K).
  - → Valence electron configurations follow the filling rule: 4*s* fills before 3*d* from scandium, $[Ar]3d^1 4s^2$, to nickel, $[Ar]3d^8 4s^2$, with the exception of chromium, $[Ar]3d^5 4s^1$.

- **Scandium, Sc, $[Ar]3d^1 4s^2$**
  - → A highly reactive metal that reacts with water as vigorously as calcium
  - → Has only a few commercial uses and is *not* essential to life
  - → In compounds, only the +3 oxidation state of scandium is found.
  - → The small, highly charged Sc^{3+} ion is strongly hydrated in water, and the aqueous ion, $Sc(H_2O)_6^{3+}$, is a weak acid (about as strong as acetic acid).

- **Titanium, Ti, $[Ar]3d^2 4s^2$**
  - → A light, strong metal passivated by an oxide coating (like Al), which masks its inherent reactivity
  - → Found in the +4 oxidation state (most common) in its ores: *rutile*, TiO_2, and *ilmenite*, $FeTiO_2$
  - → Requires strong reducing agents for extraction from its ores
  - → Metal used in jet engines, dental appliances; oxide, TiO_2, used as a paint pigment
  - → Forms series of oxides called *titanates*
  - → *Barium titanate*, $BaTiO_3$, is a *piezoelectric* material, one that develops an electrical signal when stressed.

Example 16.3a *Titanium(IV) chloride*, $TiCl_4$, is a liquid at room temperature and its melting point is −23°C. Solid $TiO_2(s)$ melts at 1750°C. Which aspects of the chemical bonding in these species help explain the dramatic melting point differences?

Solution The oxidation number of titanium is +4 in both compounds. The chloride ion, Cl^- (181 pm), is more *polarizable* than the oxide ion, O^{2-} (140 pm). The highly *polarizing* Ti^{4+} ion distorts the electrons in the chloride ions more effectively than in the oxide ions. The bonding is expected to be more covalent in $TiCl_4$ and more ionic in TiO_2. Covalent molecules form solids that are held together mainly by London forces, which are much weaker than the ionic forces that hold ionic solids together. Therefore, a much lower melting point is expected for $TiCl_4$, as is observed.

- **Vanadium, V, [Ar] $3d^34s^2$**
  - → A soft, silver-gray metal
  - → Has many commercial uses and it is thought to be essential for life
  - → Used in *ferroalloys* (Fe, V, C) to make tough steels for automotive springs
- **Compounds of vanadium**
  - → Positive oxidation states range from +1 to +5.
  - → Common oxidation states are +4 and +5.
  - → Many colored compounds, including the blue of the *vanadyl ion*, VO^{2+}, are used as ceramic glazes.
  - → *Vanadium(V) oxide*, V_2O_5, commonly called *vanadium pentoxide*, is the most important compound. The orange-yellow solid is used as an oxidizing agent and as an oxidizing catalyst in the contact process for the manufacture of sulfuric acid.

Example 16.3b The *decavanadate ion*, $V_{10}O_{28}^{6-}$, reacts in acid solution to form the *dioxovanadium ion*, VO_2^+. What is the oxidation state of vanadium in each of these species? Which oxide of vanadium has the same oxidation state as V in $V_{10}O_{28}^{6-}$? Write a balanced equation for the reaction.

Answer The oxidation state of vanadium is +5 in both species.
In *vanadium pentoxide*, $V_2O_5(s)$, vanadium has an oxidation state of +5.
The balanced equation for the reaction is

$$V_{10}O_{28}^{6-}(aq) + 16\,H_3O^+(aq) \rightarrow 10\,VO_2^+(aq) + 24\,H_2O(l)$$

- **Chromium, Cr, [Ar] $3d^54s^1$**
  - → A bright, corrosion-resistant metal
  - → Used to make stainless steel and for chromium plating
  - → An exception to the normal filling order of electrons in the periodic table
- **Compounds of chromium**
  - → Positive oxidation states range from +1 to +6.
  - → Common oxidation states are +3 and +6.
  - → *Chromium(IV) oxide*, CrO_2, is used as a *ferromagnetic* coating for "chrome" magnetic tapes. It responds better to high frequency magnetic fields than "ferric" tapes, Fe_2O_3.

→ Yellow *sodium chromate*, $Na_2CrO_4(s)$, is used as a source for many other chromium compounds. In acid solution, the *chromate ion*, CrO_4^{2-}, changes into the *dichromate ion*, $Cr_2O_7^{2-}$. The oxidation state of chromium is +6 in *both* ions.

$$2\,CrO_4^{2-}(aq) + 2\,H_3O^+(aq) \rightarrow Cr_2O_7^{2-}(aq) + 3\,H_2O(l)$$

→ Chromium compounds are used as pigments, corrosion inhibitors, fungicides, and ceramic glazes. *Chromium(III)* is essential to human health, possibly playing a role in the regulation of glucose metabolism.

Example 16.3c Is an aqueous solution of Cr^{+3} likely to be basic, neutral, or acidic?

Solution The Cr^{3+} ion is small and highly charged, and is expected to be solvated by water. The Cr–O interactions are relatively strong, and weaken the O–H bonds in the solvated water molecules. The value of pK_a is 4 for the first dissociation of $[Cr(H_2O)_6]^{3+}$, which makes Cr^{3+} a slightly stronger acid than acetic acid. The reaction is

$$[Cr(H_2O)_6]^{3+}(aq) + H_2O(l) \rightleftharpoons [Cr(H_2O)_5(OH)]^{2+}(aq) + H_3O^+(aq)$$

- **Manganese, Mn, [Ar] $3d^5 4s^2$**
  - → A gray metal that corrodes easily and is rarely used alone
  - → Important in alloys such as steel (see Table 16.2 in the text) and bronze
  - → Obtained from the ore, *pyrolusite*, mostly MnO_2
- **Compounds of manganese**
  - → Positive oxidation states range from +1 to +7.
  - → The most stable oxidation state is +2, but +4, +7, and, to some extent, +3 are also common in manganese compounds.
  - → The most important compound of manganese is *maganese(IV) oxide*, MnO_2, commonly called *manganese dioxide.* This compound is a brown-black solid used as a decolorizer to reduce the green tint of glass, as the source of other manganese compounds, and in dry cells.
  - → The *permanganate ion*, MnO_4^-, is an important oxidizing agent and a mild disinfectant. The oxidation number of manganese attains its highest value of +7 in this anion.
- **Iron, Fe, [Ar] $3d^6 4s^2$**
  - → Most abundant element on earth and most widely used of the *d*-metals
  - → Obtained from the principal ores *hematite*, Fe_2O_3, and *magnetite*, Fe_3O_4
  - → Reactive metal that readily corrodes in moist air forming rust
  - → Forms corrosion-resistant alloys, steel (see Table 16.2 in the text)
  - → A *ferromagnetic* metal, as are its oxides and many alloys
- **Compounds of iron**
  - → Positive oxidation states range from +2 to +6.
  - → Common oxidation states are +2 and +3.
  - → Salts of iron vary in color from pale yellow to dark green.

→ Colors of aqueous solutions are dominated by the $[FeOH(H_2O)_5]^{2+}$ ion, the conjugate base of $[Fe(H_2O)_6]^{3+}$.

→ *Iron(II)* is readily oxidized to *iron(III)*; the reaction is slow in acid solution and rapid in base.

→ Iron is essential for human health. It is present in ionic form in hemoglobin and in iron-containing proteins. Iron deficiency, *anemia*, results in reduced transport of oxygen to the brain and muscles; an early symptom is chronic fatigue.

- **Cobalt, Co, [Ar] $3d^7 4s^2$**

→ A silver-gray metal used mainly in iron alloys

→ Permanent magnets like those used in loudspeakers are made of *alnico steel*, an alloy of iron, nickel, cobalt, and aluminum.

→ Cobalt steels are hard and are used for drill bits and surgical tools

- **Compounds of cobalt**

→ Positive oxidation states range from +1 to +4.

→ Common oxidaiion states are +2 and +3.

→ *Cobalt(II) oxide*, CoO(s), ia a deep blue salt used to color glass and ceramic glazes.

→ Cobalt is present in the coenzyme, Vitamin B_{12}, an essential nutrient in human diets.

- **Nickel, Ni, [Ar] $3d^8 4s^2$**

→ A hard, silver-white metal used mainly for the production of stainless steel and for alloying with copper to produce *cupronickels*, the alloys used for nickel coins

→ Used as a catalyst for the *hydrogenation* of organic molecules

→ Obtained as a pure metal by the *Mond process* (the first step requires heat):

$$NiO(s,ore) + H_2(g) \rightarrow Ni(s, impure) + H_2O(g)$$
$$Ni(s, impure) + 4\,CO(g) \rightarrow Ni(CO)_4(l, pure)$$
$$Ni(CO)_4(l, pure) \rightarrow Ni(s, pure) + 4\,CO(g)$$

- **Compounds of nickel**

→ Positive oxidation states range from +1 to +4.

→ Common oxidation states are +2 and +3; +2 is most stable.

→ The green color of aqueous solutions of nickel salts is caused by $[Ni(H_2O)_6]^{2+}$ ions.

→ In nickel-cadmium (nicad) batteries, *nickel(III)* is reduced to *nickel(II)*.

- **Carbonyl compounds**

→ Many transition elements form compounds with carbon monoxide, CO, in which the *formal* oxidation state of the element is 0. This value is *not* included in the oxidation state ranges shown in Fig. 16.7 in the text. In these *carbonyl compounds*, both the metal atom and carbon monoxide molecule are regarded as *nearly* neutral species; the metal is assigned an oxidation state of 0. The CO group is not very electronegative and the number of electrons surrounding the metal in the compound is *approximately* the same as in the free metal.

→ When iron is heated in carbon monoxide, it reacts to form *iron pentacarbonyl*, $Fe(CO)_5$, a yellow molecular liquid that has a melting point of –20°C and a boiling point of 103°C. It decomposes in the presence of visible light. The shape of the molecule is *trigonal bipyramidal.*

→ Other examples of transition metal carbonyls are *nickel tetracarbonyl*, $Ni(CO)_4$, a colorless, toxic, flammable liquid that boils at 43°C and *chromium hexacarbonyl*, $Cr(CO)_6$, a colorless crystal that sublimes readily. The molecular shape of $Ni(CO)_4$ is *tetrahedral*; that of $Cr(CO)_6$ is *octahedral.*

→ Some metal carbonyl compounds can be reduced to produce anions in which the metal has a *negative* oxidation state. For example, in the $[Fe(CO)_4]^{2-}$ anion, produced by reduction of $Fe(CO)_5$, Fe has an oxidation state of –2:

$$Fe(CO)_5 \xrightarrow[\text{tetrahydrofuran}]{\text{Na}} [Fe(CO)_4]^{2-} + CO$$

Example 16.3d Nickel tetracarbonyl, $NI(CO)_4$, with a molar mass of 170.75 $g{\cdot}mol^{-1}$, melts at –25°C and is a liquid at room temperature, whereas nickel(II) bromide, $NiBr_2$, with a slightly higher molar mass of 218.53 $g{\cdot}mol^{-1}$, melts almost a thousand degrees higher at 963°C. Provide an explanation for this large difference in melting points.

Solution Nickel tetracarbonyl is composed of nearly neutral Ni atoms bonded to four CO molecules with small dipole moments (μ = 0.12 D). The molecule has *tetrahedral* shape and is *nonpolar*. Thus, $Ni(CO)_4$ molecules coalesce to form a solid by attraction derived from weak London forces, and the melting point of the liquid is expected to be low.

In contrast, nickel(II) bromide is an ionic solid with ions in the lattice held together by strong coulomb forces; consequently, its melting point is expected to be high.

16.4 Groups 11 and 12

- **Properties of the metals (see Table 16.3 in the text)**

→ All the elements of Groups 11 and 12 are metals with completely filled *d*-subshells.

→ The Group 11 metals, *copper, silver, and gold*, are called the *coinage metals*, and have valence electron configurations of $(n-1)d^{10}ns^1$. The low reactivity of the coinage metals derives partly from the poor shielding abilities of the *d*-electrons, and hence the strong attraction of the nucleus on the outermost electrons. The effect is enhanced in Period 6 by lanthanide contraction accounting for the inertness of gold.

→ The Group 12 metals, *zinc, cadmium, and mercury*, have valence electron configurations of $(n-1)d^{10}ns^2$. Zinc and, to a lesser extent, cadmium show some resemblance to beryllium or magnesium in their chemistry.

- **Copper, Cu, [Ar] $3d^{10}4s^1$**

→ Found as the free metal and in sulfide ores such as *chalcopyrite*, $CuFeS_2$

→ Used as an electrical conductor in wires

→ Forms alloys such as brass and bronze that are used in plumbing and casting of statues
→ Corrodes in moist air forming green *basic copper carbonate*

$$2\,Cu(s) + H_2O(l) + O_2(g) + CO_2(g) \rightarrow Cu_2(OH)_2CO_3(s)$$

→ Oxidized by *oxidizing* acids such as HNO_3

- **Compounds of copper**
  → Common oxidation states are +1 and +2, +2 is more stable.
  → *Copper(I)* disproportionates in water to metallic copper and copper(II).
  → *Copper(II)* is a pale blue ion in water, $[Cu(H_2O)_6]^{2+}$.
  → Copper is essential in animal metabolism. In humans, copper containing enzymes are necessary for healthy nerve and connective tissue. In the octopus and certain arthropods (such as lobsters), copper (not iron) is used to transport oxygen in the blood.

- **Silver, Ag, [Kr] $4d^{10}5s^1$**
  → Rarely found as the free metal
  → Obtained as a byproduct of the refining of copper and lead, and recycled in the photographic industry
  → Reacts with sulfur to produce a black tarnish on silver dishes and cutlery
  → Like copper, silver is oxidized by *oxidizing* acids.

$$3\,Ag(s) + 4\,H_3O^+(aq) + NO_3^-(aq) \rightarrow 3\,Ag^+(aq) + NO(g) + 6\,H_2O(l)$$

- **Compounds of silver**
  → Oxidation states are +1, +2, and +3, but +1 is most stable.
  → *Silver(I)* does *not* disproportionate in water.
  → Silver salts are insoluble in water except for $AgNO_3$ and AgF, which are soluble.
  → Silver halides are used in photographic film.

- **Gold, Au, [Xe] $4f^{14}5d^{10}6s^1$**
  → Gold is so inert that it is usually found as the metal.
  → Gold is highly malleable, easily made into thin foil (gold leaf).
  → Gold *cannot* be oxidized by nitric acid, but it dissolves in *aqua regia*, a mixture of sulfuric and hydrochloric acids.

$$Au(s) + 6\,H^+(aq) + 3\,NO_3^-(aq) + 4\,Cl^-(aq) \rightarrow AuCl_4^-(aq) + 3\,NO_2(g) + 3\,H_2O(l)$$

  → Gold is used in coins, jewelry, dental fillings, and for decorative ornamentation.

- **Compounds of gold**
  → Oxidation states are +1, +2, and +3, but +1 and +3 are most common.
  → Compounds are used to treat arthritis.
  → Gold has no known role in human health.

Example 16.3e Explain why gold metal is less reactive than silver metal.

Solution Gold and silver have the same atomic radii (144 pm, Fig. 16.4), an effect caused by lanthanide contraction. The atomic number of gold, 79, is much higher than that of silver, 47. With an additional 32 protons and similar size, gold is expected to hold its electrons more tightly than silver, making it more inert chemically.

- **Zinc, Zn, [Ar] $3d^{10}4s^2$**
  - → A silvery, reactive metal
  - → Found mainly in the sulfide ore, *sphalerite*, ZnS
  - → Obtained from the ore by *froth flotation*, followed by *smelting* with coke
  - → Used primarily for *galvanizing* iron
  - → Like copper, protected by a hard film of *basic carbonate*, $Zn_2(OH)_2CO_3$
- **Compounds of zinc**
  - → Common oxidation state is +2.
  - → An *amphoteric* metal, dissolving in both acidic and basic solutions

    Acid: $Zn(s) + 2\,H_3O^+(aq) \rightarrow Zn^{2+}(aq) + H_2(g) + 2\,H_2O(l)$

    Base: $Zn(s) + 2\,OH^-(aq) + 2\,H_2O(l) \rightarrow [Zn(OH)_4]^{2-} + H_2(g)$

    $Zn(OH)_4^{2-}$ is called the *zincate ion*. These reactions show that galvanized containers should not be used to transport either acids or alkalis.
  - → Zinc is an essential element for human health; it occurs in many enzymes. It is toxic only in very large amounts.
- **Cadmium, Cd, [Kr] $4d^{10}5s^2$**
  - → A silvery, reactive metal
  - → Obtained from its ore by *froth flotation*, followed by *smelting* with coke as is zinc
- **Compounds of cadmium**
  - → Common oxidation state is +2.
  - → *Cadmiate ions* (analogous to *zincate ions*) are known, but cadmium does not react with strong bases. Like zinc, cadmium reacts with nonoxidizing acids.
  - → Unlike zinc, cadmium salts are *deadly poisons*.
  - → Cadmium disrupts human metabolism by replacing other essential metals in the body, such as zinc and calcium, leading to soft bones and to kidney and lung disorders.
  - → Cadmium and zinc are alike chemically in many ways, but both are quite different from mercury.
- **Mercury, Hg, [Xe] $4f^{14}5d^{10}6s^2$**
  - → A volatile, silvery metal that is a liquid at room temperature (gallium and cesium are liquids on *warm* days).
  - → Found mainly in the sulfide ore, *cinnebar*, HgS
  - → Obtained from the ore by *froth flotation*, followed by *roasting* in air
  - → Vapor is poisonous.
  - → Combines with many metals to form mercury alloys called *amalgams* (an example is silver amalgams used to fill cavities in teeth)

→ The liquid temperature range of –39°C to its boiling point of 357°C is unusually wide and makes mercury well suited for use in thermometers, silent electrical switches, and high-vacuum pumps.

- **Compounds of mercury**
  → Oxidation states are +1, +2, and +3.
  → Most common oxidation states are +1 and +2.
  → *Mercury(I)* cation is a diatomic molecular ion, $[Hg{-}Hg]^{2+}$, with a covalent bond. It is usually written as Hg_2^{2+}.
  → Mercury does not react with nonoxidizing acids or with bases. It does react with oxidizing acids in a reaction similar to those for Cu and Ag.

$$3\,Hg(l) + 8\,H^+(aq) + 2\,NO_3^-(aq) \rightarrow 3\,Hg^{2+}(aq) + 2\,NO(g) + 4\,H_2O(l)$$

  → Mercury compounds, particularly organic ones, are acutely poisonous. Frequent exposure to low levels of mercury vapor results in the accumulation of mercury in the body. Damaging effects after such chronic exposure include impaired neurological function and hearing loss.

Coordination Compounds (Sections 16.5-16.7)

Key Concepts

coordination compound, complex, coordination sphere, coordination number, substitution reaction, ligand, shapes of complexes, tetrahedral complex, square-planar complex, octahedral complex, metallocene, monodentate, bidentate, polydentate, chelate, isomers, structural isomers, stereoisomers, ionization isomers, hydrate isomers, linkage isomers, coordination isomers, optical isomers, geometrical isomers, ambidentate ligands, chiral complex, enantiomers, achiral complex, optical activity, racemic mixture, racemate

Overview

- **Complex**
  → A complex is a species consisting of a central metal atom *or* ion attached to several molecules *or* ions (called *ligands*) by coordinate-covalent bonds. Examples include $[Cu(H_2O)_6]^{2+}$, $[Fe(CN)_6]^{4-}$, $Ni(CO)_4$, and $[Ag(NH_3)_2]^+$.
  → The central metal atom is regarded as a *Lewis acid* (electron pair acceptor), whereas the ligands are *Lewis bases* (electron pair donors). Complex formation is viewed as a *Lewis acid-base reaction.*
  → Complexes are formed by *d*-block and main-group metal ions. The *d*-metals form the richest array of complexes, and they are treated in Sections 16.5-16.14.
- **Coordination compounds**
  → Electrically neutral compounds, in which at least one of the ions is present as a complex, are called *coordination compounds*.

→ *Coordination compounds* may be neutral molecules or ionic solids. The molecular metal carbonyl, $Ni(CO)_4$, is an example of neutral complex species. Examples of ionic solids include $K_3[Fe(CN)_6]$, consisting of K^+ and $[Fe(CN)_6]^{3-}$ ions, $[Fe(H_2O)_5Cl]CO_3$, consisting of $[Fe(H_2O)_5Cl]^{2+}$ and CO_3^{2-} ions, and $K[Mn(CN)_5]$, consisting of K^+ and $[Mn(CN)_5]^-$ ions.

- **Goals** → An appreciation of the *nature of complexes* and *coordination compounds* is critical for understanding the chemistry and properties, such as color and magnetism, of the *d*-block elements.

16.5 The Nature of Complexes

- **Understanding complexes (terminology)**

→ *Coordination number* is the number of bonds formed by the central metal atom or ion.

→ *Ligand* is a molecule or ion attached to the metal atom or ion by a coordinate-covalent bond. A ligand *coordinates* to the metal when it attaches. If the complex is an ion, the ligands that are *directly* attached to the central atom are enclosed in *brackets*. The *charge* of the ion is displayed in the upper right corner as usual.

→ The ligands that are attached to the metal define the *coordination sphere* of the central atom or ion. Common *coordination spheres* are shown in Fig. 16.19, which is reproduced below. Almost all six-coordinate complexes are octahedral as shown in part (a). Four-coordinate complexes are either (b) tetrahedral or (c) square planar.

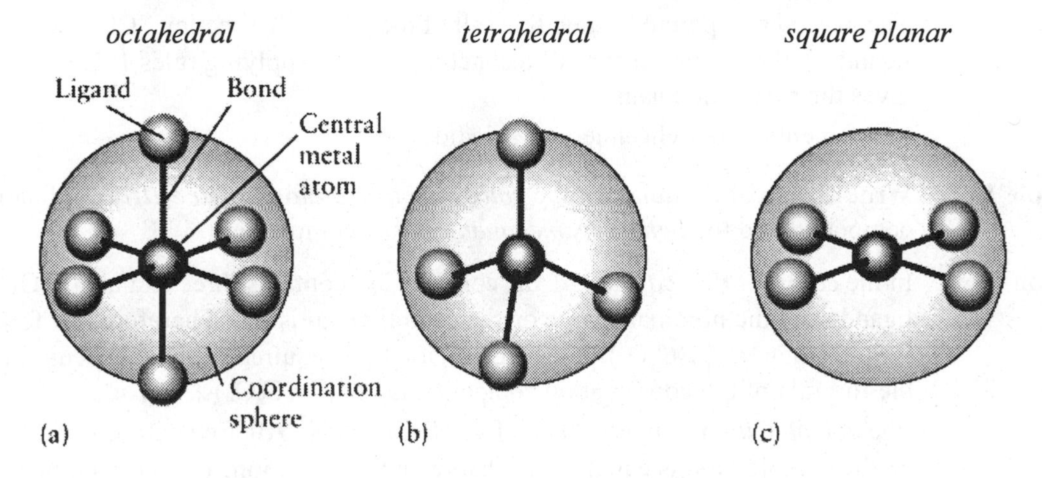

→ Examples of common ligands, showing the nonbonding electron pairs that can bond to a central metal atom, are:

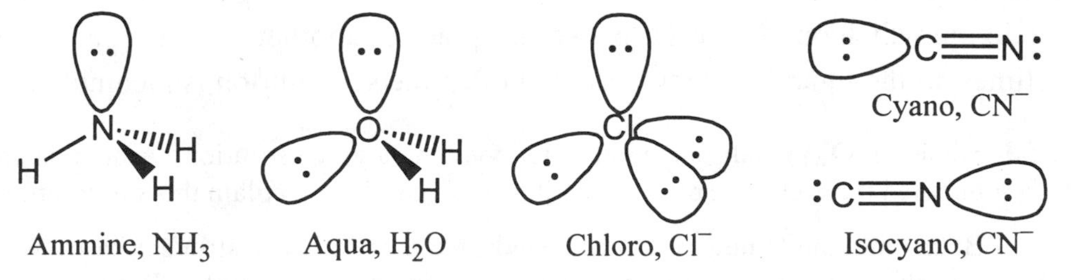

The *isocyano* ligand is attached to the metal by the electron pair on the nitrogen end.

- **Naming *d*-metal complexes and coordination compounds**
  → Use the formulas and names for the common ligands given in Table 16.4 in the text.
  → Follow the procedure outline in Toolbox 16.1 on p. 764 in the text.

Example 16.5a Name the ion, $[Fe(CN)_6]^{3-}$, and describe the properties of the complex.

Solution The *ligand* is the cyano group, CN^-, and there are six of them. The oxidation state of the metal, *iron,* is determined as follows. The *coordination number* is 6 and the total charge on all the ligands is then –6. Because the total charge on the *complex ion* is –3, the charge on iron must be +3. The name then follows from the number of ligands followed by the oxidation state of the metal, or the *hexacyanoferrate(III) ion*. The common name is the *ferricyanide* ion, because *iron(III)* is called *ferric*. A complex ion with six ligands usually has an *octahedral coordination geometry*. Note that the term *ferrate* is used for anions and *iron* for cations and neutral complexes.

Example 16.5b Give systematic names for the complex $[TiCl_2F_4]^{2-}$ and the coordination compound $[Tc(CO)_5]I$.

Solution The titanium complex ion has two *chloro* and four *fluoro* ligands, each with charge –1, for a total ligand charge of –6. To give an overall charge of –2 to the complex, Ti must be in the +4 oxidation state. Because the complex has an overall negative charge, it is given the suffix *ate*. Applying rules 1, 3, 4, 5, 7, and 8 from Toolbox 16.1 gives the systematic name:

dichlorotetrafluorotitanate(IV) ion

The neutral compound is an iodide salt of the $[Tc(CO)_5]^+$ cation. CO is a neutral ligand, so the oxidation state of technetium is +1. Applying rules 1, 2, 4, 7, and 9 gives the systematic name:

pentacarbonyltechnetium(I) iodide

Example 16.5c Write chemical formulas for the *potassium trioxalatochromate(III)* coordination compound and the *pentaamineaquanickel(II)* complex ion.

Solution In the *coordination compound*, the complex ion contains three oxalato ($C_2O_4^{2-}$) ligands and the nominal charge on the central metal is +3. Therefore, the formula for the anion is $[Cr(C_2O_4)_3]^{3-}$. Three K^+ ions are required for charge neutrality, so the formula of the coordination compound is: $K_3[Cr(C_2O_4)_3]$.

The *complex ion* has one H_2O and five NH_3 ligands. All are neutral, so the charge on the complex ion is equal to the charge on the metal ion, +2. The complex cation is: $[Ni(NH_3)_5(H_2O)]^{2+}$.

- **Ligand substitution reaction**
  → A reaction in which one Lewis base takes the place of another
  → Sometimes all the ligands are replaced, but often the substitution is incomplete.

Example 16.5d Some $NiSO_4(s)$ is dissolved in water, yielding a green solution. Ammonia gas is bubbled through the solution and it turns deep blue. Explain this observation.

Solution Both water and ammonia act as ligands, but ammonia is a stronger base. When nickel(II) sulfate is dissolved in water, the green *aquo* complex forms. When ammonia is added to the solution, it replaces (substitutes for) water as a ligand, yielding an *ammine* complex that is deep blue. The net ionic equation for the substitution reaction is

$$Ni(H_2O)_6^{2+}(aq) + 6\,NH_3(g) \rightarrow Ni(NH_3)_6^{2+}(aq) + 6\,H_2O(l)$$

Hexaaquonickel(II) green — Hexaamminenickel(II) deep blue

16.6 The Shapes of Complexes

- **Complexes** → Exhibit a variety of shapes (coordination geometries) depending on the *coordination number*
- **Coordination number 6**
  - → Complexes are usually *octahedral* in shape, with the ligands at the vertices and the metal at the center.
  - → A detailed representation is shown in Fig. 16.19a in the text and a *simplified* one is on the right.
  - → Examples are hexamminecolbalt(III), $[Co(NH_3)_6]^{3+}$, and hexacyanoferrate(II), $[Fe(CN)_6]^{4-}$.

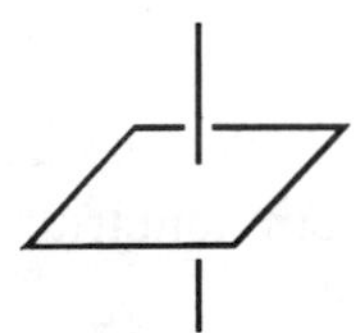

- **Coordination number 4**
  - → Two shapes are common, *tetrahedral* (Fig. 16.19b) and *square planar* (Fig. 16.19c). *Square planar* complexes are most common for metal species with the d^8 electronic configuration, such as Ni^{2+}, Pt^{2+}, and Au^{3+}.
  - → Examples of *tetrahedral* complexes are tetrahydroxozincate(II), $[Zn(OH)_4]^{2-}$ and titanium(IV) chloride, $TiCl_4$.
  - → Examples of *square planar* complexes are tetrachloroplatinate(II), $[PtCl_4]^{2-}$ and tetracyanonickelate(II), $[Ni(CN)_4]^{2-}$.
- **Coordination number 2**
  - → Complexes are *linear* in shape.
  - → Examples are dimethylmercury(0), $Hg(CH_3)_2$, and diamminesilver(I), $[Ag(NH_3)_2]^+$.
- **Metallocenes**
  - → "Sandwich" compounds with aromatic rings bonded by $2p\pi$-electrons to a central metal atom or ion. The aromatic rings serve as ligands.
  - → Examples are dibenzenechromium, $Cr(C_6H_6)_2$, and ferrocene, $[Fe(C_5H_5)_2]$.

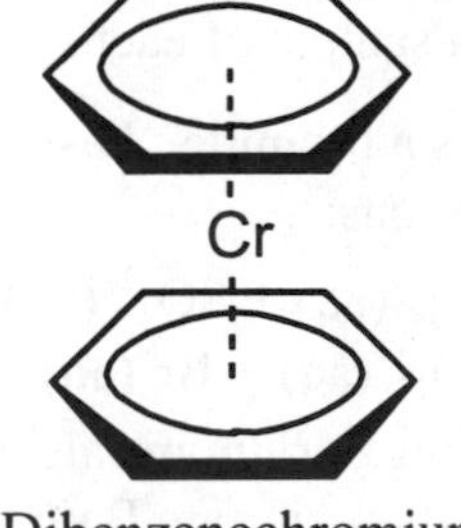

Dibenzenechromium

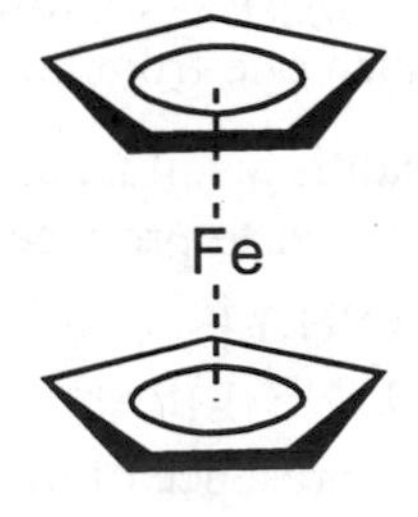

Ferrocene

- **Monodentate and polydentate ligands**
  → A *monodentate* ligand can bind to a metal at only one site. Examples are chloro, Cl^-; cyano, CN^-; carbonyl, CO; and methyl, CH_3.
  → A *polydentate* ligand can bind to a metal at more than one site simultaneously.
  → A *bidentate* ligand can bind to a metal at two sites. Examples are oxalato (ox), $C_2O_4^{2-}$, and ethylenediamine (en), $NH_2CH_2CH_2NH_2$.

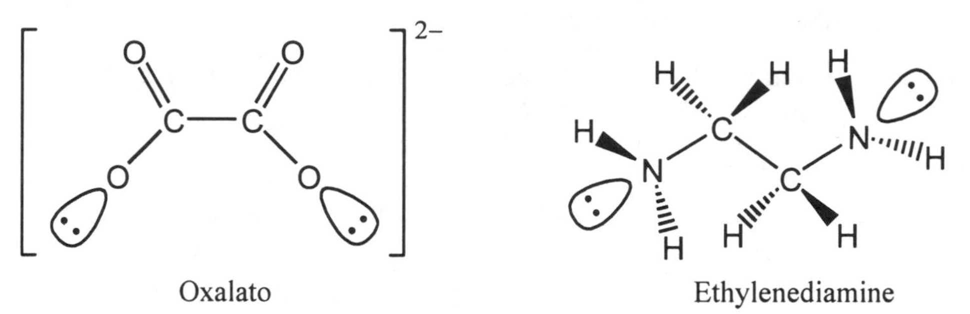

Oxalato Ethylenediamine

  → A *tridentate* ligand can bind to a metal at three sites. An example is diethylenetriamine (dien), $NH_2CH_2CH_2NHCH_2CH_2NH_2$.

16.7 Isomers

- **Isomers**
  → *Isomers* are compounds that contain the same numbers of the same atoms but in different arrangements.
  → Two major classes of isomers are *structural isomers*, in which *some* atoms have different partners (different connectivity), and *stereoisomers*, in which *all* atoms have the same partners (same connectivity) but *some* atoms are arranged differently in space.
  → *Structural isomers* are subdivided into *ionization isomers*, *hydrate isomers*, *linkage isomers*, and *coordination isomers*.
  → *Stereoisomers* are subdivided into *geometrical isomers* and *optical isomers*.

- **Ionization isomers**
  → Differ by the exchange of a ligand with an anion or molecule *outside* the coordination sphere
  → Form *different* ions in solution

Example 16.7a The coordination compounds $[CoBr(NH_3)_5]SO_4$ and $[CoSO_4(NH_3)_5]Br$ are ionization isomers. Suggest two experiments to distinguish one from the other, given a sample of each.

Answer The first salt is a sulfate and the second is a bromide. Dissolve each sample in water to produce the following ions:

$$[CoBr(NH_3)_5]SO_4(s) \rightarrow [CoBr(NH_3)_5]^{2+}(aq) + SO_4^{2-}(aq)$$
$$[CoSO_4(NH_3)_5]Br(s) \rightarrow [CoSO_4(NH_3)_5]^{+}(aq) + Br^{-}(aq)$$

Test for the presence of the anion by adding barium nitrate and silver nitrate solutions. The sulfate solution will precipitate $BaSO_4(s)$ after barium nitrate is added, but no precipitate will form if some silver nitrate is added. The bromide solution will precipitate AgBr(s) after the silver nitrate is added, but no precipitate will form if barium nitrate is added.

A second method to distinguish between the two compounds is to measure the molar conductivity of the two aqueous solutions. The sulfate solid forms an electrolyte with larger charges on each ion than the bromide solid. The sulfate solution will have the higher molar conductivity.

- **Hydrate isomers**
  - → A special type of ionization isomer
  - → *Hydrate isomers* differ by the exchange of an H_2O molecule and another ligand in the coordination sphere. For example, the solid hexahydrate of chromium(III) chloride, $CrCl_3 \cdot 6H_2O$, may be any of the following three compounds:

hexaaquachromium(III) chloride	$[Cr(H_2O)_6]Cl_3$
pentaaquachlorochromium(III) chloride monohydrate	$[CrCl(H_2O)_5]Cl_2 \cdot H_2O$
tetraaquadichlorochromium(III) chloride dihydrate	$[CrCl_2(H_2O)_4]Cl \cdot 2H_2O$

- **Linkage isomers**
  - → *Linkage isomers* differ in the identity of the atom in a given ligand that is attached to the metal atom or ion.
  - → A ligand may have more than one atom with a lone pair that can bond to the metal, but because of its size or shape, only one atom from the ligand may bond to the metal at one time. Monodentate ligands of this type are called *ambidentate ligands.*
  - → Examples from Table 16.4 in the text are: NO_2^- (M–NO_2 *nitro* and M–ONO *nitrito*), CN^- (M–CN *cyano* and M–NC *isocyano*), and SCN^- (M–SCN *thiocyanato* and M–NCS *isothiocyanato*).
  - → One pair of linkage isomers is pentaamminethiocyanatocobalt(III), $[Co(NH_3)_5SCN]^{2+}$, and pentaammineisothiocyanatocobalt(III), $[Co(NH_3)_5NCS]^{2+}$. Lewis structures for SCN^- clearly show the lone pairs on S and N that can bond to a central metal ion.

$$\left[:\ddot{S}=C=\ddot{N}:\right]^- \longleftrightarrow \left[:\underset{\cdot\cdot}{\ddot{S}}-C\equiv N:\right]^-$$

Example 16.7b A yellow solution containing the $[Co(NO_2)(NH_3)_5]^{2+}$ ion turns red when exposed to light. When the light is turned off, the color of the solution changes back to yellow. Analysis shows that the cation in the red solution and in the yellow solution have the same elemental composition and charge. Suggest structures for the two cations.

Solution Because the cations have the same elemental composition and charge, the simplest explanation consistent with the data is that the species are *linkage isomers*. Of the ligands present (NH_3 and NO_2^-), only NO_2^- is *ambidentate*. Presumably, one of the species is the pentaamminenitrocobalt(III) cation, $[Co(NO_2)(NH_3)_5]^{2+}$, and the other is the pentaamminenitritocobalt(III) cation, $[Co(ONO)(NH_3)_5]^{2+}$. Because the central metal atom is surrounded by six ligands, we expect the complexes to be octahedral. In fact, the nitro complex is yellow and the nitrito complex is red. Structures of the complexes are

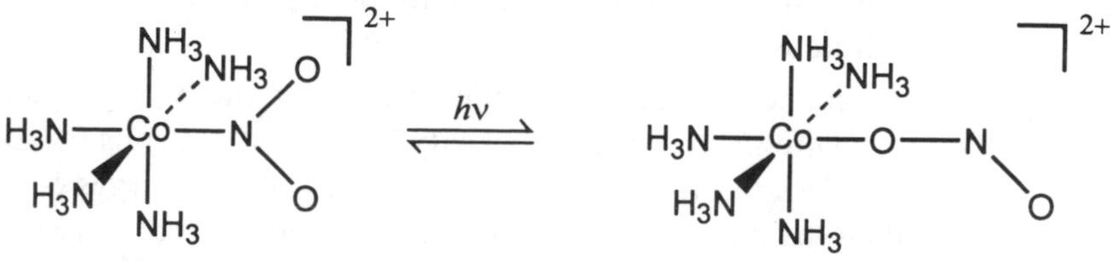

Yellow $\underset{}{\overset{hv}{\rightleftharpoons}}$ Red

- **Coordination isomers**
  - → *Coordination isomers* occur when one or more ligands are exchanged between a complex that is a cation and one that is an anion in a coordination compound.
  - → One pair of coordination isomers is tetraamminecopper(II) tetrachloroplatinate(II), $[Cu(NH_3)_4][PtCl_4]$, and tetraammineplatinum(II) tetrachlorocuprate(II), $[Pt(NH_3)_4][CuCl_4]$.

Example 16.7c Identify the types of *structural isomers* represented in the following pair:

$[FeCl(H_2O)_5]F$ and $[FeF(H_2O)_5]Cl$

Solution Between the two compounds, an anion and a ligand are exchanged. The pair represents *ionization isomers*.

Example 16.7d Identify the types of *structural isomers* represented in the following pair:

$[Cr(NH_3)_6][Fe(CN)_6]$ and $[Fe(NH_3)_6][Cr(CN)_6]$

Solution Comparison of the two compounds shows that all the ligands are exchanged between the cation and the anion. The pair represents *coordination isomers*. Note that in one isomer the chromium complex is the cation, whereas in the other isomer it is the anion.

- **Geometrical isomers**
  - → In *geometrical isomers*, atoms are bonded to the same neighbors, but have different orientations. *Geometrical isomers* are *stereoisomers*.
  - → An example is cis (adjacent) and trans (across) complexes of diamminedichloroplatinum(II), $[Pt(NH)_3Cl_2]$.

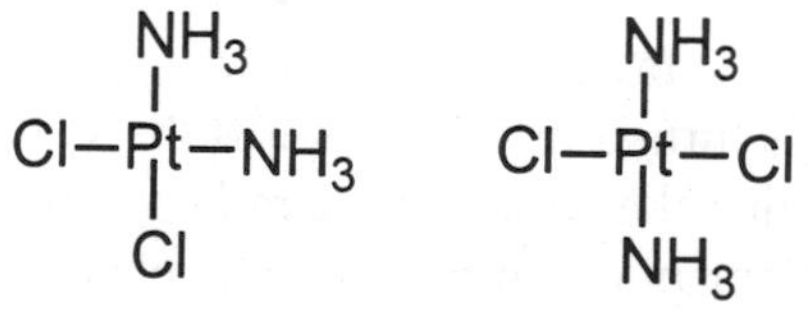

Cis Trans

- **Optical isomers**
  - → *Optical isomers* are a special type of *stereoisomers*: they are *nonsuperimposable* mirror images of each other.

→ An example is the molecule bromochlorofluoromethane, CHBrClF. The molecule has a *tetrahedral shape* and *two* optical isomers:

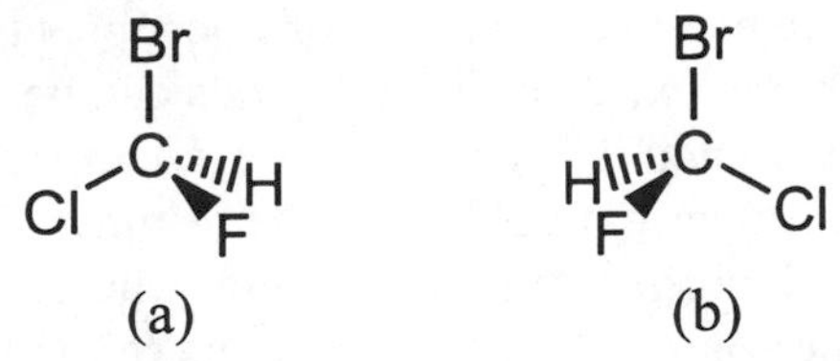

Molecules (a) and (b) are not superimposable; therefore they are different compounds. The atoms have the same arrangements in space except for "handedness." They are related to each other as the left hand is related to the right hand (*mirror images*).

- **Chirality and optical activity**
  - → A *chiral complex* or *chiral molecule* is one that is *not* identical to its mirror image. All *optical isomers* are *chiral*. An *achiral complex* or *achiral molecule* is one that is identical to its mirror image.
  - → A *chiral complex* and its mirror image form a pair of *enantiomers*. Molecules (a) and (b) shown above are *enantiomers*.
  - → *Chiral molecules* display *optical activity*, the ability to rotate the plane of polarization of light (see Box 16.2 in the text). One enantiomer of a chiral complex rotates the plane of polarization clockwise, whereas its mirror image partner rotates it by the same amount counterclockwise.
  - → A *racemic mixture* is a mixture of enantiomers in equal molar proportions. Because enantiomers rotate the plane of polarization of light in opposite directions, a racemic sample is *not* optically active.

- **Shapes of complexes**
  - → The existence of geometrical and optical isomers follows from the shape of the complex and the types of ligands that bind to the metal.
  - → In the following table, M represents a metal, A, B, C, and D represent *monodentate* ligands, and B–B represents a *bidentate* ligand such as ethylenediamine.

Shape of Complex	Complex Type	Geometrical Isomers?	Optical Isomers?
Tetrahedral	MABCD	no	yes
Square Planar	MA_2B_2	yes (cis, trans)	no
Octahedral	MA_4B_2	yes (cis, trans)	no
Octahedral	$MA_2(B\text{-}B)_2$	yes (cis, trans)	yes (cis isomer)
Octahedral	MA_3B_3	yes	no
Octahedral	$MA_2B_2C_2$	yes	yes

Example 16.7e Sketch the structures of all the distinct stereoisomers of $[Fe(en)_2Cl_2]^+$ indicating enantiomers where appropriate.

Solution The two (en) ligands are bidentate and each one occupies two cis coordination sites in an octahedral complex. The chloro ligands can be cis or trans leading to two geometric isomers (a) and (b). From (a) and (b), mirror images are generated to determine if they are superimposable with the original. The trans dichloro structure and its mirror image are superimposable, whereas the cis dichloro structure is not. So, there are three distinct isomers (a), (b) and (c). Complexes (b) and (c) are optically active and form an enantiomeric pair; they are optical isomers. Complexes (a) and (b) and (a) and (c) are geometrical isomers.

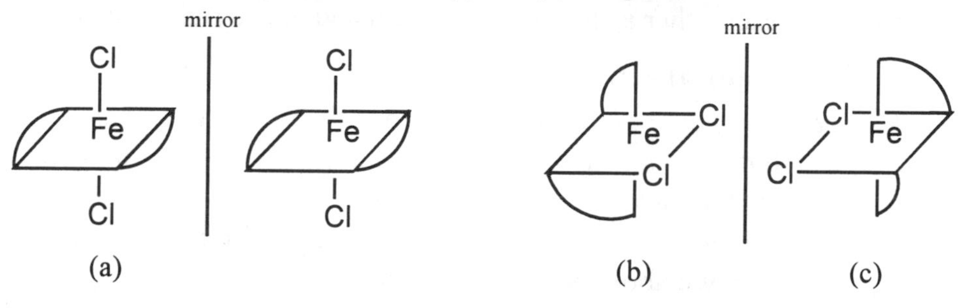

The Electronic Structures of Complexes (Sections 16.8-16.12)

Key Concepts

crystal field theory, ligand field theory, t_{2g}-orbitals, e_g-orbitals, ligand field splitting, spectrochemical series, weak-field ligands, strong-field ligands, high-spin complexes, low-spin complexes, electron configuration of a complex, colors of complexes, complementary colors, *d-d* transitions, magnetic properties of complexes

Overview

- **Topics**
  - → Electronic structure of *d*-metal complexes
  - → Nature of bonding in complexes
  - → Distribution of electrons in complexes
- **Theories**
  - → *Crystal field theory* is an electrostatic theory used to treat *d*-metal complexes. The ligands are approximated by point charges, and the metal-ligand bond is considered to be electrostatic in nature. This theory helps explain color and magnetic properties of complexes.
  - → *Ligand field theory* is an application of molecular orbital theory to help explain the nature of bonding in *d*-metal complexes. The metal and ligand electrons occupy molecular orbitals formed from the overlap of metal and ligand atomic orbitals. This theory helps explain why some ligands are strong-field and others weak-field.

16.8 Crystal Field Theory

- ***d*-Orbital splitting in an octahedral complex, ML_6**
  - → Ligands are considered to be point *negative* charges.
  - → Ligands approach the *positively charged* metal ion along the *x*-, *y*-, and *z*-axes (see Fig. 16.32 in the text). Valence electrons in the *d*-orbitals repel the negative charges on the ligands when the ligands approach.
  - → The *d*-orbitals are split into two sets: t_{2g} (three orbitals) and e_g (two orbitals).
  - → The d_{xy}-, d_{yz}-, and d_{zx}-orbitals form the t_{2g}-orbitals *lowered* in energy by (2/5)Δ_O with respect to the *d*-orbitals in the free metal ion.
  - → The $d_{x^2-y^2}$- and d_{z^2}-orbitals form the e_g-orbitals *raised* in energy by (3/5)Δ_O.
  - → The separation in energy between the t_{2g}- and e_g-orbitals is the *octahedral ligand field splitting*, Δ_O, where the O denotes octahedral.

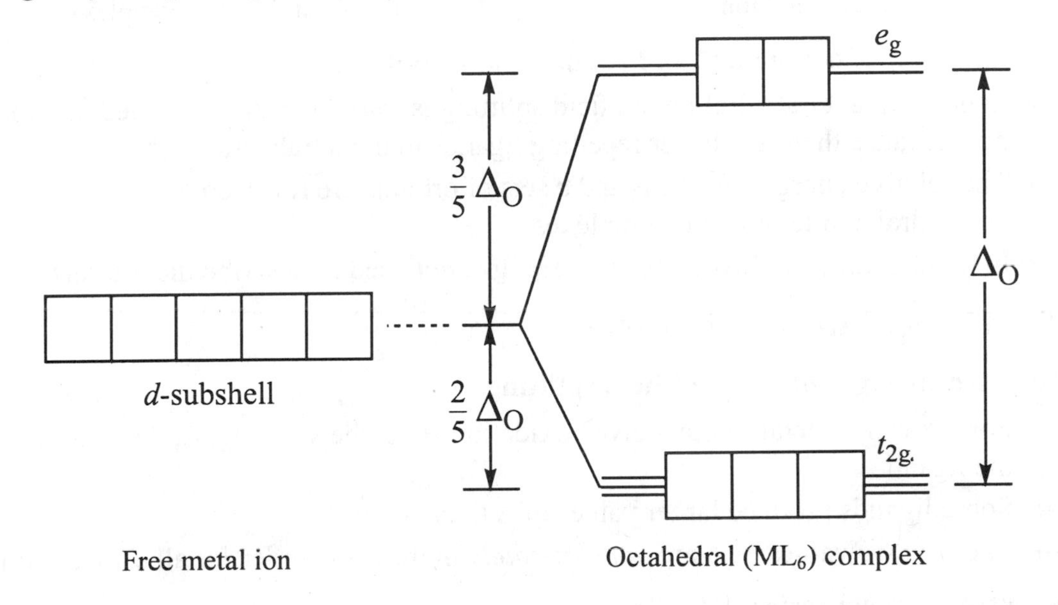

- ***d*-Orbital splitting in a tetrahedral complex, ML_4**
  - → Ligands are again approximated by point *negative* charges.
  - → Ligands approach the *positively charged* metal ion along the corner directions of a tetrahedron.
  - → The *d*-orbitals are split into two sets: t_2 (three orbitals) and *e* (two orbitals).
  - → The d_{xy}-, d_{yz}-, and d_{zx}-orbitals form the t_2-orbitals *raised* in energy by (2/5)Δ_T with respect to the *d*-orbitals in the free metal ion.
  - → The $d_{x^2-y^2}$- and d_{z^2}-orbitals form the *e*-orbitals *lowered* in energy by (3/5)Δ_T.
  - → The separation in energy between the t_2- and *e*-orbitals is the *tetrahedral ligand field splitting*, Δ_T, where the T denotes tetrahedral.

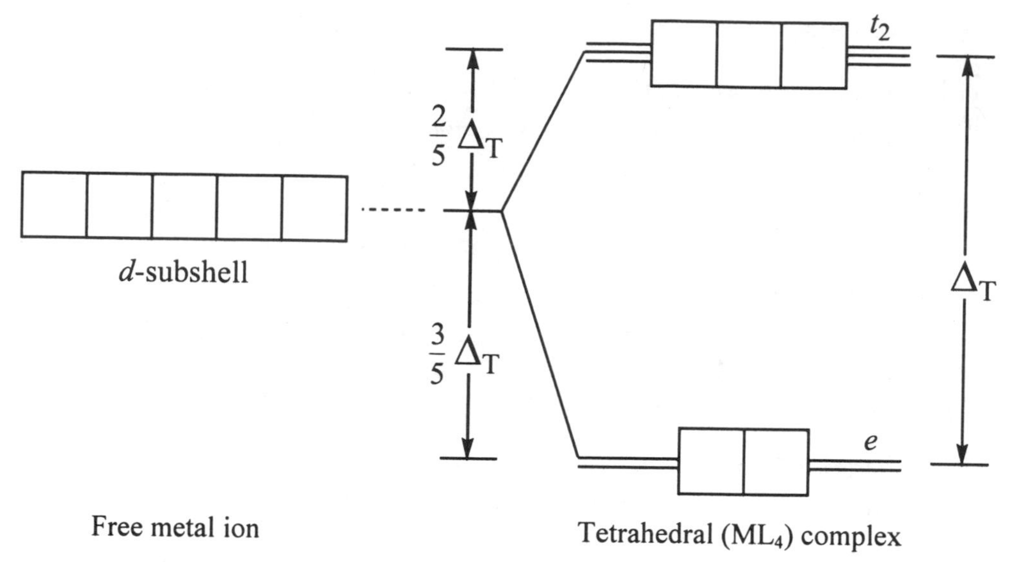

- **Comparison of octahedral and tetrahedral complexes**
  → In general, a tetrahedral ligand field splitting is *smaller* than an octahedral value, in part because there are fewer repelling ligands in the tetrahedral case.
  → The relative energies of the t_2 and e sets of orbitals are reversed in the octahedral and tetrahedral complexes.
  → In tetrahedral complexes, the subscript g is *not* used to describe the orbitals.

16.9 The Spectrochemical Series

- **Relative strength of a ligand field splitting, Δ**
  → For a given *d*-metal ion in a given oxidation state, the splitting, Δ, depends on the ligand.
  → Some ligands produce larger values of Δ than others.
  → The *relative* values of Δ are *approximately* in the same order for all *d*-metal ions.
- **Spectrochemical series**
  → Ligands are arranged according to the magnitude of the ligand field splitting that they produce (see Fig. 16.36 in the text).
  → *Weak field ligands* produce small values of Δ.

$$I^- < Br^- < Cl^- \approx SCN^- < F^- < OH^- < (ox) < H_2O$$

  → *Strong field ligands* produce large values of Δ.

$$NH_3 < (en) < NO_2^- < CN^- \approx CO$$

16.10 High-Spin and Low-Spin Complexes

- **Electron configurations in isolated atoms**
  → In an isolated atom, the five *d*-orbitals in a subshell have the same energy. According to Hund's rule, electrons occupy each orbital separately until each has one electron.
  → Hund's rule also states that the electron spins are *parallel* in different orbitals of the same energy.

- **Octahedral complexes**
  - → Similar rules apply to *d*-metal complexes; however, the five *d*-orbitals in a subshell are divided into two groups.
  - → The lower energy t_{2g}-orbitals fill first; degenerate orbitals (d_{xy}, d_{yz}, d_{zx}) fill with spins parallel in accordance with Hund's rule until the t_{2g}-orbitals are half full (d^3 complex).
  - → The electron configurations of d^4 through d^7 complexes depend on the magnitude of Δ_O compared with the energy required to pair and electron in a t_{2g}-orbital.
  - → If Δ_O is *large* compared with the spin-pairing energy (*strong field ligands*), electrons will enter the lower energy t_{2g}-orbitals with spins paired before filling the e_g-orbitals.
  - → If Δ_O is *small* compared with the spin-pairing energy (*weak field ligands*), electrons will enter the higher energy e_g-orbitals with spins parallel before pairing begins in the lower energy t_{2g}-orbitals.
  - → For d^8 through d^{10} complexes, the orbitals filled are the same for strong field and weak field ligands. Both procedures predict the same final configuration.
  - → A similar procedure is followed for *tetrahedral complexes*, but the two groups of *d*-orbitals are reversed in energy. In this case, the *e*-orbitals fill first.
  - → Electron configurations of d^n complexes are given in Table 16.5 in the text.
- **High- and low-spin complexes**
  - → A d^n complex with the maximum number of unpaired electrons is called a *high-spin complex*. High-spin complexes are expected for *weak-field ligands.*
  - → A d^n complex with the minimum number of unpaired electrons is called a *low-spin complex*. Low-spin complexes are expected for *strong-field ligands.*
  - → The d^4 through d^7 octahedral complexes may be high- or low-spin.
  - → Because Δ_T is small with respect to the energy required to pair an electron in the *e*-orbitals, t_2-orbitals are always accessible and *tetrahedral complexes* are *almost always* high-spin.

Example 16.10 The octahedral complex ion, $[MnCl_6]^{3-}$, has more unpaired spins than the octahedral complex ion, $[Mn(CN)_6]^{3-}$. Determine the electron configuration and the number of unpaired electrons of each ion.

Solution First determine the number of *d*-electrons in each complex. Use Table 16.5 to determine the electronic configuration. If the complex is d^4 through d^7, use the strength of the ligand to predict whether the complex is high- or low-spin.

Both ligands have a charge of −1, so both complexes are Mn(III). The electron configuration of an isolated Mn atom is $[Ar]3d^5 4s^2$. The electron configuration of an isolated Mn^{3+} ion is $[Ar]3d^4$; therefore, Mn(III) is a d^4 complex.

Because Cl^- is a weak-field ligand, we expect that $[MnCl_6]^{3-}$ is a high-spin complex. The electron configuration is $t_{2g}{}^3 e_g{}^1$ or in expanded form:

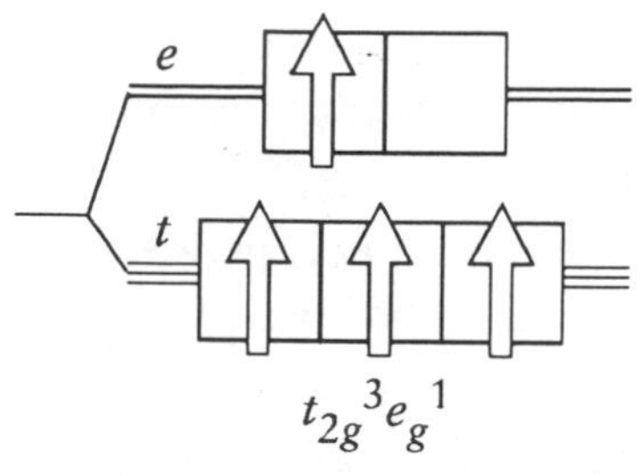

The complex has *four* unpaired electrons, drawing from Margin Figure (**19**).

Because CN^- is a strong-field ligand, we expect $[Mn(CN)_6]^{3-}$ is a low-spin complex. The electron configuration is t_{2g}^{4} or in expanded form:

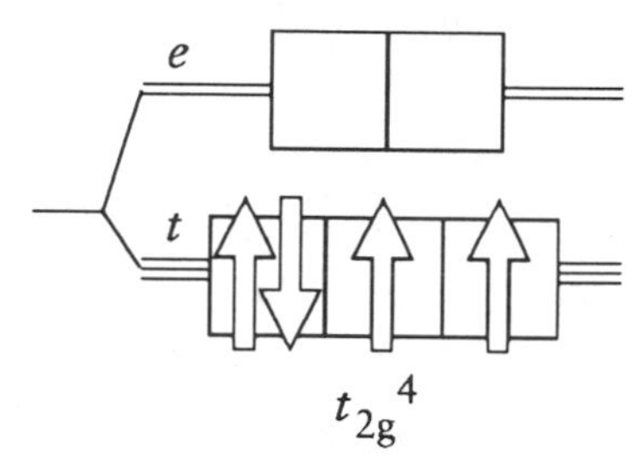

The complex has *two* unpaired electrons, drawing from Margin Figure (**18**).

16.11 The Colors of Complexes

- **Absorption of light and transmission or reflection**
  - → Visible light is made up of all wavelengths from about 400 (blue) to 800 nm (red).
  - → Transition metal complexes often *absorb* some wavelengths of visible light and *transmit* or *reflect* the rest.
  - → *Transmitted* or *reflected* light, which is the *complementary color* of the light absorbed, is the light we see and determines the color of the sample.
  - → The color wheel on p. 143 of the text provides a simple way of relating color *absorbed* to color *seen* in some systems.

Example 16.11a An aqueous solution containing $[Co(NH_3)_6]^{3+}$ appears yellow in transmitted light. What color of light is absorbed by the sample?

Solution The sample absorbs the complementary color of yellow. From the color wheel, the complementary color of yellow is violet.

Note: Prediction of color can be complicated by absorption of light by the complex in several regions of the electromagnetic spectrum. For example, chlorophyll absorbs both red and blue light, leaving only the wavelengths near green to be reflected from the leaves of plants.

- **Electronic transitions in complexes**
  - → The absorption of light in complexes is often associated with *d–d transitions.* An electron is excited from a t_{2g}- to an e_g-orbital in an *octahedral complex* or from an *e*- to a t_2-orbital in a *tetrahedral complex.*

→ *Charge-transfer transitions* may also occur in which an electron is excited from a ligand-centered orbital to a metal-centered orbital or *vice versa*. These transitions are often quite intense. An example is the deep purple color of the permanganate ion, MnO_4^-.

→ Color arises when a substance absorbs light from some portion of the visible spectrum and transmits the rest. Both *d-d* and charge-transfer transitions can occur in the visible region of the spectrum and either or both processes can contribute to the color of a transition metal complex.

- **Color, *d–d* transitions, and the spectrochemical series**

→ Complexes with *weak-field ligands* produce *small* Δ values and absorb *low-energy, long-wavelength* (*red* or *infrared*) radiation. In the absence of strong charge-transfer transitions, solutions of these complexes appear *green* (the complementary color of *red*) or *colorless* (if the absorption is in the *infrared*).

→ Complexes with *strong-field ligands* produce *large* Δ values and absorb *high-energy, short-wavelength* (*blue*, *violet* or *ultraviolet*) radiation. In the absence of complicating charge-transfer transitions, solutions of these complexes appear *orange* or *yellow* (the complementary colors of *blue* and *violet*, respectively) or *colorless* (if absorption is in the *ultraviolet*).

→ The ligand field splitting Δ can be determined from the electronic spectrum of a given complex, but it is not in most cases related simply to the color of its solution.

Example 16.11b Which of the octahedral complexes shown below is expected to have the larger ligand field splitting, Δ_O? Explain why? How could you have predicted this order if you only knew the structure of the ions and not their color?

$[Co(NH_3)_5(NO_2)]^{2+}$ $\underset{}{\overset{h\nu}{\rightleftharpoons}}$ $[Co(NH_3)_5(ONO)]^{2+}$

I (appears yellow in solution) II (appears red in solution)

Solution A complex absorbs the complementary color that its solution transmits. Complex I transmits yellow light and absorbs the complementary color of yellow, which is violet. Similarly, complex II transmits red light and absorbs green light. Photons of violet light have higher frequencies and more energy ($E = h\nu$) than photons of green light. If the color of the light arises from *d-d* transitions, we expect complex I to have a larger energy difference between its *d*-levels and therefore a higher ligand field splitting than complex II.

If only the structures of the ions were known, we could appeal to the spectrochemical series (Fig. 16.36) which shows that ligands for which oxygen is the donor atom ($C_2O_4^{2-}$, OH^-, H_2O) occur lower in the spectrochemical series than do ligands for which nitrogen is the donor atom (NH_3, en, NO_2^-). Species high in the series give larger Δ_O values so we would expect Δ_O for complex I to be higher than that for complex II.

16.12 Magnetic Properties of Complexes

- **Complexes may be diamagnetic or paramagnetic**
  - → *Diamagnetic complexes* have no unpaired electrons and are repelled by a magnetic field.
  - → *Paramagnetic complexes* have one or more unpaired electrons and are attracted into a magnetic field. The greater the number of unpaired electrons, the greater the attraction.
- **Magnetic properties of complexes**
  - → Determined from the electronic configuration of the complex
  - → Strong-field ligands produce low-spin complexes that are weakly paramagnetic.
  - → Weak-field ligands produce high-spin complexes that are strongly paramagnetic.

Example 16.12 Arrange the following octahedral species in order of increasing paramagnetism: $[Fe(CN)_6]^{4-}$, $[Ni(H_2O)_6]^{2+}$, $[CrF_6]^{4-}$, and $[Ti(H_2O)_6]^{3+}$.

Solution Determine the valence electron configuration of the ions and count the number of unpaired electrons. If the electron configuration is d^4 through d^7, care must be taken to determine whether the ion is high spin or low spin. A larger number of unpaired electrons produces an increase in paramagnetism.

$[Fe(CN)_6]^{4-}$ Fe(III) is a d^6 ion. The CN^- ligand is high in the spectrochemical series so we expect a low-spin complex. The valence electron configuration is $t_{2g}^{\ 6}$ with *no* unpaired electrons. The complex is *diamagnetic.*

$[Ni(H_2O)_6]^{2+}$ Ni(II) is a d^8 ion with the valence electron configuration $t_{2g}^{\ 6}\ e_g^{\ 2}$. The configuration has *two* unpaired electrons, one in each e_g-orbital.

$[CrF_6]^{4-}$ Cr(II) is a d^4 ion and F^- is a weak-field ligand. We expect the valence electron configuration to be $t_{2g}^{\ 3}\ e_g^{\ 1}$, with all *four* electrons unpaired.

$[Ti(H_2O)_6]^{3+}$ Ti(III) is a d^1 ion, so there is *one* unpaired electron.

Thus, in order of increasing paramagnetism, we have:

$$[Fe(CN)_6]^{4-} < [Ti(H_2O)_6]^{3+} < [Ni(H_2O)_6]^{2+} < [CrF_6]^{4-}.$$

Ligand Field Theory (Sections 16.13-16.14)

Key Concepts

ligand field theory, sigma- and pi-bonding in complexes, origin of ligand field splitting

Overview

- **Ligand field theory**
  - → Describes bonding in complexes in terms of molecular orbitals generated from metal atom *d*-orbitals and ligand orbitals.

→ More realistic than crystal field theory, which treats ligands as point charges
→ Most ideas carry over from crystal field theory; the major difference is the origin of the ligand field splitting.

16.13 Sigma-Bonding in Complexes

- **Ligand field theory**
  → Molecular orbitals are generated from the available atomic orbitals in the complex.
  → To simplify the treatment, only one atomic orbital is used on each of the ligands. In a chloride ion ligand, a Cl 3*p*-orbital is chosen, which is directed toward the metal atom. For an ammonia ligand, the sp^3 lone-pair orbital is chosen.

- **Octahedral complex**
  → For first row transition elements, the 4*s*-, 4*p*-, and 3*d*-orbitals of the central metal ion are chosen, because all these orbitals have similar energies.
  → There are 9 orbitals on the metal ion and 6 on the ligands, giving a total of 15 atomic orbitals. The 15 atomic orbitals overlap to form 15 molecular orbitals (see Fig. 16.43 in the text). The final result is that six molecular orbitals are bonding, six are antibonding, and three are nonbonding.
  → The 12 ligand electrons and the available *d* valence electrons are placed in the molecular orbitals in accordance with the building up principle to determine the ground-state electron configuration of the complex. In transition metal ions, the valence *s*- and *p*-orbitals are typically vacant, whereas the *d*-orbitals are partially occupied.
  → The first 12 electrons form six metal-ligand sigma bonds. The *d*-electrons of the metal enter the t_{2g}-orbitals and e_g-orbitals in the same way as they did in the crystal field theory. The ligand field theory, however, identifies the nature of the t_{2g}-orbitals as *nonbonding* and the e_g-orbitals as *antibonding*.

16.14 Pi-Bonding in Complexes

- **Ligand field theory**
  → The treatment of *pi-bonding* in complexes provides further insight into the *spectrochemical series*.
  → The theory explains why uncharged species such as CO are strong-field ligands, whereas negatively charge species such as Cl^- are weak-field ligands. Based on electrostatic considerations, we expect the *opposite* to be true.

- **Octahedral complex**
  → *Pi-bonding* in octahedral complexes arises from the overlap of the t_{2g}-orbitals of the metal with *p*- or π-orbitals of the ligand to form *bonding* and *antibonding* molecular orbitals, π and π^*, respectively.
  → The details of this interaction are shown in Fig. 16.44 in the text, which is reproduced on the next page.

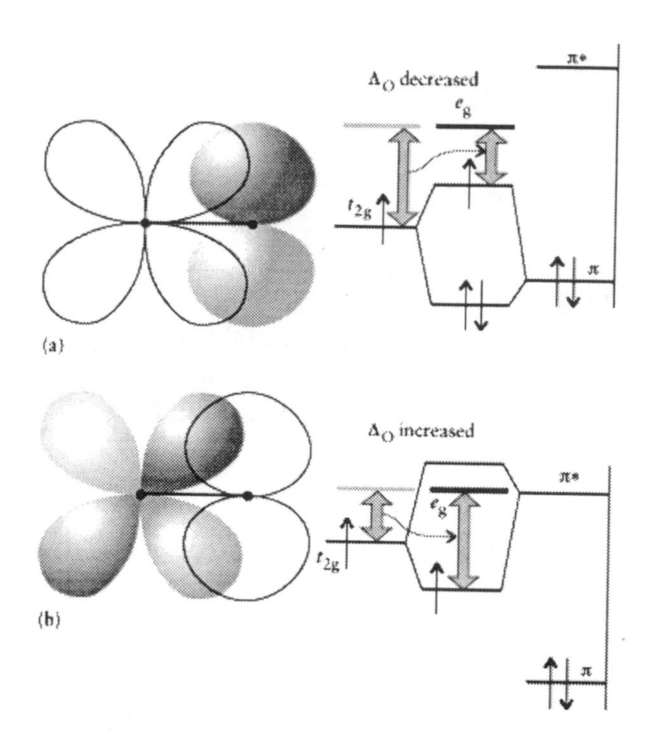

Fig. 16.44

- **Weak-field ligand (Fig. 16.44a)**
  - → The antibonding π*-orbital of the ligand is too high in energy to take part in the bonding. The bonding π-orbital of the ligand combines with a t_{2g}-orbital of the metal and the combination *raises* the energy of the molecular orbital that the metal *d*-electron occupies. The electron from the metal enters an *antibonding molecular orbital.*
  - → The energy of the e_g-orbitals is unchanged, and the splitting between the t_{2g}- and e_g-orbitals is reduced. The ligand field splitting, Δ_O, is, therefore, *decreased* by π-bonding.
  - → Weak-field ligands such as Cl^- and Br^- have a valence shell *p*-orbital occupied by two electrons that is perpendicular to the metal-ligand sigma bond, and they display this behavior.
- **Strong-field ligand (Fig. 16.44b)**
  - → The bonding π-orbital of the ligand is too low in energy to take part in the bonding. The *empty* antibonding π*-orbital of the ligand combines with a t_{2g}-orbital of the metal and the combination *lowers* the energy of the molecular orbital that the metal *d*-electron occupies. The electron from the metal enters a *bonding molecular orbital.*
  - → The energy of the e_g-orbitals is unchanged, and the splitting between the t_{2g}- and e_g-orbitals is increased. The ligand field splitting, Δ_O, is, therefore, *increased* by π-bonding.
  - → Strong-field ligands such as CO and CN^- have an unoccupied valence shell π*-orbital perpendicular to the metal-ligand sigma bond and display this behavior.

Example 16.14 Are the t_{2g}-orbitals in the following octahedral complexes best described as *bonding, nonbonding,* or *antibonding:* $[Cr(CN)_6]^{3-}$ and $[Cr(H_2O)_6]^{3+}$?

Solution Refer to Figs. 16.43 and 16.44 in the text.

$[Cr(CN)_6]^{3-}$ The CN^- ligand is isoelectronic with CO and we expect its empty π^*-orbital to combine with the t_{2g}-orbital of the complex as shown in Fig. 16.44(b). Therefore, we expect the t_{2g}-orbitals to be *bonding molecular orbitals.*

$[Cr(H_2O)_6]^{3+}$ The H_2O ligand can only form sigma bonds with metals. In the Lewis view, the water molecule has two lone pairs of electrons (doubly occupied orbitals). One of these orbitals is used to form the metal-ligand sigma bond. This type of bonding corresponds to that shown in Fig. 16.43, in which the t_{2g}-orbitals are *nonbonding molecular orbitals.* However, the electrons in the second lone pair on water can repel those in the t_{2g} set slightly, so the bonding can also be described as *slightly antibonding.*

Note: We expect the ammonia ligand to behave in a manner similar to water. In this case, there is only one lone pair of electrons, and the bonding is expected to be *nonbonding.* Ammonia and water are the *transition* ligands between weak-field and strong-field behavior.

The Impact on Materials (Sections 16.15-16.17)

Key Concepts

steel, stainless steel, pig iron, cast iron, nonferrous alloys, antiferromagnetic material, ceramic magnets, ferrofluids

Overview

- **Uses of *d*-block elements**
  - → Ferrous alloys based on iron
  - → Nonferrous alloys based on other *d*-block elements, most notably copper
  - → Fabrication of magnetic materials

16.15 Steel

- **Iron**
  - → Iron is produced by reduction of iron ores in a blast furnace. The product from the furnace is *pig iron,* which contains 3-5% C, ~2% Si, and lesser amounts of other impurities. *Cast iron* is similar to pig iron, but with some impurities removed. Both forms of iron are hard and brittle, in contrast to *pure iron,* which is malleable and more flexible.

→ In the blast furnace, limestone, $CaCO_3$, is used to remove impurities such as silica, alumina, and phosphorus(V) oxide. The following reactions occur:

$$CaCO_3(s) \rightarrow CaO(s) + CO_2(g)$$
$$CaO(s) + SiO_2(s) \rightarrow CaSiO_3(l)$$
$$CaO(s) + Al_2O_3(s) \rightarrow Ca(AlO_2)_2(l)$$
$$6\,CaO(s) + P_4O_{10}(s) \rightarrow 2\,Ca_3(PO_4)_2(l)$$

- **Steels**
  - → Steels are homogeneous alloys made from *pig iron* in two stages. In stage 1, *pig iron* is purified to *lower* the carbon content and to remove all other impurities. In stage 2, other appropriate metals are *added* to form the desired steel.
  - → Steels are solid solutions containing 2% or less carbon in iron. Their hardness, tensile strength, and ductility depends on the carbon content. Higher carbon content produces harder steel.
  - → The strength of steel can be greatly increased by heat treatment.
  - → The corrosion resistance of steel is significantly improved by alloying with other metals. For example, *stainless steel* is an alloy containing about 15% Cr by mass, and it is highly corrosion resistant.

16.16 Nonferrous Alloys

- **Alloys**
  - → Solid mixtures of metals
  - → Solutions that are *homogeneous* or *heterogeneous*
  - → Brass, bronze, gold- and silver-containing coins are *homogeneous alloys.*
  - → Tin-lead solder and mercury amalgams for dental fillings are *heterogeneous alloys.*
  - → Designed for specific needs such as titanium alloyed with tin or aluminum, producing lightweight, stiff alloys for racing bicycles
- **Substitutional alloys**
  - → Homogeneous alloys with metals of similar radius such as brass, consisting of copper and zinc (see Fig. 16.47 in the text)
  - → Typically harder than the pure metals but with lower electrical and thermal conductivity

16.17 Magnetic Materials

- **Magnetism**
  - → Arises from unpaired electrons in substances (three general types)
  - → *Paramagnetic:* Unpaired electron spins oriented *randomly*
  - → *Ferromagnetic:* Unpaired electron spins all aligned *parallel* in *domains* (see Fig. 16.6 in the text)
  - → *Antiferromagnetic:* Unpaired electron spins aligned *antiparallel* as in manganese

- **Ferromagnets**
  - → Used as permanent magnets
  - → Used for coating cassette tapes and computer disks
- **Ceramic magnets**
  - → Made from barium ferrite, $BaO \cdot nFe_2O_3$, or strontium ferrite, $SrO \cdot nFe_2O_3$
  - → Used as refrigerator magnets
  - → Brittle, relatively lightweight, and inexpensive
- **Ferrofluids**
  - → Magnetic liquids
  - → Suspensions of powdered Fe_3O_4 in a viscous oil/detergent mixture
  - → In *ferrofluids*, the Fe_3O_4 particles are attracted to the polar ends of the detergent molecules, which form micelles that are distributed throughout the oil. When a magnet approaches the fluid, the particles try to align, but are inhibited from doing so by the viscous oil. As a result, it is possible to control the flow and position of the *ferrofluid* by means of an applied magnetic field.

Chapter 17 Nuclear Chemistry

Nuclear Stability (Sections 17.1-17.5)

Key Concepts

nuclear chemistry, nuclear fission, nuclear fusion, nuclear stability, spontaneous nuclear decay, radioactivity, α particles, β particles, γ radiation, antiparticles, nuclear decay, nuclear reaction, nuclide, radioactive nuclei, parent nucleus, daughter nucleus, nuclear transmutation, α decay, nuclear equation, β decay, electron capture, positron emission, proton emission, neutron emission, band of stability, strong force, liquid drop model, independent particle model, proton-rich nuclides, radioactive series, nucleosynthesis, transuranium elements

Overview

- **Nuclei**
  → Contain *positively* charged protons and *neutral* neutrons in a tiny volume
  → Exist despite the strong repulsive forces between protons
- **Lifetime**
  → *Most* nuclei survive indefinitely.
  → *Some* nuclei decay spontaneously. The repulsive coulombic forces overcome the strong force that holds nucleons together.
- **Nuclear reactions**
  → In *nuclear fission*, large nuclei *fragment* into smaller ones.
  → In *nuclear fusion*, small nuclei *merge* into larger ones.

17.1 The Evidence for Spontaneous Nuclear Decay

- **Radioactivity**
  → Spontaneous emission of radiation or particles by a nucleus undergoing decay
  → Three types of radioactivity are the emission of α particles, β particles, and γ radiation.
  → The three types are the most common forms of radioactivity and were the first ones to be discovered.
- **Nuclear radiation (see Table 17.1 in the text):**
  α particles
  → Helium-4 nuclei, ${}^{4}_{2}He^{2+}$, traveling at speeds equal to about 10% of the speed of light, *c*
  → Deflected by electric and magnetic fields
  → When α particles interact with matter, their speed is reduced and they are neutralized.
  → Denoted by ${}^{4}_{2}\alpha$ or α

β particles

→ Rapidly moving *electrons*, e^-, emitted by nuclei at speeds less than 90% of *c*

→ Deflected by electric and magnetic fields

→ β particles are sometimes called *negatrons* (not used in the text).

→ When traveling between positively and negatively charged plates, α and β particles are deflected in opposite directions.

→ Denoted by ${}_{-1}^{0}e$, β^-, or β

γ radiation

→ High-energy *photons* (electromagnetic radiation) traveling at the speed of light

→ Uncharged and not deflected by electric and magnetic fields

→ Denoted by γ or ${}_{0}^{0}\gamma$

β^+ particles

→ A β^+ particle has the mass of an electron but carries a positive charge. It is called a *positron.*

→ When an electron and a positron, its *antiparticle*, meet, they are annihilated and completely transformed into energy, mostly γ radiation.

→ Denoted by ${}_{+1}^{0}e$ or β^+

p particles

→ A *proton* is a positively charged subatomic particle found in the nucleus. When emitted in nuclear decay, a proton typically travels at speeds equal to about 10% of *c*.

→ Denoted by ${}_{1}^{1}H^+$, ${}_{1}^{1}p$, or p

n particles

→ A *neutron* is an uncharged subatomic particle found in the nucleus. The mass of the neutron is slightly larger than the mass of the proton (see Table B.1 in the text). When emitted in nuclear decay, a neutron typically travels at speeds less than about 10% of *c*.

→ Denoted by ${}_{0}^{1}n$ or n

- **Penetrating power of nuclear radiation**

  → Penetrating power into matter decreases with increased charge and mass of the particle.

  → Uncharged γ radiation and neutrons are more penetrating than charged positrons or alpha particles.

  → Singly charged β particles (electrons) are more penetrating than the more massive, doubly charged α particles, which are virtually nonpenetrating.

17.2 Nuclear Reactions

- **Nucleons and nuclides**

  → A *nucleon* is a proton or a neutron in the nucleus of an atom.

→ A *nuclide* is an atom characterized by its atomic number, mass number, and nuclear energy state. Nuclides are often denoted by ${}^{A}_{Z}\text{E}$, where Z is the atomic number, A is the mass number (protons plus neutrons), and E is the symbol for the element. Note that $A - Z$ equals the number of neutrons in the nuclide. *A nuclide is a neutral atom and therefore is uncharged.*

→ *Nuclei* that change their structure spontaneously are called *radioactive*.

Example 17.2a How many protons, neutrons, electrons, and nucleons are present in the uranium-238 nuclide? Write the symbol for the nuclide.

Solution The number of protons is determined by the atomic number, Z. For uranium, $Z = 92$ and there are 92 protons. A nuclide is neutral, so the number of electrons must also equal 92. The mass number, A, is given in the name of the nuclide, and $A = 238$, which is also equal to the number of nucleons. The number of neutrons $= A - Z = 238 - 92 = 146$. The symbol for unranium is U and the nuclide symbol is ${}^{238}_{92}\text{U}$.

Note: The atomic number, Z, is implicit in the chemical symbol of the element, and nuclides are also written as ${}^{A}\text{E}$, which is ${}^{238}\text{U}$ in this example.

- **Nuclear reactions**

→ Any transformation that a nucleus undergoes

→ Differ from chemical reactions in three important ways

→ Isotopes of a given element often have different nuclear properties but always have similar chemical properties. Consider two isotopes of carbon, ${}^{12}\text{C}$ and ${}^{14}\text{C}$, which are difficult to separate chemically because they have similar chemical properties. The two isotopes have different nuclear stability. Carbon-12 is a stable nuclide, whereas carbon-14 decays with a half life of 5730 y (see Section 17.7).

→ Nuclear reactions often produce a different element. Titanium-44 captures one of its electrons to produce the nuclide scandium-44.

→ Nuclear reactions involve enormous energies relative to chemical reactions.

Example 17.2b Neptunium-239 decays to plutonium-239 by β decay. The plutonium-239 formed is in an *excited* nuclear energy state (indicated by *); it decays by emitting γ radiation with energy of 8.88×10^{-14} J·photon^{-1}. The relevant nuclear reactions are

$$ {}^{239}_{93}\text{Np} \rightarrow {}^{239}_{94}\text{Pu}^{*} + {}^{0}_{-1}\text{e} $$

$$ {}^{239}_{94}\text{Pu}^{*} \rightarrow {}^{239}_{94}\text{Pu} + \gamma $$

Calculate the wavelength of the γ radiation. Compare the energy of one mole of emitted photons to the molar enthalpy of combustion of methane, CH_4, which is -74.81 kJ·mol^{-1}.

Solution The emission process corresponds to high-energy electromagnetic radiation (see Section 1.2). The wavelength of the γ photon is given by

$$ \lambda = \frac{hc}{E_{\gamma}} = \frac{(6.626 \times 10^{-34}\ \text{J·s})(2.998 \times 10^{8}\ \text{m·s}^{-1})}{(8.88 \times 10^{-14}\ \text{J})} = 2.24 \times 10^{-12}\ \text{m} = 2.24\ \text{pm} $$

The energy of one mole of photons is
$E = N_A E_\gamma = (6.022 \times 10^{23}\ \text{mol}^{-1})(8.88 \times 10^{-14}\ \text{J}) = 5.35 \times 10^{10}\ \text{J·mol}^{-1}$
$= 5.35 \times 10^{7}\ \text{kJ·mol}^{-1}$

The nuclear reaction generates $(5.35 \times 10^{7})/(74.81) = 7.15 \times 10^{5}$ times as much energy as the combustion reaction.

- **Balancing nuclear reactions**
  - → In a manner similar to chemical reactions, *nuclear reactions* must be balanced with respect to both charge and mass.
  - → If ${}^{A}_{Z}\text{E}$ is used to represent one of several nuclides in a nuclear reaction, then the sum of the atomic numbers Z of the reactants must equal the sum of the Z values of the products. It also follows that the sum of the mass numbers A of the reactants must equal the sum of the mass numbers A of the products.
  - → In nuclear decay reactions, the product nucleus is called a *daughter nucleus.* The result of *most* nuclear decays is *nuclear transmutation*, the conversion of one element into another.
  - → The procedure for identifying the products of nuclear reactions is given in Toolbox 17.1 in the text. The following table lists the major types of nuclear decay, the general equations that result, and a specific example for each case.

Type	**Equation**	**Example**
α decay	${}^{A+4}_{Z+2}\text{E} \rightarrow {}^{A}_{Z}\text{E} + {}^{4}_{2}\alpha$ (or α)	${}^{232}_{90}\text{Th} \rightarrow {}^{228}_{88}\text{Ra} + {}^{4}_{2}\alpha$
β decay	${}^{A}_{Z-1}\text{E} \rightarrow {}^{A}_{Z}\text{E} + {}^{0}_{-1}\text{e}$ (or β)	${}^{3}_{1}\text{H} \rightarrow {}^{3}_{2}\text{He} + {}^{0}_{-1}\text{e}$
	$[{}^{1}_{0}\text{n} \rightarrow {}^{1}_{1}\text{p} + {}^{0}_{-1}\text{e}$ (or β)]	
Electron capture	${}^{A}_{Z+1}\text{E} + {}^{0}_{-1}\text{e} \rightarrow {}^{A}_{Z}\text{E}$	${}^{44}_{22}\text{Ti} + {}^{0}_{-1}\text{e} \rightarrow {}^{44}_{21}\text{Sc}$
	$[{}^{1}_{1}\text{p} + {}^{0}_{-1}\text{e} \rightarrow {}^{1}_{0}\text{n}]$	
Positron emission	${}^{A}_{Z+1}\text{E} \rightarrow {}^{A}_{Z}\text{E} + {}^{0}_{+1}\text{e}$ (or β^+)	${}^{22}_{11}\text{Na} \rightarrow {}^{22}_{10}\text{Ne} + {}^{0}_{+1}\text{e}$
	$[{}^{1}_{1}\text{p} \rightarrow {}^{1}_{0}\text{n} + {}^{0}_{+1}\text{e}$ (or β^+)]	
Proton emission	${}^{A+1}_{Z+1}\text{E} \rightarrow {}^{A}_{Z}\text{E} + {}^{1}_{1}\text{p}$	${}^{57}_{30}\text{Zn} \rightarrow {}^{56}_{29}\text{Cu} + {}^{1}_{1}\text{p}$
Neutron emission	${}^{A+1}_{Z}\text{E} \rightarrow {}^{A}_{Z}\text{E} + {}^{1}_{0}\text{n}$	${}^{91}_{34}\text{Se} \rightarrow {}^{90}_{34}\text{Se} + {}^{1}_{0}\text{n}$

Note: The net changes that occur are given in brackets. The neutrino, ν, is emitted in electron capture and positron emission. The antineutrino, $\tilde{\nu}$, is emitted in beta decay. Both particles have zero charge and very small mass. Proton emission and neutron emission are rare.

Example 17.2c What daughter nuclide is formed when radium-226, $^{226}_{88}\text{Ra}$, undergoes α decay?

Solution Write the general reaction for nuclear decay with radium-226 as the reactant.

$$^{226}_{88}\text{Ra} \rightarrow {}^{A}_{Z}\text{E} + {}^{4}_{2}\alpha$$

To balance charge (protons), the subscripts must have the same total value on each side of the equation: $88 = Z + 2$, or $Z = 86$. The element with atomic number 86 is radon, Rn. To balance mass (protons plus neutrons), the superscripts must have the same total value on each side of the equation: $226 = A + 4$, or $A = 222$, the mass of the daughter nuclide. The daughter nuclide is radon-222 or $^{222}_{86}\text{Rn}$.

Example 17.2d What daughter nuclide is formed when iridium-186, $^{186}_{77}\text{Ir}$, undergoes electron capture?

Solution Write the general reaction for electron capture with iridium-186 as the reactant.

$$^{186}_{77}\text{Ir} + {}^{0}_{-1}\text{e} \rightarrow {}^{A}_{Z}\text{E}$$

Balancing charge (protons): $Z = 77 - 1 = 76$, which is osmium, Os. Balancing mass (protons plus neutrons): $A = 186 + 0 = 186$. The daughter nuclide is $^{186}_{76}\text{Os}$.

Example 17.2e Write balanced nuclear equations for two processes that may transmute praseodymium-140 to cerium-140.

Solution The overall change is $^{140}_{59}\text{Pr} \rightarrow {}^{140}_{58}\text{Ce}$. The mass is balanced but the charge is reduced by one proton. A charged species is needed with zero mass number to effect the transformation. Either an electron, $^{0}_{-1}\text{e}$, is placed on the reactant side (electron capture) or a positron, $^{0}_{+1}\text{e}$, is placed on the product side (positron emission). The two balanced nuclear equations are

$$^{140}_{59}\text{Pr} + {}^{0}_{-1}\text{e} \rightarrow {}^{140}_{58}\text{Ce} \qquad \textit{electron capture}$$

$$^{140}_{59}\text{Pr} \rightarrow {}^{140}_{58}\text{Ce} + {}^{0}_{+1}\text{e} \qquad \textit{positron emission}$$

Note: The decay of praseodymium-140 occurs by both decay processes with electron capture the preferred one.

17.3 The Pattern of Nuclear Stability

- **Characterization of nuclei:**

Even-even

→ Nuclei with an even number of protons and of neutrons: *A* and *Z* are even numbers

→ In the class of even-even nuclei, 157 nuclides are stable.

→ Some members of this class are helium-4, $^{4}_{2}\text{He}$, carbon-12, $^{12}_{6}\text{C}$, oxygen-16, $^{16}_{8}\text{O}$, and neon-22, $^{12}_{10}\text{Ne}$.

Even-odd

→ Nuclei with an even number of protons and an odd number of neutrons: *Z* is an even number and *A* is an odd one

→ In the class of even-odd nuclei, 53 nuclides are stable.

→ Some members of this class are helium-3, $^{3}_{2}He$, berylium-9, $^{9}_{4}Be$, carbon-13, $^{13}_{6}C$, and neon-21, $^{21}_{10}Ne$.

Odd-even

→ Nuclei with an odd number of protons and an even number of neutrons: Z and A are both odd numbers

→ In the class of odd-even nuclei, 50 nuclides are stable.

→ Some members of this class are hydrogen-1, $^{1}_{1}H$, lithium-7, $^{7}_{3}Li$, boron-11, $^{11}_{5}B$, and nitrogen-15, $^{15}_{7}N$.

Odd-odd

→ Nuclei with an odd number of protons and of neutrons: Z is an odd number and A is an even one

→ In the class of odd-odd nuclei, only 4 nuclides are stable.

→ The members of this class are hydrogen-2 (deuterium), $^{2}_{1}H$, lithium-6, $^{6}_{3}Li$, boron-10, $^{10}_{5}B$, and nitrogen-14, $^{14}_{7}N$.

Summary

→ Even-even nuclides are the most stable and most abundant nuclides.

→ Odd-odd nuclides are the least stable and least abundant nuclides.

- **Strong force**

  → An attractive force that holds nucleons together in the nucleus

  → Overcomes the coulomb repulsion of protons

  → Acts only over a very short distance, approximately the diameter of the nucleus

- **Magic numbers**

  → Nuclei with *magic numbers* of protons and/or neutrons are more likely to be stable.

  → *Magic numbers* for *either* protons or neutrons are 2, 8, 20, 50, 82, 114, 126, and 184.

  → Analogous to the pattern of electronic stability in atoms associated with electrons in filled subshells

  → *Doubly magic* nuclides are very stable. Some members of this group are helium-4, $^{4}_{2}He$, oxygen-16, $^{16}_{8}O$, calcium-40, $^{40}_{20}Ca$, and lead-208, $^{208}_{82}Pb$.

- **Band of stability and sea of instability**

  → In Fig. 17.13 in the text, a plot of the atomic mass number of a nuclide on the y-axis and its atomic number on the x-axis is generated for all the known stable *and* unstable nuclides.

  → Stable nuclides are found in a narrow band, *band of stability*, that ends at $Z = 83$ (bismuth). All nuclides with $Z > 83$ are unstable.

  → Unstable nuclides are found in the regions above and below the *band of stability* called the *sea of instability*.

  → Nuclides *above* the band of stability are *neutron rich*, and are likely to emit a β particle (neutron → proton + β).

→ Nuclides *below* the band of stability are *proton rich*, and are likely to emit a β^+ particle (proton → neutron + β^+) or capture an electron (proton + β → neutron). Heavier nuclides below the band of stability and those with $Z > 83$ may also decay by emitting an α particle.

- **Models of nuclear structure**
  - → The three models of nuclear structure are introduced in increasing order of sophistication.
  - → In the *liquid drop model*, nucleons are packed together in the nucleus like molecules in a liquid.
  - → In the *independent particle model*, nucleons are described by quantum numbers and they occupy shells. Nucleons in the outermost shells are most easily lost as a result of radioactive decay.
  - → In the *collective model*, nucleons occupy quantized energy levels and interact with each other by the strong force and the electrostatic (coulomb) force.

17.4 Predicting the Type of Nuclear Decay

- **Massive nuclei with $Z > 83$**
  - → Must lose protons to reduce their atomic number and generally need to lose neutrons as well
  - → Decay in a stepwise manner and give rise to a *radioactive series*, a characteristic sequence of nuclides (see Fig. 17.16 in the text for the uranium-238 series).
  - → First step is the ejection of an α particle, then another α particle or β^- particle and so on. The final step is the formation of a stable isotope of lead (magic number 82). The uranium-238 series ends at lead-206, the uranium-235 series ends at lead-207, and the thorium-232 series ends at lead-208. The uranium-235 series is sometimes called the actinium series.
- **Neutron rich nuclides**
  - → Nuclides that have high neutron to proton (n/p) ratios compared to those nuclides in the band of stability
  - → Tend to decay by reducing the number of neutrons
  - → Commonly undergo β decay, which decreases the neutron to proton ratio
- **Proton rich nuclides**
  - → Nuclides that have high proton to neutron (p/n) ratios compared to those nuclides in the band of stability
  - → Tend to decay by reducing the number of protons
  - → Commonly undergo electron capture, positron emission, or proton emission (least likely), which decreases the proton to neutron ratio
- **Decay modes are summarized in the following table**

Nuclide type	Common decay mode(s)	Result of decay
too massive ($Z > 83$)	α decay	A decreases by 4, Z decreases by 2
neutron rich	β decay	Z increases, number of neutrons decreases
proton rich	electron capture or positron emission	Z decreases, number of neutrons increases
proton rich	proton emission	proton plus daughter nuclide with decreased Z

Example 17.4a Xenon-140 lies outside the *band of stability*. Predict the most likely mode of decay. Write a balanced nuclear equation for the transformation and identify the daughter nuclide.

Solution In the periodic table, the atomic mass of Xe ($Z = 54$) is 131.30, suggesting that the stable isotopes of Xe have mass numbers equal to or near 131. The mass number of 140 is much larger than 131, which suggests that xenon-140 is neutron rich. If so, the tendency is one of decreasing the number of neutrons in the nuclide. The most likely process is β decay. The unbalanced equation is

$$^{140}_{54}\text{Xe} \rightarrow {}^{A}_{Z}\text{E} + {}^{0}_{-1}\text{e}$$

Comparison of superscripts yields $A = 140$, and that of subscripts gives $Z = 55$, which is the daughter nuclide, the element cesium, Cs. The final balanced nuclear equation is

$$^{140}_{54}\text{Xe} \rightarrow {}^{140}_{55}\text{Cs} + {}^{0}_{-1}\text{e}$$

Note: The stable nuclide of Xe with the highest mass number is xenon-136. Xenon-140 is unstable and decays by β particle emission.

Example 17.4b Actinium-225 is an unstable nuclide. Predict the most likely mode of decay. Write a balanced nuclear equation for the transformation and identify the daughter nuclide.

Solution Elements with $Z > 83$ often decay by α particle emission. Actinium has $Z = 89$ and is expected to decay by α emission. The unbalanced equation is

$$^{225}_{89}\text{Ac} \rightarrow {}^{A}_{Z}\text{E} + {}^{4}_{2}\alpha$$

Comparison of superscripts yields $A = 221$, and that of subscripts gives $Z = 87$, which is the daughter nuclide, the element francium, Fr. The final balanced nuclear equation is

$$^{225}_{89}\text{Ac} \rightarrow {}^{221}_{87}\text{Fr} + {}^{4}_{2}\alpha$$

Note: Actinium-225 decays by α particle emission. The daughter nuclide, francium-221, is also unstable; it decays by α particle emission to given astatine-217.

17.5 Nucleosynthesis

- **Formation of the elements**
  - → Elements are formed by *nucleosynthesis.*
  - → Occurs when particles collide vigorously with one another, as in stars
  - → *Transmutation* of elements is the conversion of one element into another.
  - → First artificial transmutation occurred in 1919 by Rutherford. A rapidly moving alpha particle collides with a nitrogen-14 nuclide to produce a proton and the oxygen-17 nuclide. The velocity must be fast enough to overcome the repulsive coulomb force between positively charged nuclei. The balanced nuclear equation is

$$^{14}_{7}\text{N} + {}^{4}_{2}\alpha \rightarrow {}^{17}_{8}\text{O} + {}^{1}_{1}\text{p}$$

Nucleosynthesis reactions are commonly written in shorthand as

target (incoming species, ejected species) product

In the Rutherford example, the shorthand notation is

$$^{14}_{7}\text{N}(\alpha,\text{p})^{17}_{8}\text{O}$$

  - → Nucleosynthesis reactions can be accomplished using a variety of projectile species including γ radiation, protons, neutrons, deuterium nuclides, α particles, and heavier nuclides. Neutrons, because they are uncharged, can travel relatively slowly and still react with nuclides.

Example 17.5a In *neutron-induced transmutation*, a uranium-238 nuclide absorbs a neutron and emits a β particle. What is the product of this reaction?

Solution Write an unbalanced nuclear equation. The reactants are uranium-238 and a neutron, and the products are a daughter nuclide and a β particle.

$$^{238}_{92}\text{U} + {}^{1}_{0}\text{n} \rightarrow {}^{A}_{Z}\text{E} + {}^{0}_{-1}\text{e}$$

Comparison of superscripts yields $A = 239$, and that of subscripts gives $Z = 93$, which is the daughter nuclide, the element neptunium, Np. The final balanced nuclear equation is

$$^{238}_{92}\text{U} + {}^{1}_{0}\text{n} \rightarrow {}^{239}_{93}\text{Np} + {}^{0}_{-1}\text{e} \quad \textit{or} \quad {}^{238}_{92}\text{U}(\text{n},\beta){}^{239}_{93}\text{Np}$$

Note: The reaction is a two step process:

$$^{238}_{92}\text{U} + {}^{1}_{0}\text{n} \rightarrow {}^{239}_{92}\text{U}$$

$$^{239}_{92}\text{U} \rightarrow {}^{239}_{93}\text{Np} + {}^{0}_{-1}\text{e}$$

Example 17.5b In *γ-induced transmutation*, a nuclide absorbs γ radiation and then undergoes β decay to produce the neon-20 nuclide. What is the original nuclide?

Solution Write an unbalanced nuclear equation. The reactant nuclide that absorbs γ radiation is unknown, the products are neon-20 and a β particle.

$$^{A}_{Z}\text{E} + \gamma \rightarrow {}^{20}_{10}\text{Ne} + {}^{0}_{-1}\text{e}$$

Comparison of superscripts yields $A = 20$, and that of subscripts gives $Z = 9$, which is the nuclide for the element fluorine, F. The final balanced nuclear equation is

$$^{20}_{9}\text{F} + \gamma \rightarrow {}^{20}_{10}\text{Ne} + {}^{0}_{-1}\text{e} \quad \textit{or} \quad {}^{20}_{9}\text{F}(\gamma,\beta){}^{20}_{10}\text{Ne}$$

- **Transuranium elements**
  - → Elements following uranium ($Z = 92$) in the periodic table
  - → Elements from rutherfordium (Rf, $Z = 104$) to meitnerium (Mt, $Z = 109$) were given official names in 1997.
- **Transmeitnerium elements**
  - → Elements following meitnerium, only a few have been discovered
  - → Temporary names are based on short hand numbers derived from Latin (see Table 17.2 in the text).

Example 17.5c Give the systematic name and symbol for elements with $Z = 114$ and $Z = 126$.

Solution The element with 114 protons has the systematic name of un(1)un(1)quad(4)ium or ununquadium. The symbol for the element is Uuq.

The element with 126 protons has the systematic name of un(1)bi(2)hex(6)ium or unbihexium. The symbol for the element is Ubh.

Example 17.5d Determine the atomic number of the elements with the systematic symbols Uto and Bun.

Solution The element, Uto, is Untrioctium with $Z = 138$.

The element, Bun, is Biunnilium with $Z = 210$.

Nuclear Radiation (Sections 17.6-17.8)

Key Concepts

ionizing radiation, absorbed dose, rad (radiation absorbed dose), gray, relative biological effectiveness, dose equivalent, roentgen equivalent man (rem), sievert, activity, becquerel, curie, law of radioactive decay, decay constant, nuclear fallout, isotopic dating, radiocarbon dating, radioisotopes, tracers

Overview

- **Nuclear radiation**
  - → Includes α, β, and β^+ particles and γ radiation; sometimes called *ionizing radiation*
  - → Ejects electrons from atoms to form ions
  - → Breaks chemical bonds in molecules in tissue
  - → Used to kill cancer cells, but may damage healthy tissue

17.6 The Biological Effects of Radiation

- **Penetrating power**
  - → Alpha, beta, and gamma particles are forms of *ionizing radiation* with different *penetration power* and require different amounts of shielding to prevent harmful effects (see Table 17.3 in the text).

→ Alpha particles have the least penetrating power and are absorbed by paper and the outer layers of skin. If inhaled or ingested, alpha particles are much more damaging.
→ Beta particles are 100 times more penetrating than alpha particles and are absorbed by 1 cm of flesh or 3 mm of aluminum, for example.
→ Gamma radiation is 100 times more penetrating than beta particles and can pass through buildings and bodies leaving a trail of ionized and damaged molecules. Gamma radiation is absorbed by lead bricks or thick concrete.
→ The order of *increasing* penetrating power is $\alpha < \beta < \gamma$.

- **Radiation damage to tissue**

→ Depends on the *strength* of the source of radiation, the *type* of radiation, the *length* of exposure, and the *extent* to which the radiation can reach sensitive tissue
→ For example, Pu^{4+} is an α emitter and, if ingested, replaces Fe^{3+} in the body resulting in the inhibition of the production of red blood cells.

- **Absorbed dose of radiation**

→ Energy deposited in a sample, such as a human body when exposed to radiation
→ Units of the absorbed dose of radiation are

1 rad = amount of radiation that deposits 0.01 J of energy per kg of tissue
1 gray (Gy) = an energy deposit of 1 $J{\cdot}kg^{-1}$ (SI unit)
1 Gy = 100 rad

Note: The name *rad* stands for "radiation absorbed dose."

- **Dose equivalent**

→ Actual dose modified to account for different destructive powers of radiation
→ Relative destructive power is called the *relative biological effectiveness*, Q. For β^- and γ radiation, $Q \equiv 1$. For α radiation, $Q \approx 20$.
→ The natural unit of dose equivalent is the rem (*roentgen equivalent man*).
→ Definition: Dose equivalent in rem = $Q \times$ absorbed dose in rad
→ SI Units: Dose equivalent in sievert (Sv) = $Q \times$ absorbed dose in gray (Gy)
1 Sv = 100 rem

- **Common radiation dose and harmful effects**

→ Typical annual human dose equivalent from natural sources is 0.2 $rem{\cdot}y^{-1}$, a number with a wide range depending upon habitat and lifestyle.
→ Typical chest x-ray gives a dose equivalent of about 7 millirem.
→ 30 rad (30 rem) of γ radiation may cause a reduction in white blood cell count.
→ 30 rad (600 rem) of α radiation causes death.

Example 17.6 What is the dose equivalent in rem and in sievert when 5.0 J of radiation is delivered to 0.55 kg of tissue by α radiation?

Solution Calculate the energy delivered per kg, the dose in rad, and the dose in rem using the relative biological effectiveness factor Q for α radiation. Finally, convert the unit of rem to the SI unit of Sv, recalling that 1 Sv= 100 rem.

Energy per kilogram $= \dfrac{5.0\ \text{J}}{0.55\ \text{kg}} = 9.09\ \text{J}\cdot\text{kg}^{-1}$

Conversion to rad: $\left(\dfrac{9.09\ \text{J}\cdot\text{kg}^{-1}}{0.01\ \text{J}\cdot\text{kg}^{-1}}\right)\text{rad} = 909\ \text{rad}$

Conversion to rem: Dose equivalent = (909 rad)(Q = 20) = 1.8×10^4 rem

Conversion to Sv: (1.8×10^4 rem)/(100 rem·Sv^{-1}) = 1.8×10^2 Sv

Note: An older unit for absorbed dose was the *roentgen*. It was defined as the quantity of γ or x-radiation that produces a certain amount of ionization in a certain quantity of air.

17.7 Measuring the Rate of Nuclear Decay

- **Activity of a sample**
  - → Number of nuclear disintegrations per second
  - → One nuclear disintegration per second is called 1 becquerel, Bq (SI unit).
  - → The curie, Ci, an older unit for activity, is equal to 3.7×10^{10} Bq, which is equal to the radioactive output of 1 g of radium-226.
  - → Because the curie represents a large value of decay, the activity of typical samples has units expressed in the millicurie, mCi, or the microcurie, μCi.

  Note: See Table 17.4 in the text for a list of the radiation units.

Example 17.7a The activity of a certain cobalt-60 source is 45.0 μCi. What is the value of the activity in the unit of becquerel and disintegrations per second?

Solution Because the becquerel is defined as one disintegration per second, the only conversion we need is the one between the curie and the becquerel. Recall that 1 Ci = 3.7×10^{10} Bq and 1 μCi = 1×10^{-6} Ci.

$$45.0\ \mu\text{Ci}\left(\frac{1\times10^{-6}\ \text{Ci}}{1\ \mu\text{Ci}}\right)\left(\frac{3.7\times10^{10}\ \text{Bq}}{1\ \text{Ci}}\right) = 1.66\times10^{6}\ \text{Bq} = 1.66\ \text{MBq}$$

The answer is the same for the unit of disintegrations per second.

- **Nuclear decay**
  - → The decay of a nuclide (isotope) can be written as

    Parent nuclide → daughter nuclide + radiation
  - → This form is the same as an *elementary reaction* that is *unimolecular*. An unstable nuclide takes the place of an excited molecule.
  - → The rate of nuclear decay depends only on the identity of the isotope, not on its chemical form or temperature.
- **Law of radioactive decay**
  - → The rate of nuclear decay and the number N of radioactive nuclei present is given by

    $$\boxed{\text{Activity} = \text{rate of decay} = k \times N}$$
  - → k = *decay constant* (or rate constant for the reaction)

    The units of the *decay constant* are the same as a *first order* rate constant, $(\text{time})^{-1}$.

- **Integrated rate law for nuclear decay**

→ Exponential form: $N = N_0 e^{-kt}$ Logarithmic form: $\ln\left(\frac{N}{N_0}\right) = -kt$

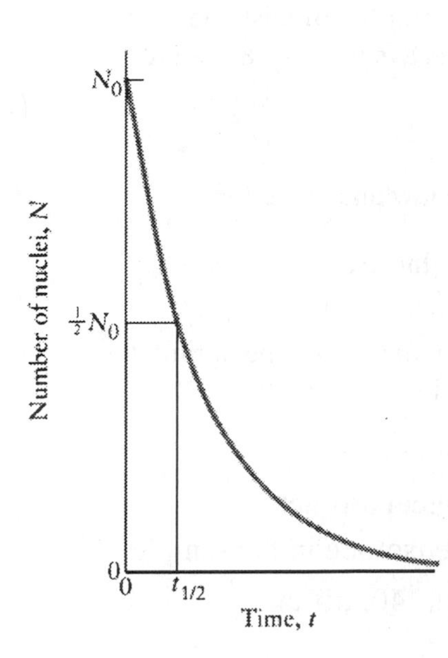

On the left is a plot of N *vs.* t for nuclear decay. N_0 is the number of radioactive nuclei present at $t = 0$. N is the number of radioactive nuclei present at a later time t. The time required for the nuclei to decay to half their initial number is equal to the half-life, $t_{1/2}$. At this time, $N = \frac{1}{2}N_0$. Substituting this value into the integrated rate law yields

$$t_{1/2} = \frac{\ln 2}{k}$$

Half-lives of radioactive nuclides span an enormous range from picoseconds (^{215}Fr, 120 ps) to billions of years (^{238}U, 4.5×10^9 y). See Table 17.5 in the text for a list of half-lives of common radioactive isotopes.

Example 17.7b The half-life of the β emitter, carbon-14, is 5.73×10^3 y. Determine the decay constant for this nuclide in y^{-1} and min^{-1}. What percentage of the original activity of a given sample of carbon-14 remains after 1.00×10^4 y?

Solution Use the half-life relation to determine the decay constant in the original units.
$k = (\ln 2)/(5.73 \times 10^3 \text{ y}) = 1.21 \times 10^{-4} \text{ y}^{-1}$
Use the factor method to convert from reciprocal years to reciprocal minutes.

$$k = 1.21 \times 10^{-4}\left(\frac{1}{y}\right)\left(\frac{1 \text{ y}}{365 \text{ d}}\right)\left(\frac{1 \text{ d}}{24 \text{ h}}\right)\left(\frac{1 \text{ h}}{60 \text{ min}}\right) = 2.30 \times 10^{-10} \text{ min}^{-1}$$

The activity of a sample is proportional to the number of nuclides, N, in the sample. Comparing the activity at $t = 0$ with the value at future time t gives

$$\frac{(\text{Activity})_t}{(\text{Activity})_0} = \frac{kN_t}{kN_0} = \frac{N_t}{N_0} = \text{fraction of the original activity remaining at time } t$$

$$\frac{N_t}{N_0} = e^{-kt} = e^{-(1.21\times10^{-4}\text{ y}^{-1})(1.00\times10^{4}\text{ y})} = e^{-1.21} = 0.298$$

The fraction of the original activity remaining is 0.298 and the percentage remaining is the same number multiplied by 100% or 29.8%.

Example 17.7c A 26 mg sample of thorium-215 with $t_{1/2} = 1.2$ s decays by α emission. What is the daughter nuclide? How much time will elapse for 18 mg of the sample to decay?

Solution The unbalanced nuclear equation is

$$^{215}_{90}\text{Th} \rightarrow {}^{A}_{Z}\text{E} + {}^{4}_{2}\alpha$$

Comparison of subscripts yields $A = 211$, and that of subscripts gives $Z = 88$, corresponding to the nuclide for the element radium, Ra. The daughter nuclide is $^{211}_{88}\text{Ra}$.

The mass of thorium-215 is proportional to the number of unstable nuclei. Therefore, the mass that remains after 18 mg decays is 26 – 18 = 8 mg.

$$\frac{N_t}{N_0} = \frac{m_t}{m_0} = \frac{8 \text{ mg}}{26 \text{ mg}} = 0.3077$$

Use the logarithmic form of the first-order rate law and solve for t.

$$t = -\frac{1}{k}\ln\left(\frac{N_t}{N_0}\right) = -\left(\frac{t_{1/2}}{\ln 2}\right)\ln\left(\frac{N_t}{N_0}\right) = -\left(\frac{1.2 \text{ s}}{0.6931}\right)\ln(0.3077) = 2.0 \text{ s}$$

Note: In these examples, we identify the number of nuclei remaining at time t with the symbol N_t, a notation identical to the one used in Chapter 13.

- **Isotopic dating**
  - → Used to determine the ages of rocks and of archeological artifacts
  - → Carried out by measuring the activity of a radioactive isotope in the sample
  - → Useful isotopes for dating are uranium-238, potassium-40, tritium, $^{3}_{1}\text{H}$, and carbon-14.
- **Radiocarbon dating**
  - → Most important example of isotopic dating
  - → Uses the beta decay of carbon-14, for which the half-life is 5730 y.
- **Radioactive ^{14}C**
  - → Formed in the atmosphere by neutron bombardment of ^{14}N

$$^{14}_{7}\text{N} + {}^{1}_{0}\text{n} \rightarrow {}^{14}_{6}\text{C} + {}^{1}_{1}\text{p}$$

  - → Enters living organisms as $^{14}CO_2$ through photosynthesis and digestion
  - → Leaves living organisms by excretion and respiration
  - → Achieves a steady-state concentration in living organism with an activity of 15 disintegrations per minute per gram of *total* carbon
  - → Shows decreasing activity in dead organisms because they no longer ingest ^{14}C from the atmosphere
  - → Using the integrated rate law, we can estimate the time of death of an organism by measuring the activity of the sample.

Example 17.7d A sample of charcoal from an archeological dig undergoes 7.80×10^3 disintegrations of carbon-14 nuclides per gram of total carbon in a 15-h period. Determine the age of the tree from which the charcoal was derived.

Solution Calculate the number of disintegrations per gram of carbon expected over the same 15-h period from living material and compare the value to that observed for the dead object. For one gram of carbon in a living organism, we find

$(15 \text{ disintegrations}\cdot\text{min}^{-1})(60 \text{ min}\cdot\text{h}^{-1})(15 \text{ h}) = 1.35 \times 10^4$ disintegrations

$$t = -\left(\frac{t_{1/2}}{\ln 2}\right)\ln\left(\frac{(\text{Activity})_t}{(\text{Activity})_0}\right) = -\left(\frac{5730 \text{ y}}{0.6931}\right)\ln\left(\frac{7.80\times10^3}{1.35\times10^4}\right) = 4.54\times10^3 \text{ y}$$

The tree died approximately 4500 years ago.

17.8 Uses of Radioisotopes

- **Radioisotopes**
  - → Used to cure disease (see Box 17.2 in the text), preserve food, and trace the mechanisms of chemical reactions
  - → Used to power spacecraft, locate sources of water, and determine the age of nonliving materials including wood (^{14}C dating), rocks ($^{238}U/^{206}Pb$ ratio), and ground water (tritium dating using the $^1H/^3H$ ratio)
- **Radioactive tracers**
  - → Radioactive isotopes are used to track changes and locations. For example, phosphorus-32, a β emitter with $t_{1/2} = 14.28$ d, can be incorporated into phosphate-containing fertilizer to follow the mechanism of plant growth.
  - → In chemical reactions, tracers help determine the mechanism of the reaction. Nonradioactive isotopes are also used for this purpose as in the following example.

Example 17.8 A chemist suggests that when acetone is placed in water, a small amount of acetone hydrate is formed. The amount is small because the equilibrium lies far to the reactant side as indicated in the following diagram:

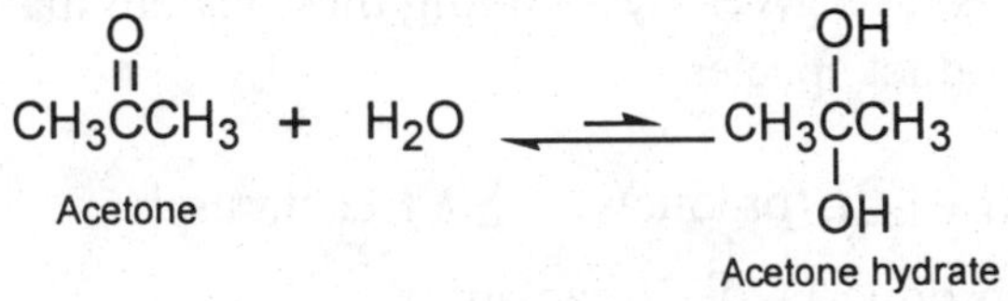

$$CH_3C(=O)CH_3 + H_2O \rightleftharpoons CH_3C(OH)_2CH_3$$

Acetone — Acetone hydrate

A skeptic does not believe this reaction occurs at all. Propose an experiment using a nonradioactive isotope that supports the formation of acetone hydrate in water.

Solution Use water that is isotopically enriched with nonradioactive oxygen-18. Determine the extent to which ^{18}O is incorporated into the acetone. This can be determined by mass spectroscopy of the vaporized liquid. Ordinary acetone has a molar mass of 58 g·mol^{-1}, whereas ^{18}O-substituted acetone, $C_3H_6{}^{18}O$, has a molar mass of 60 g·mol^{-1} and is easily detected. Finding a substantial amount of oxygen-18 in the acetone supports the proposed equilibrium and the existence of acetone hydrate.

Nuclear Energy (Sections 17.9-17.12)

Key Concepts

mass-energy conversion, nuclear binding energy, electronvolt, nuclear fission, spontaneous nuclear fission, induced nuclear fission, fissionable nuclei, fissile nuclei, critical mass, supercritical mass, subcritical mass, moderator, breeder reactor, nuclear fusion, thermonuclear explosion, plasma, isotope enrichment

Overview

- **Nuclear reactions**
  → Tend to release huge amounts of energy
  → Widely used to provide electrical power
  → Free of chemical pollution compared with combustion reactions
  → Generate radioactive waste and/or radioactive pollution

17.9 Mass-Energy Conversion

- **Nuclear binding energy**
  → Energy *released* when protons and neutrons join together to form a *nucleus*

$$(Z)\,{}^{1}_{1}\mathrm{p} + (A-Z)\,{}^{1}_{0}\mathrm{n} \rightarrow {}^{A}_{Z}\mathrm{E}^{Z+}$$

  → Since *nuclides* are *neutral* atoms, the energy released is also given *approximately* by

$$(Z)\,{}^{1}_{1}\mathrm{H} + (A-Z)\,{}^{1}_{0}\mathrm{n} \rightarrow {}^{A}_{Z}\mathrm{E}$$

  → The binding energy is then given by applying the Einstein mass-energy equation to each reactant and product species:

$$\begin{aligned} E_{\text{bind}} &= |\Delta E| = |\Sigma E(\text{products}) - \Sigma E(\text{reactants})| \\ &= |E(\text{product}) - \Sigma E(\text{reactants})| \\ &= |mc^2(\text{product}) - \Sigma mc^2(\text{reactants})| \\ &= |m(\text{product}) - \Sigma m(\text{reactants})| \times c^2 \end{aligned}$$

$$\boxed{E_{\text{bind}} = |\Delta m| \times c^2}$$

  → *Binding energy* is a measure of the stability of the nucleus.
  → The *binding energy per nucleon* is defined as E_{bind}/A and is a better measure of the stability of the nucleus.

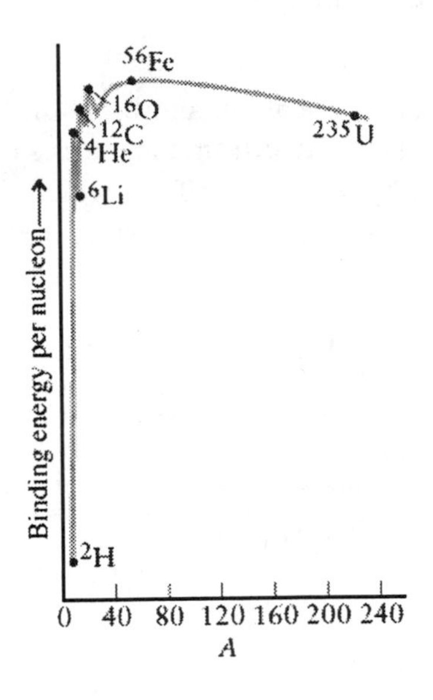

A plot of the binding energy per nucleon *vs.* atomic mass number is shown on the left (Fig. 17.20 in the text). Nucleons are bound together most strongly in the elements near iron and nickel, which accounts for the high abundance of iron and nickel in meteorites and on planets similar to the earth. From the figure, it appears that light nuclei become more stable when they "fuse" together to form heavier ones. Heavy nuclei become more stable when they undergo "fission" and split into lighter ones. Also, note the increased stability of the nuclei with magic numbers, particularly oxygen-16, which is *doubly magic*.

- **Units used in nuclear calculations**
  - → *Mass* is usually reported in *atomic mass units*, u.
  - → One atomic mass unit is defined as exactly $\frac{1}{12}$ the mass of one atom (nuclide) of ^{12}C:

$$\boxed{1\ \text{u} = 1.660\,54 \times 10^{-27}\ \text{kg}}$$

  - → *Binding energy* is usually reported in electronvolts, eV, or millions (mega) of electronvolts, MeV. An electronvolt is the change in potential energy of an electron (charge of $1.602\,18 \times 10^{-19}$ C) when it is moved through a potential difference of 1 V:

$$\boxed{1\ \text{eV} = 1.60218 \times 10^{-19}\ \text{J}}$$

Example 17.9 Calculate the binding energy of nitrogen-14 in MeV and kJ·mol^{-1} given the following masses: $m_H = 1.0078$ u (hydrogen atom), $m_n = 1.0087$ u (neutron), and $m_N = 14.0031$ u (nitrogen-14 nuclide). Determine the binding energy per nucleon in MeV.

Solution Nitrogen-14 contains 7 protons, 7 neutrons, and 7 electrons.

Calculate $|\Delta m|$ in atomic mass units.

$|\Delta m| = |m_N - 7\,m_H - 7\,m_n| = |14.0031 - 7(1.0078) - 7(1.0087)| = |-0.1124\ \text{u}|$
$= 0.1124$ u

Convert atomic mass units to kilograms.

$|\Delta m| = (0.1124\ \text{u})(1.66054 \times 10^{-27}\ \text{kg·u}^{-1}) = 1.8664 \times 10^{-28}$ kg

Calculate the binding energy in J, MeV, and kJ·mol^{-1}.

$E_{bind} = |\Delta m| \times c^2 = (1.8664 \times 10^{-28}\ \text{kg})(2.997\,92 \times 10^8\ \text{m})^2 = 1.6774 \times 10^{-11}$ J

$= (1.6774 \times 10^{-11}\ \text{J})/(1.602\,18 \times 10^{-19}\ \text{J·eV}^{-1}) = 1.047 \times 10^8\ \text{eV} = 104.7$ MeV

$= (1.6774 \times 10^{-11}\ \text{J})(10^{-3}\ \text{kJ·J}^{-1})(6.022\,14 \times 10^{23}\ \text{mol}^{-1}) = 1.010 \times 10^{10}\ \text{kJ·mol}^{-1}$

Calculate the binding energy per nucleon in MeV·nucleon^{-1}.
E_{bind}/A = (104.7 MeV)/14 = 7.479 MeV·nucleon^{-1}

Note: Typical energies associated with chemical reactions are about a million times smaller. If the binding energy of nitrogen-14 is calculated from the masses of the nuclei rather than the nuclides, the result differs by about 2×10^{-6} u, which corresponds to 2×10^{-3} MeV. This is a relatively small error that arises from the binding energy of the electrons to the nuclei.

- **Masses of particles** → The following table lists selected values of particle mass.

Name	Symbol	Mass
electron	${}^{0}_{-1}e$	5.485799×10^{-4} u
neutron	${}^{1}_{0}n$	1.008665 u
proton	${}^{1}_{1}p$	1.007276 u
hydrogen (atom)	${}^{1}_{1}H$	1.007825 u
deuteron (nucleus)	${}^{2}_{1}d$	2.013553 u
deuterium (atom)	${}^{2}_{1}H$	2.014102 u
triton (nucleus)	${}^{3}_{1}t$	3.015501 u
tritium (atom)	${}^{3}_{1}H$	3.016049 u
alpha particle (nucleus)	${}^{4}_{2}\alpha$	4.001506 u
helium-4 (atom)	${}^{4}_{2}He$	4.002603 u

Note: The free neutron has a half-life of 10.3 min. Tritium has a half-life of 12.3 y.

17.10 Nuclear Fission

- **Nuclear fission**
  → Occurs when a nucleus breaks into two or more smaller nuclei
  → Releases a large amount of energy
  → May be *spontaneous* or *induced*

- **Spontaneous nuclear fission**
  → Occurs when oscillations in heavy nuclei lead them to break into two smaller nuclei of similar mass
  → Yields a variety of products for a given nuclide (see Fig. 17.22 in the text)
  → An example is the spontaneous nuclear fission of americium-244. Two of the daughter nuclides formed are iodine-134 and molybdenum-107. The reaction is

$$ {}^{244}_{95}Am \rightarrow {}^{134}_{53}I + {}^{107}_{42}Mo + 3\,{}^{1}_{0}n $$

- **Induced nuclear fission**
  → Caused by bombarding a heavy nucleus with neutrons

→ Yields a variety of daughter nuclides for a given parent

→ An example is the induced fission occurring by the neutron bombardment of plutonium-239. A number of products are formed. Two important reactions are

$$^{239}_{94}\text{Pu} + ^{1}_{0}\text{n} \rightarrow ^{98}_{42}\text{Mo} + ^{138}_{52}\text{Te} + 4\,^{1}_{0}\text{n}$$

$$^{239}_{94}\text{Pu} + ^{1}_{0}\text{n} \rightarrow ^{100}_{43}\text{Tc} + ^{135}_{51}\text{Sb} + 5\,^{1}_{0}\text{n}$$

- **Energy changes in fission reactions**

→ Energy released during fission is calculated by using Einstein's equation.

→ For a balanced nuclear reaction,

$$\boxed{\Delta m = \sum m(\text{products}) - \sum m(\text{reactants})} \quad \text{and} \quad \boxed{\Delta E = \Delta m \times c^2}$$

Example 17.10 Determine the energy released in MeV and in kJ·mol^{-1} for the fission reaction

$$^{235}_{92}\text{U} + ^{1}_{0}\text{n} \rightarrow ^{97}_{40}\text{Zr} + ^{137}_{52}\text{Te} + 2\,^{1}_{0}\text{n}$$

The masses of the particles are $^{1}_{0}\text{n}$, 1.0087 u; $^{235}_{92}\text{U}$, 235.0439 u; $^{97}_{40}\text{Zr}$, 96.9110 u; $^{137}_{52}\text{Te}$, 136.9254 u.

Solution The change in mass associated with the reaction is

$\Delta m = m(\text{Zr}) + m(\text{Te}) + 2m_n - \text{m(U)} - m_n = m(\text{Zr}) + m(\text{Te}) + m_n - m(\text{U})$

$= 96.9110 + 136.9254 + 1.0087 - 235.0439 = -0.1988$ u

Convert change in mass to kilograms.

$\Delta m = -(0.1988\ \text{u})(1.66054 \times 10^{-27}\ \text{kg·u}^{-1}) = -3.301 \times 10^{-28}$ kg

Calculate the energy change for the reaction in J, MeV and kJ·mol^{-1}.

$\Delta E = -(3.301 \times 10^{-28}\ \text{kg})(2.9979 \times 10^{8}\ \text{m·s}^{-1})^2 = -2.967 \times 10^{-11}$ J

$= -(2.967 \times 10^{-11}\ \text{J})/(1.60218 \times 10^{-19}\ \text{J·eV}^{-1}) = -1.852 \times 10^{8}$ eV

$= -185.2$ MeV

The energy released when one uranium-235 nuclide undergoes fission is the negative of this value or 185.2 MeV.

$= -(2.967 \times 10^{-11}\ \text{J})(10^{-3}\ \text{kJ·J}^{-1})(6.02214 \times 10^{23}\ \text{mol}^{-1})$

$= -1.787 \times 10^{10}$ kJ·mol^{-1}

Similarly, the energy released is 1.787×10^{10} kJ·mol^{-1}.

- **Fissionable and fissile nuclei**

→ *Fissionable nuclei* are nuclei that can undergo induced fission.

→ *Fissile nuclei* are fissionable nuclei that can undergo induced fission with slow-moving neutrons.

→ Examples of fissile nuclei are $^{235}_{92}\text{U}$, $^{233}_{92}\text{U}$, and $^{239}_{94}\text{Pu}$, which are the fuels of nuclear power plants. The nuclide, $^{238}_{92}\text{U}$, is fissionable by fast-moving neutrons but is *not* fissile.

- **Nuclear branched chain reactions**

→ Neutrons are *chain carriers* in a branched chain reaction (see Section 13.12).

→ Chain branching occurs when an induced fission reaction produces two or more neutrons.

→ An example is the reaction, ${}^{235}_{92}U + {}^{1}_{0}n \rightarrow {}^{97}_{40}Zr + {}^{137}_{52}Te + 2\,{}^{1}_{0}n$, in which two neutrons are produced for each uranium-235 nuclide that reacts. The two neutrons can either escape into the surroundings or be captured by (react with) other uranium-235 nuclides.

→ A *critical mass* is the mass of fissionable material above which so few neutrons escape from the sample that the fission chain reaction is sustained. The critical mass for pure plutonium of normal density is about 15 kg. A sample of plutonium with mass greater than 15 kg is *supercritical*; the reaction is self-sustaining and may result in an explosion. A *subcritical* sample is one that has less than the critical mass for its density.

- **Controlled and uncontrolled (explosive) nuclear reactions**

→ Explosive nuclear reactions occur when a *subcritical* amount of fissile material is made *supercritical* so rapidly that the chain reaction occurs uniformly throughout the material.

→ Controlled nuclear reactions occur in nuclear reactors. They are not explosive because the fuel is subcritical. The rate of reaction is controlled by a *moderator* such as graphite rods, which slow the emitted neutrons down so that a greater fraction can induce fission.

→ The difference between a fission explosion and a controlled fission is shown in the figure. Each symbol ⊗ represents a fission event. The neutrons from the fission are shown but products are not.

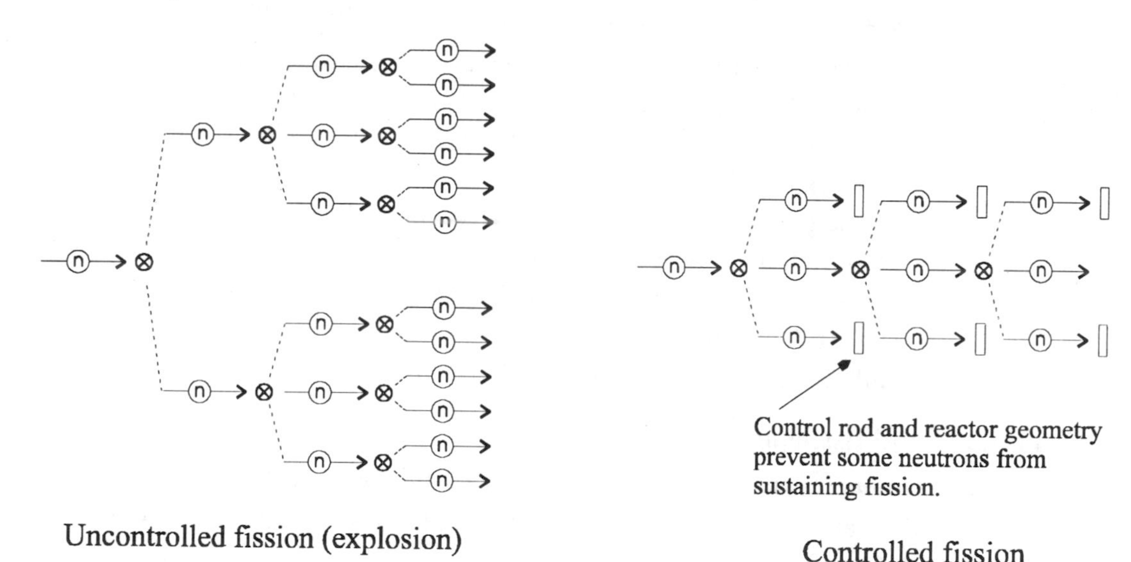
n
Control rod and reactor geometry prevent some neutrons from sustaining fission.

Uncontrolled fission (explosion) Controlled fission

Note: The number of neutrons released in an event is variable as is the identity of the products (see Fig. 17.22 in the text).

- **Breeder reactors**

→ Nuclear reactors used to synthesize fissile nuclides for fuel and weapon use are called *breeder reactors*.

→ Breeder reactors run very hot and fast because no moderator is present. As a result, they are more dangerous than nuclear reactors used for power generation.

17.11 Nuclear Fusion

- **Nuclear fusion**
  - → Occurs when lighter nuclei fuse together to form a heavier nucleus
  - → Accompanied by a large release of energy if light nuclei fuse, because the heavier nuclei have larger binding energies per nucleon (see Fig. 17.20 in the text)
  - → Occurs in stars and in the hydrogen bomb, in which the fusion reaction is uncontrolled
  - → Difficult to achieve because charged nuclei must collide at tremendous speeds to fuse together
  - → Fusion is essentially free of radioactive waste.
  - → Fusion has not yet been sustained for an appreciable time in a controlled reaction on earth. A safe, sustained fusion reaction could provide an almost unlimited source of energy.

- **Energy changes in fusion reactions**
  - → Energy released during a fusion reaction is calculated by using Einstein's equation.
  - → The procedure is the same as for fission reactions (see Section 17.10).

Example 17.11 Calculate the energy released when 1.00 g of deuterium forms helium-3 in the fusion reaction

$$^{2}_{1}\text{H} + ^{2}_{1}\text{H} \rightarrow ^{3}_{2}\text{He} + ^{1}_{0}\text{n}$$

Compare the value to 69.7 kJ, which is the energy released when 1.00 g of deuterium is oxidized at 298K to form liquid deuterium oxide.
The masses of the particles are $^{1}_{0}\text{n}$, 1.0087 u; $^{2}_{1}\text{H}$ (deuterium), 2.0141 u; $^{3}_{2}\text{He}$, 3.0160 u.

Solution The change in mass associated with the reaction is
$\Delta m = m(\text{He}) + m_n - 2m(\text{H}) = 3.0160 + 1.0087 - 2(2.0141) = -0.0035$ u
Convert change in mass to kilograms.
$\Delta m = -(0.0035\ \text{u})(1.66054 \times 10^{-27}\ \text{kg·u}^{-1}) = -5.81 \times 10^{-30}$ kg
Calculate the energy change for the reaction in J and kJ·mol^{-1}.
$\Delta E = -(5.81 \times 10^{-30}\ \text{kg})(2.9979 \times 10^{8}\ \text{m·s}^{-1})^2 = -5.22 \times 10^{-13}$ J
$= -(5.22 \times 10^{-13}\ \text{J})(10^{-3}\ \text{kJ·J}^{-1})(6.02214 \times 10^{23}\ \text{mol}^{-1})$
$= -3.14 \times 10^{8}\ \text{kJ·mol}^{-1}$
This amount is for two mol of deuterium. The energy change for one mole of deuterium is
$\Delta E = -(3.14 \times 10^{8}\ \text{kJ})/2\ \text{mol} = -1.57 \times 10^{8}\ \text{kJ·mol}^{-1}$
The molar mass of deuterium is 2.0141 g·mol^{-1}. The energy change for 1.00 g of deuterium is
$\Delta E = -(1.57 \times 10^{8}\ \text{kJ·mol}^{-1})/(2.0141\ \text{g·mol}^{-1}) = -7.80 \times 10^{7}\ \text{kJ·g}^{-1}$
The energy produced is the negative of this value, $7.80 \times 10^{7}\ \text{kJ·g}^{-1}$.
The chemical combustion of 1.00 g of deuterium produces only 69.7 kJ.
The energy "advantage" for the nuclear fusion reaction is

$$\frac{7.80 \times 10^7}{69.7} = 1.12 \times 10^6$$ or a million times increase in energy production.

Note: Nuclear fusion clearly has the potential to provide enormous amounts of energy, and it could replace fossil fuels. One difficulty in developing this energy source is the tremendously high temperatures required (on the order of 100 million degrees kelvin) to maintain fusion reactions. At such high temperatures, no vessel can contain the reactants. One solution is to heat an ionized gas (plasma) that is contained by magnetic fields.

17.12 The Chemistry of Nuclear Power

- **Chemistry**
  - → The key to the safe use of nuclear power
  - → Used to prepare nuclear fuel
  - → Used to recover important fission products
  - → Used to dispose of nuclear waste safely
- **Uranium**
  - → The fuel of nuclear reactors
  - → Obtained primarily from the ore pitchblende, UO_2, by reducing the oxide to the metal and then *enriching* or increasing the fraction of the fissile nuclide, uranium-235
  - → To be useful as a fuel, the percentage of ^{235}U must be increased from its natural abundance of 0.7% to about 3%.
- **Enrichment**
  - → Exploits the mass difference between ^{235}U and ^{238}U
  - → Separation is accomplished by repeated effusion of $^{235}UF_6$ and $^{238}UF_6$ vapor (see Graham's law of effusion in Section 4.11).

$$\frac{\text{Rate of effusion of } ^{235}UF_6}{\text{Rate of effusion of } ^{238}UF_6} = \sqrt{\frac{M_{238}}{M_{235}}} = \sqrt{\frac{352.1}{349.0}} = 1.004$$

Note: Because the ratio is close to one, repeated effusion steps (hundreds) are required to get the desired separation.

Example 17.12a Calculate the theoretical number of stages needed to enrich ^{235}U from its natural abundance of 0.72% to 3.0% by vapor effusion, using a uranium hexafluoride mixture containing $^{235}UF_6(g)$ and $^{238}UF_6(g)$. Assume that the enrichment factor for a single stage is the ideal value of 1.004.

Solution Let N_{235}/N_{238} represent the ratio of the number of molecules of $^{235}UF_6$ and $^{238}UF_6$ in the gas. According to the natural abundance of 0.72%, the initial ratio is

$$\left(\frac{N_{235}}{N_{238}}\right)_{\text{initial}} = \frac{0.72}{99.28} = 7.252 \times 10^{-3}$$

The final ratio desired is

$$\left(\frac{N_{235}}{N_{238}}\right)_{\text{final}} = \frac{3.00}{97.00} = 3.093 \times 10^{-2}$$

For each stage of effusion, the ratio of the light to heavy isotope is multiplied by a factor of 1.004. Therefore, we write the following

$$\left(\frac{N_{235}}{N_{238}}\right)_{\text{initial}} \times (1.004)^x = \left(\frac{N_{235}}{N_{238}}\right)_{\text{final}}$$

$$(7.252 \times 10^{-3})(1.004)^x = (3.093 \times 10^{-2})$$

$$(1.004)^x = 4.265$$

Solve by using logarithms as follows

$$\ln (1.004)^x = x \ln (1.004) = \ln (4.265)$$

$$(3.992 \times 10^{-3})\, x = 1.450$$

$$x = 363$$

Therefore, 363 stages of effusion are required to purify ^{235}U to an abundance of 3% assuming ideal separation efficiency.

Note: In this example, we ignored the presence of uranium-233, which has a natural abundance of 0.0056%. This isotope is also enriched in the final mixture.

Example 17.12b Why is it necessary to enrich the ^{235}U content from the 0.72% natural abundance value to 3% or more in order to use uranium as a fuel in fission reactors?

Solution Enrichment is necessary for a sustained fission reaction to occur. For a fission reaction to be self-sustaining, a certain fraction of the neutrons released during a fission event must induce additional events by collisions with another fissile nuclide. If the density of uranium-235 nuclei is not high enough, too large a fraction of the neutrons released in the fission events will escape from the fuel and the reaction will not be sustained.

- **Nuclear waste**
  - → Spent nuclear fuel called *nuclear waste* is still radioactive. It is a mixture of uranium and fission products.
  - → Nuclear waste must be stored safely for about 10 half-lives, and it is generally buried underground.
  - → For underground burial, incorporation of the highly radioactive fission (HRF) products into a glass or ceramic material is better than placing it in metal storage drums. Storage drums can corrode, allowing the waste to seep into aquifers and/or contaminate large areas of soil.

Chapter 18 Organic Chemistry I: The Hydrocarbons

Aliphatic Hydrocarbons (Sections 18.1-18.7)

Key Concepts

hydrocarbons, types of hydrocarbons, condensed structural formula, stick structure, aliphatic hydrocarbon, aromatic hydrocarbon, saturated hydrocarbon, unsaturated hydrocarbon, nomenclature of hydrocarbons, alkanes, cycloalkanes, alkenes, alkynes, isomers, structural isomers, connectivity, stereoisomers, geometrical isomers, optical isomers, chiral molecule, enantiomers, achiral molecule, racemic mixture, properties of alkanes, alkane substitution (mechanism), properties of alkenes, elimination reaction, dehydrohalogenation reaction, electrophilic addition to alkenes (mechanism), electrophile

Overview

- **Hydrocarbons**
  - → Compounds containing only carbon and hydrogen (*Study Guide*, Section 3.7)
  - → Hydrocarbons without benzene rings are *aliphatic.*
  - → Saturated hydrocarbons with no multiple bonds are *alkanes*.
  - → Unsaturated hydrocarbons with at least one double bond are *alkenes* and with at least one triple bond are *alkynes.*
  - → Hydrocarbons with benzene rings are *aromatic.*
  - → Complex molecules often have "aromatic parts" and "aliphatic parts."

18.1 Types of Hydrocarbons

- **Structural formula**
  - → Shows chemical connectivity, how atoms are attached to one another
  - → For example, butane and methylpropane have the same *empirical* (C_2H_5)and *molecular* (C_4H_{10}) formulas, respectively, but different *structural formulas*

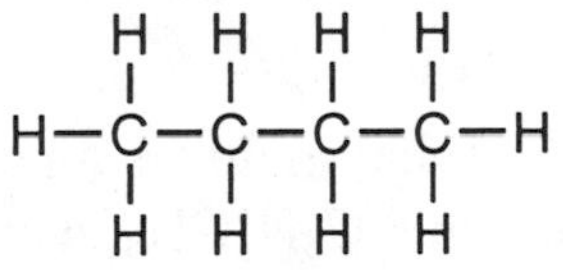

```
    H  H  H  H            H CH3 H
    |  |  |  |            |  |  |
  H-C--C--C--C-H        H-C--C--C-H
    |  |  |  |            |  |  |
    H  H  H  H            H  H  H
       Butane           Methylpropane
```

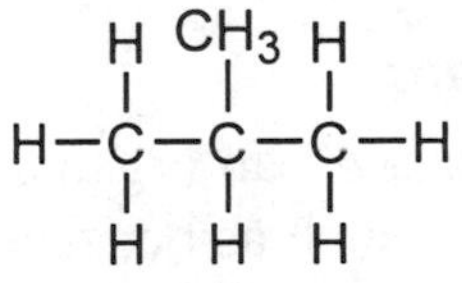

In the structure on the right, an abbreviated notation for the methyl group (CH_3–) attached to the propane chain is used (the three C–H bonds are *not* shown). A structure with all three terminal methyl groups in abbreviated form is

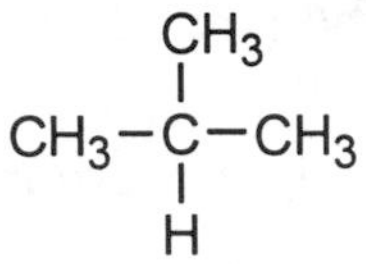

```
        CH3
         |
   CH3 - C - CH3
         |
         H
```

- **Condensed structural formula**
  - → Simple way to represent structures of complicated organic molecules (see Section C)
  - → Shows how the atoms are grouped together in an abbreviated form
  - → *Condensed structural formula* for butane is $CH_3CH_2CH_2CH_3$ or $CH_3(CH_2)_2CH_3$, the latter often written when the chain is lengthy.
  - → *Condensed structural formula* for methylpropane is $CH_3CH(CH_3)CH_3$ or $(CH_3)_3CH$, because the methyl groups are equivalent. The first formula is the one chosen to name the compound (methylpropane *not* trimethylmethane).
- **Stick structure**
  - → A *stick structure* represents a chain of carbon atoms as a zigzag line. The end of each short line represents a carbon atom. The C–H bonds and H atoms are not shown, but are added "mentally" to the structure. Other atoms, such as oxygen and nitrogen atoms, are written explicitly, as well as any hydrogen atoms to which they are bonded.
  - → The zigzag *structural formula* and the *stick structure* of methylpropane are

CH3
CH
CH3 CH3

  - → A benzene ring is represented by a circle inside a hexagon or by one of the Kekulé structures. The presence of a hydrogen atom attached to each carbon atom is implied. These forms of the benzene ring are also used in structural formulas. The respective *structural formulas* and *stick structures* of ethylbenzene, $C_6H_5CH_2CH_3$, are

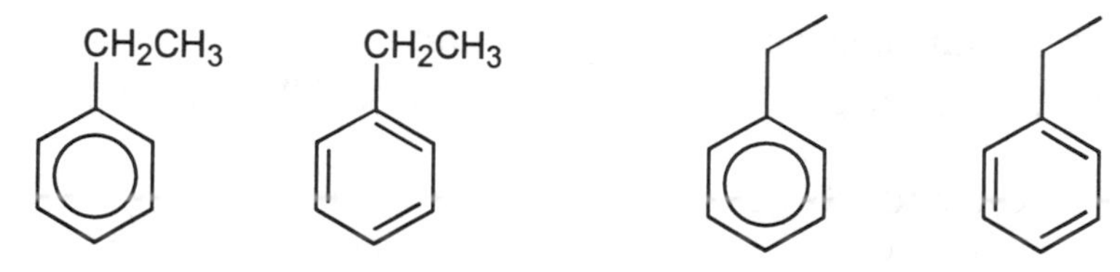

Example 18.1a Write the *structural formula* and the *stick structure* of $NH_2CH_2CH_2OH$.

Answer $NH_2–CH_2–CH_2–OH$

NH2 OH

- **Types of hydrocarbons**
  - → *Aromatic hydrocarbon:* an organic hydrocarbon that includes a benzene ring as part of its structure, such as benzene, C_6H_6, toluene, $C_6H_5CH_3$, naphthalene, $C_{10}H_8$:

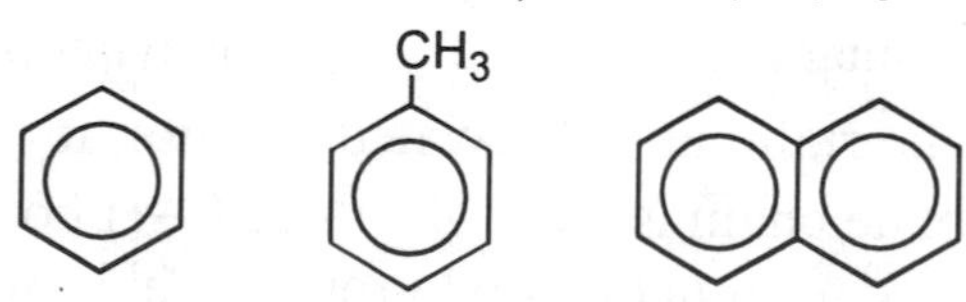

  - → *Aliphatic hydrocarbon:* a hydrocarbon that does not have an aromatic ring in its structure.
  - → *Saturated hydrocarbon:* an aliphatic hydrocarbon with no carbon-carbon multiple bonds, such as butane, $CH_3CH_2CH_2CH_3$.

→ *Unsaturated hydrocarbon:* an aliphatic hydrocarbon with at least one carbon-carbon multiple bond, such as butene, $CH_2{=}CHCH_2CH_3$, or butyne, $CH{\equiv}CCH_2CH_3$.

Example 18.1b Write *stick structures* for butane, 1-butene, and 1-butyne. The number indicates the position of the multiple bond.

Answer

Butane 1-Butene 1-Butyne

18.2 Nomenclature of Hydrocarbons

- **Alkanes**
  - → Saturated hydrocarbons
  - → In naming single chain *alkanes*, we add a prefix denoting the number of carbon atoms in the chain to the suffix *-ane* (see Table 18.1 in the text). The first three members of the alkane family are methane, CH_4, ethane, CH_3CH_3, and propane, $CH_3CH_2CH_3$.
  - → If a *hydrogen atom* is removed from an *alkane*, the name of the *alkyl* group that results is derived by replacing the suffix *-ane* with *-yl*. The first three members of the alkyl family are the methyl, CH_3-, ethyl, CH_3CH_2-, and propyl, $CH_3CH_2CH_2-$, groups.
  - → Methylpropane, $CH_3CH(CH_3)CH_3$, is a propane molecule in which an H atom on the middle C atom is replaced by a CH_3 group.
  - → *Cycloalkanes* are alkanes that contain *rings* of $-CH_2-$ units. The first three members are cyclopropane, C_3H_6, cyclobutane, C_4H_8, and cyclopentane, C_5H_{10}.

- **Properties of alkanes**
  - → Structures may be written as linear, but they are actually linked tetrahedral units. Rotation about C–C bonds is fairly unrestricted, so the chains often roll up into a ball in gases and liquids to maximize the attraction between different parts of the chain.
  - → Electronegativity values of C and H are similar, so hydrocarbon molecules are best regarded as nonpolar. The dominant interaction between the molecules is therefore the London interaction (see Section 5.4). The strength of London interactions increases with the number of electrons in the molecule; therefore, alkanes should be less volatile with increasing molar mass (see Fig.18.1 in the text).
  - → Many alkane molecules appear in forms with the same atoms bonded together in different arrangements such as butane and methylpropane. *Isomers* are different compounds with the same molecular formula.

- **Alkenes**
  - → Ethene or ethylene, C_2H_4 or $H_2C{=}CH_2$, is the simplest unsaturated hydrocarbon.

→ *Alkenes* are a series of compounds with formulas derived from $H_2C{=}CH_2$ by inserting CH_2 groups. The next member of the family is propene, $H{-}CH_2{-}CH{=}CH_2$ ($CH_3CH{=}CH_2$).

→ Names of alkenes are formed in an analogous manner to that of alkanes except the ending is *-ene*. The double bond is located by numbering the carbon atoms in the chain and writing the lower of the two numbers of the two carbons joined by the double bond (see Toolbox 18.1 in the text).

- **Alkynes**

→ *Alkynes* are aliphatic hydrocarbons that have at least one carbon-carbon triple bond.

→ Ethyne or acetylene, $HC{\equiv}CH$, is the simplest alkyne.

→ Alkynes are named like the alkenes but with the suffix *-yne*.

- **Nomenclature rules** → The nomenclature rules for aliphatic hydrocarbons are given in Toolbox 18.1 in the text.

- **Alkane nomenclature**

→ If the carbon atoms in an alkane are bonded in a row so that the formula is of the form $CH_3(CH_2)_mCH_3$, the compound is named according to the number of carbon atoms present. These compounds are called *unbranched alkanes*. Examples of such compounds are given in Table 18.1 in the text. All are named with a stem name and the suffix *-ane*. A cyclic (ring) alkane is designated by the prefix *cyclo-*.

→ A *branched alkane* has one or more carbon atoms that are not in a single row of carbon atoms, and these carbons constitute *side chains*. Butane is an *unbranched alkane*, and methylpropane is a *branched alkane* with a CH_3 *side chain*. The side chains are treated as *substituents*, which are atoms or groups of atoms that have been substituted for a hydrogen atom.

→ A systematic procedure for naming alkanes is

Step 1. Identify the *longest* chain of carbon atoms in the molecule.

Step 2. Number the carbon atoms on the longest chain so that the carbon atoms with substituents have the *lowest* possible numbers.

Step 3. Identify each substituent and the number of the carbon atom in the chain on which it is located.

Step 4. Name the compound by listing the substituents in *alphabetical order*, with the numbered location of each substituent preceding its name. The name(s) of the substituent(s) are followed by the parent name of the alkane, which is determined by the number of carbon atoms in the longest chain. As an example, heptane with a methyl group at the number 3 carbon is named 3-methylheptane.

Step 5. If two or more of the same substituents are present (such as two methyl groups), a Greek prefix such as *di-*, *tri-*, or *tetra-* is attached to the name of the group. Numbers of the carbon atoms to which groups are attached, separated by commas, are include in the name. As an example, heptane with a methyl group at carbon atom 3 and another one at carbon atom 4 is named 3,4-dimethylheptane.

Step 6. Numbers in the name are always separated from letters by a hyphen; numbers are separated from each other by commas.

Example 18.2a What is the name of the alkane $CH_3(CH_2)_8CH_3$?

Solution The alkane is *unbranched* with a total of 10 carbon atoms. The stem name is obtained by combining the prefix for ten (*dec-*) with the alkane suffix *-ane* to get the name *decane*.

Example 18.2b What is the name of the *branched* alkane $CH_3(CH_2)_2C(CH_3)_2CH_2CH_3$?

Solution The first step is to number the carbons of the longest carbon chain so that the substituents appear on the lowest numbered carbons.

```
                    CH3
                    |
CH3–CH2–CH2–C –CH2–CH3
 6    5    4   3|  2    1
                    CH3
```

The longest carbon chain contains six carbons, so the compound is a hexane. There are two methyl groups on carbon number 3. Combining all the information gives the name 3,3-dimethylhexane.

Example 18.2c What is the name of the alkane $CH_3CH_2CH(CH_3)CH(CH_2CH_3)CH(CH_3)CH_3$?

Answer 3-ethyl-2,4-dimethylhexane

- **Alkene nomenclature**
  - → Alkenes are named by using the stem name of the corresponding alkane with a number that specifies the location of the double bond. The number is obtained by numbering the longest carbon chain so that the double bond is associated with the lowest numbered carbon possible and assigning to the double bond the lower number of the two double-bonded carbons.

```
  6    5    4    3     2  1
CH3CH2CH2CH=CHCH3
```

2-Hexene (not 3-hexene)

  - → Other substituents are named as described earlier for alkanes, with a number specifying the location of the group and an appropriate prefix denoting how many groups of each kind are present.
  - → Numbering the longest chain so that the double bond has the lowest number takes precedence over keeping the substituent numbers low.
  - → If more than one double bond is present, the number of these bonds is indicated by a Greek prefix. The suffix is *-ene.*

Example 18.2d What is the name of the *branched* alkene $CH_3(CH_2)_2C(CH_3){=}CHCH_3$?

Solution The first step is to number the carbons of the longest carbon chain so that the double bond appears on the lowest numbered carbons.

```
                    CH3
                    |
CH3–CH2–CH2–C =CH–CH3
 6    5    4   3   2    1
```

The longest carbon chain contains six carbons, so the compound is a hexene. There is one methyl group on carbon number 3. Combining all the information gives the name 3-methyl-2-hexene.

Note: The location of the double bond is not numbered in the two unambiguous cases ethene, $CH_2{=}CH_2$, and propene $CH_3CH{=}CH_2$. The molecule 1-butene, $CH_2{=}CHCH_2CH_3$, *is* different than 2-butene, $CH_3CH{=}CHCH_3$.

Example 18.2e What is the name of the alkene $CHCl_2CH_2C(CH_3){=}CHCH_3$?

Answer 5,5-dichloro-3-methyl-2-pentene
(low number for the double bond takes precedence)

Example 18.2f What is the name of the alkene $CH_3CH{=}CHCH_2CH{=}CH_2$?

Solution The molecule contains two double bonds and six carbon atoms in a chain, so this compound is a hexadiene. Two numbers are required to locate the double bonds. The name with the lowest numbers is 1,4-hexadiene.

- **Alkyne nomenclature**
  - → Alkynes are named by using the stem name of the corresponding alkane with a number that specifies the location of the triple bond. The numbering convention is the same as the one used for double bonds. If more than one triple bond is present, the number of these bonds is indicated by a Greek prefix. The suffix is *-yne.*
  - → When numbering atoms in the chain, the lowest numbers are given preferentially to (a) functional groups named by suffixes (see Toolbox 19.1), (b) double bonds, (c) triple bonds, and (d) groups named by prefixes.

Example 18.2g What is the name of the *branched* alkyne $CH_3CH_2CH(CH_3)C{\equiv}CCH_3$?

Solution The first step is to number the carbons of the longest carbon chain so that the triple bond appears on the lowest numbered carbons.

$$\underset{6}{CH_3}-\underset{5}{CH_2}-\underset{4}{\overset{\overset{\displaystyle CH_3}{|}}{CH}}-\underset{3}{C}\equiv\underset{2}{C}-\underset{1}{CH_3}$$

The longest carbon chain contains six carbons, so the compound is a hexyne. There is one methyl group on carbon number 4. Combining all the information gives the name 4-methyl-2-hexyne.

Note: The location of the triple bond is not numbered in the two unambiguous cases ethyne, $CH{\equiv}CH$, and propyne $CH_3C{\equiv}CH$. The molecule 1-butyne, $CH{\equiv}CCH_2CH_3$, *is* different than 2-butyne, $CH_3C{\equiv}CCH_3$.

Example 18.2h What is the name of the alkyne $CHCl_2CH(CH_3)C{\equiv}CCH_3$?

Answer 5,5-dichloro-4-methyl-2-pentyne
(low number for the triple bond takes precedence)

Example 18.2i What is the name of the alkyne $CH_3C{\equiv}CCH_2C{\equiv}CH$?

Solution The molecule contains two triple bonds and six carbon atoms in a chain, so this compound is a hexadiyne. Two numbers are required to locate the triple bonds. The name with the lowest numbers is 1,4-hexadiyne.

Note: A molecule with a double and triple bond is 1-butene-3-yne, $CH_2{=}CH{-}C{\equiv}CH$, with the lowest number given preferentially to the location of the double bond.

18.3 Isomers

- **Structural isomers**
  → Molecules that are *structural isomers* are constructed from the same atoms but the atoms are connected differently; that is, the molecules have a different *connectivity* (see Section 16.7).
  → The molecules have the *same* molecular formula, but *different* structural formulas.
  → Butane, $CH_3(CH_2)_2CH_3$, and methylpropane, $CH_3CH(CH_3)CH_3$, are two structural isomers with the same molecular formula C_4H_{10}. The molecules have a different *connectivity*.

Example 18.3a Without referring to the text, draw all the structural isomers that have the formula C_5H_{12}.

Solution First draw the different possible arrangements of five carbon atoms then add the required hydrogen atoms. Start with the five carbons bonded in a straight chain, and then move one or more of the carbons to branched positions. Arrangements that may be interconverted by rotating the carbon framework are *not* structural isomers. The three possible carbon frameworks are

```
                        C              C
                        |              |
C-C-C-C-C         C-C-C-C            C-C-C
                                       |
                                       C
```

Three other structures may be drawn that are equivalent to the center one. These are *not* different structural isomers. They are formed by rotating the center structure. The three "rotated" structures shown are ignored.

```
                     C
                     |
C-C-C-C           C-C-C-C           C-C-C-C
    |                                 |
    C                                 C
```

To complete the drawings of the three structural isomers, add the 12 hydrogen atoms and recall that each carbon atom must have four bonds. For clarity, the branched C–C bonds are stretched.

```
                                   H               H
                                   |               |
                                 H-C-H           H-C-H
  H H H H H                  H H   |   H         H   |   H
  | | | | |                  | |   |   |         |   |   |
H-C-C-C-C-C-H              H-C-C---C---C-H     H-C---C---C-H
  | | | | |                  | |   |   |         |   |   |
  H H H H H                  H H   H   H         H   |   H
                                                   H-C-H
                                                     |
                                                     H
```

Example 18.3b The carbon frameworks for six alkanes with the formula C_6H_{14} are shown in the figure. Which are distinct structural isomers and which represent the same compound?

```
  C                   C
  |                   |
C-C-C-C-C         C-C-C-C-C          C-C-C-C-C
                                           |
                                           C
     I                II                  III
```

```
  C  C            C                 C
  |  |            |                 |
C-C--C-C      C---C---C---C     C---C---C---C
                  |                     |
                  C                     C
   IV             V                    VI
```

Answer I, II, IV, and V are distinct; I and III are the same; IV and VI are the same.

- **Stereoisomers**
  - → If two molecules have the same connectivity but some of their atoms are arranged differently in space, they are called *stereoisomers.*
  - → *Geometrical isomers* are *stereoisomers* that have different arrangements in space on either side of a double bond, or above and below the ring of a cycloalkane.
  - → *Optical isomers* are *stereoisomers* in which each isomer is the mirror image of the other, *and* the images are nonsuperimposable.
- **Geometrical isomers**
  - → Compounds with the same molecular formula and with atoms bonded to the same neighbors, but with a different arrangement of atoms in space
  - → In the case of molecules with a double bond, the effective lack of rotation around the double bond makes this type of isomer possible.
  - → *Geometrical isomers* of organic compounds are distinguished by the italicized prefixes *cis* (on this side) and *trans* (across).
  - → An example is 2-butene, for which two geometrical isomers are possible:

In the trans isomer, the two methyl groups are across the double bond from each other. In the cis isomer, they are on the same side of the double bond. Rotation around the double bond does not occur under normal conditions because of the rigid π-bond, and the two isomers are distinct compounds with different chemical and physical properties. If sufficient energy is added to a pure sample of either compound, a *cis-trans isomerization* reaction may occur in which the π-bond is partially broken and part of the sample is converted to the other isomer.

Example 18.3c Draw stick structures for the two geometrical isomers of 3-hexene.

Solution The double bond in 3-hexene is located on carbon 3. There is an ethyl group attached to carbon 3 and another to carbon 4. These two groups may be arranged in either a cis or trans fashion. The two isomers are

Note: Remember that the restriction on rotation applies only to the C=C bond. The C–C single bonds are *not* restricted and relatively free rotation about them can occur. Also, if three of the four groups (or all four) bonded to the double-bonded carbon atoms are the same, then cis-trans isomers are not possible.

- **Optical isomers**
  → *Optical isomers* are *nonsuperimposable* mirror images.
  → A *chiral* molecule has a mirror image that is nonsuperimposable. Your hand is a chiral object; its mirror image is not superimposable. An organic molecule is chiral if at least one of its central carbon atoms has four *different* groups attached to it. Such a carbon atom is called a *stereogenic center*. Note that this terminology is *not* used in the text.
  → A pair of *enantiomers* is made up of a chiral molecule and its mirror image. Enantiomers have identical chemical properties, except when they react with other chiral molecules. Enantiomers differ in only one physical property; chiral molecules display *optical activity*, the ability to rotate the plane of polarization of light (see Section 16.7 and Toolbox 16.2). Mixtures of enantiomers in equal proportions are called *racemic mixtures,* which are not optically active.
  → A molecule that has a superimposable mirror image is said to be *achiral.*

Example 18.3d Test each of the following organic molecules for chiral properties.

```
     H  H  Cl H          H  Br H  Cl          H  Br H  H
     |  |  |  |          |  |  |  |           |  |  |  |
   H-C--C--C--C-H      H-C--C--C--C-H       H-C--C--C--C-H
     |  |  |  |          |  |  |  |           |  |  |  |
     H  H  Br H          H  Br H  H           H  H  H  H
          I                  II                   III
```

Solution In molecule I, the second carbon atom from the right has four different groups attached to it (ethyl, methyl, chloro, and bromo) and constitutes the *stereogenic center*. Molecule I is *chiral*. In molecule II, each carbon atom has at least two of the same groups attached. Molecule II is *achiral*. In molecule III, the second carbon atom from the left has four different groups attached to it (methyl, ethyl, bromo, and a hydrogen atom). Molecule III is *chiral.*

Example 18.3e Draw a stick structure for the chiral molecule, 3-methylhexane, and identify the *stereogenic center*.

Solution The stick structure is 3

Carbon atom 3 has four different groups attached to it (ethyl, propyl, methyl, and a hydrogen atom), and constitutes the *stereogenic center*.

- **Summary** → A pictorial summary of the three different types of isomerism is displayed in Fig. 18.2 in the text (reproduced here).

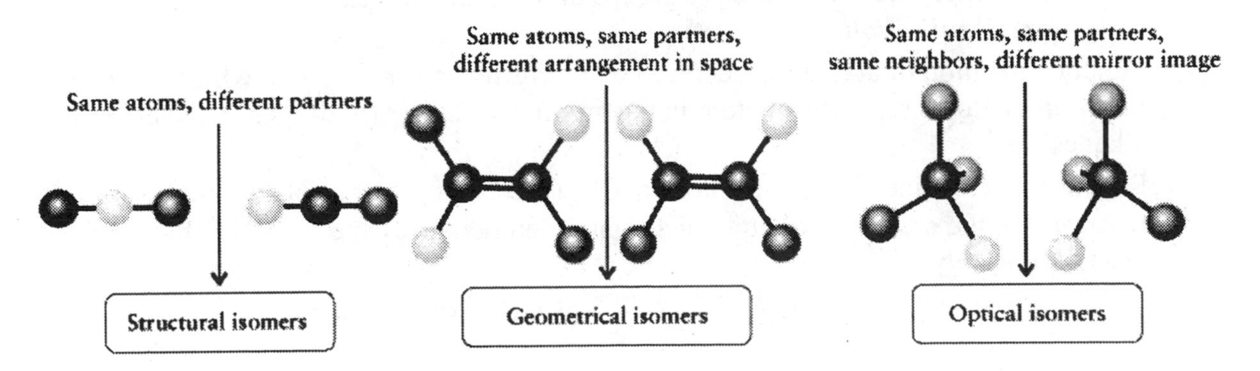

18.4 Properties of Alkanes

- **Physical properties**
  - → An *alkane* with a long, unbranched chain tends to have a higher melting point, boiling point, and heat of vaporization than its branched structural isomer.
  - → Molecules with unbranched chains can get closer together than molecules with branched chains. As a result, molecules with branched chains have *weaker* intermolecular forces than their isomers with unbranched chains.
  - → As an example, butane containing an unbranched chain has a higher boiling point (–0.5°C) than its structural isomer, methylpropane, containing a branched chain (–11.6°C).
- **Chemical properties**
  - → *Alkanes* are not very reactive chemically. They were once called *paraffins*, derived from the Latin "little affinity."
  - → *Alkanes* are unaffected by concentrated sulfuric acid, by boiling nitric acid, by strong oxidizing agents such as potassium permanganate, and by boiling aqueous sodium hydroxide.
  - → The C–C bond enthalpy, 348 $kJ \cdot mol^{-1}$, and the C–H bond enthalpy, 412 $kJ \cdot mol^{-1}$, are large, so there is little energy advantage in replacing them with most other bonds. Notable exceptions are C=O, 743 $kJ \cdot mol^{-1}$, C–OH, 360 $kJ \cdot mol^{-1}$, and C–F, 484 $kJ \cdot mol^{-1}$.
- **Combustion (oxidation) reactions**
  - → *Alkanes* are used as fuels, because their enthalpies of combustion are high (see Section 6.18). The products of combustion are carbon dioxide and water. Strong C–H bonds are replaced by even stronger O–H bonds in H_2O, and the O=O bonds are replaced by two strong C=O bonds in CO_2.
  - → As an example, the combustion of one mole of octane releases 5471 kJ of heat. Recall that $\Delta H_r°$ for a reaction depends on the stoichiometric coefficients. The balanced equation for the combustion of octane is

$$2\,C_8H_{18}(l) + 25\,O_2(g) \rightarrow 16\,CO_2(g) + 18\,H_2O(l) \qquad \Delta H_r° = -10\,942\ kJ \cdot mol^{-1}$$

- **Substitution reactions**
  - → *Alkanes* are used as raw materials for the synthesis of many reactive organic compounds.
  - → Organic chemists introduce reactive groups into alkane molecules in a process called *functionalization.*
  - → Functionalization of alkanes is achieved by a *substitution reaction*, in which an atom or group of atoms replaces an atom in the original molecule (hydrogen in the case of alkanes).
  - → Reaction of methane, CH_4, and chlorine, Cl_2, is an example of a substitution reaction. In the presence of ultraviolet light or temperatures above 300°C, the gases react explosively:

$$CH_4(g) + Cl_2(g) \xrightarrow{\text{light or heat}} CH_3Cl(g) + HCl(g)$$

Chloromethane, CH_3Cl, is one of four products; the others are dichloromethane, CH_2Cl_2, trichloromethane, $CHCl_3$, and tetrachloromethane, CCl_4. The properties and uses of the chloromethanes are

Formula	Common name	Boiling point	Dipole moment	Uses
CH_3Cl	methyl chloride	–24.2°C	1.87 D	refrigerant
CH_2Cl_2	methylene chloride	40°C	1.60 D	dry cleaning agent
$CHCl_3$	chloroform	61.7C	1.01 D	anesthetic
CCl_4	carbon tetrachloride	76.8°C	0	nonpolar solvent

Note: Tetrachloromethane is carcinogenic and its use is restricted. It once was used as a nonflammable fluid in certain fire extinguishers.

18.5 Mechanism: Alkane Substitution

- **Radical chain mechanism**
  - → Kinetic studies of the rate of the chlorine substitution reaction, as affected by concentration, catalyst, solvent, and temperature, suggest that alkane substitution reactions proceed by a *radical chain mechanism* (see Section 13.12).
  - → The *initiation step* is the dissociation of chlorine:

    $$Cl_2 \xrightarrow{\text{light or heat}} 2\,Cl\cdot$$
  - → Chlorine atoms proceed to attack methane molecules and extract a hydrogen atom:

    $$Cl\cdot + CH_4 \rightarrow HCl + \cdot CH_3$$

    Because one of the products is a radical, this reaction is a *propagation step*.
  - → In a second propagation step, the methyl radical may react with a chlorine molecule:

    $$Cl_2 + \cdot CH_3 \rightarrow CH_3Cl + Cl\cdot$$

    The chlorine atom formed may take part in the other *propagation step* or attack a CH_3Cl molecule to eventually form CH_2Cl_2.
  - → A termination step occurs when two radicals combine to form a nonradical product, as in the reaction:

    $$Cl\cdot + \cdot CH_3 \rightarrow CH_3Cl$$

    The substitution reaction is not very clean and the product is usually a mixture of compounds. One may limit the production of the more highly substituted alkanes by using a large excess of the alkane.

18.6 Properties of Alkenes

- **Double bond**
  - → The carbon-carbon double bond, C=C, consists of a σ-bond and a π-bond.
  - → Each carbon atom is sp^2 hybridized and one of the hybrid orbitals is used to form the σ-bond.
  - → The unhybridized *p*-orbitals on each atom overlap with each other to form a π-bond.
  - → All four atoms attached to the C=C group lie in the same plane and are fixed into that arrangement by the resistance to twisting of the π-bond (see Fig. 18.7 in the text).

→ Alkene molecules cannot roll up into a compact arrangement as alkanes can, so alkenes have lower melting points.
→ In C=C, the π-bond is weaker than the σ-bond (see Section 3.7). A consequence of this weakness is the reaction most common in alkenes, the replacement of the π-bond by two new σ-bonds.

- **Formation of alkenes by elimination reactions**

→ In the petrochemical industry, abundant alkanes are converted to more reactive alkenes by a catalytic process called *dehydrogenation:*

$$CH_3CH_3(g) \xrightarrow{Cr_2O_3} CH_2{=}CH_2(g) + H_2(g)$$

This is an example of an *elimination reaction*, a reaction in which two groups or two atoms on neighboring carbon atoms are removed from a molecule, leaving a multiple bond (see Fig. 18.8 in the text).

→ In the laboratory, alkenes are produced by *dehydrohalogenation* of haloalkanes, the removal of a hydrogen atom and a halogen atom from neighboring carbon atoms:

$$CH_3CH_2Br \xrightarrow{CH_3CH_2O^- \text{ in ethanol at } 70°C} CH_2{=}CH_2 + HBr$$

This elimination reaction is carried out in hot ethanol, using sodium ethoxide, CH_3CH_2ONa, as the reagent.

→ *Dehydrohalogenation* occurs by the attack of the ethoxide ion on a hydrogen atom attached to the carbon atom next to the one containing the bromine atom. The hydrogen atom is removed from the molecule as a proton and $CH_3CH_2O{-}H$ is formed. When the carbon atom forms a second bond to its neighbor, the Br^- ion departs.

18.7 Mechanism: Electrophilic Addition to Alkenes

- **Addition reaction**

→ Characteristic chemical reaction of alkenes, in which atoms supplied by the reactant form σ-bonds to the two carbon atoms joined by the π-bond, which is broken (see Fig. 18.9 in the text)
→ Almost all addition reactions are exothermic.
→ A *hydrogenation* reaction is the addition of two hydrogen atoms at a double bond:

$$CH_3CH{=}CHCH_3 + H_2 \rightarrow CH_3CH_2{-}CH_2CH_3$$

→ A *halogenation* reaction is the addition of two halogen atoms at a double bond:

$$CH_3CH{=}CHCH_3 + Cl_2 \rightarrow CH_3CHCl{-}CHClCH_3$$

→ A *hydrohalogenation* reaction is the addition of a hydrogen atom and a halogen atom at a double bond:

$$CH_3CH{=}CHCH_3 + HCl \rightarrow CH_3CH_2{-}CHClCH_3$$

- **Estimating the reaction enthalpy of an addition reaction**

→ Use bond enthalpies in Tables 6.6 and 6.7 in the text for the bonds that are broken and formed. Note that the bond enthalpies strictly apply to gas-phase reactions.

→ Recall:

$$\Delta H_r° \approx \sum_{\text{reactants}} n\Delta H_B(\text{bonds broken}) - \sum_{\text{products}} n\Delta H_B(\text{bonds formed})$$

Example 18.7 Estimate the reaction enthalpy for the *hydrobromination* reaction:

$$CH_3CH{=}CHCH_3 + HBr \rightarrow CH_3CH_2{-}CHBrCH_3$$

Solution The following bonds are broken: C=C and H–Br.
$\Sigma n\Delta H_B$(bonds broken) = 612 + 366 = 978 kJ·mol^{-1}
The following bonds are formed: C–C, C–H , and C–Br.
$\Sigma n\Delta H_B$(bonds formed) = 348 + 412 + 276 = 1036 kJ·mol^{-1}
The final result is: $\Delta H_r° \approx 978 - 1036 = -58$ kJ·mol^{-1}

- **Mechanism of addition reactions**
  - → Double bonds contain a high density of high energy electrons associated with the π-bond. This region of high electron density represents an accumulation of negative charge that is attractive to positively charged reactants.
  - → An *electrophile* is a reactant that is attracted to a region of high electron density. It may be a positively charged species or one that has or can acquire a *partial* positive charge during the reaction.
  - → An example treated in the text is the bromination of ethene to form dibromoethane. Bromine molecules are polarizable; a partial positive charge builds up on the bromine atom closest to the double bond. Bromine acts as an *electrophile*. The partial charge becomes a full charge and a bromine cation attaches to the double bond, leaving a Br^- ion behind. The cyclic intermediate is called a *bromonium ion*.

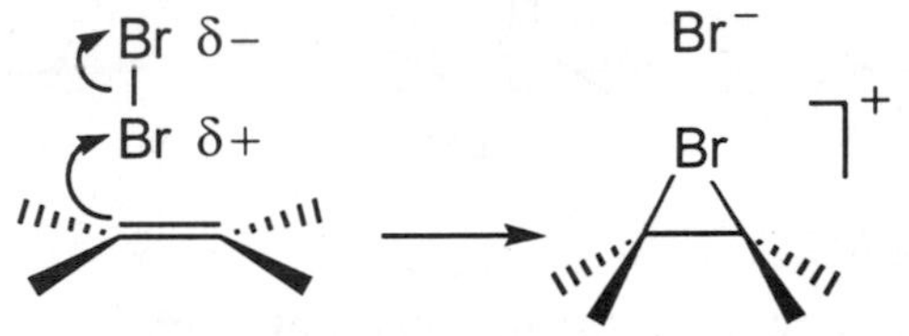

Br δ−
Br δ+
Br^-
Br +

The Br^- ion (or a different Br^- ion) is attracted by the positive charge of the bromonium ion. It forms a bond to one carbon atom, and the bromine atom in the cyclic ion forms another bond, giving 1,2-dibromoethane.

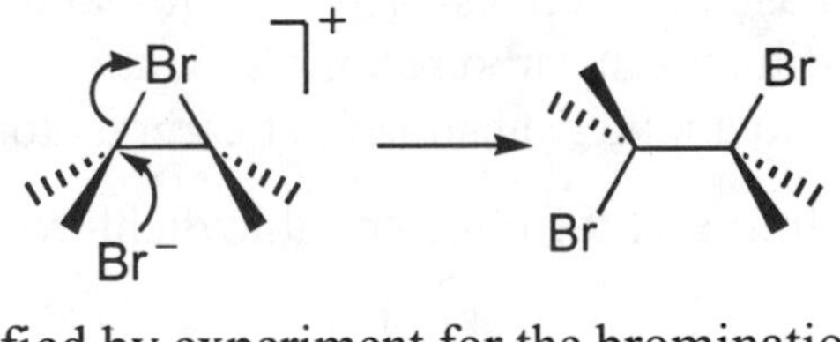

Br +
Br^-
Br
Br

  - → The mechanism is verified by experiment for the bromination of *cis*-2-butene as explained in the text (see Margin Figures **31**, **32**, and **33**). The stereochemistry of the mechanism restricts the 2,3-dibromobutane isomers that may form.

Aromatic Compounds (Sections 18.8-18.9)

Key Concepts

arenes, aromatic hydrocarbons, benzene, phenyl group, naphthalene, anthracene, phenanthrene, ring numbering, electrophilic substitution, mechanism

Overview

- **Definition**
  - → Aromatic hydrocarbons are called *arenes*.
  - → *Benzene*, C_6H_6, is the parent compound. The benzene ring is sometimes called the *phenyl* group, as in 2-phenyl-*trans*-2-butene, $CH_3C(C_6H_5){=}CHCH_3$.

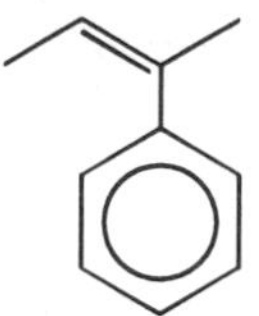

  - → Aromatic compounds include those with fused benzene rings, such as naphthalene, $C_{10}H_8$, anthracene, $C_{14}H_{10}$, and phenanthrene, $C_{14}H_{10}$.

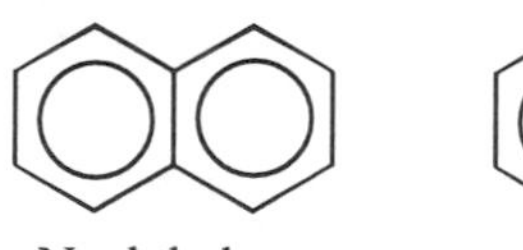

Naphthalene

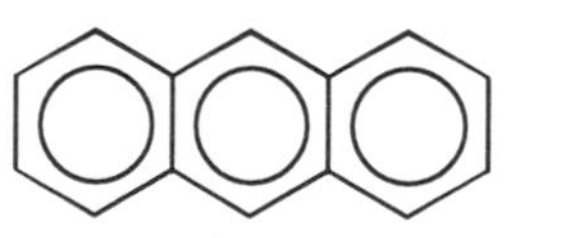

Anthracene

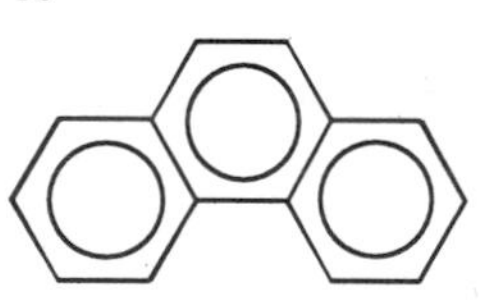

Phenanthrene

- **Benzene ring numbering**
  - → Substituents on the benzene ring are designated by numbering the carbon atoms from 1 to 6 around the ring. Compounds are named by counting around the ring in the direction that gives the smallest numbers to the substituents. It is also common to use the prefixes, *ortho*-, *meta*-, and *para*- to denote substituents at carbon atoms 2, 3, and 4, respectively, relative to another substituent at carbon atom 1.

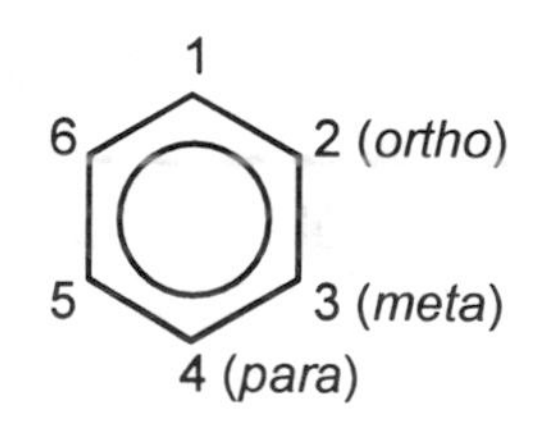

  - → An example is the naming of the three possible dichlorobenzene molecules:

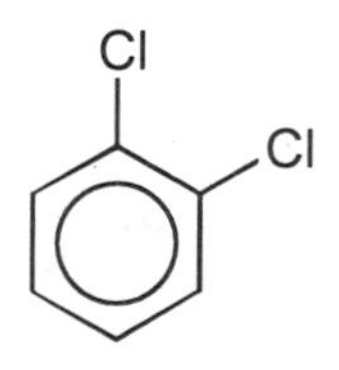

1,2-Dichlorobenzene
ortho-Dichlorobenzene

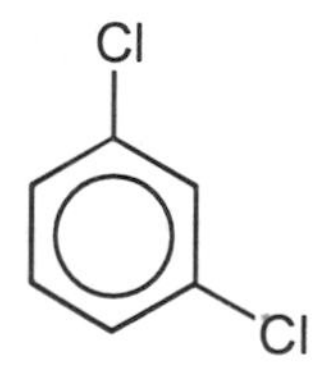

1,3-Dichlorobenzene
meta-Dichlorobenzene

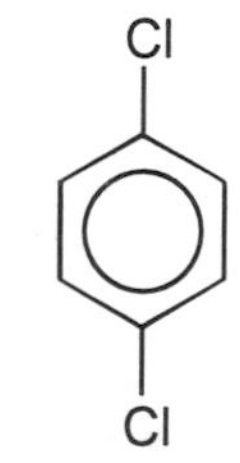

1,4-Dichlorobenzene
para-Dichlorobenzene

Note: Numbering carbon atoms in fused ring systems is somewhat more complicated and is not covered in the text.

18.8 Reactions of Arenes

- **Substitution reactions**
  → *Arenes* have delocalized π-electrons; but, unlike alkenes, arenes undergo predominantly *substitution* reactions, with the π-bonds of the ring unaffected.
  → The reaction of benzene, C_6H_6, with chlorine produces chlorobenzene when one chlorine atom substitutes for a hydrogen atom:

$$C_6H_6 + Cl_2 \xrightarrow{Fe} C_6H_5Cl + HCl$$

18.9 Mechanism: Electrophilic Substitution

- **Electrophilic substitution** → Mechanism of substitution of a benzene ring is an electrophilic attack, so the reaction is called *electrophilic substitution*.
- **Halogenation**
  → In the halogenation of benzene, an iron catalyst is used. The iron is converted to iron(III) halide, FeX_3. Iron(III) halide reacts further to polarize the X_2 molecule.
  → The commonly accepted mechanism for aromatic halogenation, X_2, is then

$$X_2 + FeX_3 \rightleftharpoons X_3Fe\cdots X^{\delta-}–X^{\delta+} \qquad \textit{fast equilibrium}$$

$$X_3Fe\cdots X^{\delta-}–X^{\delta+} + C_6H_6 \rightarrow FeX_4^- + C_6H_6X^+ \qquad \textit{slow}$$

$$C_6H_6X^+ + FeX_4^- \rightarrow C_6H_5X + HX + FeX_3 \qquad \textit{fast}$$

  → In the last step, the hydrogen atom is easily removed from the ring, for in that way the delocalization of π-electrons, lost in the slow step, is regained.

- **Nitration**
  → A mixture of nitric acid and concentrated sulfuric acid slowly converts benzene into nitrobenzene.
  → The nitronium ion, NO_2^+, is the electrophile and the nitrating agent. Nitronium salts such as nitronium perchlorate and nitronium fluoroborate are known. The nitronium ion survives when the salt is dissolved in nonaqueous solvents such as nitromethane or acetic acid.
  → The commonly accepted mechanism for aromatic nitration is:

$$HNO_3 + 2\,H_2SO_4 \rightleftharpoons NO_2^+ + H_3O^+ + 2\,HSO_4^-$$

$$NO_2^+ + C_6H_6 \rightarrow C_6H_6NO_2^+ \qquad \textit{slow}$$

$$C_6H_6NO_2^+ + HSO_4^- \rightarrow C_6H_5NO_2 + H_2SO_4 \qquad \textit{fast}$$

  → In the last step, the hydrogen atom is removed from the ring to restore delocalization of π-electrons.

- **Ortho- and para-directing activators**
  → Electrons are donated into the delocalized molecular orbitals of the ring (*resonance effect)*.
  → The reaction rate is *much* faster than in unsubstituted benzene, and the substituent is called an *activator*.
  → Products of a substitution reaction favor the *ortho* and *para* positions of the ring, because these locations have more electron density than the *meta* positions.
  → Examples include, –OH, $-NH_2$, and substituted amines.
  → The atom bonded to the benzene ring has a *nonbonding* pair of electrons.

  Note: Alkyl groups do not have a nonbonding pair of electrons, but they are also ortho- and para-directing activators. The mechanism in this case is a different one (see any organic chemistry text).

- **Ortho- and para-directing deactivators**
  → Electrons are donated into the delocalized molecular orbitals of the ring (*resonance effect*), but the substituent is very electronegative.
  → The reaction rate is *slightly* slower than in unsubstituted benzene, and the substituent is called a *deactivator*. Deactivation occurs when the substituent is *highly* electronegative and withdraws some electron density from the ring.
  → Products of a substitution reaction still favor the *ortho* and *para* positions of the ring, because these positions have *relatively* more electron density than the *meta* positions.
  → The *only* examples are –F, –Cl, –Br, and –I.
  → The atom bonded to the benzene ring has a *nonbonding* pair of electrons.

- **Meta-directing deactivators**
  → Highly electronegative substance which can withdraw electrons partially and/or a substituent that *removes* electrons by resonance
  → The reaction rate is *much* slower than in unsubstituted benzene, and the substituent is called a *deactivator*.
  → Products of a substitution reaction favor the *meta* position of the ring, because the *ortho* and *para* positions have greatly decreased electron density.
  → Examples include –COOH, $-NO_2$, $-CF_3$, and –C≡N.
  → *All* electron pairs on the atom bonded to the benzene ring are *bonding* pairs.

Chapter 19 Organic Chemistry II: Functional Groups

Functional Groups (Sections 19.1-19.8)

Key Concepts

functional groups, haloalkane, nucleophilic substitution, nucleophile, hydrolysis reaction, S_N1 reaction, heterolytic dissociation, carbocation, leaving group, inversion of configuration, S_N2 reaction, alcohol, hydroxyl group, primary alcohol, secondary alcohol, tertiary alcohol, diol, ether, crown ether, phenol, aldehyde, ketone, carbonyl group, carboxylic acid, carboxyl group, ester, amine, amino group, amino acid, amide

Overview

- **Organic chemistry**
  - → Foundation of the pharmaceutical industry and of medicine
  - → Basis of biochemistry and molecular biology
  - → Understanding organic compounds through the presence of groupings of atoms with characteristic functions
- **Functional groups**
  - → Predicting the properties of many organic molecules, including large biomolecules, carbohydrates, and genetic material as well as polymeric materials, through the properties of their *functional groups*
  - → Identification of *functional groups* and their corresponding organic reaction mechanisms has led to the systemization of organic chemistry.
  - → Common *functional groups* are listed in Table 19.1 in the text.

19.1 Haloalkanes

- **Haloalkanes**
  - → Alkanes in which a *halogen atom*, X, replaces one or more hydrogen atoms
  - → *Haloalkanes* are insoluble in water. Some are highly toxic and environmentally unfriendly, such as the chlorofluorocarbon (CFC) 1,2-dichloro-1-fluoroethane, $HFClC–CClH_2$, which is partly responsible for the depletion of the ozone layer.
  - → The C–X bonds in haloalkanes are *polar*; the carbon atom carries a partial positive charge and the halogen, X, a partial negative one. The polarity of the C–X bond governs much of the chemistry of haloalkanes.
  - → Haloalkanes are susceptible to *nucleophilic substitution*, in which a reactant that seeks out centers of positive charge in a molecule, the *nucleophile*, replaces a halogen atom.
  - → *Nucleophiles* include anions such as OH^- and CN^-, as well as neutral Lewis bases with lone pairs such as H_2O and NH_3.

Example 19.1 Iodomethane reacts with water in a substitution reaction to form methanol and hydroiodic acid. Write a balanced equation for the reaction and identify the nucleophile.

Solution The balanced equation is: $CH_3I + H_2O \xrightarrow{\text{slow}} CH_3OH + HI$
The carbon atom of CH_3I has a partial positive charge to which the lone pair electrons on water are attracted. Water is the *nucleophile* in this *hydrolysis reaction,* in which a C–I bond is replaced by a C–O bond. The reaction may also be viewed as a *nucleophilic substitution* reaction, in which a hydroxyl group has formally replaced the iodine atom.

19.2 Mechanisms: Nucleophilic Substitution

- **Nucleophilic substitution reactions**
  - → Proceed by one of two basic mechanisms, designated S_N1 or S_N2; S stands for substitution, N for nucleophilic, and 1 or 2 indicates the molecularity of the *rate-determining step* (see Section 13.11)
  - → A haloalkane may react by *either* mechanism.
- **S_N1 reaction in haloalkanes**
  - → Characterized by a unimolecular rate-determining step
  - → The reaction rate law is first order overall, first order in the haloalkane and zero order in the nucleophile.
  - → In the *rate determining step,* a C–X bond is broken in a process called *heterolytic dissociation.* A *carbocation*, R_3C^+, and a halide ion, X^-, which is called the *leaving group*, are produced:

    $R_3C{-}X \rightarrow R_3C^+ + X^-$ R is an alkyl functional group attached to C.

    The nucleophile then attacks the *carbocation* at the C^+ site *usually* on either side of the carbocation.
  - → In a pure S_N1 reaction (leaving group departs rapidly and neighboring groups do not interfere), a *stereogenic carbon center* under attack will produce a *racemic* mixture of enantiomers as products. If the leaving group departs slowly, it continues to protect one side of the carbocation. The nucleophile may then attach preferentially to the opposite side, leading to *inversion* of configuration. The leaving group may depart quickly, but a substituent may block the path of the entering nucleophile. In this case, one side of the molecule is protected, and the stereochemistry of the reactant may be partially retained in a process called *neighboring group participation* (see Box 19.1 in the text).
  - → The S_N1 reaction mechanism is favored in haloalkanes with several bulky groups, R. An example of a haloalkane with three bulky R groups is 2-chloro-2-methylpropane, $(CH_3)_3CCl$.

Note: *Chirality* is a property of a molecule; *stereogenicity* is a property of an atom. In 2-chlorobutane, $CH_3CH_2CHClCH_3$, the carbon atom at position 2 is bonded to four different groups; the carbon atom is *stereogenic* and the molecule is *chiral.*

- **S_N2 reaction in haloalkanes**
  → Characterized by a bimolecular rate-determining step
  → The reaction rate law is second order overall, first order in the haloalkane and in the nucleophile.
  → Reaction occurs in a single, concerted step through the formation of a *bimolecular complex*. A schematic S_N2 reaction between an entering nucleophile, N^-, and a haloalkane, RCH_2X, is

$$N^- + RCH_2X \longrightarrow [N\cdots C(R)(H)(H)\cdots X]^- \longrightarrow NCH_2R + X^-$$

Bimolecular complex

  → The reaction proceeds with an *inversion* of configuration, meaning that an optically active reactant gives rise to an optically active product with the *reverse* stereochemical configuration at the site of attack. The process of inversion may be viewed as similar to an umbrella turning inside out.
  → The S_N2 reaction mechanism is favored in haloalkanes with an unobstructed reaction center. An example is a haloalkane with a single alkyl group R, such as RCH_2X.

Example 19.2 Is the hydrolysis reaction of the haloalkane, 3-bromo-3-methylhexane, expected to proceed by an S_N1 or S_N2 mechanism? If the reactant is optically pure, is the product expected to be optically pure as well? Explain.

Solution In the structure of 3-bromo-3-methylhexane, a central carbon atom is bonded to four different groups; hence, the central carbon atom is *stereogenic* and the molecule is *chiral*.

$$CH_3CH_2CH_2C(Br)(CH_3)CH_2CH_3$$

The methyl, ethyl, and propyl alkyl groups are bulky, and they are expected to inhibit nucleophilic attack on the side of the molecule opposite the bromo group, as would occur in an S_N2 mechanism. Because of the bulky groups, an S_N1 mechanism is expected; the product will be a *racemic* mixture of enantiomers if the bromide ion departs quickly *and* the neighboring groups do not participate. Participation of a neighboring group or the slow departure of a leaving group could lead to an excess of one enantiomer over the other.

$$CH_3CH_2CH_2C(OH)(CH_3)CH_2CH_3 \qquad CH_3CH_2CH_2C(CH_3)(OH)CH_2CH_3$$

19.3 Alcohols

- **Definition**
  → The *hydroxyl group* is an –OH group covalently bonded to a carbon atom.
  → An *alcohol* is an organic compound that contains a *hydroxyl* group not directly bonded to an aromatic ring or to a carbonyl group, $>C=O$.

→ Alcohols are named by adding the suffix *–ol* to the stem of the parent hydrocarbon, as in methanol for CH_3OH (methyl alcohol) or ethanol for CH_3CH_2OH (ethyl alcohol). Note that common names are given in parenthesis. When the location of the –OH group needs to be identified, the number of the carbon atom to which it is attached is given, as in 1-propanol for $CH_3CH_2CH_2OH$ (*n*-propyl alcohol) and 2-propanol for $CH_3CH(OH)CH_3$ (isopropyl alcohol). A *diol* is an organic compound with two hydroxyl groups, such as 1,2-ethanediol for $HOCH_2CH_2OH$ (ethylene glycol).

- **Classes of alcohols**

→ A *primary alcohol* has the form RCH_2–OH, where R can be any group. Examples of primary alcohols are methanol, ethanol, and 1-propanol.

→ A *secondary alcohol* has the form R_2CH–OH, where the R groups can be the same or different. Examples of secondary alcohols are 2-propanol, $(CH_3)_2CH$–OH (*same* R groups, CH_3–), and 2-butanol, $CH_3CH_2CH(–OH)CH_3$ (*different* R groups, CH_3CH_2– and CH_3–).

→ A *tertiary alcohol* has the form R_3C–OH, where the R groups can be the same or different. Examples of tertiary alcohols are 2-methyl-2-propanol, $(CH_3)_3C$–OH (*same* R groups, CH_3–), and 3-methyl-3-pentanol, $CH_3CH_2C(CH_3)(–OH)CH_2CH_3$ (*different* R groups, CH_3CH_2– and CH_3–). A structure for the latter is:

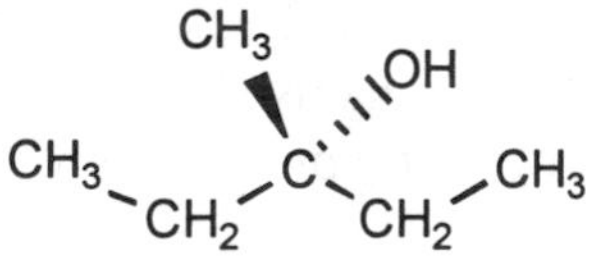

- **Properties of alcohols**

→ Alcohols are polar molecules that can lose the –OH proton in certain solvents, but typically not in water.

→ Alcohols have relatively high boiling points and low volatility for their molar mass because of hydrogen bond formation through the –OH group. In this way, they are similar to water. The normal boiling point of ethanol with a molar mass of 46.07 $g{\cdot}mol^{-1}$ is 78.2°C, whereas that of pentane with a molar mass of 72.14 $g{\cdot}mol^{-1}$ is 36.0°C.

Example 19.3 Which one of the following species is an alcohol?

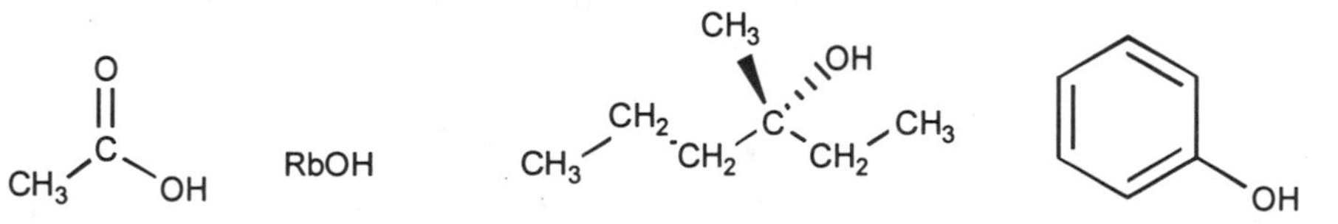

I II III IV

Solution Recall that an alcohol is an organic compound with a hydroxyl group *not* directly bonded to an aromatic ring or a carbonyl group.

In species I, the hydroxyl group is bonded to a carbonyl group, so it is *not* an alcohol (it is a carboxylic acid, see Section 19.7).

Species II, rubidium hydroxide, is an ionic solid, which is inorganic and therefore, *not* an alcohol.

Species III is an organic compound with a hydroxyl group not bonded to an aromatic ring or carbonyl and is therefore an alcohol.
Species IV has an –OH group bonded to an aromatic ring; it is phenol, not an alcohol.

19.4 Ethers

- **Ethers**
  - → Organic compounds of the form R–O–R, where R is any alkyl group and the two R groups may be the same or different
  - → *Ethers* are more volatile than alcohols with the same molar mass because ethers do not form hydrogen bonds in the pure state. They can, however, act as hydrogen-bond acceptors by using the lone pairs of electrons on the oxygen atom.
  - → Because ethers have low polarity and low reactivity, they are useful solvents for other organic compounds. However, ethers are quite flammable and must be handled with care.
  - → Examples are diethyl ether, CH_3CH_2–O–CH_2CH_3 (*same* R groups, CH_3CH_2–), and 1-butyl methyl ether, $CH_3CH_2CH_2CH_2$–O–CH_3 (*different* R groups, $CH_3CH_2CH_2CH_2$– and CH_3–).
- **Crown ethers**
  - → Cyclic polyethers of formula $\text{(}CH_2CH_2-O\text{)}_n$
  - → Name reflects the crownlike shape of the molecules.
  - → Bind strongly to alkali metal ions such as Na^+ and K^+ and allow inorganic salts to be dissolved in organic solvents
  - → The common oxidizing agent potassium permanganate, $KMnO_4$, is insoluble in nonpolar solvents such as benzene, C_6H_6. In the presence of [18]-crown-6, it dissolves in benzene according to the following reaction:

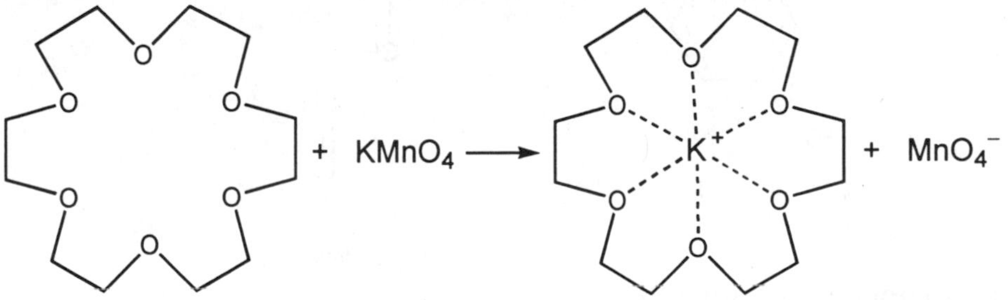

$$\text{[18]-crown-6} + KMnO_4 \longrightarrow [K(\text{18-crown-6})]^+ + MnO_4^-$$

[18]-Crown-6

Example 19.4 Determine which one of the following compounds is an ether and name it.

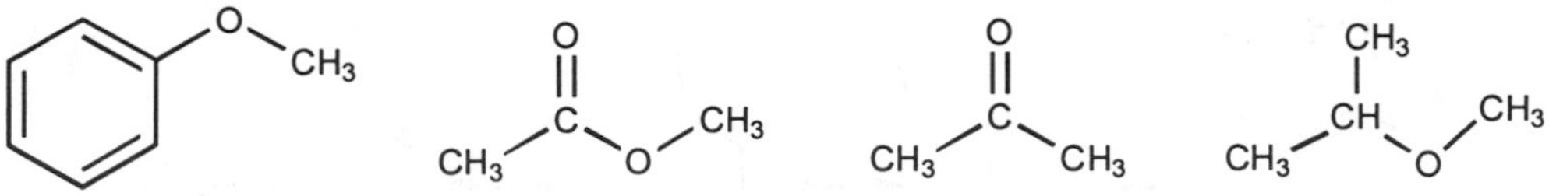

I: C_6H_5–O–CH_3

II: CH_3–C(=O)–O–CH_3

III: CH_3–C(=O)–CH_3

IV: CH_3–CH(CH_3)–O–CH_3

Solution An ether is any organic compound of the form R–O–R, where R is any alkyl group.
In compound I, one of the R groups attached to O is a phenyl ring, which is not an alkyl group, so I is not an ether.
Neither O atom in compound II has two alkyl groups attached to it, so II is not an ether (it is an ester, see Section 19.7).
There is only one group attached to the O atom in compound III, and it is not an ether (it is a ketone, see Section 19.6).
Compound IV has the requisite R–O–R form and is an ether. Its name is 2-propyl methyl ether or isopropyl methyl ether.

19.5 Phenols

- **Phenols**
  - → Organic compounds in which a *hydroxyl group* is attached *directly* to an aromatic ring
  - → The parent compound, phenol, is a white, crystalline, molecular solid.

OH OH

Phenol, C_6H_5OH: or Melting point: 40.9°C

  - → Substituted phenols occur naturally, and some are responsible for the fragrances of plants. *Thymol* is the active ingredient in oil of thyme, and *eugenol* provides the scent and flavor in oil of cloves.

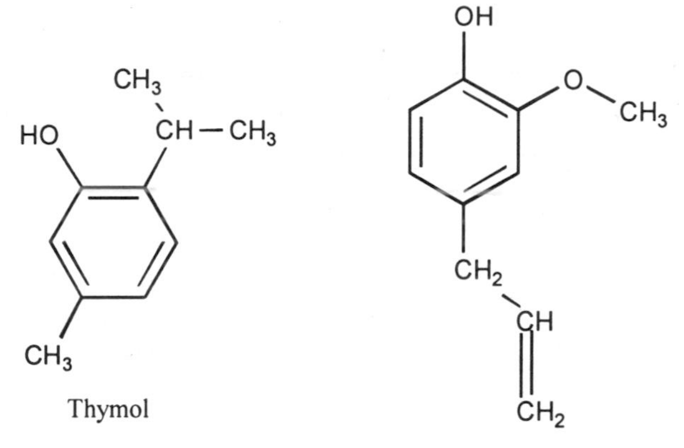

Thymol

Eugenol

- **Acid-base properties of phenols**
  - → Phenols are generally weak acids in contrast to alcohols, which typically are not acidic.
  - → Acidity of phenols can be understood on the basis of resonance stabilization of the negative charge of the conjugate base of phenol, $C_6H_5O^-$.

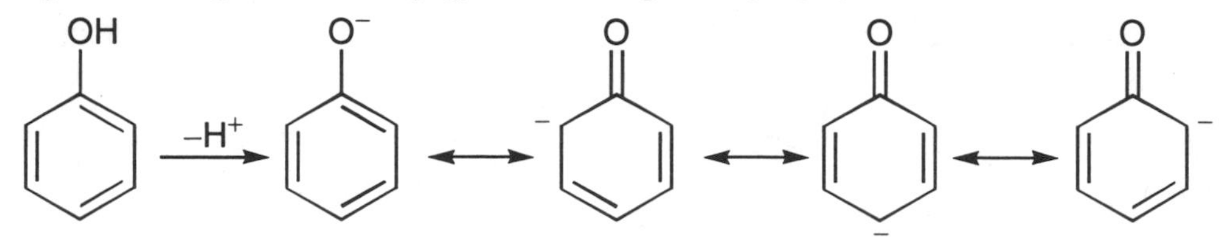

→ Because of resonance stabilization, the phenoxide ($C_6H_5O^-$) anion is a weaker conjugate base than the corresponding conjugate base of typical alcohols. For example, the ethoxide ($CH_3CH_2O^-$) anion, the conjugate base of ethanol, is a stronger base than the phenoxide anion.

19.6 Aldehydes and Ketones

- **Aldehydes**

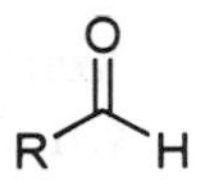

→ *Aldehydes* are organic compounds of the form shown on the right, where R is a hydrogen atom, an aliphatic group, or an aromatic group.

→ The group characteristic of aldehydes is written as –CHO, as in formaldehyde, HCHO, the first member of the family.

→ The names are given systematically by replacing the ending *–e* by *–al*, but many common names exist, such as formaldehyde for methanal and acetaldehyde for ethanal. The carbon atom on the carbonyl group is *included* in the count of carbon atoms when determining the alkane from which the aldehyde is derived.

→ A few common aldehydes are shown below.

Ethanal　　Benzaldehyde　　Methanal

Benzaldehyde provides part of the aroma of almonds and cherries. Aldehydes generally contribute to the flavor of fruits and nuts, and to the odors of plants.

→ Aldehydes can be prepared by the mild oxidation of *primary* alcohols. The industrial preparation of formaldehyde is by the oxidation of methanol with silver catalyst, which functions as a mild oxidizing medium preventing the further oxidation of formaldehyde to formic acid, HCOOH. The overall reaction is

$$2\,CH_3OH(g) + O_2(g) \xrightarrow{600°C,\ Ag} 2\,HCHO(g) + 2\,H_2O(g)$$

- **Ketones**

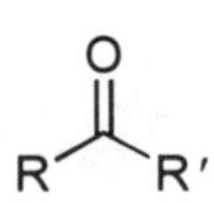

→ *Ketones* are organic compounds of the form shown on the right, where the R groups (alkyl or aryl) may be the same or different.

→ The *carbonyl group*, –C=O, characteristic of ketones, is written –CO, as in propanone, CH_3COCH_3 (acetone), the first member of the family.

→ The names are given systematically by replacing the ending *-e* by *-one*, but many common names exist such as acetone or dimethyl ketone for propanone and methyl ethyl ketone for butanone. The carbon atom on the carbonyl group is *included* in the count of carbon atoms when determining the alkane from which the ketone is derived. A number is used to locate the carbon atom in the carbonyl group to avoid ambiguity.

→ A few common ketones are shown below:

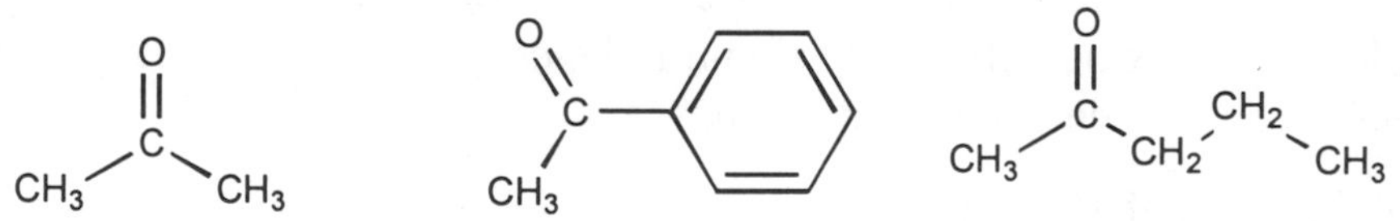

Propanone (acetone) Benzophenone 2-Pentanone

→ *Ketones* contribute to flavors and fragrances. An example shown below is *carvone*, which is the essential oil in spearmint.

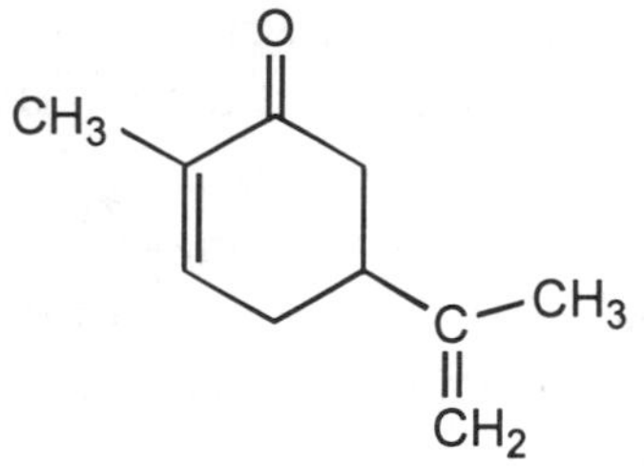

→ Ketones can be prepared by the oxidation of *secondary* alcohols. There is less risk of further oxidation than in aldehydes, so stronger oxidizing agents are used. Dichromate oxidation of secondary alcohols produces ketones with little excess oxidation and in good yields. An example is the oxidation of 2-propanol to propanone (acetone) shown schematically below:

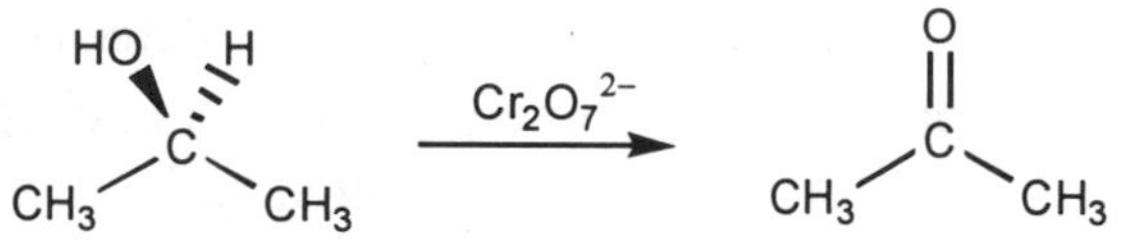

$Cr_2O_7^{2-}$

Example 19.6 Aldehydes are easier to oxidize than ketones, and this feature may be used to distinguish between them. *Tollens reagent* (a solution of Ag_2O in aqueous ammonia) will oxide an aldehyde to its corresponding carboxylic acid, but will *not* react with a ketone. The oxidizing agent in Tollens reagent is Ag^+, which is reduced to silver metal forming a mirror that is easy to detect.

Write a reaction for the oxidation of butanal using Tollens reagent.

Solution Butanal is oxidized to the corresponding carboxylic acid, butanoic acid, and Ag^+ is reduced to Ag(s), forming a mirror. The overall (unbalanced) reaction is

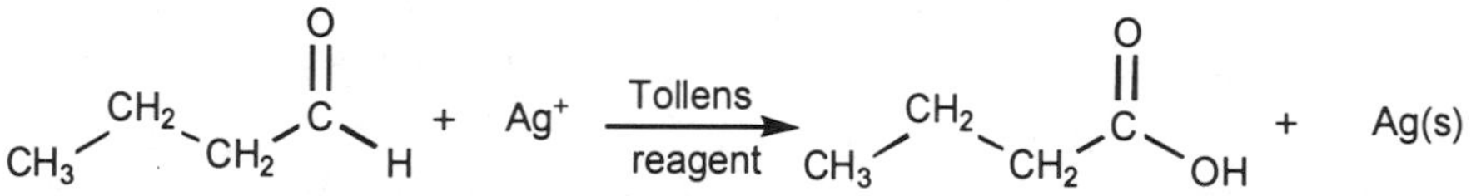

$$CH_3CH_2CH_2CHO + Ag^+ \xrightarrow[\text{reagent}]{\text{Tollens}} CH_3CH_2CH_2COOH + Ag(s)$$

Often in organic chemistry, reactions are written by specifying only *key* reactants and products, and the conditions and reagents required for the reaction to occur.

19.7 Carboxylic Acids

- **Carboxylic acids**

→ *Carboxylic acids* are organic compounds of the form shown on the right, where R is a hydrogen atom, an alkyl or an aryl group.

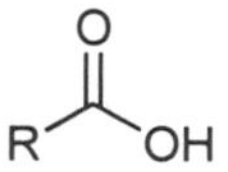

→ The *carboxyl group* characteristic of carboxylic acids is written –COOH as in formic acid, HCOOH, the first member of the family. The *carboxyl group* is composed of a *hydroxyl group*, –OH, attached to a *carbonyl group*, –C=O.

→ The names are given systematically by replacing the ending *-e* by *-oic acid*, but many common names are used such as formic acid for methanoic acid and acetic acid for ethanoic acid. The carbon atom on the carboxyl group is included in the count of carbon atoms to determine the parent hydrocarbon molecule.

→ A few common carboxylic acids are shown below:

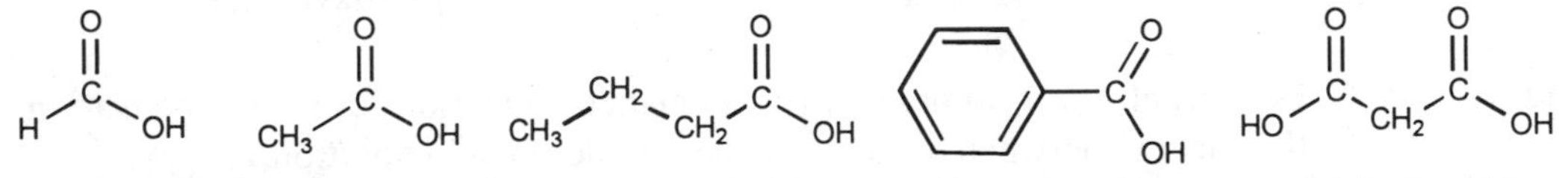

Methanoic acid (Formic acid) Ethanoic acid (Acetic acid) Butanoic acid Benzoic acid Malonic acid

Note: Malonic acid contains two –COOH groups; it is called a *diacid.*

→ Carboxylic acids contain hydroxyl groups, –OH, which can take part in hydrogen bonding.

→ Carboxylic acids can be prepared by oxidation of aldehydes or of *primary* alcohols in an acidified solution containing a strong oxidizing agent such as potassium permanganate, $KMnO_4$, or sodium dichromate, $Na_2Cr_2O_7$:

Aldehyde Primary alcohol

→ In some cases, alkyl groups can be oxidized *directly* to carboxyl groups. An important industrial example is the oxidation of the methyl groups on *p*-xylene by a cobalt(III) catalyst to form *terephthalic acid*, which is used in the production of artificial fibers.

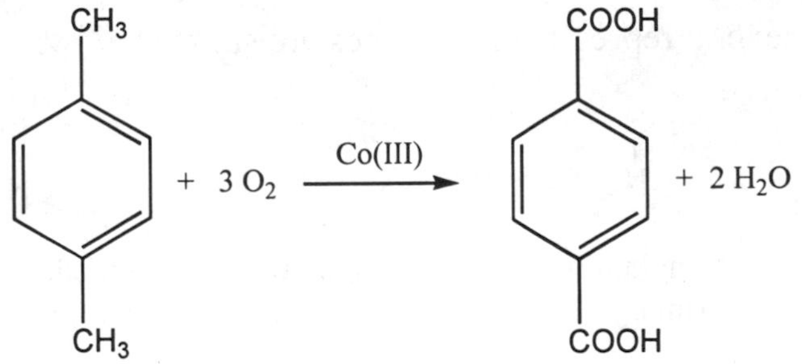

$$\text{+ } 3\,O_2 \xrightarrow{\text{Co(III)}} \text{+ } 2\,H_2O$$

- **Esters**

→ An *ester* is the product of a *condensation* reaction (defined below) between a carboxylic acid and an alcohol. A water molecule is also produced. *Esterification* reactions can be acid catalyzed:

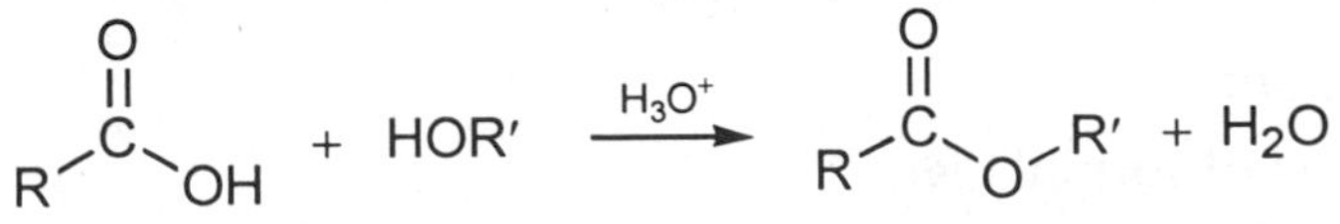

$$RCOOH + HOR' \xrightarrow{H_3O^+} RCOOR' + H_2O$$

Carboxylic acid Alcohol Ester Water

→ A *condensation* reaction is one in which two molecules combine to from a large one and a small molecule is eliminated.

→ Many *esters* have fragrant aromas and contribute to flavors of fruits. The esters *n*-amyl acetate and *n*-octyl acetate are responsible for the aromas of bananas and oranges, respectively.

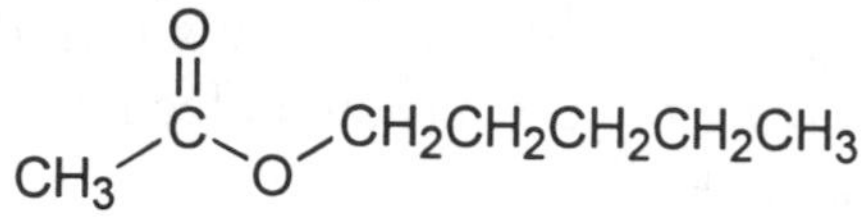

$$CH_3-C(=O)-O-CH_2CH_2CH_2CH_2CH_3$$

n-Amyl acetate

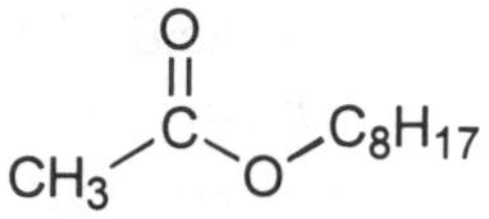

$$CH_3-C(=O)-O-C_8H_{17}$$

n-Octyl acetate

Example 19.7 Use structural formulas to write a balanced equation for the condensation reaction that occurs between acetic acid and ethanol in acid solution.

Solution The condensation reaction is between a carboxylic acid and an alcohol to yield an ester. As shown below, the product is the ester *ethyl acetate*. The atoms in the box are the ones that form the product water molecule.

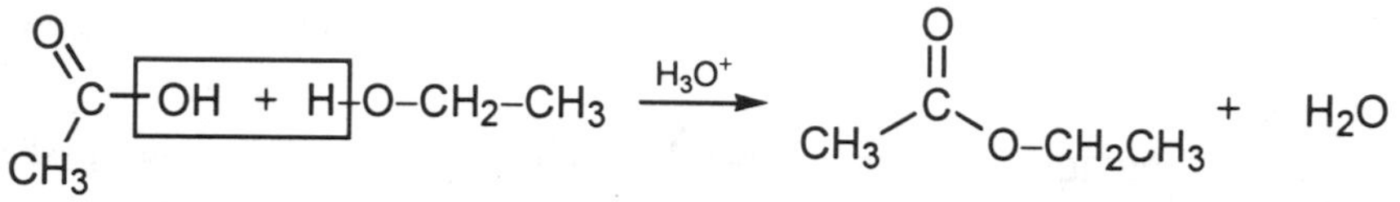

$$CH_3-C(=O)-\boxed{OH + H}-O-CH_2-CH_3 \xrightarrow{H_3O^+} CH_3-C(=O)-O-CH_2CH_3 + H_2O$$

Acetic acid, Ethanol, Ethyl acetate, Water

19.8 Amines, Amino Acids, and Amides

- **Amines**

→ Derivatives of *ammonia*, NH_3, formed by replacing one or more hydrogen atoms with organic groups, R; the *amino group*, $-NH_2$, is the functional group of *amines*

→ Named by specifying the groups attached to the nitrogen atom, N, alphabetically followed by the suffix *amine*

→ Designated as *primary, secondary,* or *tertiary*, depending on the number of R groups attached to the N atom

→ Ammonia and several representative amines are shown below:

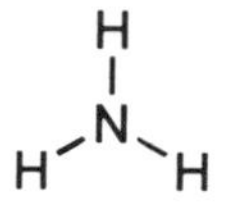

NH_3	H_2N-CH_3	$H-N(CH_2CH_3)-CH_3$	$CH_3-N(CH_2CH_3)-CH_3$
Ammonia	Methylamine (primary: 1 R)	Ethylmethylamine (secondary: 2 R)	Dimethylethylamine (tertiary: 3 R groups)

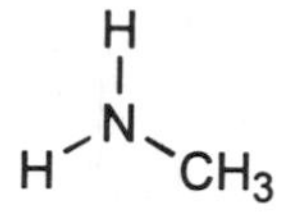

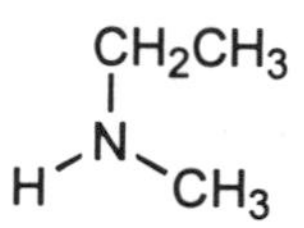

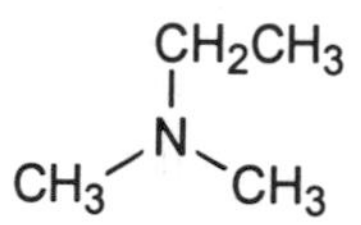

- **Properties of amines**

→ Characterized by four sp^3 hybrid orbitals on the nitrogen atom; three participate in three single bonds and the fourth has a lone pair of electrons

→ *Amines* are widespread in nature, often have disagreeable odors (for example *putrescine*, $NH_2(CH_2)_4NH_2$), and, similar to ammonia itself, are weak bases.

- **Quaternary ammonium ions**

→ Tetrahedral ions of formula R_4N^+

→ Four groups bonded to the central nitrogen atom, may be the same or different

→ Negligible acid or base properties and little effect on pH
→ Isolated as quaternary ammonium salts
→ Dimethylethylpropylammonium chloride, with four R groups, is an example of a quaternary ammonium salt.

$$CH_3-\overset{+}{N}(CH_2CH_3)(CH_3)(CH_2CH_2CH_3)\quad Cl^-$$

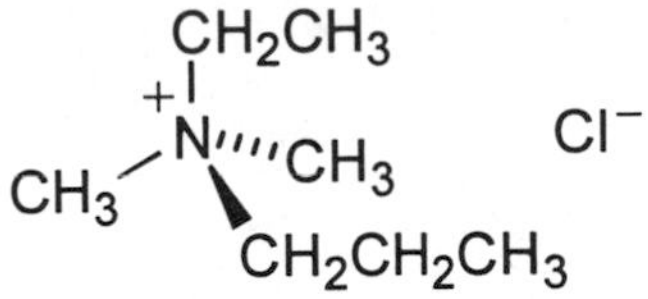

- **Amino acids**
  → An *amino acid* is a carboxylic acid that contains an *amino* group as well as a *carboxyl* group separated by *one* carbon atom. A more precise name is an *aminocarboxylic acid.*
  → General formula of an *amino acid* is $R(NH_2)CH(COOH)$.
  → Examples of *amino acids* are *glycine*, *methionine*, and *phenylalanine.*

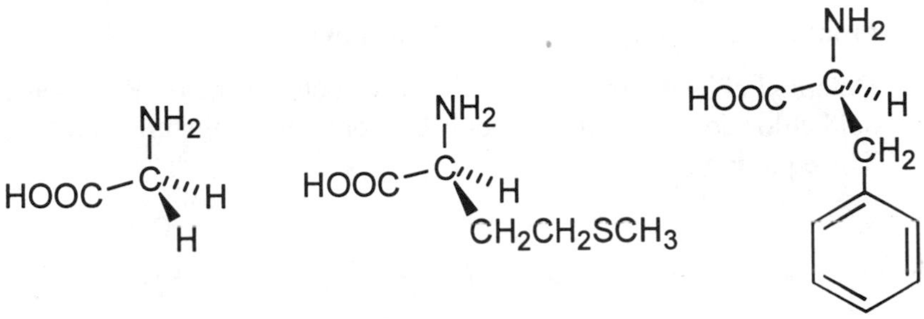

$$HOOC-CH(NH_2)-H \qquad HOOC-CH(NH_2)-CH_2CH_2SCH_3 \qquad HOOC-CH(NH_2)-CH_2-C_6H_5$$

Glycine   Methionine   Phenylalanine

Note: In amino acids, the central carbon atom is *stereogenic* and the molecule is *chiral* except for the amino acid *glycine.*

- **Properties of amino acids**
  → Building blocks of proteins (see Section 19.13)
  → Essential for human health
  → Form *hydrogen bonds* with *both* the *amino* and *carboxyl* groups
  → Form *double ions* (*zwitterions*) by transferring a proton from the *carboxyl* to the *amino* group as shown for a general amino acid

$$HOOC-CH(NH_2)-R \qquad {}^{-}OOC-CH(NH_3^{+})-R$$

Amino acid   Double ion form

Note: The term *zwitterion* comes from the German, *Zwitter*, hermaphrodite. In solution, all amino acids exist virtually exclusively as *double ions.*

- **Amides**
  → Molecules described by the general formula R–(CO)–NH–R
  → *Amides* are formed by the condensation reaction of carboxylic acids or acid chlorides with amines. In the reaction of carboxylic acids, water is eliminated. The

reaction mixture near room temperature forms an ammonium salt that reacts further to form the amide at higher temperature. In the reaction of acid chlorides, HCl is formed and removed by strong base added to neutralize the reaction mixture, or by the addition of excess amine.

→ The general reaction for the formation of an amide by a carboxylic acid and an amine is

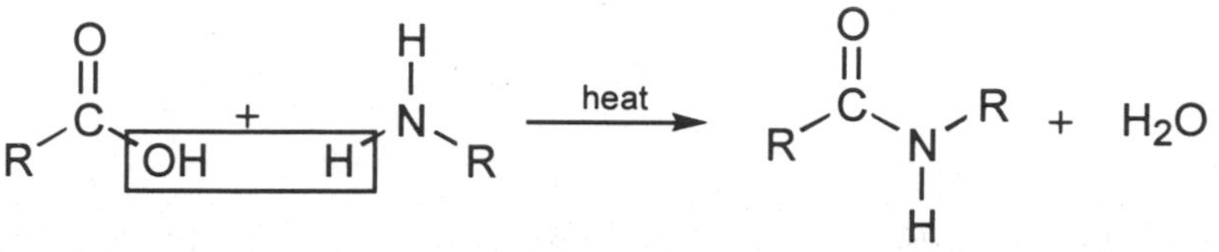

$$R\text{-}C(=O)\text{-}\boxed{OH + H}\text{-}N(H)\text{-}R \xrightarrow{\text{heat}} R\text{-}C(=O)\text{-}N(H)\text{-}R + H_2O$$

Note: The amide shown above has an N–H group that may participate in *intermolecular* hydrogen bonding. The atoms that form the product water molecule are shown in a box.

Example 19.8a Write a balanced equation using structural formulas for the principal products formed in the reaction of octanoyl chloride with 2-propylamine.

Solution Octanoyl chloride is the acid chloride of octanoic acid. Reaction of an acid chloride and an amine yields the corresponding amide and HCl. The equation is

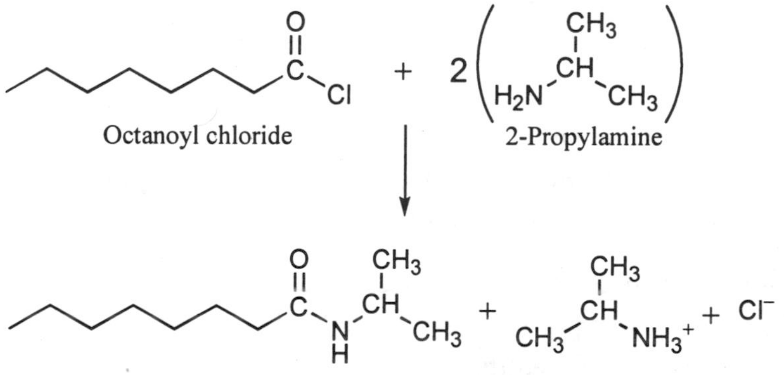

$$CH_3(CH_2)_6C(=O)Cl + 2\,(H_2N\text{-}CH(CH_3)\text{-}CH_3) \longrightarrow CH_3(CH_2)_6C(=O)\text{-}NH\text{-}CH(CH_3)\text{-}CH_3 + CH_3\text{-}CH(CH_3)\text{-}NH_3^+ + Cl^-$$

Octanoyl chloride; 2-Propylamine

Note: Two equivalents of the base 2-propylamine are required for the reaction to go to completion. One equivalent reacts with the acid chloride to form the amide, and the second with the HCl product to form an ammonium chloride.

Example 19.8b Write a balanced equation using structural formulas for the principal products formed in the reaction of acetic acid with triethylamine.

Solution Normally, we expect the reaction to yield an amide and water. Triethylamine, however, is a *tertiary* amine and has no hydrogen atom to react with the acid. An amide is *not* formed in the reaction. Rather, the acid reacts with the basic amine to form the ammonium salt.

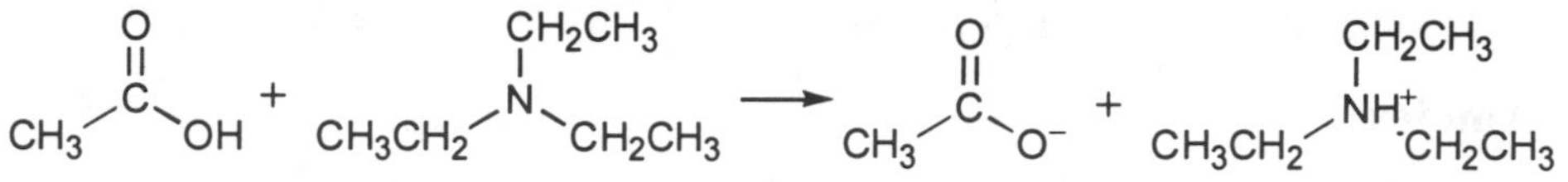

$$CH_3\text{-}C(=O)\text{-}OH + CH_3CH_2\text{-}N(CH_2CH_3)\text{-}CH_2CH_3 \longrightarrow CH_3\text{-}C(=O)\text{-}O^- + CH_3CH_2\text{-}NH^+(CH_2CH_3)\text{-}CH_2CH_3$$

Summary of Functional Groups

Class of compound	Functional group	Generic formula	Example	Condensed structural formula
alcohol	–OH	R–OH	ethanol	CH_3CH_2OH
ether	–O–	R–O–R	dimethyl ether	CH_3OCH_3
aldehyde	–C(=O)–H	R–(CO)–H	ethanal	CH_3CHO
ketone	–C(=O)–	R–(CO)–R	butanone	$CH_3CH(O)CH_2CH_3$
carboxylic acid	–C(=O)–OH	R–COOH	ethanoic acid (acetic acid)	CH_3COOH
ester	–C(=O)–O–	R–COO–R	ethyl acetate	$CH_3C(O)OCH_2CH_3$
amine	$–NH_2$			
primary		RNH_2	ethylamine	$CH_3CH_2NH_2$
secondary		R_2NH	ethylmethylamine	$CH_3CH_2NHCH_3$
tertiary		R_3N	trimethylamine	$(CH_3)_3N$
quaternary		R_4N^+	tetramethyl-ammonium ion	$(CH_3)_4N^+$
amide	–C(=O)–N(–)–			
primary		$R–(CO)–NH_2$	ethanamide (acetamide)	CH_3CONH_2
secondary		R–(CO)–NH–R	N-methyl propanamide	$CH_3CH_2CONH(CH_3)$
tertiary		$R–(CO)–NR_2$	N,N-dimethyl ethanamide	$CH_3CON(CH_3)_2$

The Impact on Materials (Sections 19.9-19.12)

Key Concepts

polymers, addition polymerization, monomer, repeating unit, radical polymerization, Ziegler-Natta catalyst, syndiotactic polymer, atactic polymer, isotactic polymer, latex, condensation polymerization, condensation polymers, polyamides, polyesters, copolymers, alternating copolymers, block copolymers, random copolymer, graft copolymer, composite material, elasticity

Overview

- **Polymers**
  - → Compounds of high molar mass consisting of smaller repeating *subunits* called *monomers*
  - → Used in many modern synthetic materials
  - → *Polymers* may be synthetic or naturally occurring. Examples of synthetic polymers are polystyrene, Teflon, polyethylene, Plexiglas, Dacron, and Orlon. Examples of natural polymers are nucleic acids and carbohydrates.
  - → Polymers can be formed by *addition* or *condensation* polymerization.

19.9 Addition Polymerization

- **Addition polymers**
  - → Formed when *alkene* monomers react with themselves with no net loss of atoms to form *polymers*
  - → Prepared by *radical polymerization*, a radical chain reaction
  - → Radical polymerization reactions are started by using an *initiator* such as an organic peroxide, R–O–O–R, which decomposes when heated to form two free radicals.
  - → The radical polymerization mechanism is shown for ROOR and $CH_2{=}CHX$:

Initiation: $$R{-}O{-}O{-}R \xrightarrow{\text{heat}} R{-}O\bullet + \bullet O{-}R$$

Propagation: $$\underset{\text{Initiator}}{R{-}O\bullet} + \underset{\text{Monomer}}{H_2C{=}CHX} \longrightarrow R{-}O{-}CH_2{-}\dot{C}HX$$

$$R{-}O{-}CH_2{-}\dot{C}HX \xrightarrow{n\,(H_2C{=}CHX)} R{-}O{-}\left(CH_2{-}CHX\right)_{n+1}\bullet$$

Termination: $2\left(R{-}O{-}CH_2{-}\overset{H}{\underset{X}{C}}\bullet\right) \longrightarrow R{-}O{-}CH_2CHXCHXCH_2{-}O{-}R$

→ Radical polymerization reactions feature *initiation* and *propagation* steps as shown on the preceding page. The reaction terminates when all the monomer is consumed or when two radical chains of any length react to form a single diamagnetic (nonradical) species. One possible *termination* step is shown in the scheme above.

Example 19.9a Name the polymers produced in the preceding radical polymerization for X = H and X = CH_3.

Solution For X = H, the monomer is ethene (ethylene) and the polymer is *polyethylene*. For X = CH_3, the monomer is propene (propylene) and the polymer is *polypropylene*.

Example 19.9b Sketch three repeating units of the polymer formed by radical polymerization of phenylethene.

Solution A sketch of the monomer phenylethene, commonly known as *styrene*, is shown along with the repeating unit formed from it in a radical polymerization reaction. The three repeat unit is formed by linking three repeating units together. The polymer is called *polystyrene*.

$CH{=}CH_2$ $-\S-CH{-}CH_2-\S-$ $-\S-CH{-}CH_2{-}CH{-}CH_2{-}CH{-}CH_2{-}\S-$

Styrene monomer   Repeating unit   Sketch of three repeating units of polystyrene

Note: The squiggly lines here and elsewhere are used as boundaries for a unit abstracted from a larger polymer chain.

- **Stereoregular polymer**
  - → Each unit or pair of repeating units in the polymer has the same relative orientation.
  - → These polymers pack well together, and are relatively strong and impact resistant.
- **Isotactic polymer**
  - → A *stereoregular* polymer in which all substituents are found on the same side of the extended carbon chain
  - → An example of an *isotactic* configuration is a portion of a polypropylene chain in which the methyl side groups are oriented as follows:

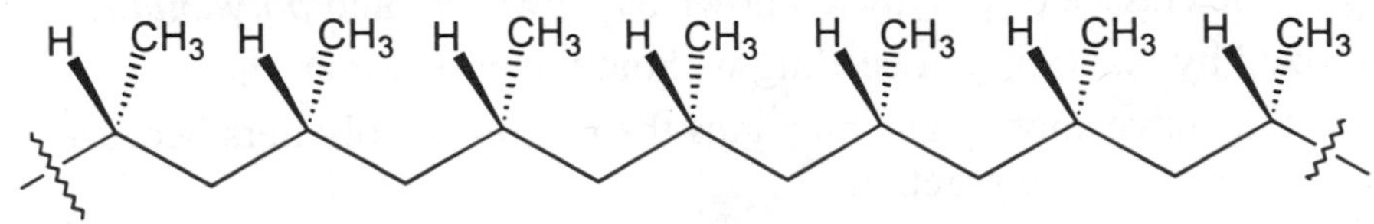

Isotactic configuration – same side

- **Syndiotactic polymer**
  → A *stereoregular* polymer in which substituents alternate regularly on either side of the extended carbon chain
  → An example of a *syndiotactic* configuration is a portion of a polypropylene chain with the following orientation of the methyl side groups:

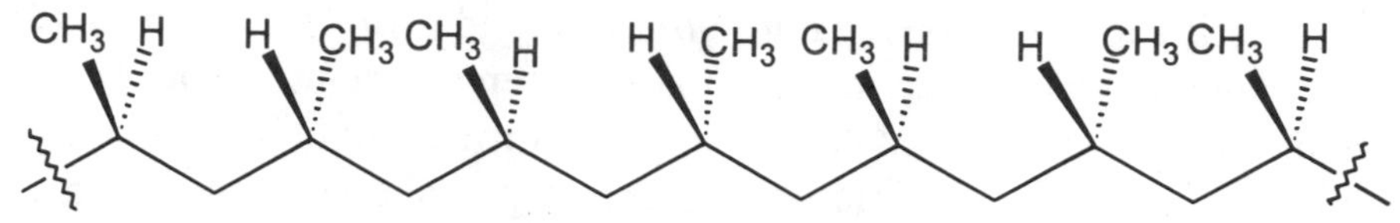

Syndiotactic configuration – alternating sides

- **Atactic polymer**
  → A polymer in which the substituents are randomly oriented with respect to the extended carbon chain
  → *Atactic* polymers are *not* stereoregular, but amorphous and less suited for many applications.
  → An example of an *atactic* configuration is a portion of the polypropylene chain with the following random orientation of the methyl side groups:

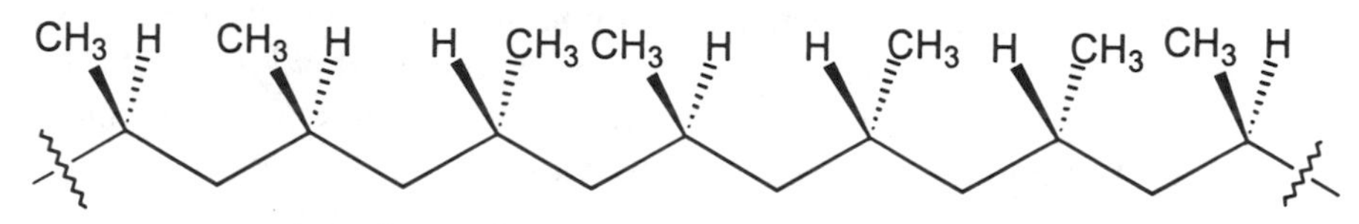

Atactic configuration – random sides

- **Ziegler-Natta catalysts**
  → Consist of aluminum- and titanium-containing compounds, such as titanium tetrachloride, $TiCl_4$, and triethyl aluminum, $(CH_3CH_2)_3Al$
  → Used in the production of stereoregular polymers, including synthetic rubber
  → Chemists were unable to synthesize rubber with useful mechanical properties until the discovery of Ziegler-Natta catalysts in 1953.

19.10 Condensation Polymerization

- **Condensation reaction** → In polymer chemistry, a condensation reaction is *also* one in which two molecules combine to form a larger one, and a small molecule is eliminated.

- **Condensation polymers**
  → Polymers formed by a series of *condensation* reactions
  → Formed from reactants each of which has *two* functional groups
  → Formed from stoichiometric amounts of the reactants
  → Encompass the classes of polymers known as *polyesters* and *polyamides*
  → Characterized by reactions proceeding without the need for an initiator
  → Typically, they have shorter chain lengths than addition polymers because each monomer can initiate the reaction.

- **Polyesters**
  → Formed from the condensation of a *diacid* with a *diol*
  → Kodel polyester is formed from the esterification of terephthalic acid and 1,4-bis(hydroxymethyl)-cyclohexane as shown:

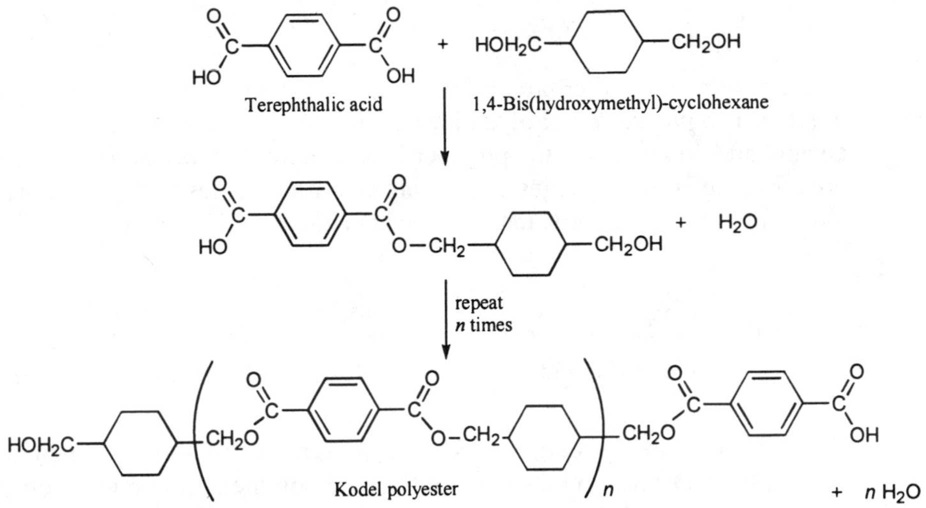

Terephthalic acid + $HOH_2C-C_6H_{10}-CH_2OH$ (1,4-Bis(hydroxymethyl)-cyclohexane)

↓

$HO-C(=O)-C_6H_4-C(=O)-O-CH_2-C_6H_{10}-CH_2OH$ + H_2O

↓ repeat *n* times

$HOH_2C-C_6H_{10}-\left(CH_2O-C(=O)-C_6H_4-C(=O)-O-CH_2-C_6H_{10}\right)_n-CH_2O-C(=O)-C_6H_4-C(=O)-OH$ + $n\,H_2O$

Kodel polyester

- **Polyamides**
  → Formed from the condensation of a *diacid* with a *diamine*
  → Nylon-66 forms from the condensation of 1,6-hexanedioic acid and 1,6-hexanediamine. The reactants first form a salt by proton transfer, which then condenses to form the polyamide as shown:

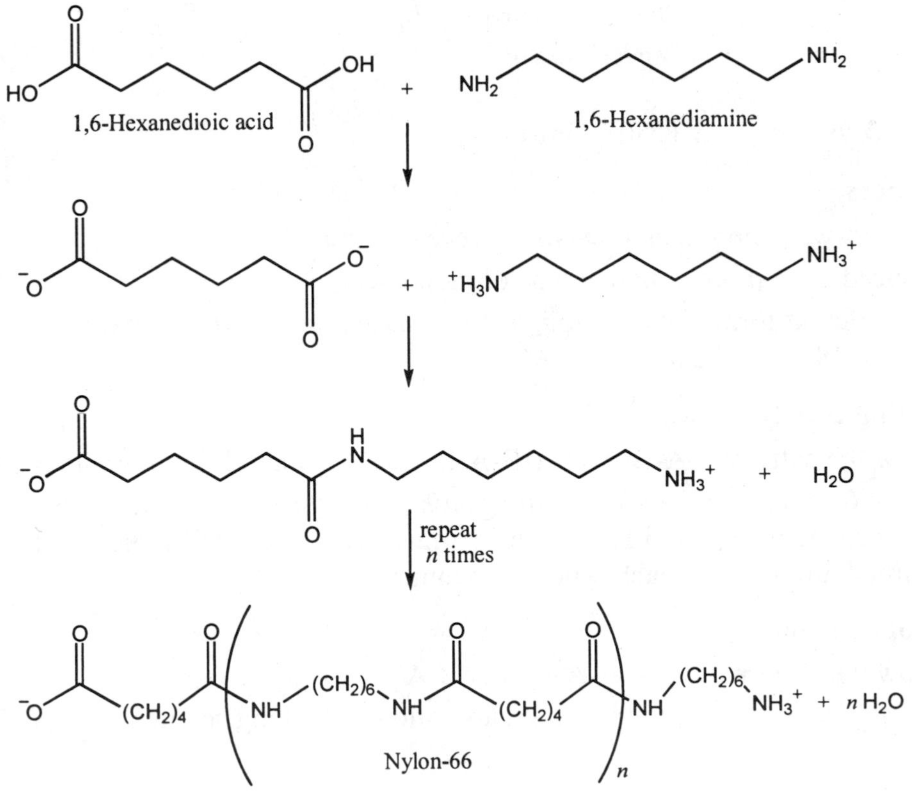

1,6-Hexanedioic acid + 1,6-Hexanediamine

$HO-C(=O)-(CH_2)_4-C(=O)-OH$ + $NH_2-(CH_2)_6-NH_2$

↓

$^-O-C(=O)-(CH_2)_4-C(=O)-O^-$ + $^+H_3N-(CH_2)_6-NH_3^+$

↓

$^-O-C(=O)-(CH_2)_4-C(=O)-NH-(CH_2)_6-NH_3^+$ + H_2O

↓ repeat *n* times

$^-O-C(=O)-(CH_2)_4-C(=O)\left(NH-(CH_2)_6-NH-C(=O)-(CH_2)_4-C(=O)\right)_n NH-(CH_2)_6-NH_3^+$ + $n\,H_2O$

Nylon-66

Example 19.10 Write structural formulas for the monomers used to form the following polymers. If the polymer is a *polyester* or a *polyamide*, identify it as such.

I. 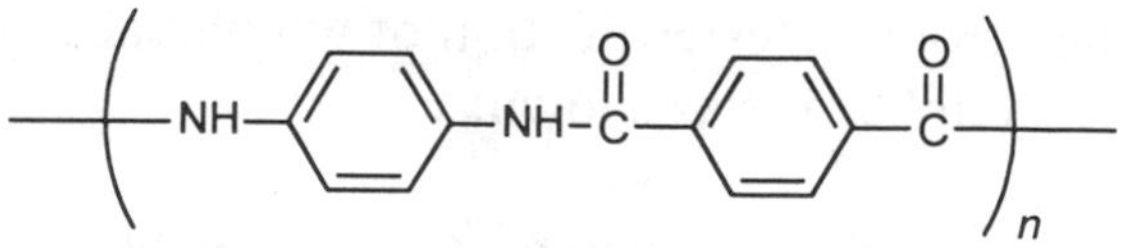 $-\!\left(-NH-C_6H_4-NH-\overset{O}{\overset{\|}{C}}-C_6H_4-\overset{O}{\overset{\|}{C}}-\right)_n\!-$ II. 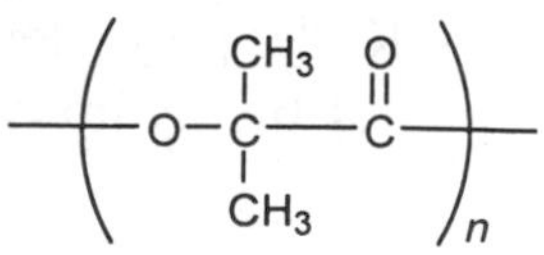 $-\!\left(-O-\underset{CH_3}{\overset{CH_3}{C}}-\overset{O}{\overset{\|}{C}}-\right)_n\!-$

Solution The amide functional group, $-\overset{O}{\overset{\|}{C}}-\overset{|}{N}-$, is clearly present in structure I, so the polymer is a *polyamide*. Polyamides are formed by the repeated condensation of an acid and an amine, so the polymer is formed from a *diacid* and a *diamine*. By breaking the amide linkages and adding water molecules to the structure, we obtain the two monomers, terephthalic acid and 1,4-diaminobenzene:

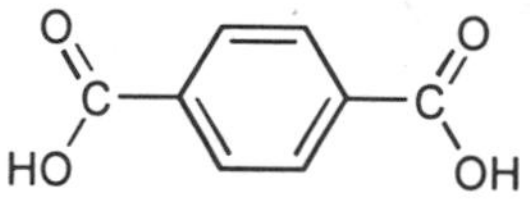

$HO-\overset{O}{\overset{\|}{C}}-C_6H_4-\overset{O}{\overset{\|}{C}}-OH$ Terephthalic acid

$H_2N-C_6H_4-NH_2$ 1,4-Diaminobenzene

The ester functional group, $-\overset{O}{\overset{\|}{C}}-O-$, is present in structure II, which is difficult to visualize. Draw two repeating units of the polymer, and you will only see ester functional groups. The polymer is a *polyester* made from a single type of monomer, 2-hydroxy-2-methylpropionic acid, which contains both carboxylic acid and alcohol functional groups. A polymer with one type of repeating unit is called *simple* (see Fig. 19.14a in the text).

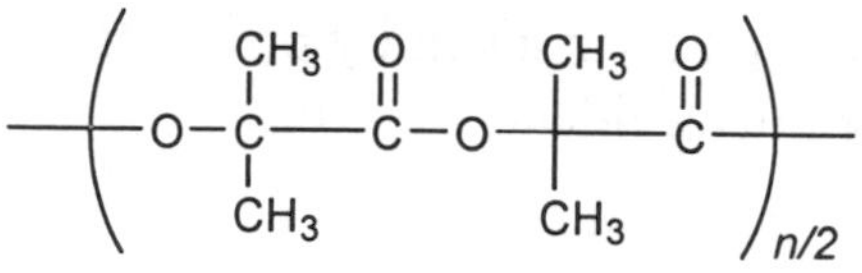

$-\!\left(-O-\underset{CH_3}{\overset{CH_3}{C}}-\overset{O}{\overset{\|}{C}}-O-\underset{CH_3}{\overset{CH_3}{C}}-\overset{O}{\overset{\|}{C}}-\right)_{n/2}\!-$ Two repeating units

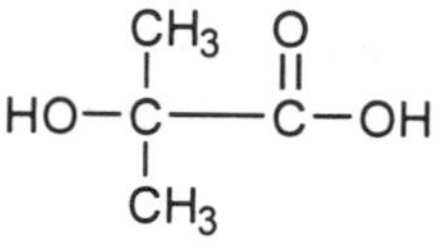

$HO-\underset{CH_3}{\overset{CH_3}{C}}-\overset{O}{\overset{\|}{C}}-OH$ Monomer

19.11 Copolymers and Composites

- **Copolymers**
  - → Polymers with more than one type of repeating unit
  - → Produced from more than one type of monomer
  - → Four different forms: *alternating*, *block*, *random*, and *graft* copolymers (see Fig. 19.14 in the text)
- **Alternating copolymers**
  - → Follow the pattern: –A-B-A-B-A-B-A-B–, where A and B are monomer units
  - → Nylon-66 (preceding page) is an *alternating* copolymer formed from the monomers 1,6-hexanedioic acid and 1,6-hexanediamine. The index (66) indicates the number of carbon atoms (6) in each type of monomer.
- **Block copolymers**
  - → Follow the pattern: –A-A-A-A-B-B-B-B-A-A-A–
  - → Long segments of one monomer, A, are followed by long segments of the other, B.

→ One example is *high-impact polystyrene*, which is a block copolymer of *styrene* and *butadiene*:

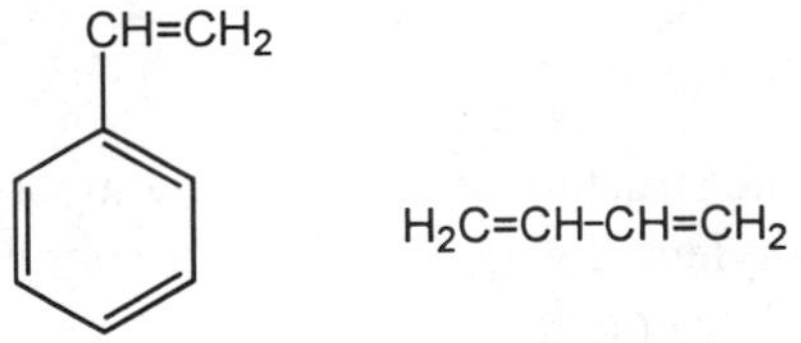

$CH=CH_2$ (on benzene ring)

Styrene

$H_2C=CH-CH=CH_2$

Butadiene

- **Random copolymers**
  → Follow no particular pattern: –A-A-B-A-B-A-B-B-A-B-A-B-B-B-A–
  → One example is the radical polymerization of the monomers styrene and 3-methylstyrene, which are expected to form a random copolymer:

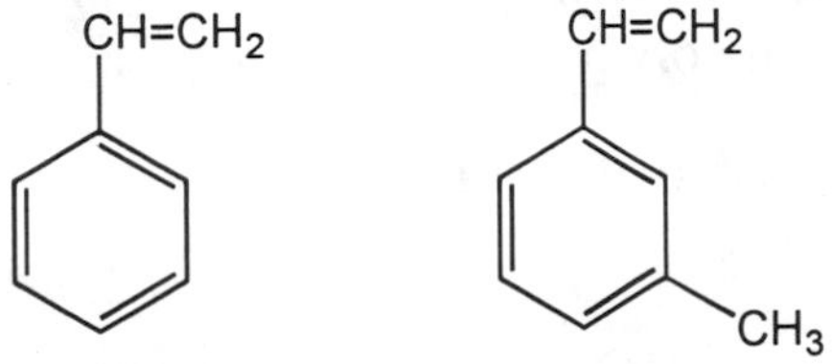

Styrene   3-Methylstyrene

- **Graft copolymers**
  → Consists of long chains of one monomer, A, with pendant chains of the second monomer, B, as in the sketch:

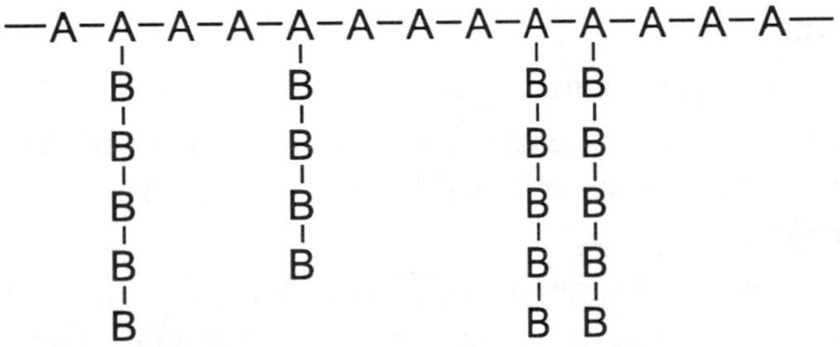

—A–A–A–A–A–A–A–A–A–A–A–A–A—

- **Properties of copolymers**
  → Different types of copolymers extend the range of physical properties obtainable for materials.
  → One example is the formation of soft polyurethane foams from *diisocyanates* and *glycols*, which are used for insulation and for furniture stuffing. The formation of a *polyurethane* is shown below:

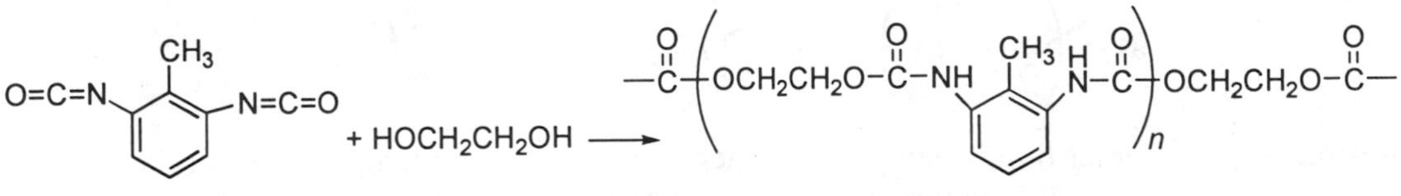

O=C=N–(C₆H₃CH₃)–N=C=O + $HOCH_2CH_2OH$ ⟶ $-\overset{O}{\overset{\|}{C}}\left(OCH_2CH_2O-\overset{O}{\overset{\|}{C}}-NH-C_6H_3(CH_3)-NH-\overset{O}{\overset{\|}{C}}\right)_n OCH_2CH_2O-\overset{O}{\overset{\|}{C}}-$

Toluene-2,6-diisocyanate Ethylene glycol   A polyurethane

Example 19.11 What type of copolymer is the following?

$$\sim CH_2-CH_2-CF_2-CF_2-CH_2-CH_2-CF_2-CF_2-CH_2-CH_2-CF_2-CF_2\sim$$

Solution Examination of the structure reveals that the monomers are ethylene, $H_2C{=}CH_2$, and tetrafluoroethylene, $F_2C{=}CF_2$. The polymer has the form –A-B-A-B-A-B– and is an *alternating copolymer.*

- **Composite materials**
  - → Consist of two or more materials solidified together
  - → Combine the advantages of the component materials
  - → Properties are superior to those of the component materials.
  - → One example is fiberglass, a material of great strength and flexibility, which consists of inorganic materials in a polymer matrix.

19.12 Physical Properties of Polymers

- **Synthetic polymers**
  - → Synthetic polymers have no definite molar mass, only an *average* value.
  - → Upon heating, synthetic polymers tend to soften gradually; they have no definite melting point.
  - → Polymers have chains of various lengths mixed together. For a given polymer, longer *average* chain length leads to higher viscosity and a higher softening point.
- **Properties of polymers**
  - → Depend upon the *average* chain length
  - → Depend upon the *polarity* of the functional groups; polar groups are associated with stronger intermolecular forces and tend to increase both softening points and mechanical strength.
  - → Properties depend on the manner in which chains pack. Long unbranched chains form crystalline regions that lead to strong, dense materials. Branched-chain polymers exhibit more tangled arrangements; they are less likely to form crystalline regions and tend to be weaker, less dense materials.

Example 19.12a Which of the following polymers would you expect to have the highest softening point when heated: a) *polyethylene*, b) *polyacrylate*, or c) *polypropylene*?

a) $\left(CH_2-CH_2 \right)_n$ b) $\left(CH_2-CH_2(CN) \right)_n$ c) $\left(CH_2-CH(CH_3) \right)_n$

Solution Other things being equal (such as average chain length), polyacrylate, with its polar side group, would be expected to have the highest softening point.

- **Elasticity**
  - → Ability of a polymer to return to its original shape after being stretched

→ *Elastomers* are materials that return easily to their original shape after stretching.
→ *Elasticity* of natural rubber can be improved by vulcanization (heating with sulfur), which forms disulfide (–S–S-) cross-links between polymer chains, increasing the resilience of the material (see Fig. 19.16 in the text). Extensive cross-linking provides a rigid network of interlinked polymer chains and leads to very hard materials.

Example 19.13b Glyptal, a strong, rigid polyester used for electronic parts, is a polymer of terephthalic acid (I) and glycerol (II). Draw a segment of the polymer and explain why it is so strong.

I.  HOOC–C_6H_4–COOH

II. 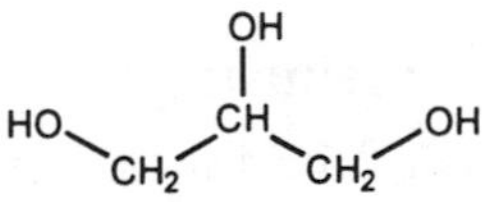 HO–CH_2–CH(OH)–CH_2–OH

Solution We expect the acid and the alcohol to form a polyester by condensation and elimination of water. We tend to associate rigid, strong polymers with *extensive cross-linking*. Notice that glycerol has *three* functional groups capable of forming a polymer chain with terephthalic acid. If two form the chain in the usual way, the third is left dangling and can react further with additional acid to form a highly branched, cross-linked polymer, a section of which is shown below. Cross-linking gives the polymer its strength.

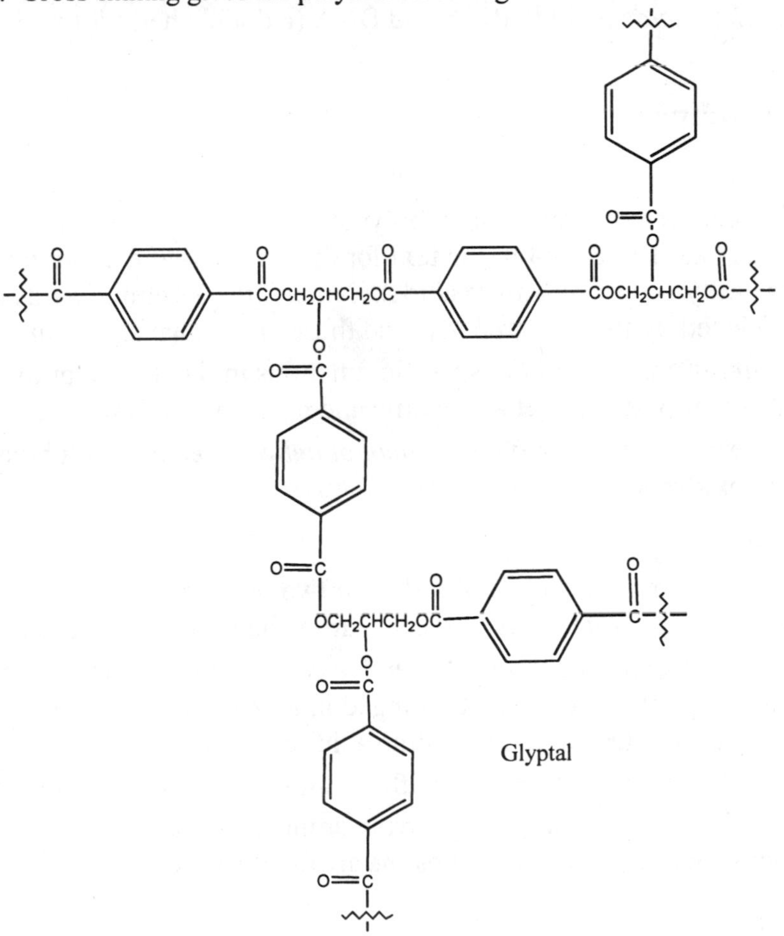

COCH₂CHCH₂OC — COCH₂CHCH₂OC — OCH₂CHCH₂OC

Glyptal

The Impact of Biology (Sections 19.13-19.15)

Key Concepts

proteins, essential amino acids, peptides, peptide bond, residue, oligopeptide, dipeptide, primary structure, secondary structure, ♀ helix, †-pleated sheet, tertiary structure, disulfide link, quaternary structure, denaturation, carbohydrates, polysaccharides, DNA, RNA, nucleoside, nucleotide, polynucleotides, nucleic acid

Overview

- **Organic chemistry**
  - → Basis of the functioning of all living things
  - → Organic compounds participating in reactions in organisms range from small molecules to huge polymers (*biopolymers*).
- **Biopolymers**
  - → *Cellulose* is found in wood, cotton, and silk.
  - → *Proteins* are found in skin, hair, muscle, and as *enzymes*.
  - → *Carbohydrates* include polymeric sugars (polysaccharides), starches, and cellulose.
  - → *Nucleic acids* include RNA and DNA (a double helix formed from two nucleic acid strands).

19.13 Proteins

- **Proteins**
  - → *Proteins* are condensation copolymers of *up to* 20 naturally occurring amino acids (see Table 19.4 in the text for the formulas and abbreviations of the naturally occurring amino acids). Nine of the 20 amino acids cannot be produced by the human body, and these are known as *essential amino acids.*
  - → *Proteins* perform highly specific functions in the human body. Enzymes (globular proteins) act as specific and efficient catalysts.
  - → An example of an enzyme is *alcohol dehydrogenase*, a globular protein that oxidizes ethanol to ethanal.
- **Peptides**
  - → *Peptides* are molecules formed from two or more amino acids. The –CO–NH– link is called a *peptide bond,* and each amino acid in a peptide is called a *residue.*
  - → Typical proteins contain *polypeptide chains* of more than a hundred residues joined through peptide bonds and arranged in a particular order. *Oligopeptides* are molecules with only a few amino acid residues.
  - → Peptides are named starting with the amino acid on the left (N-terminus).
  - → Reaction of the naturally occurring amino acids, aspartic acid (Asp) and phenylalanine (Phe), produces the artificial sweetener aspartame (Asp-Phe).

N-terminus H$_2$N–CH(CH$_2$COOH)–C(=O)–NH–CH(CH$_2$C$_6$H$_5$)–C(=O)–OH C-terminus

Aspartame, a dipeptide

Notice the peptide link, –CO–NH–, shown in the box. Aspartame is a *dipeptide* formed from *two* amino acids, and contains two *residues*.

Example 19.13 Write a chemical equation showing the formation of the dipeptide, Ala-Gly, from its constituent amino acids.

Solution The constituent amino acids are alanine and glycine, and the structures are given in Table 19.4 in the text. Because it is named first, alanine constitutes the N-terminus. The reaction is

$$H_2N\text{–}CH(CH_3)\text{–}C(=O)\text{–}OH + H\text{–}NH\text{–}CH_2\text{–}C(=O)\text{–}OH \longrightarrow H_2N\text{–}CH(CH_3)\text{–}C(=O)\text{–}NH\text{–}CH_2\text{–}C(=O)\text{–}OH + H_2O$$

Alanine Glycine Ala-Gly

The atoms that are eliminated to form a water molecule are shown in the box.

- **Structure of proteins:**

Primary structure

→ Sequence of residues in the peptide chain

→ The *primary* structure of *aspartame* is Asp-Phe. The *dipeptide* with the N- and C-terminus interchanged is Phe-Asp, a *different* oligopeptide.

Secondary structure

→ Describes the shape of the polypeptide chain

→ Controlled by *intramolecular* interactions between different parts of the peptide chain

→ Common *secondary* structures include the α *helix* (see Fig. 19.18 in the text) and the β*-pleated sheet* (see Fig. 19.19). The α *helix* is a helical portion of a polypeptide held together by hydrogen bonding. The β*-pleated sheet* is a portion of a polypeptide in which the segments of the chain lie side by side to form nearly flat sheets.

Tertiary structure

→ *Tertiary* structure describes the overall three-dimensional shape of the polypeptide, including the way in which α helix and β-pleated sheet regions fold together to shape the macromolecule. Folding is a consequence of the interactions between residues lying in different parts of the primary structure. One important link responsible for *tertiary* structure is the *disulfide link* (–S–S–) between amino acids containing sulfur.

→ Proteins are classified as *globular* (ball shaped) or *fibrous* (hair shaped with long chains of polypeptides that occur in bundles). Globular proteins are soluble in water; fibrous proteins are not. Essentially all enzymes are globular proteins.

→ *Hemoglobin* (responsible for oxygen transport in blood) and *cytochrome c* (a component of electron transport chains in mitochondria and bacteria) are globular proteins; *fibroin* (the protein of silk) is a fibrous protein.

Quaternary structure

→ Describes the arrangement of *subunits* in proteins with more than one polypeptide chain (or *subunit*)

→ Not all proteins contain more than one subunit, so not all have a *quaternary* structure.

→ *Hemoglobin* (see Fig. 19.19) contains four polypeptide units and has a *quaternary* structure.

- **Denaturation**

→ Loss of structure of proteins

→ Occurs when a protein loses quaternary, tertiary, or secondary structure (disruption of *noncovalent* interactions)

→ *Denaturation* is caused by heating and other means, is often irreversible, and is accompanied by loss of function of the protein.

19.14 Carbohydrates

- **Carbohydrates**

→ Most abundant class of naturally occurring organic compounds

→ Constitute more than 50% of the Earth's biomass

→ Members include starches, cellulose, and sugars.

→ Often have the empirical formula CH_2O (hence the name *carbohydrates*)

- **D-Glucose**

→ Most abundant carbohydrate

→ Forms starch and cellulose by polymerization reactions

→ Has the molecular formula $C_6H_{12}O_6$

→ Classified as both an alcohol *and* an aldehyde (*pentahydroxyl aldehyde*)

→ Exists in an acyclic (open-chain) and in two cyclic (closed-chain) forms called *anomers*. In water, the cyclic structures, also called *glucopyranose*, are favored.

- **Structure of D-glucose**
  - → The straight-chain structure of D–glucose and the interconversion to the two cyclic *anomers* are shown below. The percentages indicate proportions of the several species present in water solution. The anomers, α-D-glucose and β-D-glucose, differ in the stereochemistry at the carbon atom derived from the aldehyde C atom.

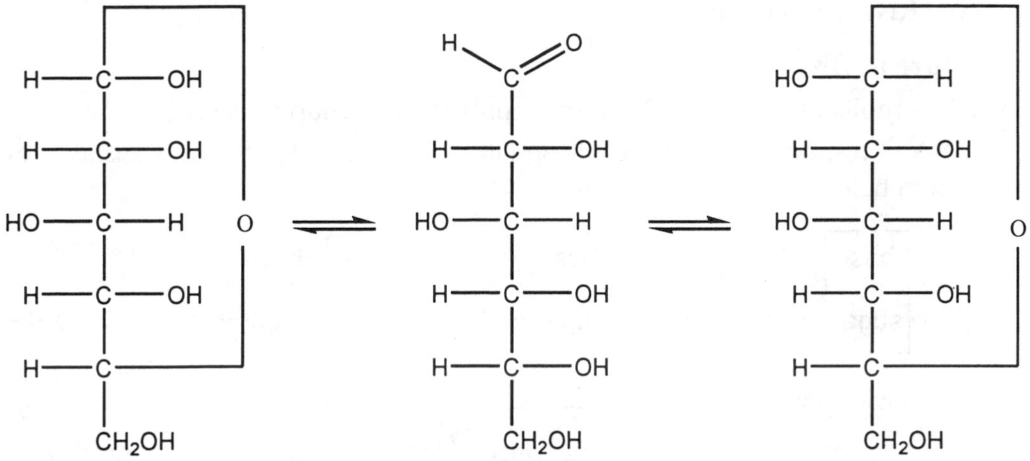

α-D-Glucose
36%

D-Glucose
0.02%

β-D-Glucose
64%

  - → A more realistic representation of the cyclic forms of D-glucose is:

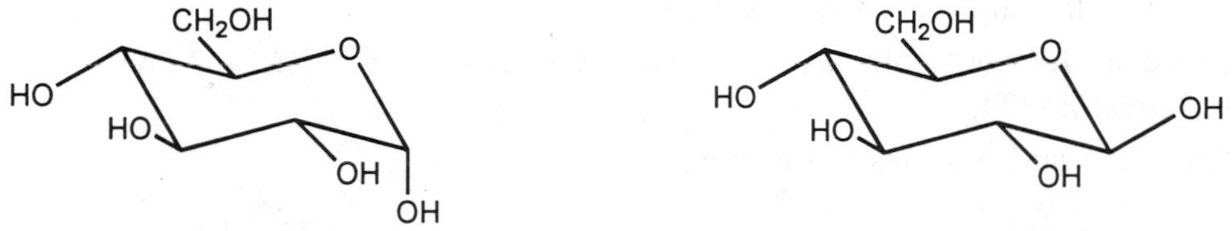

α-D-Glucose

β-D-Glucose

Note: L-Glucose, not found in nature, is the mirror image of D-glucose, and together they form a pair of enantiomers, which differ in the direction of rotation of polarized light. The symbols D and L refer to a configuration relationship to the compound D-glyceraldehyde, and not necessarily to the sign of rotation of light.

- **Polysaccharides**
  - → Polymers of glucose
  - → Include starch (digestible by humans) and cellulose (not digestible by humans)
  - → An example is cellulose, the structural material of plants and wood, which is a condensation polymer of β-D-glucose. A three subunit chain of cellulose is shown below. Note that the polymer is formed formally by elimination of H_2O from glucose monomers.

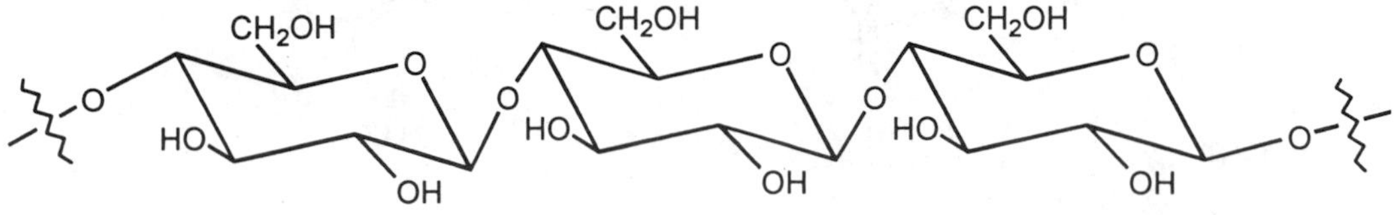

19.15 DNA and RNA

- **DNA**
  → The abbreviation for *deoxyribonucleic acid*
  → DNA molecules carry genetic information from generation to generation, control the production of proteins, and serve as the template for the synthesis of RNA (see RNA subsection).
- **Structure of DNA**
  → DNA molecules are condensation copolymers of enormous size.
  → DNA is composed of a sugar phosphate backbone and pendent bases as shown in the sketch below:

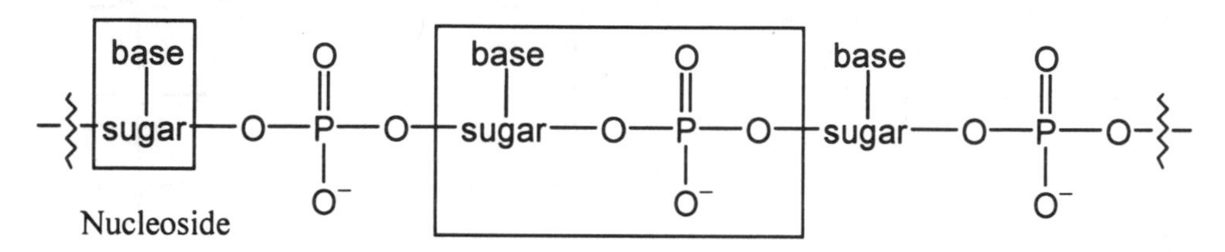

  → A base-sugar unit is called a *nucleoside.* A base-sugar-phosphate grouping is called a *nucleotide.*
- **Sugar and bases**
  → In DNA, the sugar is *deoxyribose.*
  → There are four possible bases: *cytosine* (C), *guanine* (G), *adenine* (A), and *thymine* (T).
  → The sugar and four bases are shown below:

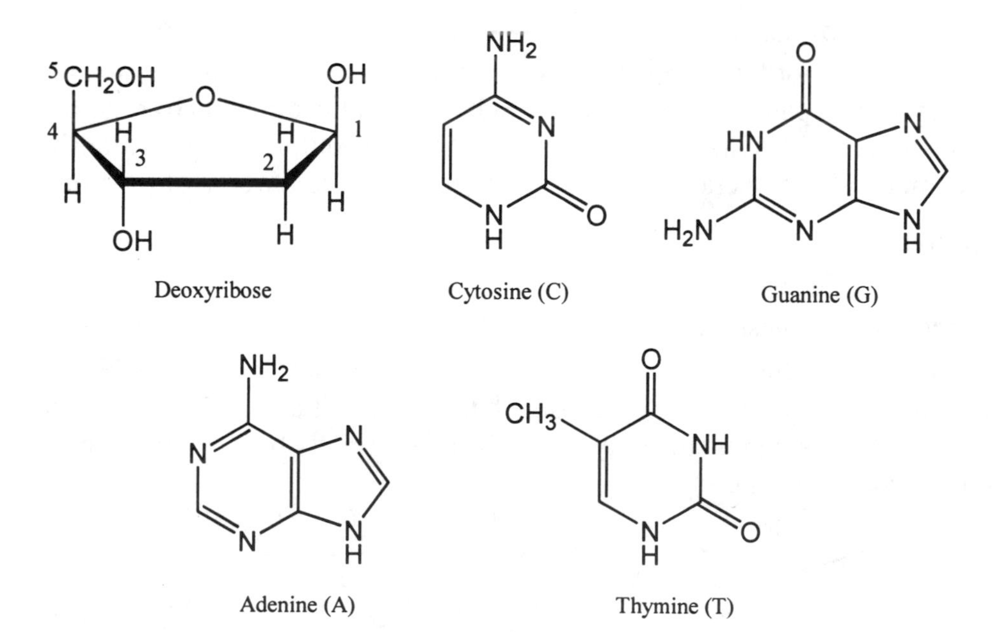

- **Nucleosides** → Formed in DNA by replacement of the OH group at carbon 1 in deoxyribose with the appropriate amine of the base:

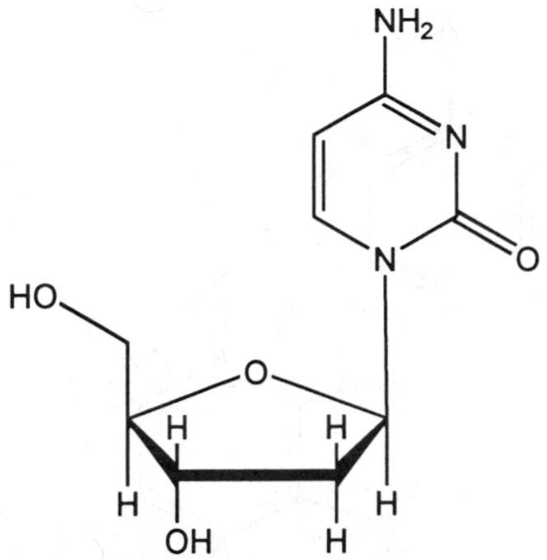

Cytosine nucleoside

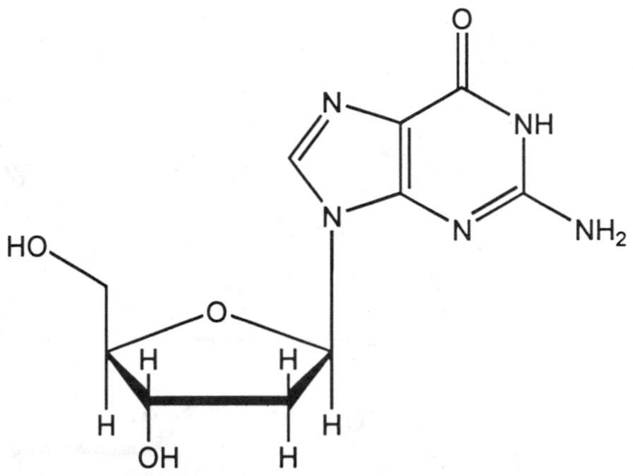

Guanine nucleoside

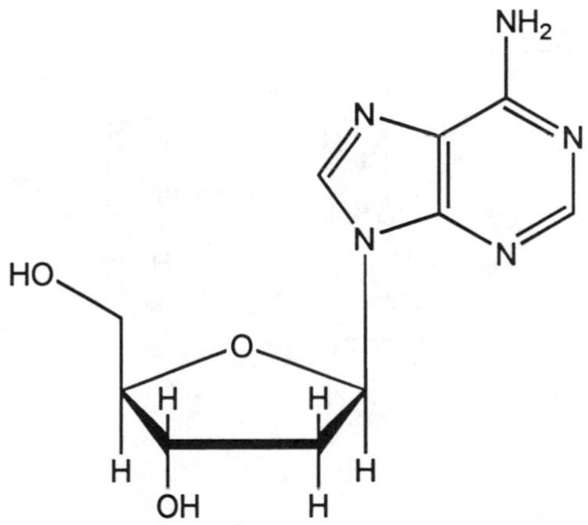

Adenine nucleoside

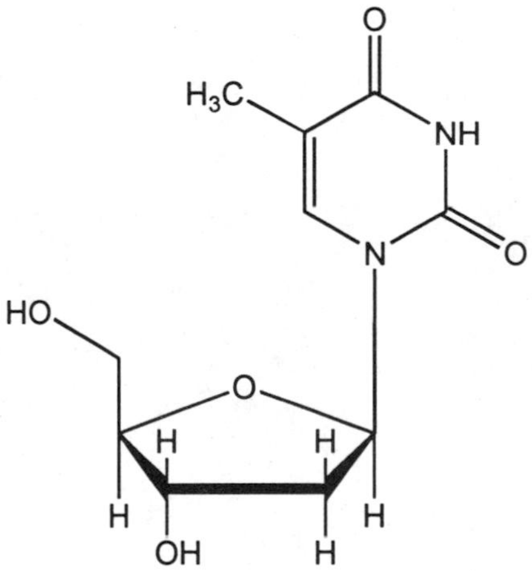

Thymine nucleoside

- **Nucleotides** → Condense at carbon 3 and carbon 5, eliminating water, to form the DNA molecule (nucleic acid), a *polynucleotide*; a trinucleotide segment of a polynucleotide is shown below:

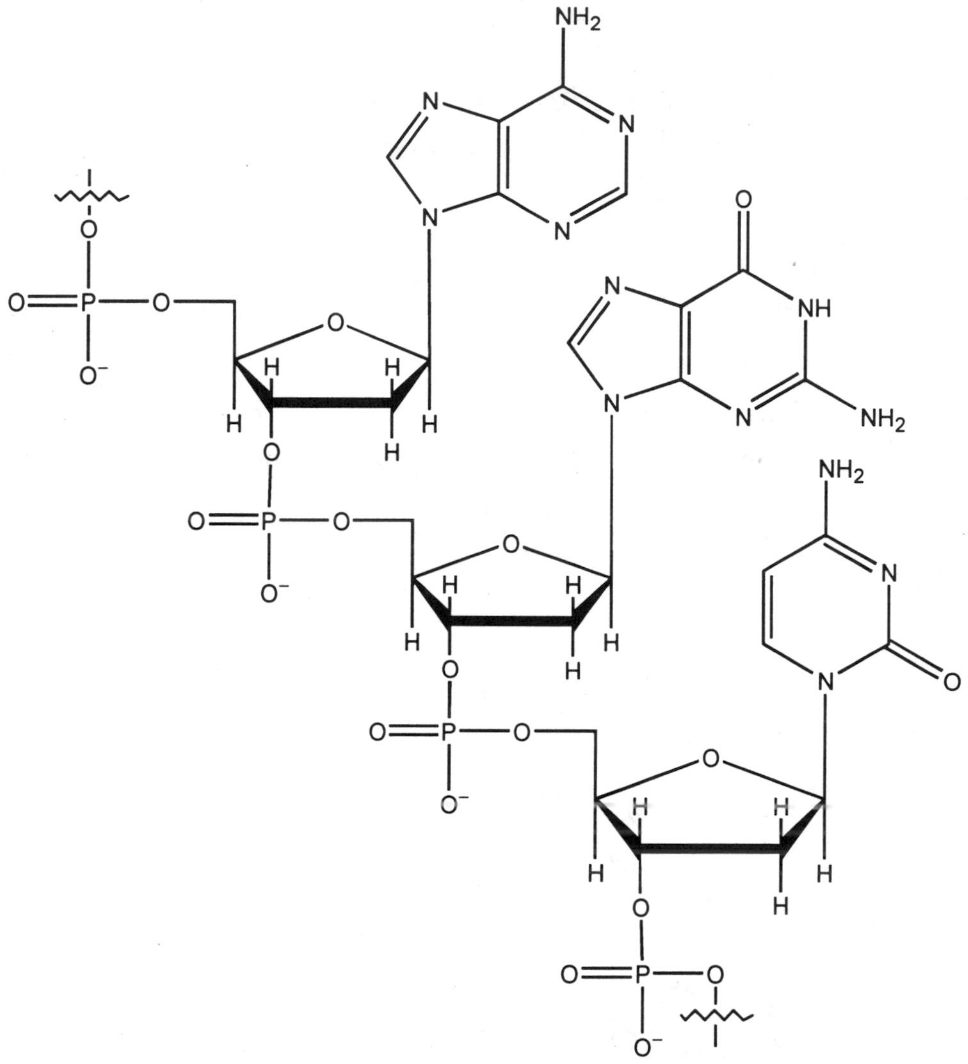

- **DNA double helix**
  → The polynucleotide (nucleic acid) strands link to each other in pairs to form the well-known double helix structure (see Fig. 19.28 in the text).
  → Association between strands of two polynucleotides occurs when hydrogen bonds between bases on *different* strands are formed.
  → In DNA, only two types of base pairs (G with C and A with T) occur, which are shown below with the hydrogen bonds indicated by dashed lines:

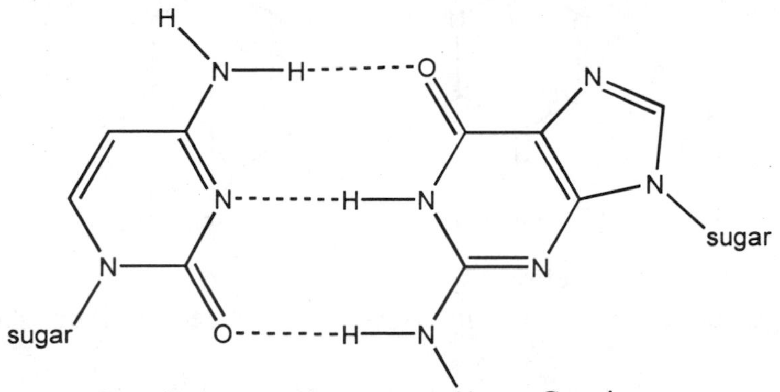

Cytosine Guanine

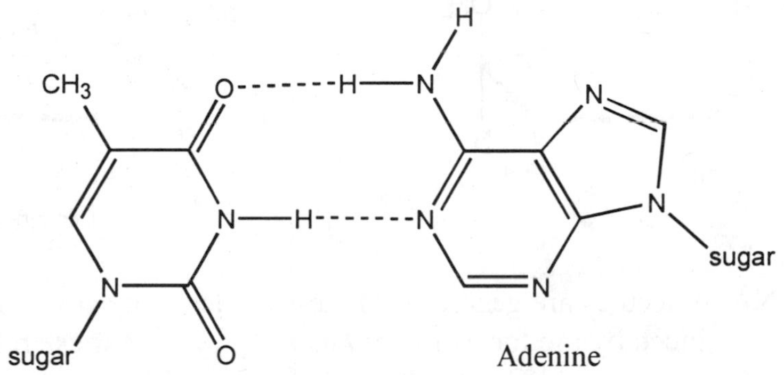

Thymine Adenine

Note: The CG and AT base pairs have approximately the same size and shape, which reduces distortions in the double helix structure.

- **RNA**
  → The abbreviation for *ribonucleic acid*
  → A polynucleotide, like DNA
  → Shorter than DNA, and generally single-stranded

- **Three types of RNA**
  → *Messenger RNA* (mRNA) carries genetic information from a cell nucleus to the cytoplasm where translation to protein takes place.
  → *Ribosomal RNA* (rRNA) appears to act as a catalyst in the biosynthesis of proteins.
  → *Transfer RNA* (tRNA) is used to transport amino acids during protein synthesis.

- **Structure of RNA**

  → RNA contains four base pairs as in DNA, but *uracil* (U) substitutes for *thymine* (T).

  → *Uracil* lacks the methyl group of *thymine* at the carbon-5 position. The structures of thymine and uracil are

Thymine (T) Uracil (U)

Note: The position of the methyl group on thymine does not interfere with its hydrogen bonding with adenine (A) (see the preceding figure showing an AT base pair).

→ The sugar in RNA is ribose (deoxyribose shown for comparison):

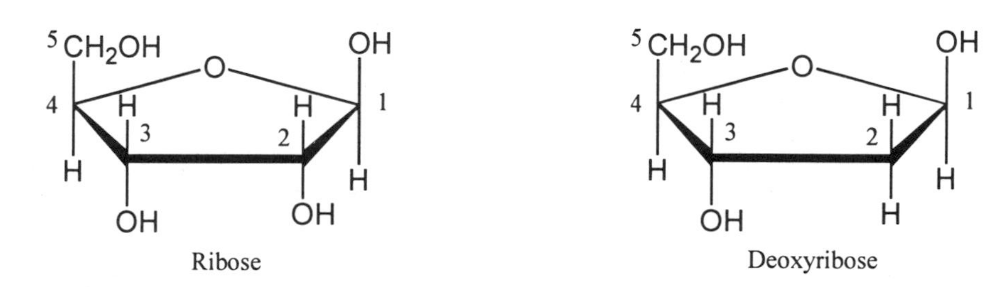

Ribose Deoxyribose

→ Although RNA molecules are generally single-stranded, they contain regions of a double helix produced by the formation of *hairpin loops*. In these regions, the usual base pairing is A with U, and G with C. Imperfections in base pairing are common, however, in RNA.

- **Summary**

  → Nucleic acids are copolymers of four nucleotides joined by phophate ester links.

  → The uracil-ribose nucleotide in RNA replaces the thymine-deoxyribose nucleotide in DNA.

  → The nucleotide sequence stores and carries all genetic information.